Hyperon Resonances – 70

edited by

Earle C. Fowler

Duke University

Moore Publishing Company
Durham, North Carolina

ISBN — 0-87716-031-7

HYPERON RESONANCES – 70

Conference held at Duke University, Durham, North Carolina, April 24 and 25, 1970.

Organized by:

Earle C. Fowler, Robert D. Tripp, and William J. Willis

With the assistance of:

Rodney L. Cool, David B. Cline, Malcolm Derrick, Bernard Feld, Gerson Goldhaber, and Riccardo Levi-Setti

Sponsored by:

Duke University, Durham, North Carolina with additional assistance from the U. S. Atomic Energy Commission.

Discussion Leaders

Rodney L. Cool
David Cline
Riccardo Levi-Setti
Malcolm Derrick
Gerson Goldhaber
Bernard Feld

Reviewers

Gerald Lynch
Richard Dalitz
Robert Tripp
David Plane
Angela Barbaro-Galtieri
David Miller
George Kalbfleisch
Sydney Meshkov

Scientific Secretaries

Peter Haberler
Dwight Carpenter
Carl Rose
Morris Binkley
Nelson Lipshutz
Moo Han
Lawrence Biedenharn

FOREWORD

Since the probable existence of the first hyperon resonance was reported by M. L. Good on behalf of the Alvarez group at the Rochester Conference in 1960, the subject has expanded explosively. At our general meetings, it has been increasingly difficult to bring together experimenters with results which ought to be compared directly. Many of us have felt for some time that a single meeting devoted to hyperon resonances was needed. Finally, with the encouragement of William Willis and Robert Tripp, we at Duke have undertaken the task of bringing together experimenters from most of the world's laboratories who specialize in the study of these entities.

The resulting conference was held on the campus of Duke University, Durham, North Carolina, on the 24th and 25th of April, 1970. The table below shows the organization of the conference after the initial division of topics and tentative assignment of available time.

Session	Discussion Leader	Reviewer	Scientific Secretary
I. Total Cross Sections	Rodney L. Cool	Gerald Lynch	Peter Haberler
II. Lower Energy Formation	David Cline	Richard Dalitz Robert Tripp	Dwight Carpenter
III. Moderate Energy Formation	Riccardo Levi-Setti	David Plane	Carl Rose
IV. Higher Energy Formation	Riccardo Levi-Setti	Angela Barbaro-Galtieri	Carl Rose
V. Y^* Production	Malcolm Derrick	David Miller	Morris Binkley
VI. Ξ^* Production	Malcolm Derrick	George Kalbfleisch	Nelson Lipshutz
VII. Z^* Production	Gerson Goldhaber	Contributors: Yokosawa, Hughes, Wagner, Goldhaber	Moo Han
VIII. Theoretical Standpoint	Bernard Feld	Sydney Meshkov	Lawrence Biedenharn

After completion of the choice of key reviewers the organization of each session was left in the hands of the Discussion Leaders. Short contributions to the meeting were accepted whether or not formal talks were given. The Scientific Secretaries not only assisted the Discussion Leaders but also were responsible for completing the written record of the discussions.

First, I want to thank the Organizers, the Discussion Leaders, the Reviewers, the Speakers, the Contributors, and the Participants for their efforts to optimize the exchange of information and to maximize understanding. We at Duke found that everyone was most enthusiastic and cooperative in their efforts to make the conference successful. Our special thanks go to everyone who has made written contributions to the proceedings not only for their arduous technical work but also for adhering to a relatively complex set of rules in order to provide us with standardized manuscripts ready for the cameras of our publisher. Second, I want to thank all those at Duke who contributed mightily to the planning of the details of the meetings themselves and associated functions. The brunt of this was carried by the Scientific Secretaries and their wives, our Duke graduate students and their wives in both experimental and theoretical high energy physics, and Dr. Lloyd Fortney who recorded the discussions on tape. Help from other parts of the physics department was always provided when needed. I want to give special acknowledgement to the technical staff in our high energy physics group; particularly Mary Moses, Frances Pendergrass, Karen Spalek, and Judy Butcher. Third, I want to express my appreciation to Duke University and to the U. S. Atomic Energy Commission for their support of this endeavor in supplying both facilities and funds. Fourth, as the final deadline for submission of the manuscript of these proceedings to the publisher has approached I have relied more and more on Dr. Dwight Carpenter and Mary Moses to complete the final details and incorporate the latest contributions and modifications. Without their help there would be no publication.

Finally, I wish to thank my wife not only for her constant support of my activities in this endeavor, but also for organizing the greater share of the auxiliary activities that contributed so much to the pleasant atmosphere of the conference.

We believe that most of the participants found the conference both enjoyable and profitable. We are hopeful that these proceedings have retained the spirit of the meetings and will help open the way to the reconciliation of apparently conflicting results.

Earle C. Fowler
June 23, 1970
Durham, North Carolina

CONTENTS

CONTRIBUTIONS

SESSION V: Y^* PRODUCTION

TALKS

CONTRIBUTION

CONTRIBUTIONS

TALKS

CONTRIBUTION

Note: The underlines in the table of contents indicate the names of the persons who presented the talks at the conference.

SESSION I

TOTAL CROSS SECTIONS

Discussion Leader: Rodney Cool

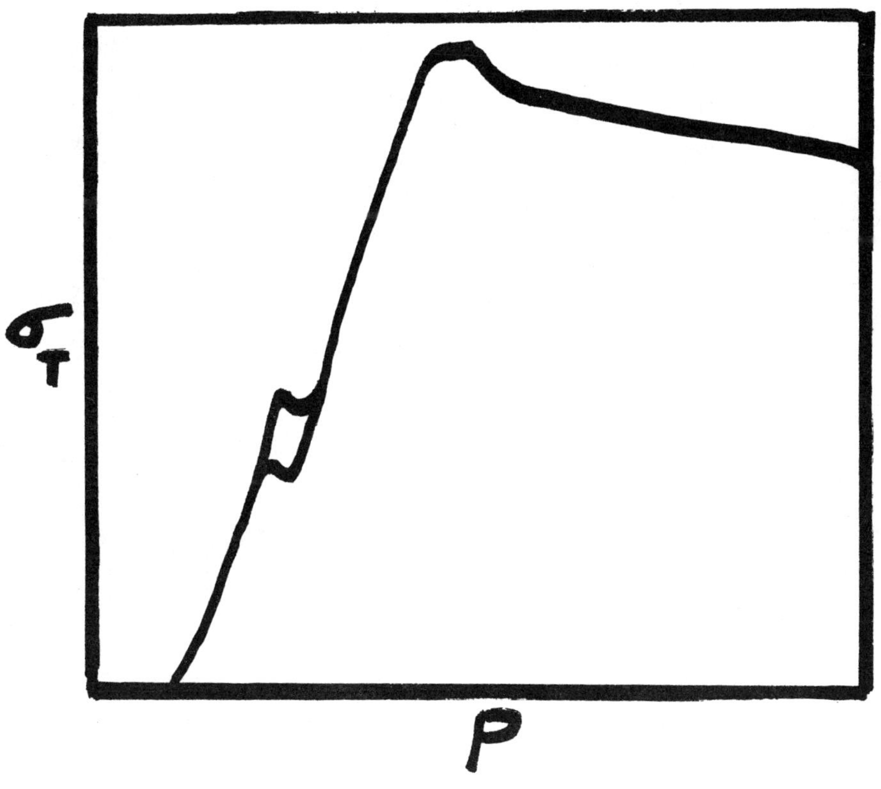

KAON-NUCLEON TOTAL CROSS SECTIONS FROM 0.36 TO 0.72 GeV/c

T. Bowen, P.K. Caldwell, F. Ned Dikmen, E.W. Jenkins,
R. M. Kalbach, D.V. Petersen, and A.E. Pifer
Department of Physics, University of Arizona,
Tucson, Arizona 85721

The K^+ and K^- total cross sections were measured on both hydrogen and deuterium at a series of momenta between 360 MeV/c and 720 MeV/c by the transmission counter technique. The experiment was performed at the Berkeley Bevatron using a low momentum separated kaon beam. The kaons were identified by use of two threshold lucite Cerenkov counters, by time-of-flight, and by dE/dx in the beam counters. Three 18 inch targets, one each of hydrogen, deuterium, and empty were alternately placed in the beam. The amount of incident beam transmitted through the targets was measured by five transmission counters mounted on a movable stand. The smallest transmission counter subtended a constant momentum transfer squared of $-t = 2.95 \times 10^{-3}$ $(\mathrm{GeV/c})^2$. The data were corrected for single coulomb (Rutherford) scattering, coulomb-nuclear interference scattering and decay effects, and the pure isospin states extracted.

The measured total cross sections for K^+p and K^+d are listed in Table I and are shown in Figures 1 and 2, respectively, with the results of the experiment by Bugg et al.[1] It may be noted that there is a significant systematic disagreement in the K^+d data of this experiment and those of Bugg et al. while the hydrogen data agree quite well. This same disagreement appears in the K^-d data. This suggests a fundamental difficulty in the treatment of the deuterium data in one of the experiments, but at the present time the disagreement cannot be accounted for.

The measured total cross sections for K^-p and K^-d are listed in Table II and are shown in Figures 3 and 4, respectively, with the data of Bugg et al.[1], Watson et al.[2], and Armenteros et al.[3]

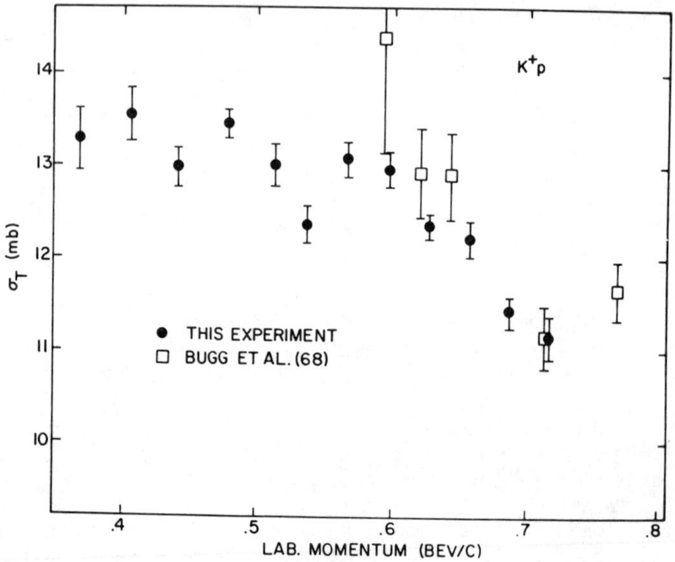

Figure 1 K^+ p measured total cross sections

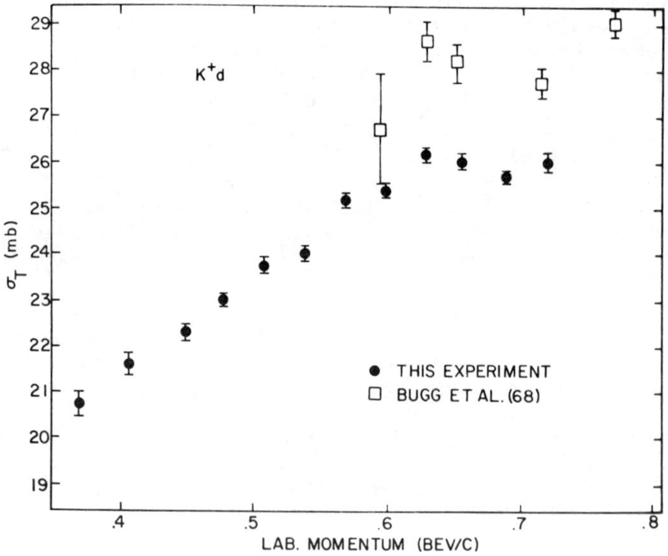

Figure 2 K^+ d measured total cross sections

TABLE I

K^+p and K^+d measured total cross sections

P_{lab} (Mev/c)	σ_{K^+p} (mb)	increased error	σ_{K^+d} (mb)	increased error
717	11.14 ± .25	0.50	26.02 ± .24	0.41
686	11.40 ± .18	0.40	25.68 ± .14	0.36
657	12.20 ± .20	0.63	26.03 ± .17	0.33
627	12.33 ± .14	0.39	26.15 ± .17	0.40
596	12.94 ± .19	0.34	25.40 ± .16	0.20
566	13.05 ± .19	0.25	25.17 ± .16	0.18
536	12.35 ± .20	0.57	24.01 ± .17	0.49
506	12.99 ± .23	0.28	23.77 ± .19	0.19
475	13.44 ± .17	0.29	22.99 ± .16	0.20
440	12.97 ± .23	0.51	22.31 ± .20	0.36
405	13.55 ± .31	0.48	21.61 ± .26	0.32
366	13.28 ± .35	0.64	20.74 ± .30	0.48

TABLE II

K^-p and K^-d measured total cross sections

P_{lab} (Mev/c)	σ_{K^-p} (mb)	increased error	σ_{K^-d} (mb)	increased error
717	34.01 ± .37	0.38	56.50 ± .34	0.35
686	35.29 ± .39	0.42	56.45 ± .34	0.46
657	35.53 ± .33	0.33	55.22 ± .32	0.44
627	36.41 ± .28	0.37	55.80 ± .26	0.26
596	37.76 ± .33	1.16	58.34 ± .29	0.96
566	39.29 ± .34	0.39	60.04 ± .29	0.73
536	42.08 ± .38	0.38	63.07 ± .33	0.38
506	44.94 ± .45	0.57	67.93 ± .39	0.46
475	48.96 ± .47	0.56	74.53 ± .42	0.49
440	58.42 ± .57	1.16	87.57 ± .50	0.92
405	75.54 ± .68	0.71	96.74 ± .64	0.64
385	83.75 ± .83	0.87	- - -	- - -
366	77.24 ± .79	1.04	102.93 ± .70	1.02

The measured cross sections quoted in these tables then have been unfolded for the experimental momentum spread from energy loss in the targets. In addition the deuterium results have been unfolded for the fermi motion of the deuteron. The Glauber-Wilkin[4] screening correction has been applied to the deuteron and the pure I=0 and I=1 isospin states computed. These values are listed in Tables III and IV. The errors quoted for these unfolded cross section values include the experimental cross section error plus the contribution from our unfolding procedure as estimated by Monte Carlo calculations.

These new data show that the K^+ p I=1 cross section is without resonant structure in this energy range although the presence of a small dip at 700 MeV/c is confirmed. The K^+ I=0 cross section is smoothly rising in this energy range. Above our energy range one is not sure of the K^+ I=0 cross section due our systematic disagreement with previous K^+ - d cross sections.

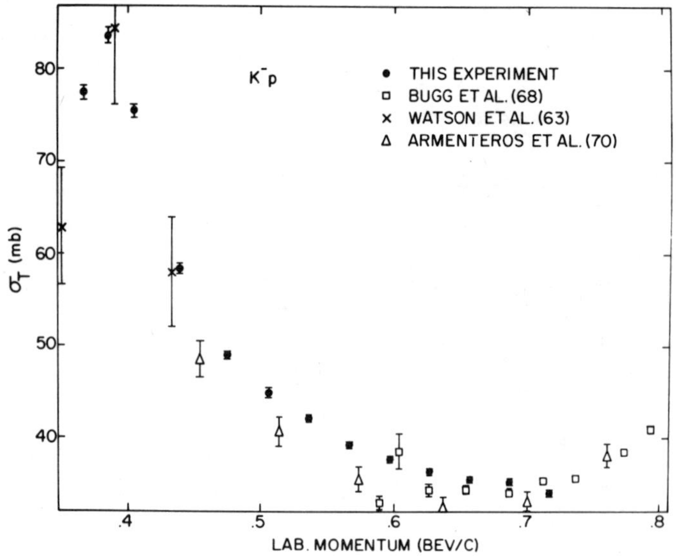

Figure 3 K^-p measured total cross sections

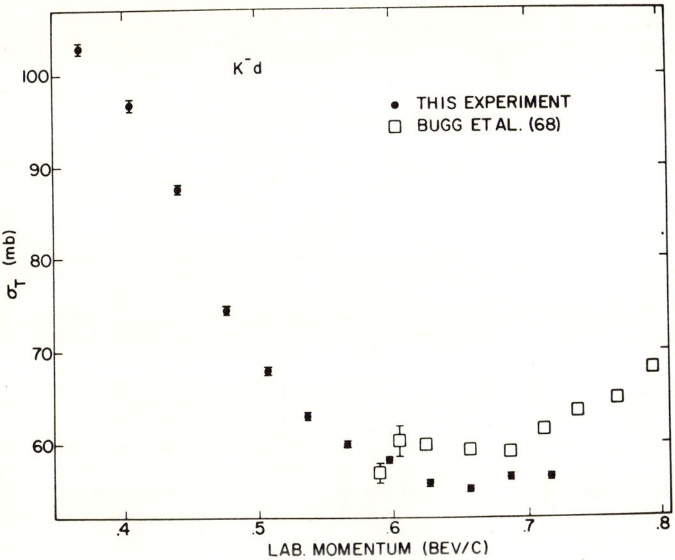

Figure 4 K^-d measured total cross sections

TABLE III

K^+N Momentum Unfolded Total Cross Sections

P_{lab} (Mev/c)	σ_p (mb)	σ_d (mb)	σ_1 (mb)	σ_0 (mb)
717	11.10 ∓.25	24.97 ∓.47	11.10 ∓.25	17.48 ∓ 1.20
686	11.25 ∓.18	24.61 ∓.42	11.25 ∓.18	16.27 ∓ 1.00
657	12.43 ∓.20	26.95 ∓.39	12.43 ∓.20	17.59 ∓ 0.98
627	12.18 ∓.14	27.57 ∓.35	12.18 ∓.14	19.60 ∓ 0.82
596	13.04 ∓.19	26.28 ∓.33	13.04 ∓.19	14.42 ∓ 0.87
566	13.20 ∓.19	25.65 ∓.30	13.20 ∓.19	12.65 ∓ 0.83
536	12.09 ∓.20	24.64 ∓.28	12.09 ∓.20	13.90 ∓ 0.82
506	13.02 ∓.23	24.09 ∓.27	12.02 ∓.23	9.99 ∓ 0.88
475	13.58 ∓.17	24.41 ∓.27	13.58 ∓.17	8.98 ∓ 0.74
440	12.81 ∓.23	23.06 ∓.28	12.81 ∓.23	8.54 ∓ 0.89
405	13.69 ∓.31	21.89 ∓.30	13.69 ∓.31	3.44 ∓ 1.11
366	13.20 ∓.35	21.64 ∓.33	13.20 ∓.35	4.47 ∓ 1.24

TABLE IV

K^-N Momentum Unfolded Total Cross Sections

P_{lab} (MeV/c)	σ_p (mb)	σ_d (mb)	σ_1 (mb)	σ_0 (mb)
686	35.33 ±.39	53.52 ±.79	19.34 ±.88	51.32 ± 1.41
657	35.40 ±.33	50.63 ±.73	16.01 ±.80	54.79 ± 1.23
627	36.36 ±.28	51.45 ±.67	15.92 ±.73	56.80 ± 1.07
596	37.80 ±.33	55.42 ±.63	18.89 ±.71	56.71 ± 1.17
566	39.10 ±.34	56.29 ±.61	18.37 ±.70	59.83 ± 1.19
536	42.15 ±.38	55.64 ±.61	13.88 ±.72	70.42 ± 1.29
506	44.71 ±.45	58.03 ±.61	13.46 ±.76	75.96 ± 1.48
475	48.45 ±.47	63.26 ±.62	14.83 ±.78	82.07 ± 1.54
440	58.41 ±.57	78.49 ±.65	20.58 ±.87	96.24 ± 1.83
405	69.80 ±.68	93.75 ±.71	25.36 ±.98	114.24 ± .16
385	100.19 ±.83	- - - - -	(26.5 ±1.0)	(173.9 ± 2.5)
366	69.32 ±.79	94.98 ±.79	28.29 ±1.12	110.35 ± 2.50

List of References

1. D. V. Bugg, R. S. Gilmore, K. M. Knight, D. C. Salter, G. H. Stafford, E. J. N. Wilson, J. D. Davies, J. D. Dowell, P. M. Hattersley, R. J. Homer, A. W. O'Dell, A. A. Carter, R. J. Tapper, and K. F. Riley, Phys. Rev. 168, 1466 (1968).

2. M. B. Watson, M. Ferro-Luzzi, and R. D. Tripp, Phys. Rev. 131, 2248 (1963).

3. R. Armenteros, P. Baillon, C. Bricman, M. Ferro-Luzzi, E. Pagiola, J. O. Petersen, D. E. Plane, N. Schmitz, E. Burkhardt, N. Filthuth, E. Kluge, H. Oberlack, R. R. Ross, R. Barloutaud, P. Granet, J. Meyer, J. P. Porte, and J. Prevost, CERN Report/ D. Ph. II/PHYS. 70-7.

4. C. Wilkin, Phys. Rev. Letters 17, 561 (1966).

REVIEW OF THE KN AND $\bar{K}N$ TOTAL CROSS SECTIONS BELOW 3.5 GeV/c*

Gerald Lynch

Lawrence Radiation Laboratory

Berkeley, California 94720

INTRODUCTION

Accurate K^+ and K^- total cross-section measurements have been made in hydrogen and deuterium for kaon laboratory momenta from 400 MeV/c to 3.4 GeV/c. Many of the resonances that are presently tabulated in this region (including all of those with masses greater than 2.2 GeV) have these total cross-section data as their only evidence, and some other resonances depend greatly upon the total cross-section data for their determination. It is the purpose of this paper to present these total cross-section data and also present the derived cross sections in the I = 0 and I = 1 channels in such a way that one can get a feeling for how well the different experiments agree concerning various features of the data and to what extent some of the conclusions are placed in doubt by uncertainties that are introduced during the analysis.

K^+ CROSS-SECTION DATA

Figure 1 shows the K^+p total cross-section data. The most precise data come from three experimental groups: the Brookhaven group, the Rutherford group, and the Arizona group. The square symbols (◫ and ⊠) are for the two Brookhaven experiments of Abrams et al. and Cool et al.[1] The cross entries (✳) represent the Rutherford experiment of Bugg et al.,[2] and the diamond points (◇) are for the data of the Arizona group[3] that have been presented at this conference by Ed Jenkins. These data of the Arizona group are the only new data to be added in the past year and as yet they are unpublished. The points shown by the symbol △ are those of Burrows et al.,[4] those denoted by ▽ are from Cook et al.,[5] and the points shown by the double triangle (⧖) at low momentum are

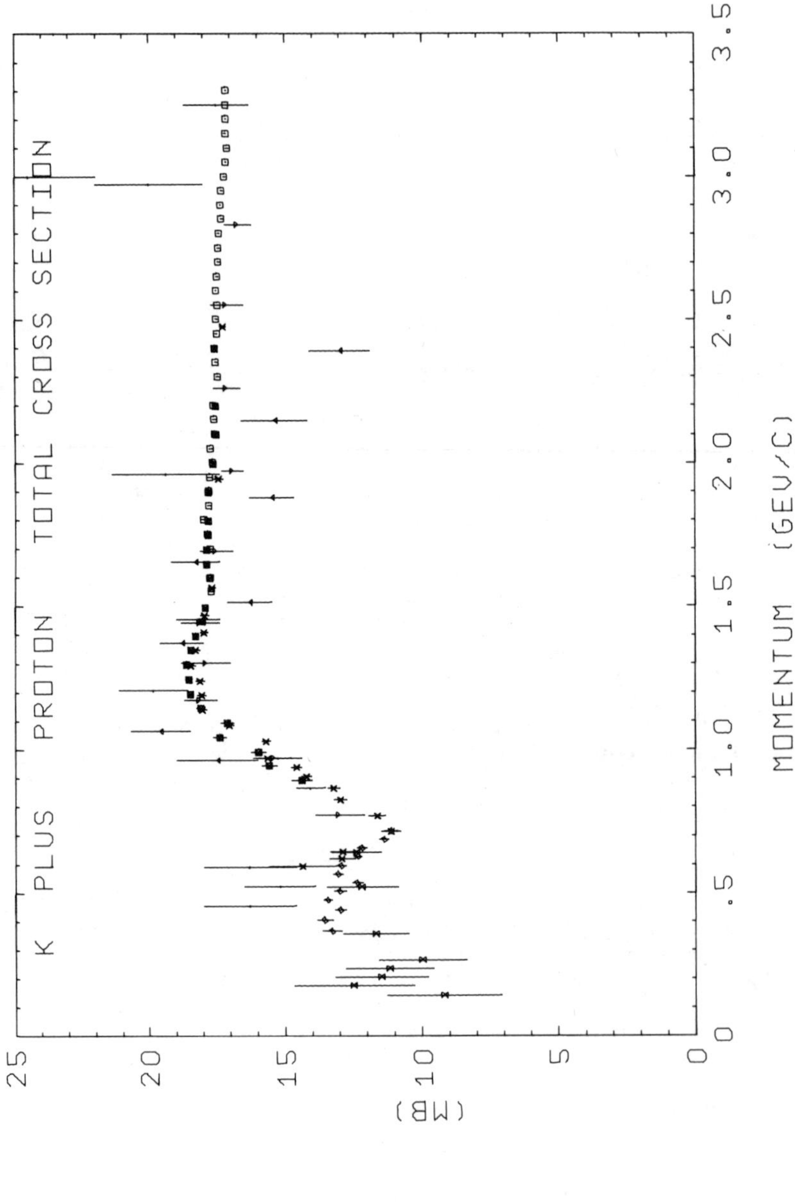

FIG. 1. The K^+p total cross-section data from 0 to 3.5 GeV/c.

those of Goldhaber et al.[6] The points denoted by + are from six other experiments, none of which covered a broad momentum range. Figures 2 and 3 show somewhat magnified representations of these data, with the less precise experiments omitted. The solid curve that is drawn here represents no model, but is merely a freehand curve drawn through the data. The purpose of this curve is not to prejudice the eye but rather to show what curve has been used for making corrections to the data, corrections for the Glauber effect, the finite momentum resolution of the experiments, and the effect of the internal momentum of the deuteron.

Since the K^+p system has I = 1, these figures (1 through 3) represent the I = 1 K-nucleon cross section. They show the peak at about 1.25 GeV/c (corresponding to a mass of 1.91 GeV) that has been called the Z_1^* since the time that it was first observed by Cool et al.,[1] in 1966. We can also see that the valley in the cross sections at 0.7 GeV/c that was indicated by the data of Bugg et al.[2] is confirmed by the data of the Arizona group. Figure 4 shows the difference between the measured values and the smooth curve. The absolute values of these differences merely reflect where I have chosen to draw the smooth curve. What is meaningful on this graph is the differences between different experiments. It should be pointed out that the errors shown on these plots are statistical errors only. This is the procedure that has been customary in presenting these total cross-section data. One should realize, however, that all of these experiments may have systematic errors of at least 1%, which is five times the statistical error on some of the higher-energy cross-section measurements. These systematic errors should not affect any local structure indicated by one experiment, but must be considered when comparing different experiments.

Figure 5 shows the K^+d total cross-section data. The symbols are the same as those used for the K^+p data. In the more magnified plot of these data shown in Fig. 6 one can see that there is a discrepancy of about 7% between the Arizona data and the Rutherford data in the region in which they overlap. This systematic error is a serious one. Since, as we shall see later, this same effect occurs in the K^- data, one is led to believe that somewhere there is a systematic error that affects the Kd data but not the Kp data. One possible contribution to this is that there is some uncertainty concerning the determination of the deuterium density. But since this uncertainty seems to be only about 1% and it is independent of the beam momentum it is probably not the main source of this discrepancy. There is another possible explanation for this effect; this arises from the difference in the extrapolation procedures used by the two groups. In any transmission-experiment measurement an extrapolation to zero angle must be made to account for those interactions resulting in particles produced at very small angles. This procedure is more straightforward for the Kp

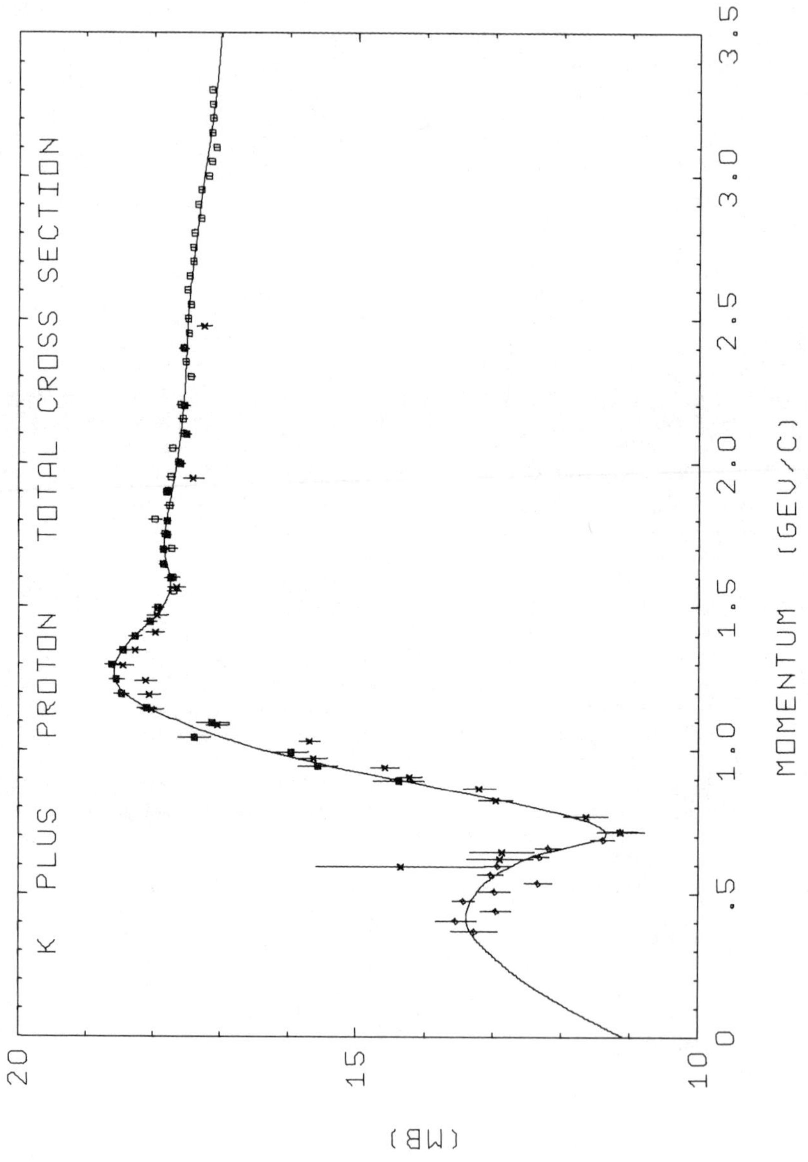

FIG. 2. The K^+p total cross-section data from 0 to 3.5 GeV/c with a smooth curve.

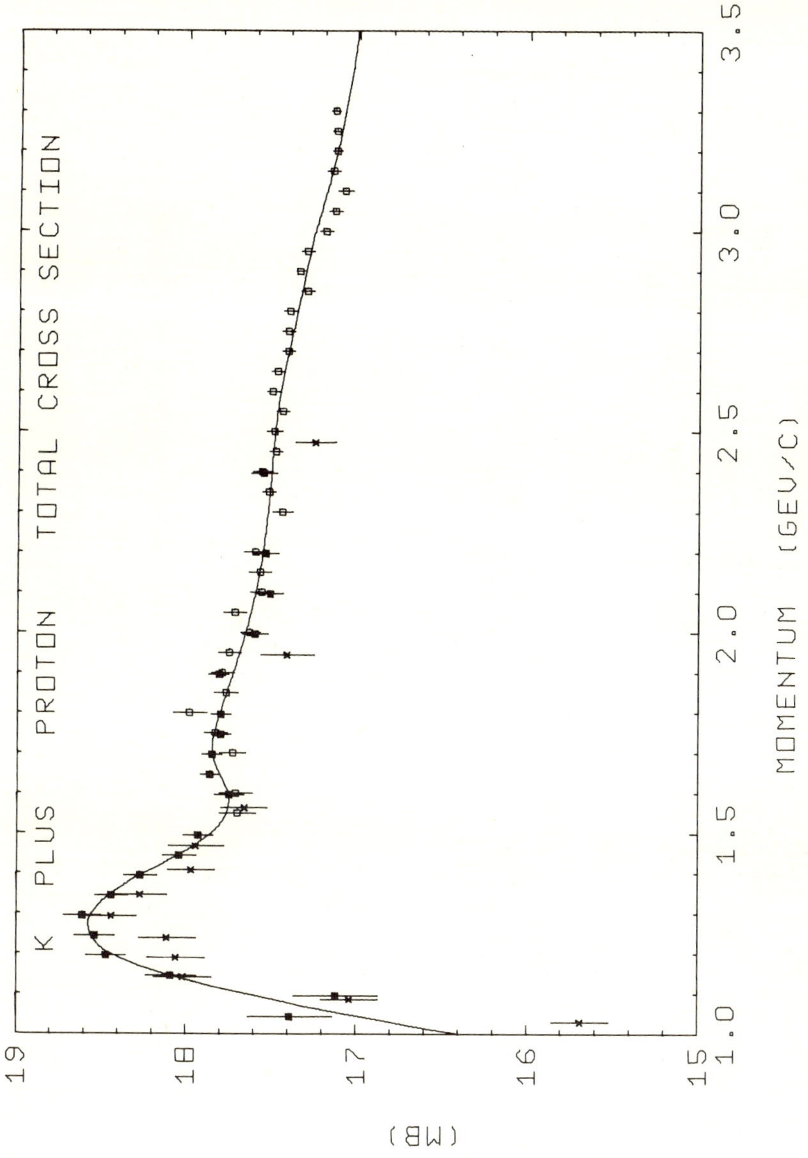

FIG. 3. The K^+p total cross-section data from 1.0 to 3.5 GeV/c with a smooth curve.

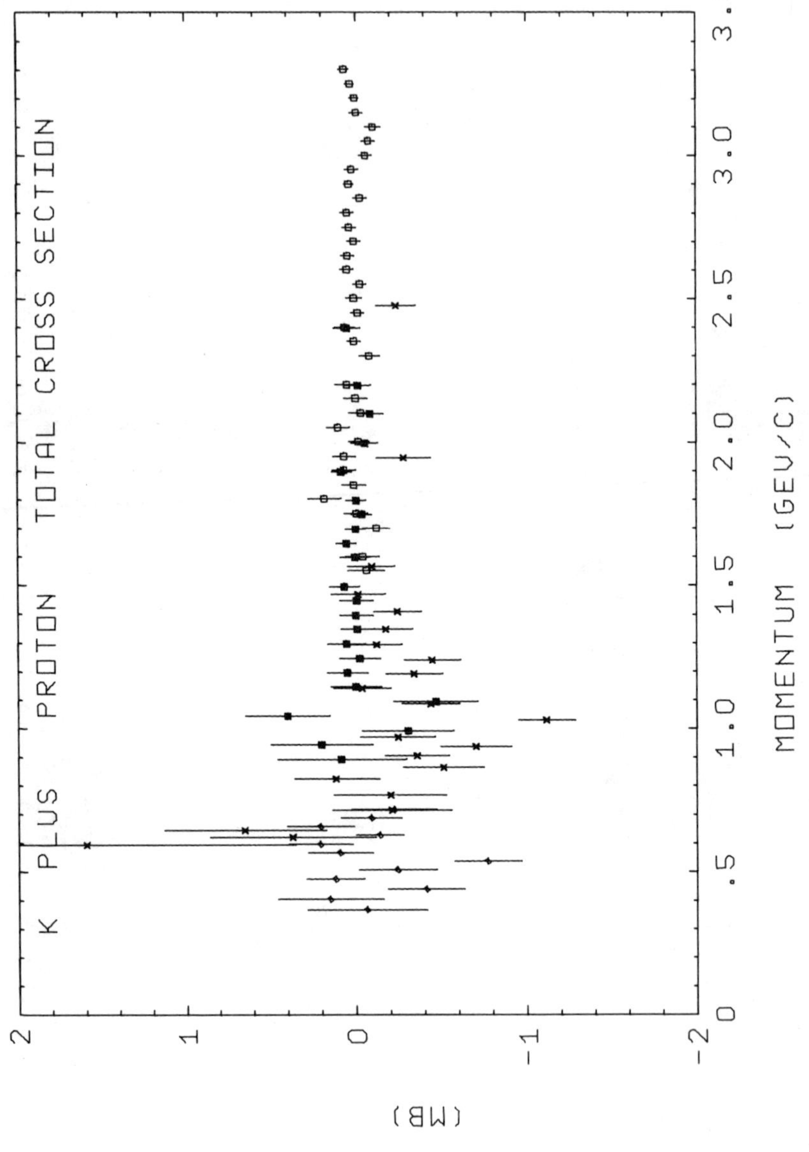

FIG. 4. The difference between the K^+p total cross-section data and a smooth curve.

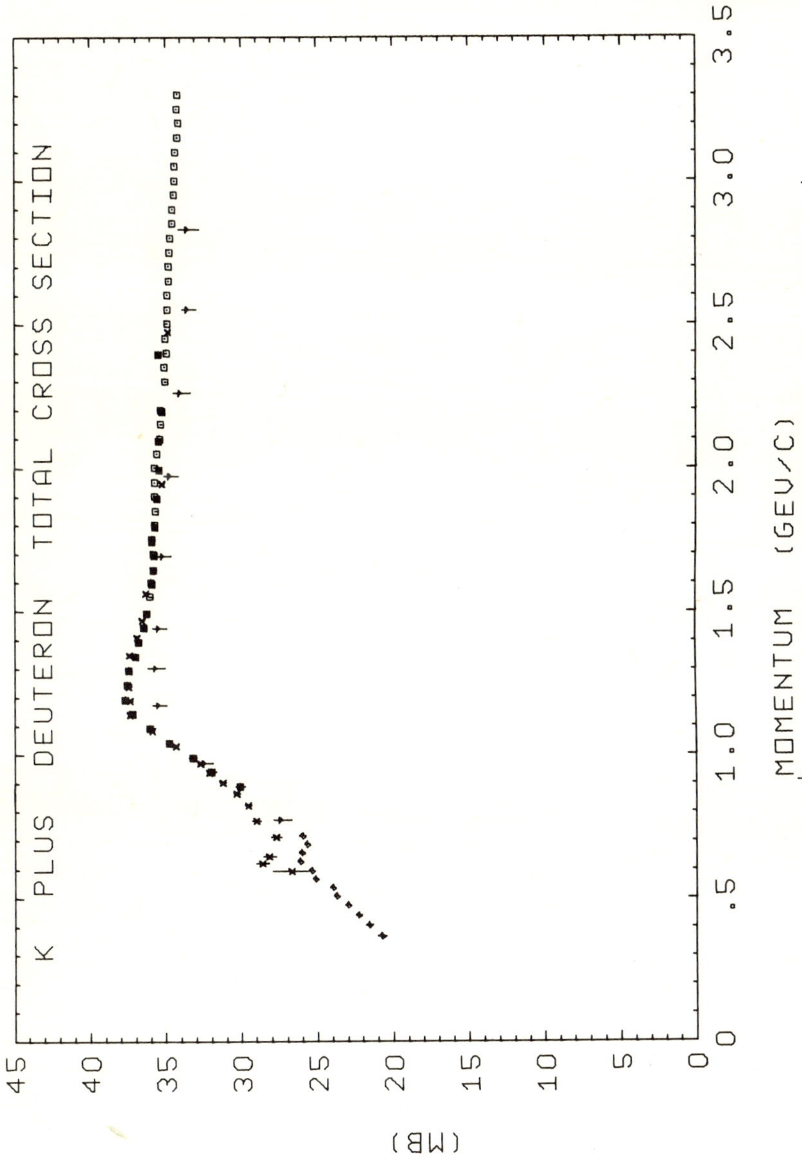

FIG. 5. The K^+d total cross-section data from 0 to 3.5 GeV/c.

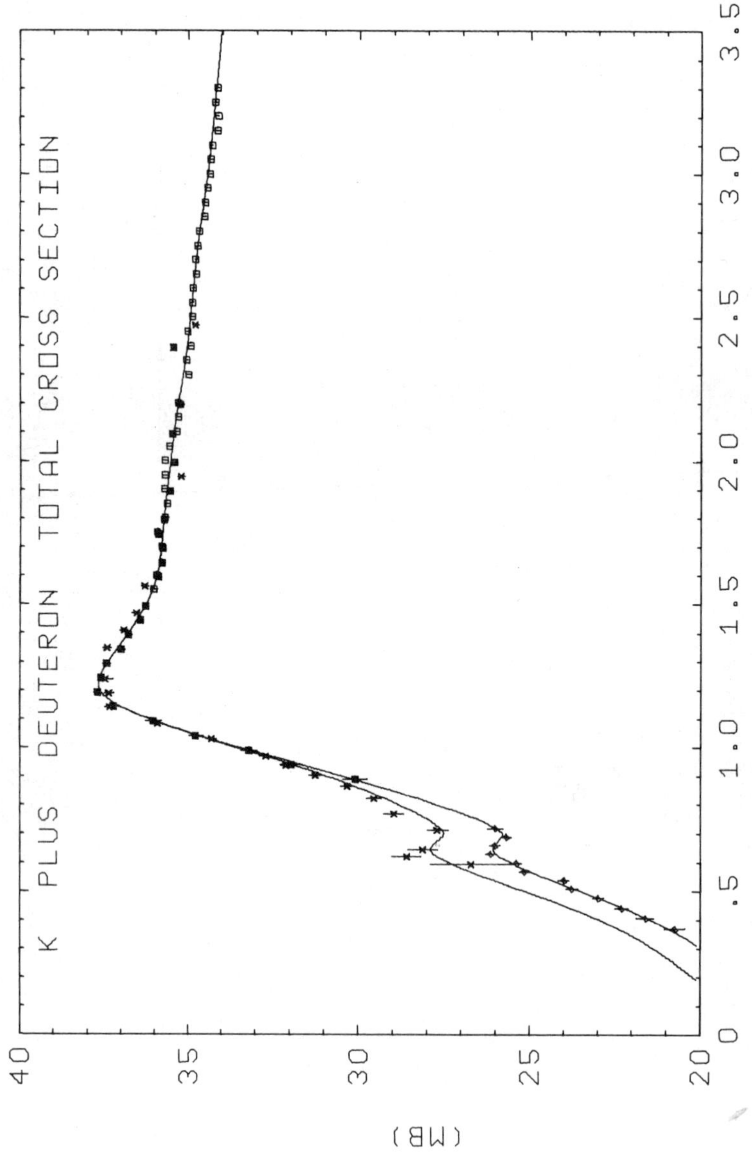

FIG. 6. The K^+d total cross section data from 0 to 3.5 GeV/c with smooth curves.

cross section where the elastic-scattering angular distribution is very nearly linear on a semilog plot than it is for the Kd cross section, for which the plot is not linear but has a peak at small angles due to coherent production. The Arizona group used a linear extrapolation whereas the Rutherford group used a quadratic extrapolation and found that the quadratic term is negligible for hydrogen but several standard deviations for deuterium. It would be worthwhile to investigate how much of the discrepancy can be explained by this effect.

Two smooth curves have been drawn in Fig. 6. The lower one connects the data of the Brookhaven group with those of the Arizona group, following the shape of the Rutherford data in between. The higher curve follows the data of the Rutherford group and is exactly 7% above the lower curve for all momenta less than 650 MeV/c. Another possible curve that can be drawn through these data is shown in Fig. 7, which was taken from an analysis of the K^+ data recently performed by Dowell.[8] This interpretation explains the disagreement between the two experiments in terms of poorly measured points by both experiments in the region in which they overlap, rather than in terms of a systematic difference between the experiments.

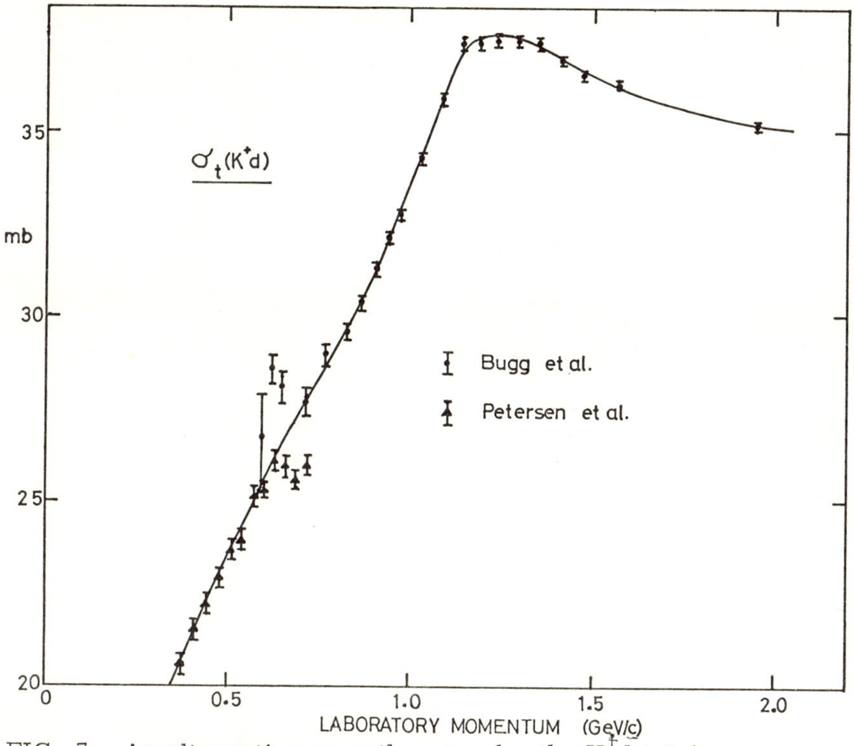

FIG. 7. An alternative smooth curve for the K^+d total cross-section data.

Figure 8 is a magnified plot of the K^+d data above 1 GeV/c, and Fig. 9 shows the differences between the data points and the lower curve of Fig. 6. Again let me emphasize that the fact that some of the points deviate from zero is no reflection on that experiment but rather a reflection of my somewhat arbitrary choice of where to draw the smooth curve.

K^- CROSS SECTION DATA

The K^- total cross-section data are shown in Fig. 10. The symbols on this plot are the same as those on the previous plots except that the double triangle symbol (⧗) represents the data of the CHS (CERN-Heidelberg-Saclay) collaboration.[9] These CHS data, along with the data of the Arizona group, are the results that are new in the past year. The first three points in Fig. 10 show the peak in the cross section caused by the Y^* (1520).

In Figs. 11 and 12 one can see the other structure in the K^-p cross section. In particular, the small but significant peak at 800 MeV/c shows up here and its detection does not depend upon the K^-d data. Figure 13 shows how well the experiments agree. Around 1.1 GeV/c the Brookhaven data are higher than the Rutherford data. This seems to be due to a small disagreement in the momentum determination. At the low-momentum end, the Arizona data are systematically higher than the CHS data.

The K^-d cross-section data are shown in Figs. 14, 15, and 16. The smooth curves in Fig. 15 were chosen in the same way that the curves for the K^+d data were chosen, with the upper curve 7% above the lower one for momenta less than 650 MeV/c. In Fig. 17 one can see that there are quite a few systematic differences between the experiments for the K^-d cross-section data.

DERIVED CROSS SECTIONS

The purpose of measuring both the Kp and Kd cross sections is to obtain from them the cross sections for I = 0 and I = 1. If the deuteron were merely composed of a proton and a neutron at rest, then we could calculate these cross sections from the deuteron and proton cross sections as follows:

for K^+, $\sigma_1 = \sigma_p$,
$\sigma_0 = 2\sigma_d - 3\sigma_p$,

and for K^-, $\sigma_1 = \sigma_d - \sigma_p$,
$\sigma_0 = 3\sigma_p = \sigma_d$.

There are a number of corrections to be made to these relations. One correction is the shadowing correction, often called the Glauber correction.[10] One replaces σ_d in the above formulas by $\sigma_d + \sigma_g$, where the correction term σ_g is approximately equal to $\langle r^{-2}\rangle \sigma_n\sigma_p/4\pi \approx \sigma_n\sigma_p/400$ mb, where $\langle r^{-2}\rangle$ is the average inverse square of the neutron-proton separation in the deuteron ground state

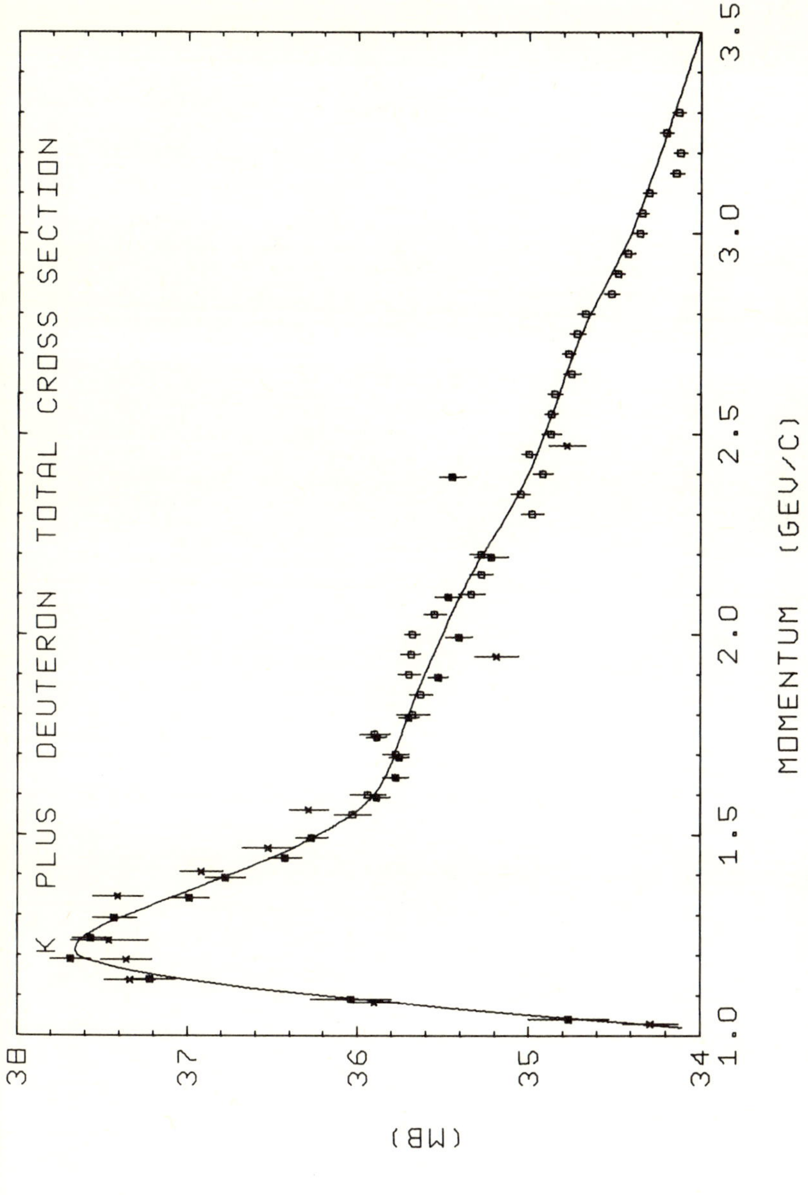

FIG. 8. The K^+d total cross-section data from 1.0 to 3.5 GeV/c with a smooth curve.

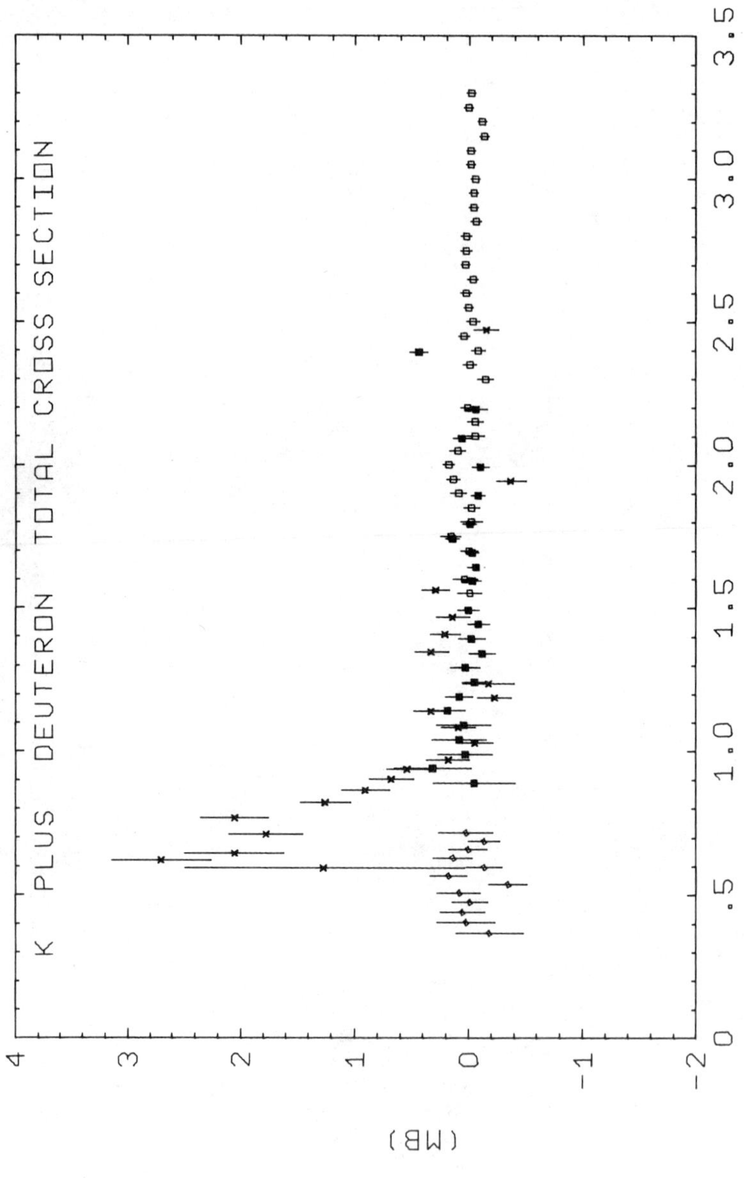

FIG. 9. The difference between the K^+d total cross-section data and the lower smooth curve.

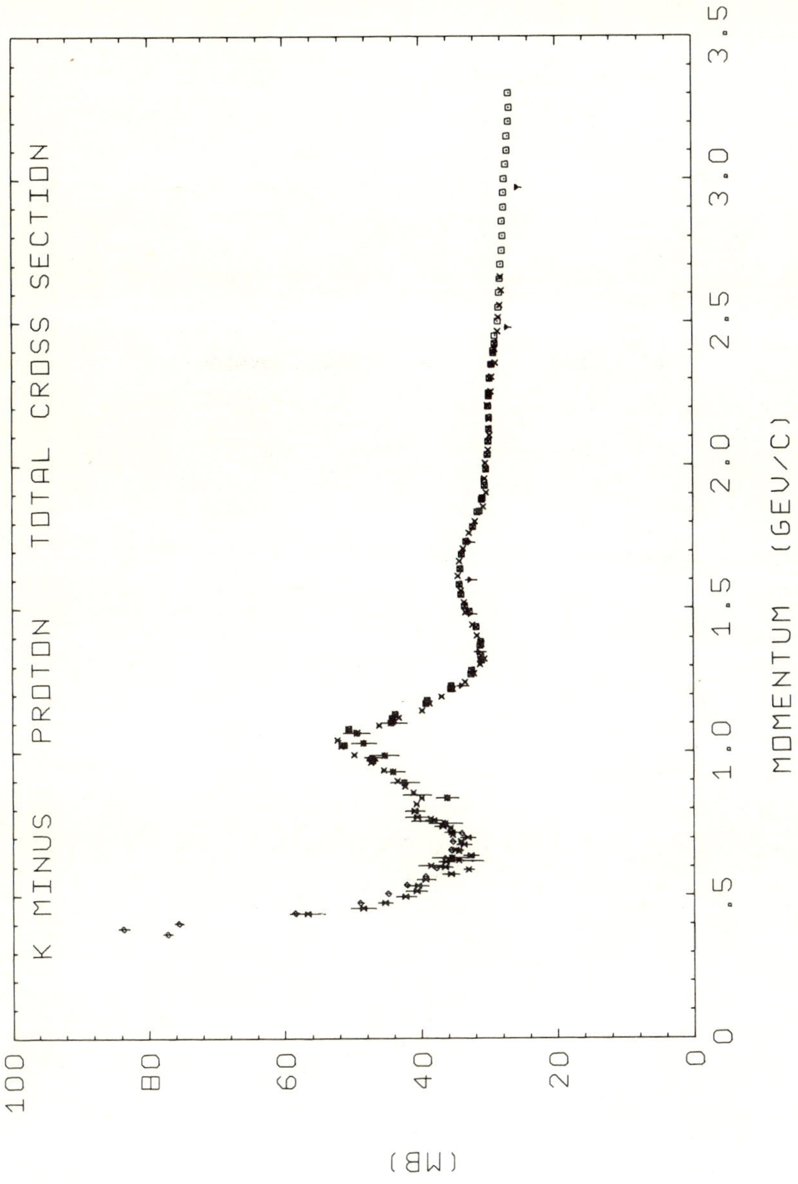

FIG. 10. The K^-p total cross-section data from 0 to 3.5 GeV/c.

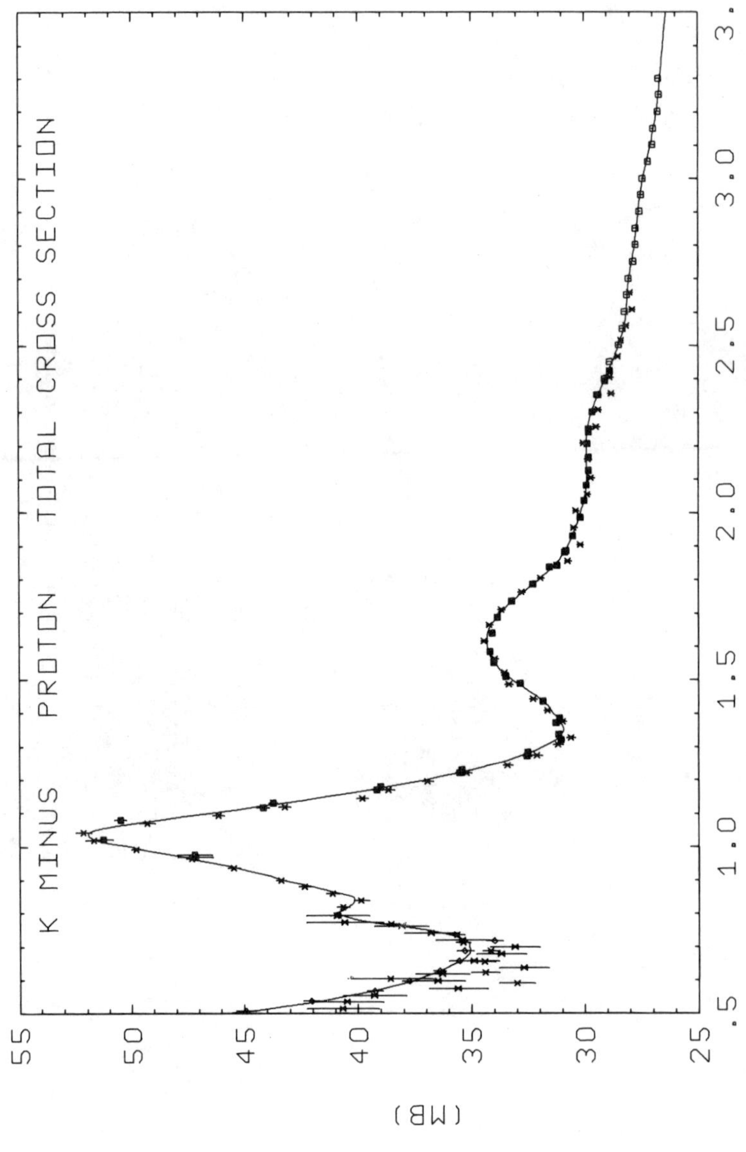

FIG. 11. The K^-p total cross-section data from 0.5 to 3.5 GeV/c with a smooth curve.

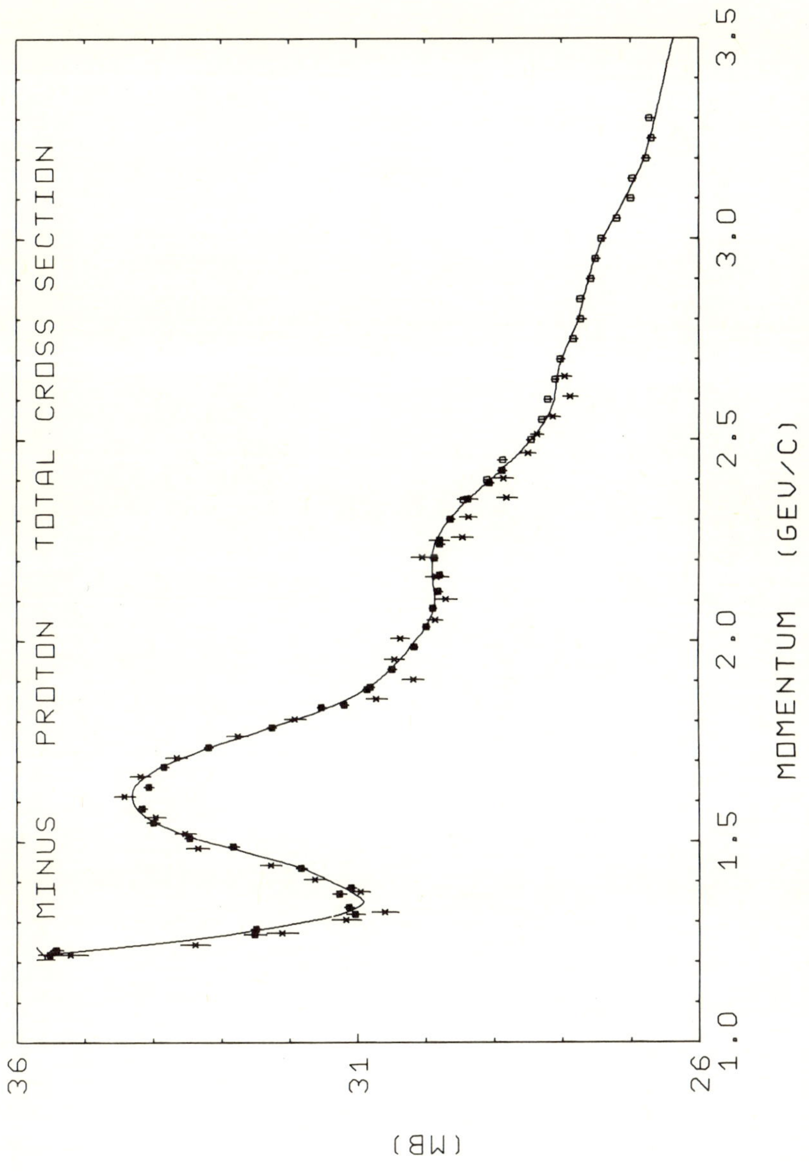

FIG. 12. The K^-p total cross-section data from 1.0 to 3.5 GeV/c with a smooth curve.

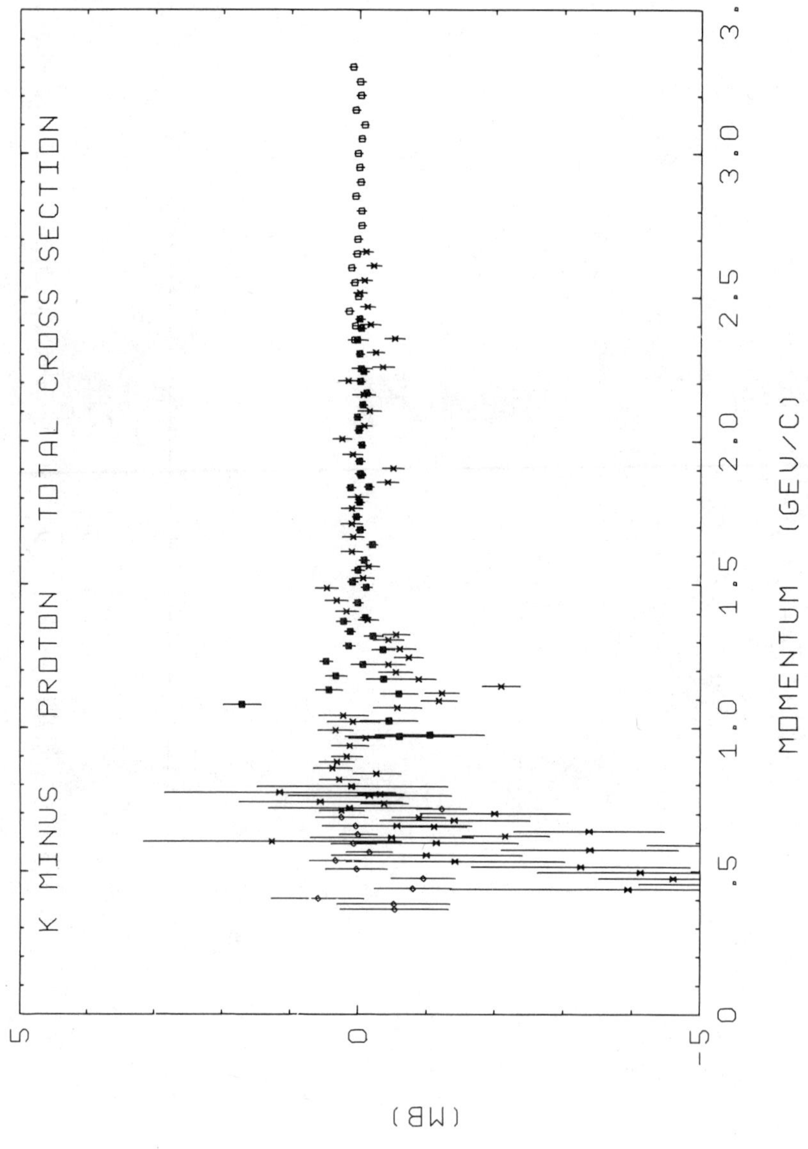

FIG. 13. The difference between the K^-p total cross-section data and the smooth curve.

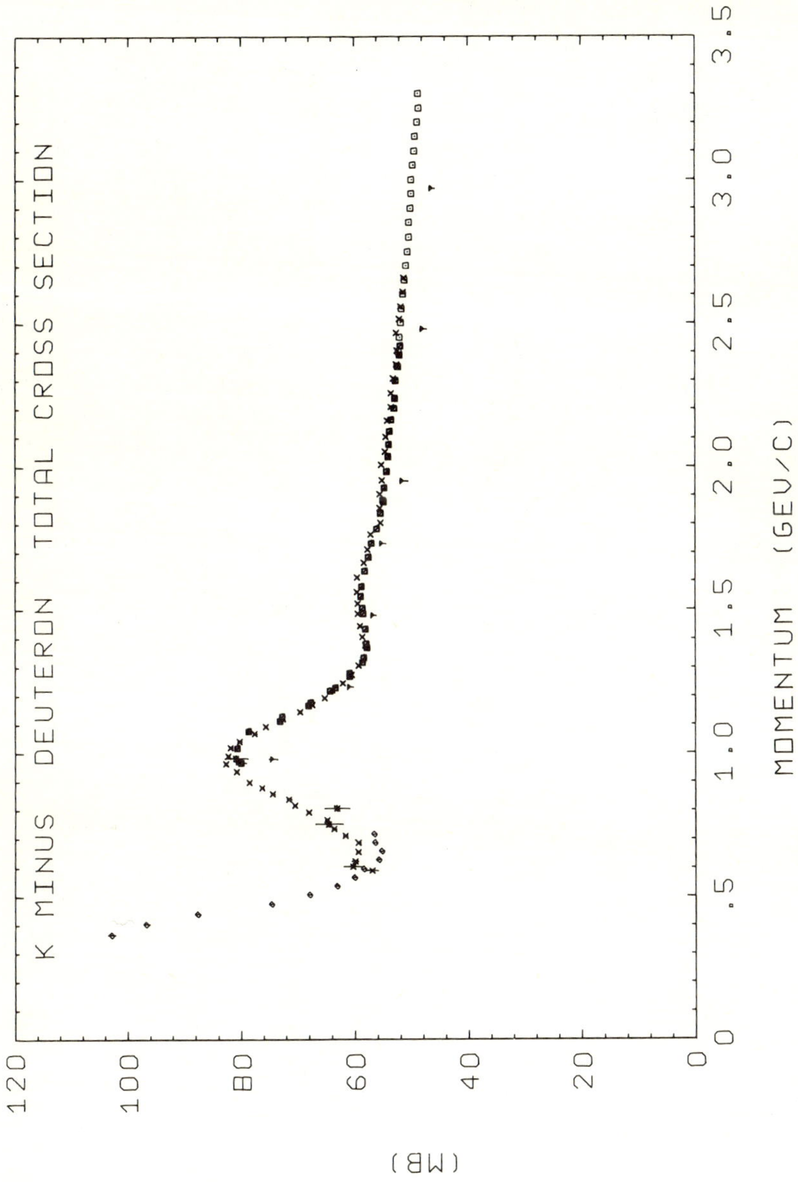

FIG. 14. The K^-d total cross-section data from 0 to 3.5 GeV/c.

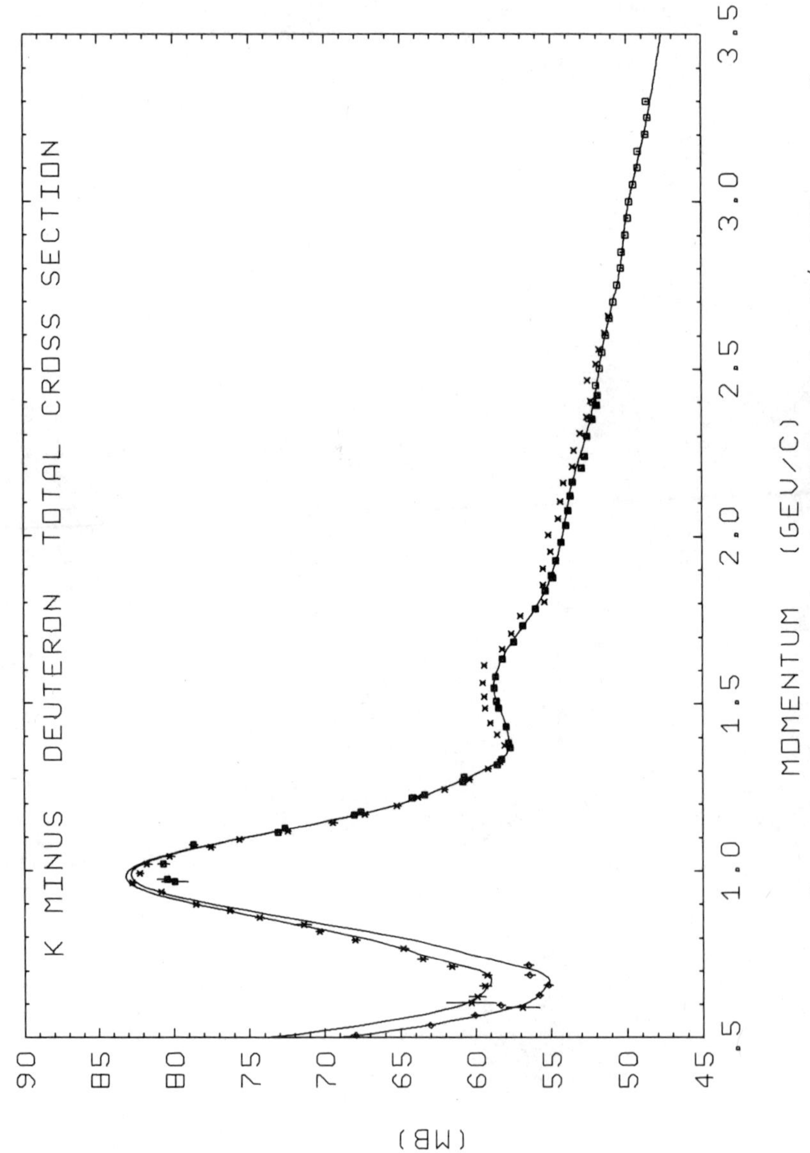

FIG. 15. The K^-d total cross-section data from 0.5 to 3.5 GeV/c with a smooth curve.

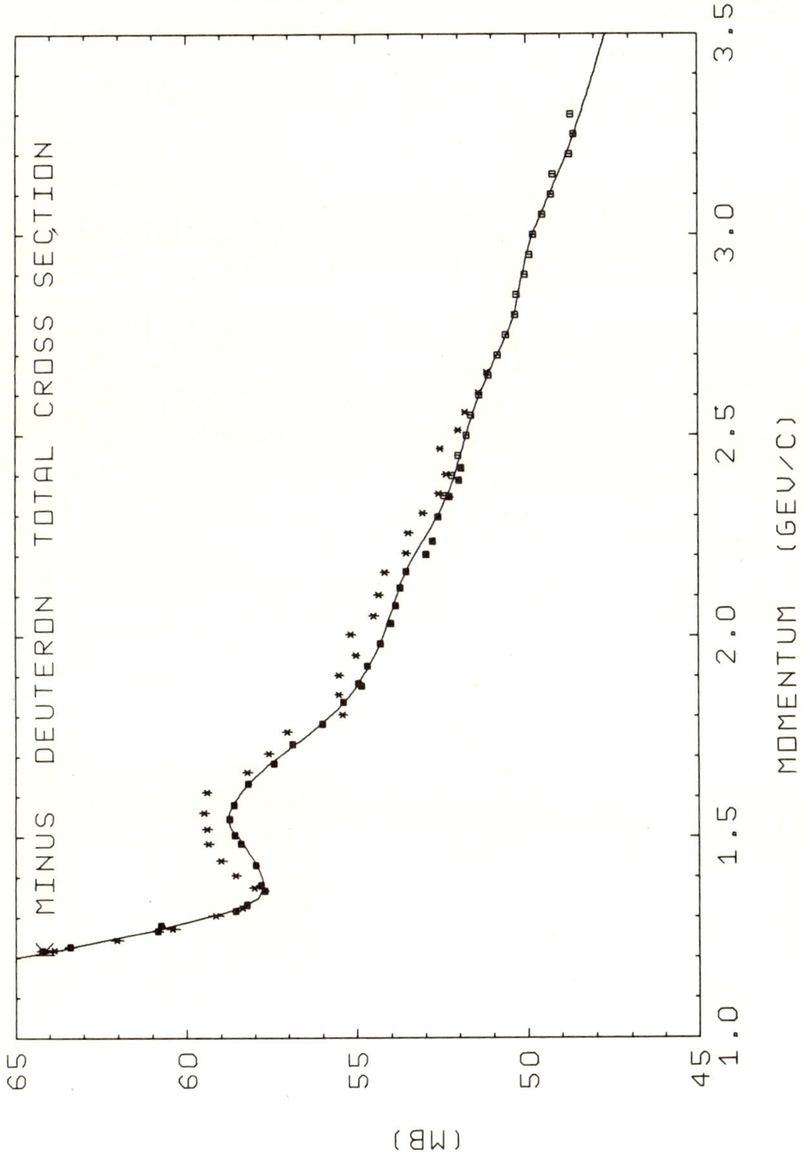

FIG. 16. The K^-d total cross-section data from 1.0 to 3.5 GeV/c with a smooth curve.

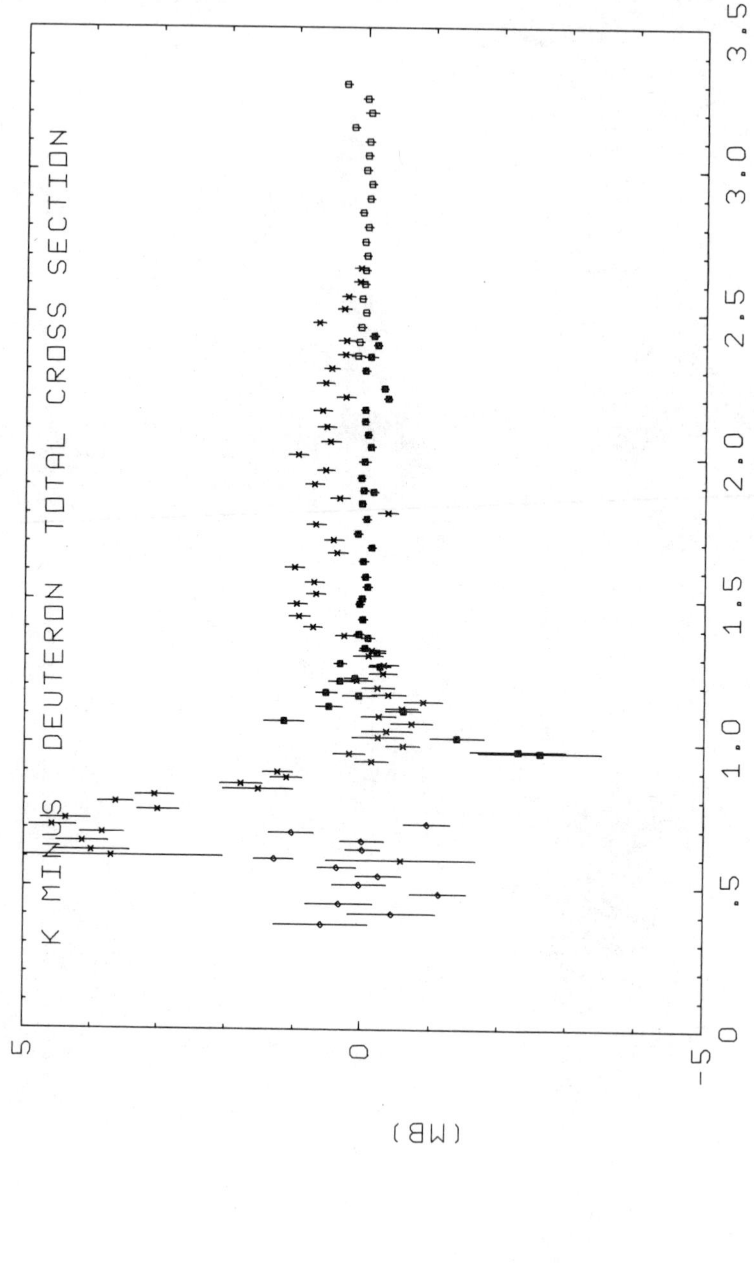

FIG. 17. The difference between the K^-d total cross-section data and the lower smooth curve.

In practice a more complicated form for σ_g has been used that makes it satisfy charge independence and tries to take into account the real parts of the scattering amplitudes.[11]

A second correction is the correction for the effect of the internal momentum of the nucleons in the deuteron. To make this correction one must "unfold" the deuteron cross section; that is, find that neutron cross section which, when added to the proton cross section, smeared out by the internal momentum of the deuteron, and corrected for the Glauber correction produces the observed deuteron cross section. As long as the nucleon cross sections have no fine structure this process is straightforward and adds no significant additional errors to the final result. However, if there is structure with widths on the order of the effect of the smearing caused by the deuteron internal momentum, then the process of unfolding does add additional uncertainties. The full width at half height of the effective resolution caused by this smearing is about 28 MeV in center-of-mass energy at a beam momentum of 500 MeV/c and 46 MeV/c at 1000 MeV/c, and 82 MeV at 2.5 GeV/c.

There is a third correction that has been made to these data, a correction for the finite momentum resolution of the experiments. This correction is very small for the K^+ data, and for the K^- data is significant only at the lowest momenta. It affects both the proton and the deuteron cross sections. In order to make the corrections to the data to obtain σ_0 and σ_1, calculations were made using the smooth curves that were drawn through the data. But in order to present a result that is nearly independent of the smooth curve that was used, one can present the individual data points modified by this correction. This procedure is facilitated by the fact that each of these experiments has measured the hydrogen and deuterium cross sections at the same, or very nearly the same, momenta.

THE I = 0 KN CROSS SECTION

The data points for σ_0 for the KN system are shown Fig. 18. Clearly there is a broad peak from 700 to 1200 MeV. It is unclear whether this broad peak is separated into two distinct peaks, and this determination depends upon which data points are correct in this region. The next graph (Fig. 19) shows these same data points along with the smooth curve that resulted from the original smooth curves assumed, which in the case of the deuteron data was a curve that is exactly midway between the two smooth curves shown in Fig. 6. This smooth curve in Fig. 19 should not be favored over any other smooth curve that could be drawn through these data. Indeed it does not seem like a very natural curve to draw in this case. What is important is that the position of the points on Figs. 18 and 19 would differ very little from those shown if the corrections had been made using any reasonably smooth curve drawn through the raw data.

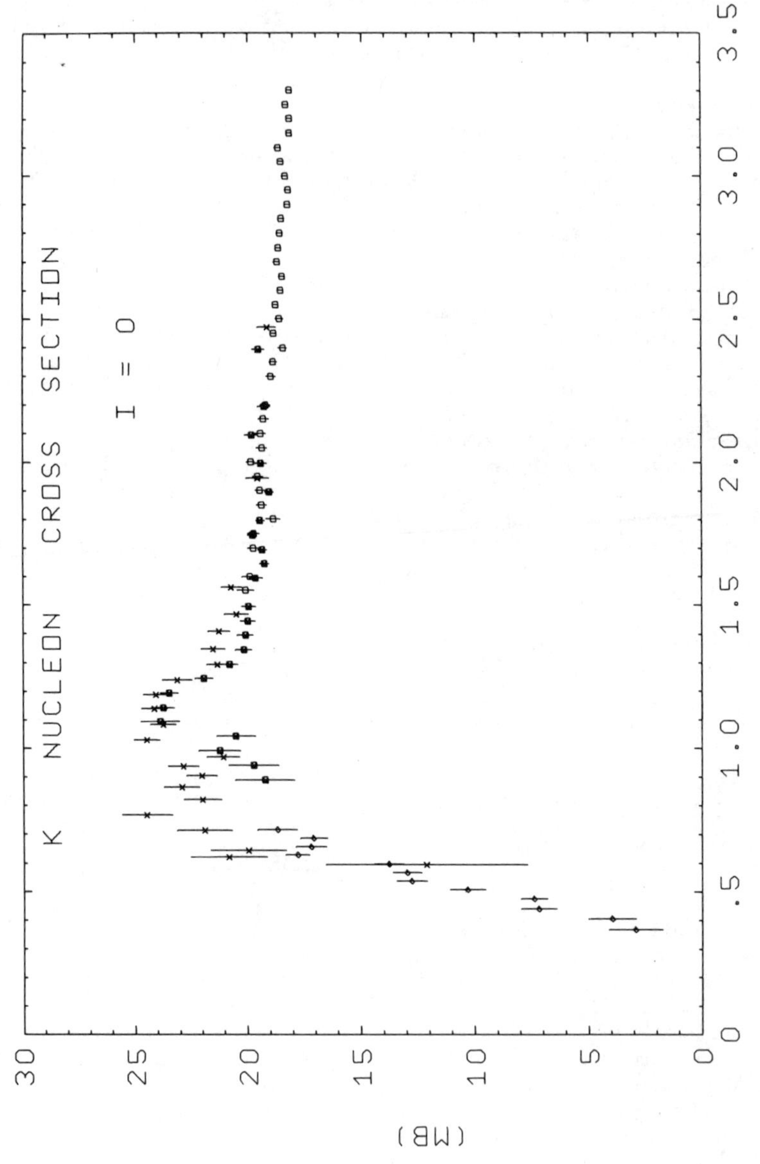

FIG. 18. The data points for the I = 0 KN cross section.

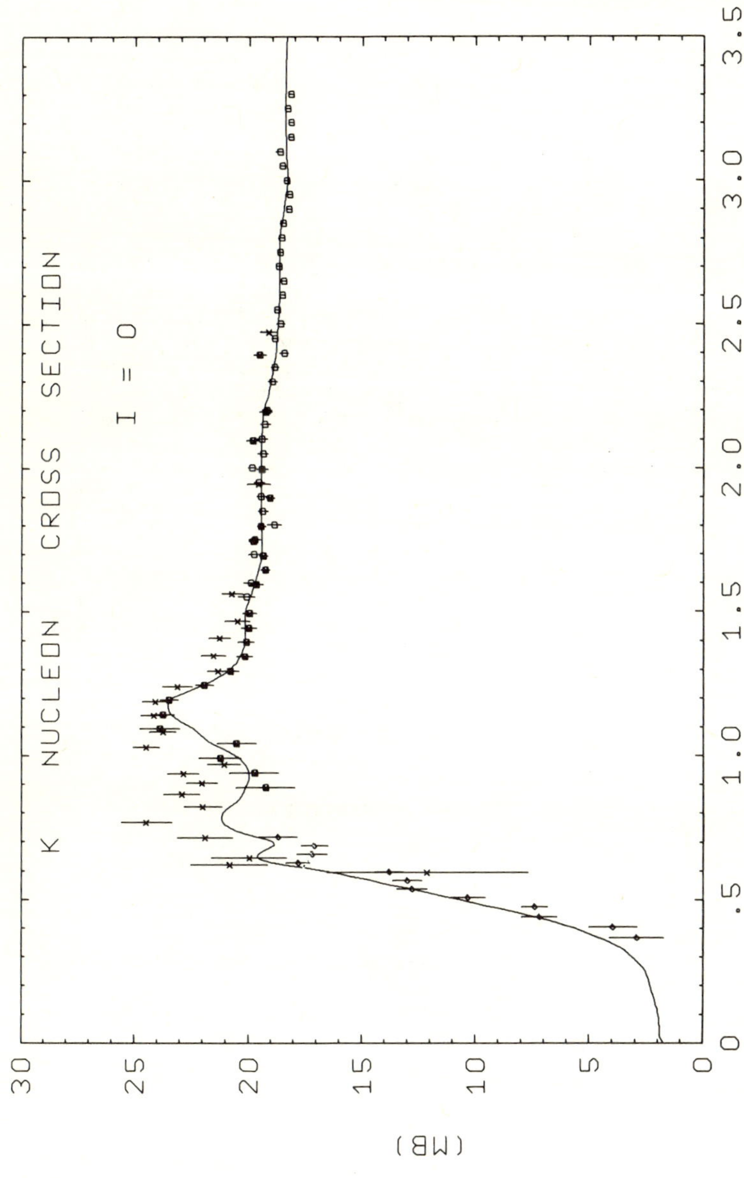

FIG. 19. The data points and smooth curve for the I = 0 KN cross section.

The next few figures are presented in an attempt to give one a feeling for the effect of various factors involved in the calculation of σ_0. Figure 20 compares the same solid curve that is in Fig. 19 with the result obtained if the internal momentum of the deuteron is ignored or if a different deuteron wave function is used. In doing the calculation a two-parameter Moravcsik deuteron wave function[12] was used. Here we see that the result is negligibly different if the Hulthén wave function is used. We can see that the unfolding process accentuates the structure that is there but for the K^+ data does not make the structure qualitatively different.

The effect of the Glauber correction is illustrated in Fig. 21. Clearly this Glauber correction has a considerable effect on the magnitude of the cross section but has essentially no effect on its structure. As I said before, the Glauber-correction formula involves the real part of the KN scattering amplitude. Figure 21 shows that whether or not this effect is included is unimportant. Actually the form of the Glauber correction used is derived from a high energy approximation that may be violated at the lower energies of this experiment. Above one GeV/c it is probably very good.

The methods used to do the unfolding and the Glauber correction have been checked for the pion-nucleon data by both the Brookhaven and Rutherford groups.[13] Since the $I = 1/2$ pion-nucleon cross section can be calculated from $\pi^- p$ and $\pi^+ p$ cross sections as well as from $\pi^- p$ and $\pi^- d$ cross sections, the method can be checked. The Brookhaven group found that above 900 MeV/c this method checks very well. But the Rutherford group found that at lower momenta the self-consistency is much poorer, to the extent that the Glauber correction seems to have the wrong sign. It is very difficult to estimate how much uncertainty is placed into the final derived cross sections owing to the inadequacy of the assumptions made: that one can use an impulse approximation to calculate the effect of the internal momentum of the nucleons in the deuteron, and that one can use at low momenta the Glauber correction that is valid in the diffraction region.

Previous attempts at unfolding the cross sections have neglected the flux-factor effect that arises from the fact that the measured rate depends upon not only the cross section but the relative velocities of the beam and the target. The result of neglecting this effect, as well as the result of neglecting the binding energy of the dueteron, is shown in Fig. 22. Each of these effects causes a slight shifting of the positions of peaks, but the magnitude of the effect is not important at the present level of accuracy.

THE I = 0 AND I = 1 $\bar{K}N$ CROSS SECTIONS

Before I present the results for the $\bar{K}N$ cross sections in the form of data points, let us look at how the results in this case

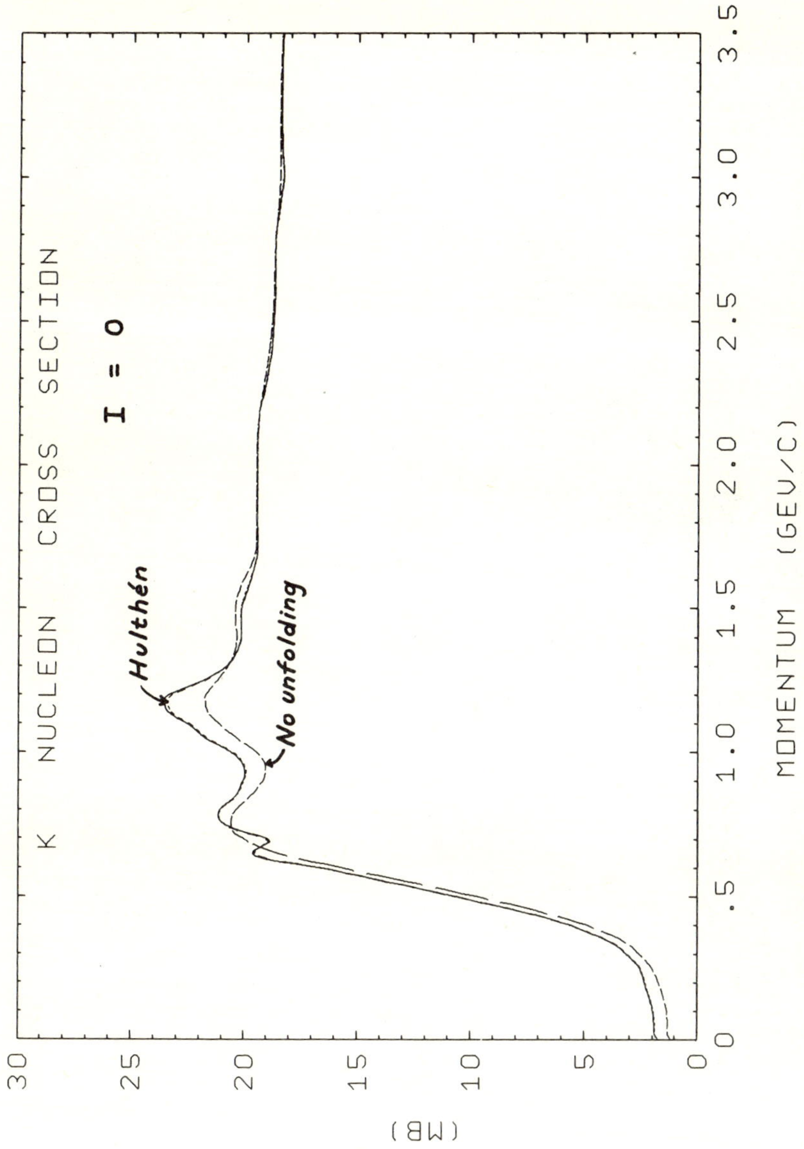

FIG. 20. The effect on σ_0 of neglecting unfolding and of using a the Hulthén wave function.

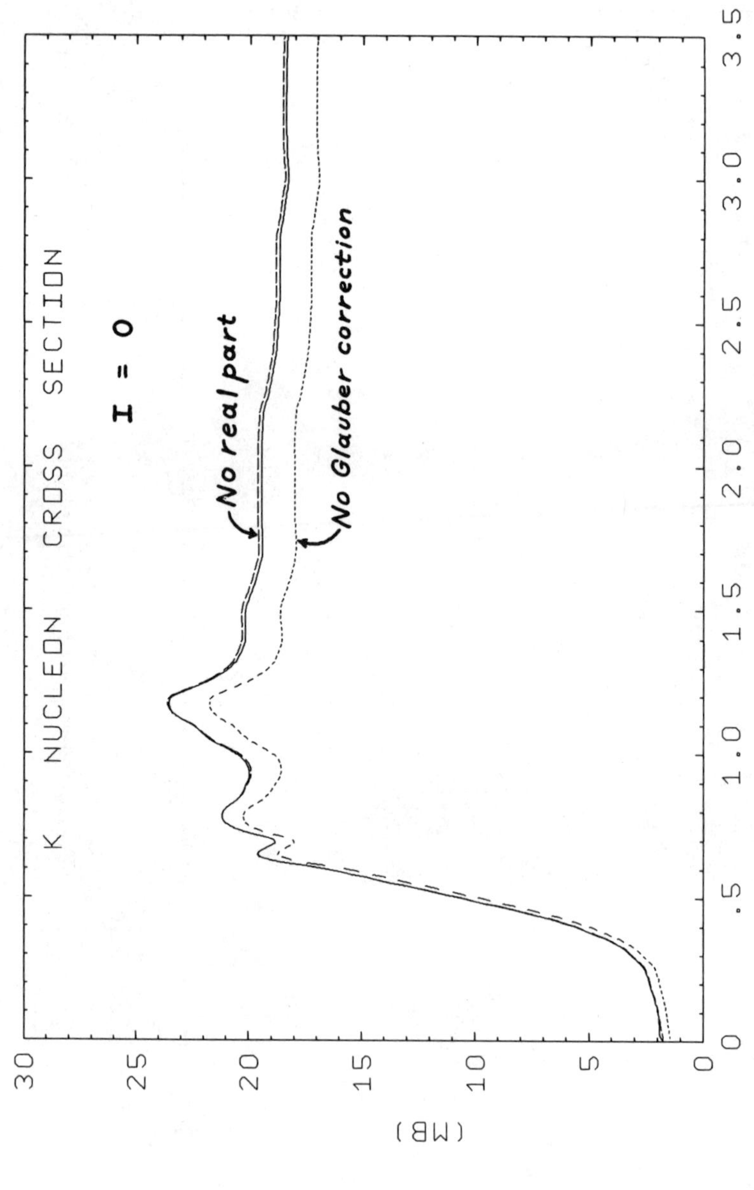

FIG. 21. The effect on σ_0 of neglecting or modifying the Glauber correction.

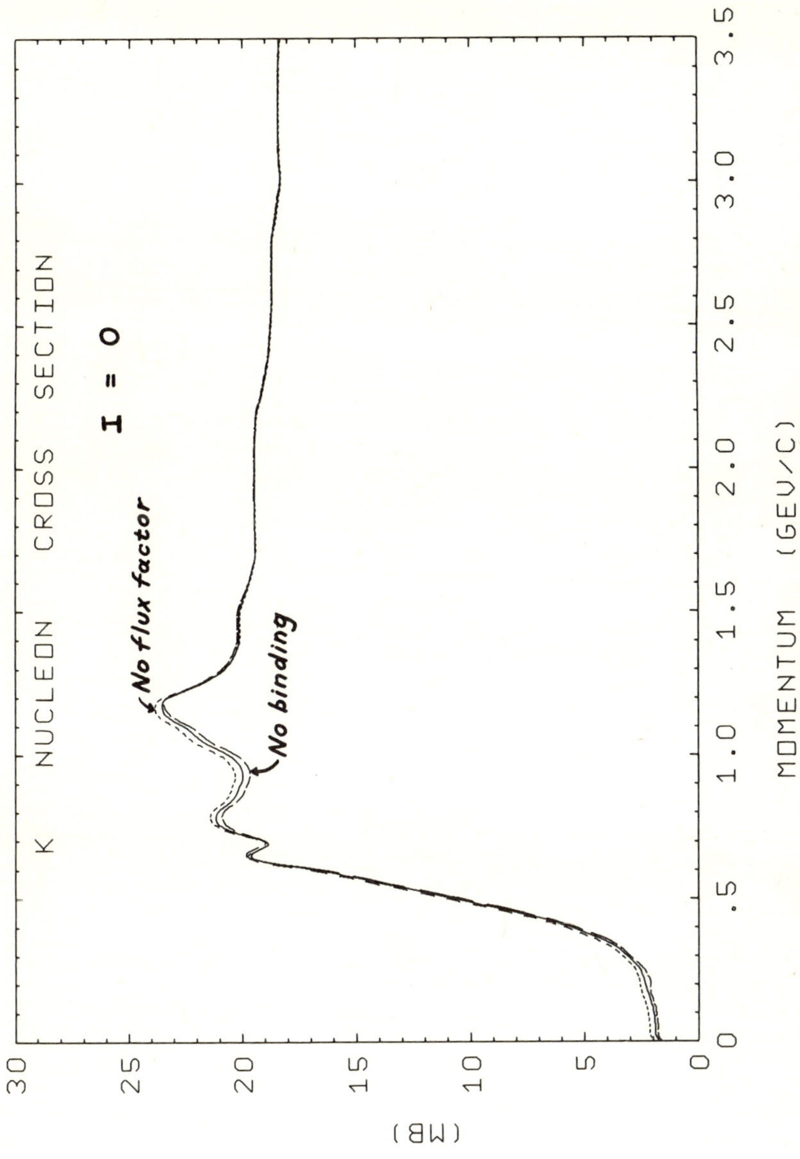

FIG. 22. The effect on σ_0 of neglecting the flux factor or the deuteron binding energy.

depend upon various things. The effect of the Glauber correction is shown in Fig. 23, and Fig. 24 shows the effect of neglecting the unfolding altogether. One can see that many of the features in the I = 1 cross section are not visible, or are hardly present before the unfolding is done. In particular the peak at 1.26 GeV/c, which has been often been identified as the recurrence of the Σ, is not present before unfolding is done. These features of the derived cross section are suspect since in this case, in contrast to the K^+ case, there are non-negligible errors introduced by the unfolding process itself, because there are resonances such as the $Y_0^*(1520)$ and the $Y_0^*(1815)$ that are narrow enough compared with the effect of the internal momentum to complicate the unfolding. There are many different schemes that have been used to do the unfolding. One rather elegant scheme is to write down a set of linear equations that express how each measured data point depends upon the unknown unfolded cross-section points. Then one obtains the answer by solving this set of simultaneous equations. A difficulty with this method is that there is no way of imposing smoothness on the final answer; it may give an answer that has high frequency fluctuations. Therefore most attempts at unfolding have used an iterative method. By starting with a guess of the unfolded cross section and correcting this by the amount that the resulting folded cross section disagrees with the measured data, one can iterate toward an answer. Usually this process does not converge but first approaches a self-consistent result and then develops instabilities that cause it to diverge. To avoid this trouble the Brookhaven group stopped after three iterations. My approach was to smooth each new guess by setting each point to the value of the best-fit parabola through the five points centered about that point (for a point spacing of 20 MeV/c). This method converged to a very self-consistent solution in four interactions for the K^+d data, but took about 10 iterations to converge to a not nearly so self-consistent result for the K^-d data. This latter method produced results that have slightly more pronounced structure than the results of the former method, but the differences are so small that on graphs such as the ones that I have been presenting here the differences are difficult to see above 500 MeV/c.

Since the solution is not very self-consistent, one is tempted to use points every 10 MeV/c rather than every 20 MeV/c. This amounts to relaxing the smoothness criterion. The result of doing this is shown in Fig. 25. We see that this solution has considerably more exaggerated structure than the solution using 20-MeV/c intervals. Thus the solutions that we get are somewhat sensitive to the extent to which the solutions are forced not to have fine structure. The differences between the two curves in Fig. 25 is a fair measure of the size of the systematic error due to this effect above 500 MeV/c. Below 500 MeV/c the systematic errors are larger because of the presence of the large narrow $Y^*(1520)$ and because this region is strongly influenced by the poorly known but large and rapidly varying cross section below 360 MeV/c.

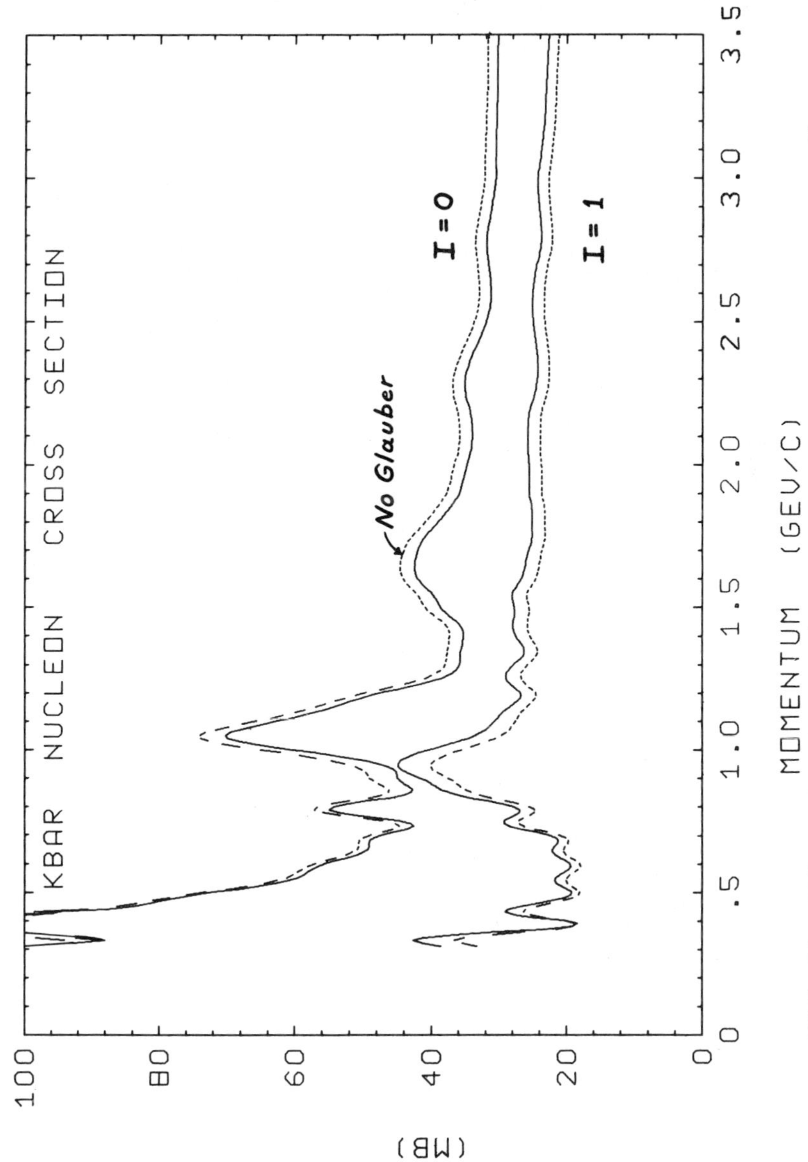

FIG. 23. Effect of neglecting **the** Glauber correction on the derived $\bar{K}N$ total cross sections.

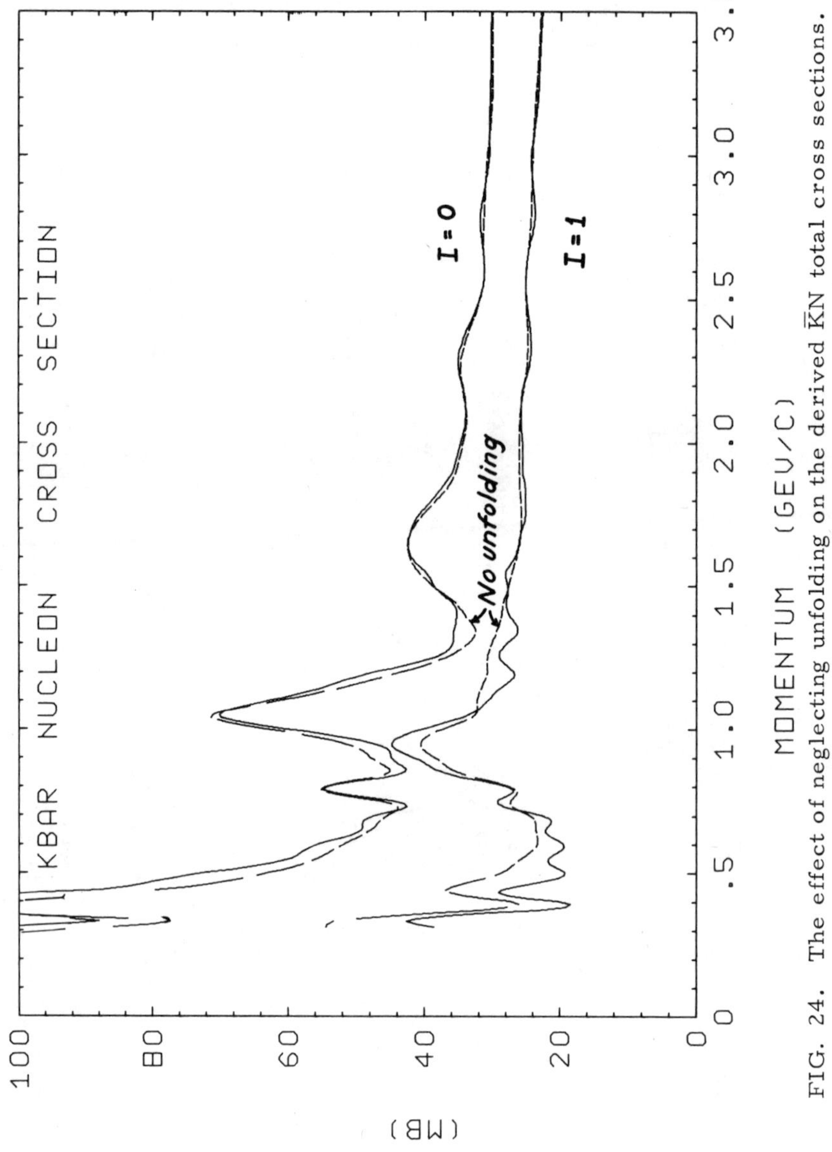

FIG. 24. The effect of neglecting unfolding on the derived $\bar{K}N$ total cross sections.

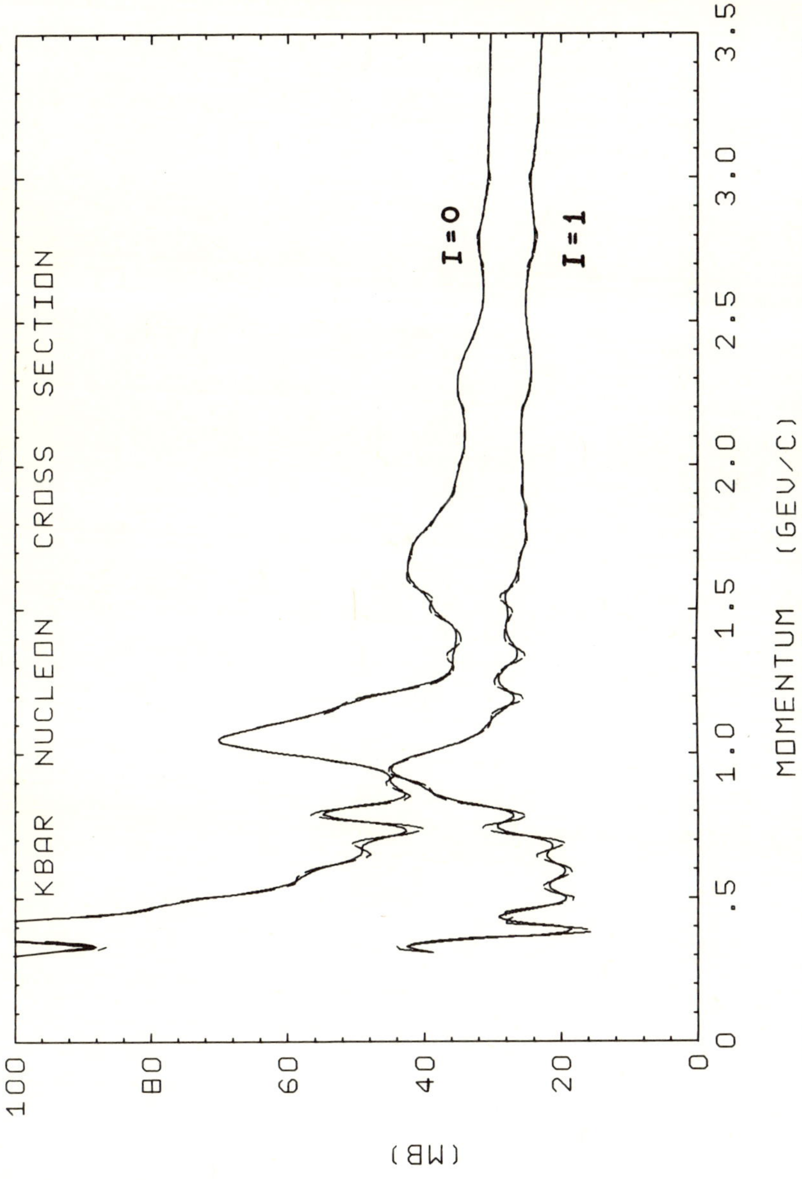

FIG. 25. Effect of using 10- rather than 20-MeV/c intervals on the derived $\overline{K}N$ total cross sections.

The sensitivity to the amount of smoothing is much greater than the sensitivity to other reasonable modifications, such as changing the deuteron wave function, neglecting the real part of the scattering amplitude, or neglecting the flux factor. It also makes little difference whether one finds the neutron cross section by substracting the proton cross section from the unfolded deuteron cross section (which I did) or if one unfolds the difference between the deuteron cross section and the folded proton cross section (which the Brookhaven group did).

The data points for the I = 0 cross section are shown in Figs. 26 and 27. As we have seen, the statistical errors do not reflect the entire uncertainties in this cross section. Nevertheless all of the peaks in this plot are unquestionably correct, including the $Y^*(1520)$ at 390 MeV/c, the $Y^*(1690)$ at 800 MeV/c, the $Y^*(1815)$ at 1050 MeV/c, the $Y^*(2100)$ at 1700 MeV/c, and the bump at 2300 MeV/c.

The situation for I = 1, as seen in Figs. 28 and 29, is much less clear. The apparent peak at 440 MeV/c is almost surely not real, but is a result of the inability to adequately unfold the $Y^*(1520)$ peak in the K^-d cross section. The structure in the smooth curve between 500 and 800 MeV/c is not demanded by the data. The $Y^*(1765)$ is clearly visible at 950 MeV/c. The $Y^*(1915)$ at 1260 MeV/c is, as we have seen, entirely a product of the unfolding, and because of this must be considered as much more suspect than the resonances indicated in the I = 0 state. Nevertheless the $Y^*(1915)$ emerges from the unfolding no matter what plausible method is used. The $Y^*(2030)$ shows up here at 1500 MeV/c. Its magnitude is not as great as the magnitude indicated by the analyses of the charge-exchange data[14] in this region. There is a hint here that this object is a double rather than a single structure.

CONCLUSION

The I = 0 cross section for the KN system can be calculated in a way that depends very little upon the method of analysis that is chosen. The greatest uncertainty in these results comes from the disagreement between the different experiments. Yet another experiment will be needed before the detailed shape of the cross section will be known in the region from 700 MeV/c to 1 GeV/c.

For the K^- data the uncertainties introduced by the unfolding are more serious than the disagreement between the experiments, and the derived cross sections should not be taken as accurate cross sections for the purpose of testing models. The total cross-section data are very precise data and should be used in phase-shift analyses. But it is my advice that anyone who does phase-shift analyses should use the raw data themselves rather than the derived cross section, that is, he should fold and correct his assumed solutions to compare with the measured Kd cross section rather than compare his solution with an unfolded corrected derived cross section.

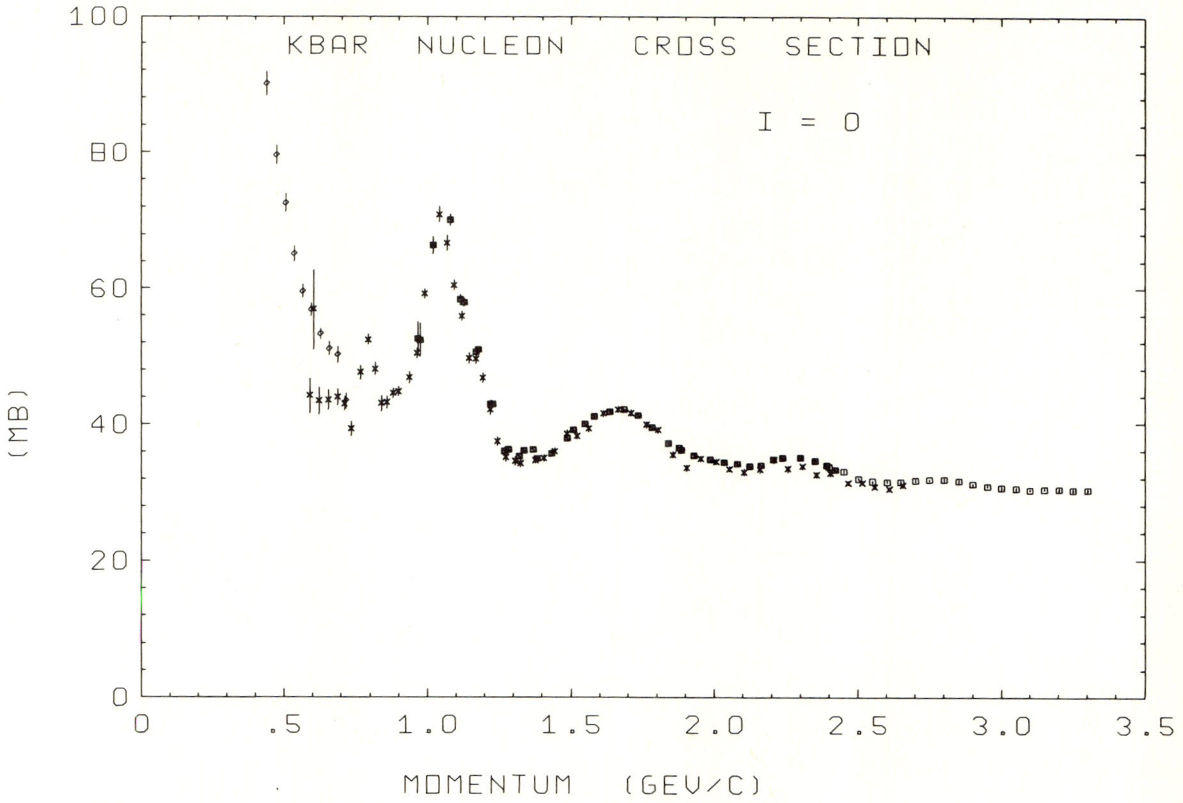

FIG. 26. The data points for the I = 0 $\overline{K}N$ cross sections.

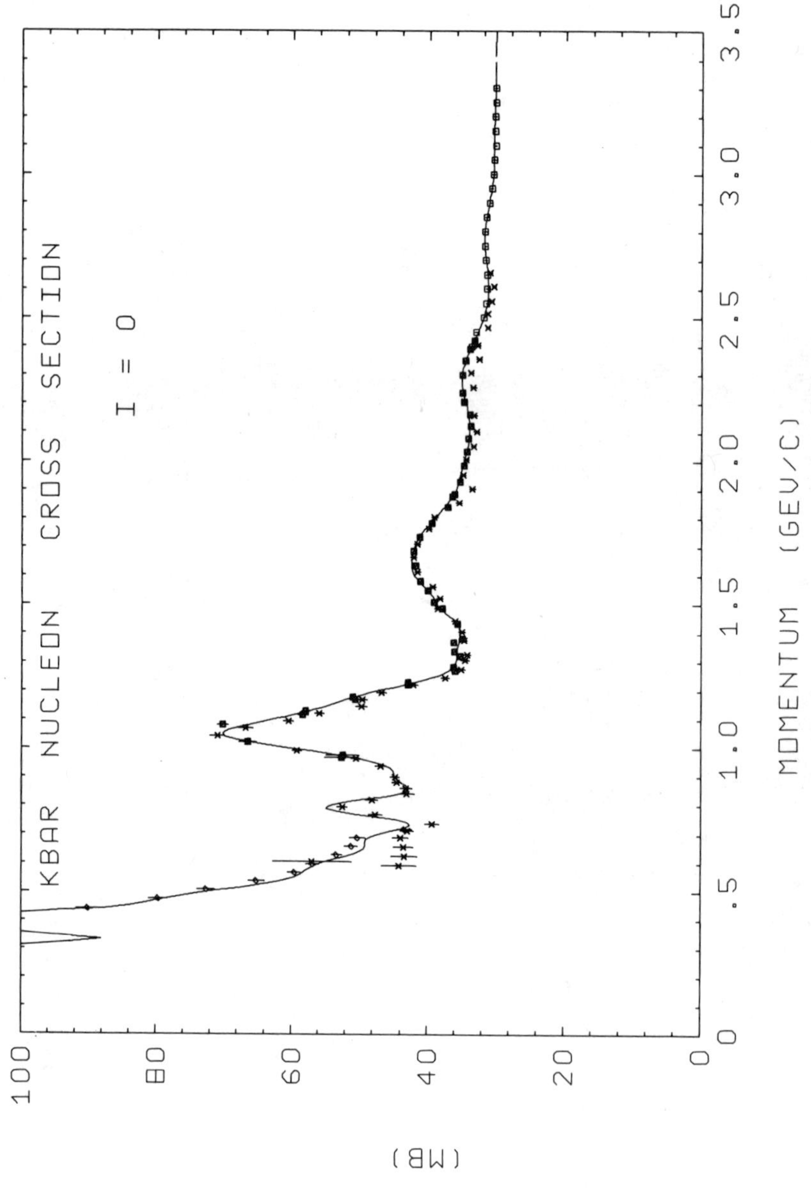

FIG. 27. The data points and smooth curve for the $I = 0$ $\bar{K}N$ cross section.

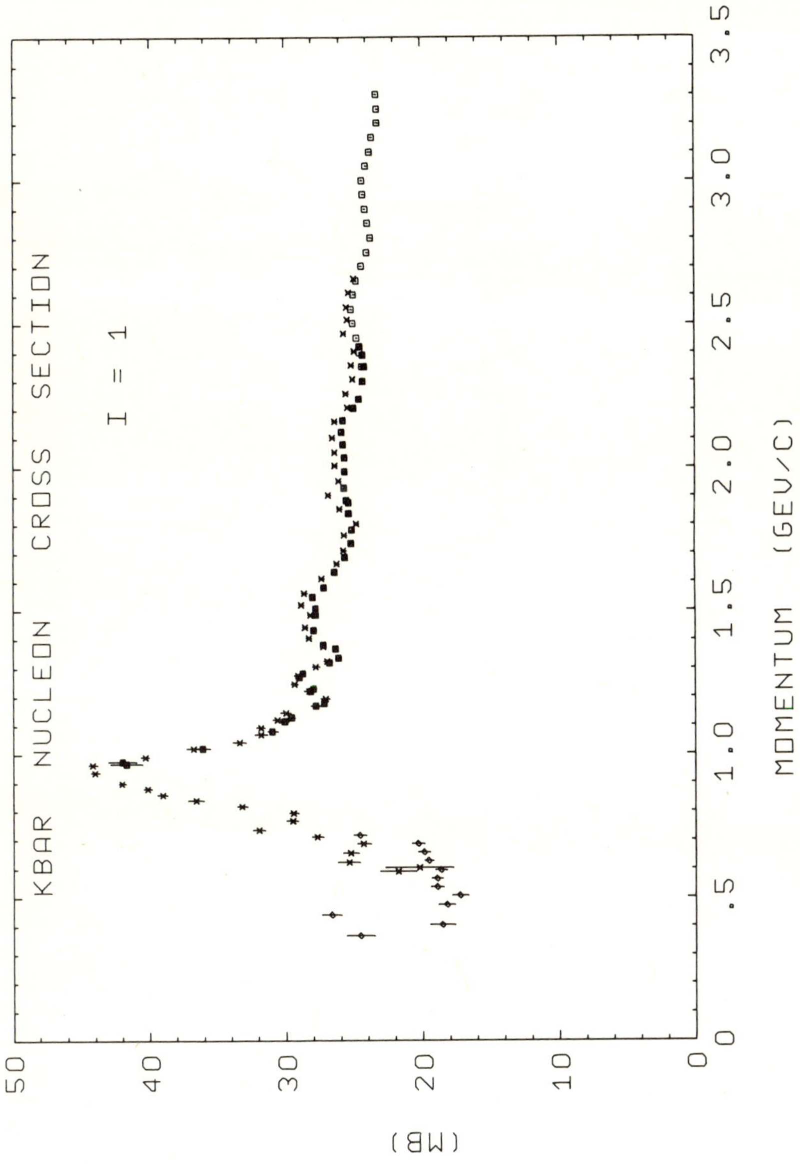

FIG. 28. The data points for the I = 1 $\overline{K}N$ cross section.

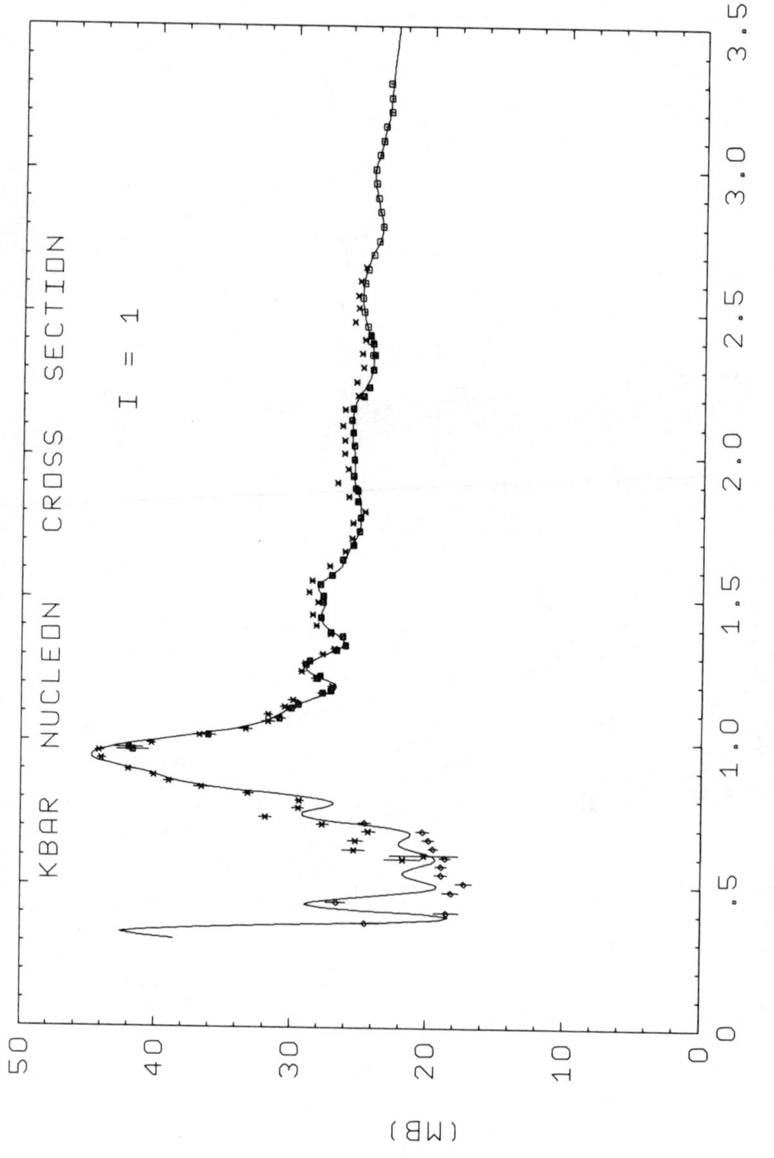

Fig. 29. The data points and smooth curve for the I = 1 $\bar{K}N$ cross section.

FOOTNOTE AND REFERENCES

*Work done under the auspices of the U. S. Atomic Energy Commission.

1. R. J. Abrams, R. L. Cool, G. Giacomelli, T. F. Kycia, B. A. Leontic, K. K. Li, and D. N. Michael, Phys. Rev. Letters 18, 1209 (1967); 19, 259 (1967) and 19, 678 (1967). An article by the same authors has been submitted to the Physical Review [BNL 14046 (1969)].
   R. L. Cool, G. Giacomelli, T. F. Kycia, B. A. Leontic, K. K. Li, A. Lundby, and J. Teiger, Phys. Rev. Letters 16, 1228 (1966); 17, 102 (1966). An article by the same authors plus C. Wilkin has been submitted to the Physical Review [BNL 14047 (1969)].
2. D. V. Bugg, R. S. Gilmore, K. M. Knight, D. C. Salter, G. H. Stafford, E. J. N. Wilson, J. D. Davies, J. D. Powell, P. M. Hattersley, R. J. Homer, A. W. O'Dell, A. A. Carter, R. J. Tapper, and K. F. Riley, Phys. Rev. 168, 1466 (1968).
3. E. Jenkins (private communication) and D. V. Petersen, Ph.D. Thesis, University of Arizona.
4. H. C. Burrows, D. O. Caldwell, D. H. Frisch, D. A. Hill, D. M. Ritson, and R. A. Schluter, Phys. Rev. Letters 2, 117 (1952).
5. V. Cook, D. Keefe, L. T. Kerth, P. G. Murphy, W. A. Wenzel, and T. F. Zipf, Phys. Rev. Letters 7, 182 (1961).
   V. Cook, B. Cork, T. F. Koang, D. Keefe, L. T. Kerth, W. A. Wenzel, and T. F. Zipf, Phys. Rev. 123, 320 (1961).
6. S. Goldhaber, W. Chinowsky, G. Goldhaber, W. Lee, T. O'Halloran, T. F. Stubbs, G. M. Pjerrou, D. H. Stork, and H. K. Ticho, Phys. Rev. Letters 9, 135 (1962).
7. T. F. Kycia, L. T. Kerth, and R. G. Baender, Phys. Rev. 118, 553 (1960).
   R. W. Bland et al., Lawrence Radiation Report UCRL-18758 (1969).
   W. F. Baker, R. L. Cook, E. W. Jenkins, T. F. Kycia, R. H. Phillips, and A. L. Read, Phys. Rev. 129, 2285 (1963).
   G. von Dardel, D. H. Frisch, R. Mermod, R. H. Milburn, P. A. Piroue, M. Kivargent, G. Weber, and K. Winter, Phys. Rev. Letters 5, 333 (1960).
   A. Bettini, M. Cresti, S. Limentani, L. Peruzzo, R. Santangelo, D. Locke, D. J. Crennell, W. T. Davies, and P. B. Jones, Physics Letters 16, 83 (1965).
   W. Chinowsky, G. Goldhaber, S. Goldhaber, T. O'Halloran, and B. Schwarzchild, Phys. Rev. 139B, 1411 (1965).
   P. Sällström, G. Otter, and G. Ekspong, Nuovo Cimento, 49A, 348 (1967).
   O. Chamberlain, K. M. Crowe, D. Keefe, L. T. Kerth, A. Lemonick, Tin Maung, and T. F. Zipf, Phys. Rev. 125, 1696 (1962).

P. L. Bastien and J. P. Berge, Phys. Rev. Letters 10, 188 (1963).
A. N. Diddens, E. W. Jenkins, T. F. Kycia, and K. F. Riley, Phys. Rev. 132, 2721 (1963).
8. J. D. Dowell, Isotopic Spin Analysis of K^+p and K^+d Total Cross Sections (to be published).
9. R. Armenteros et al., (to be submitted to Nuclear Physics B.
10. R. J. Glauber, Phys. Rev. 100, 242 (1955).
R. J. Glauber and V. Franco, Phys. Rev. 142, 1195 (1966).
11. The formulation is that of C. Wilkin, Phys. Rev. Letters 17, 561 (1966).
12. M. Moravcsik, Nuclear Physics 7, 113 (1958).
13. A. A. Carter, K. F. Riley, R. J. Tapper, D. V. Bugg, R. S. Gilmore, K. M. Knight, D. C. Salter, G. H. Stafford, E. J. Wilson, J. D. Davies, J. D. Dowell, P. M. Hattersley, R. J. Homer, and A. W. O'Dell, Phys. Rev. 168, 1457 (1968).
14. C. W. Wohl, F. T. Solmitz, and M. L. Stevenson, Phys. Rev. Letters 17, 107 (1966).

COMMENTS ON THE SCATTERING OF K^+ MESONS IN THE ISOSPIN ZERO STATE

Rodney L. Cool

Brookhaven National Laboratory, Upton, New York

I would like at this time to exercise my prerogative as discussion leader to take a few minutes for remarks on the isospin zero K^+ cross sections and their interpretation. Some months ago we published a reanalysis of our data, combined with the Rutherford data, of the I=0 total cross section.[1] The combined data showed the probable existence of two low energy peaks; one near 0.79 GeV/c, the other near 1.14 GeV/c. A satisfactory fit could be made to our data alone, for which the lowest point was 0.9 GeV/c, with a single peak at 1.15 GeV/c.[2] The lower peak, if interpreted as due to resonant behavior, gave the parameters $E_r \simeq 1.78$ GeV with $\Gamma \cong 0.56$ GeV. New data of Jenkins, et al. at lower energies has been presented here. This data for K^+p agrees with the Rutherford data but does not agree well with the K^+d data. The isospin zero cross sections derived from the K^+p and K^+d data then depend to some extent on which of these discrepant data is accepted as correct. Dowell has submitted a paper to this conference in which he has made a fit to the Rutherford data and the Arizona data as shown in Fig. 1 which essentially averages the data in the region of most serious disagreement. At the higher energies, he has used the Rutherford data alone since it is in excellent agreement with our observations. The I=0 cross section which has been deduced is shown in Fig. 2. From this fit of the data a clear two-peaked structure results. However, as Lynch has clearly pointed out earlier in this session, the exact choice of which data points to accept as being correct can appreciably influence the shape of the I=0 cross section.

We have also made a new analysis based on all of the data now available. For this analysis, we have again used the method we developed earlier[1] and which was described here by Lynch. This method allows us to display the I=0 points which are deduced from the data, together with their errors, essentially independent of the choice of the curves used to fit the original data.

The I=0 points thus derived are shown in Fig. 3. This figure now makes clear where the uncertainties lie. Since we do not seem to be in a position to choose which data are correct, the exact I=0 energy dependence is unclear.

But I wish to point out that some features are still qualitatively clear, even allowing for the uncertainties in the data. They can be enumerated as follows:

1. All analyses show a peak of $\sim$ 3-4 mb at about 1.15 GeV/c.

2. There is definite structure between 0.6 and 0.9 GeV/c which is, at minimum, a shoulder and, at maximum, a definite peak in I=0.

3. As emphasized by Lynch, the uncertainties arising from the nuclear physics of the deuteron, namely the Glauber correction, the internal momentum spectrum, the contribution of the real amplitude and the like, do not alter the qualitative features of the energy dependence of the cross section.

Figure 4 shows the behavior of the total cross section from our analysis (and the limits of uncertainty) together with the elastic and inelastic contributions as taken from bubble chamber data. Clearly, despite the uncertainty in σ_T(I=0), the elastic cross section must have a large peak near 0.7 GeV/c. This behavior is due simply to the fact that σ_T rises sharply at energies 0.4 to 0.5 GeV/c below the rise of the inelastic cross section. If the peak is as large as indicated by some of the data, the elastic cross section reaches $4\pi\lambda\!\!\!^{-2}$ (see Fig. 2). It is worth noting that it is very difficult to achieve such a large elastic cross section in a low angular momentum state without resonance or near-resonance behavior. Taken literally it suggests a $P_{1/2}$ nearly elastic resonance.

The cause of the peak near 1.2 GeV/c remains obscure. The interpretation of these structures must await better data, particularly elastic scattering angular distributions and polarization measurements which will permit phase shift analyses.

REFERENCES

1. R.J. Abrams, et al., Physics Letters 30B, 564 (1969).

2. R.L. Cool, et al., Phys. Rev. Letters 17, 102 (1966).

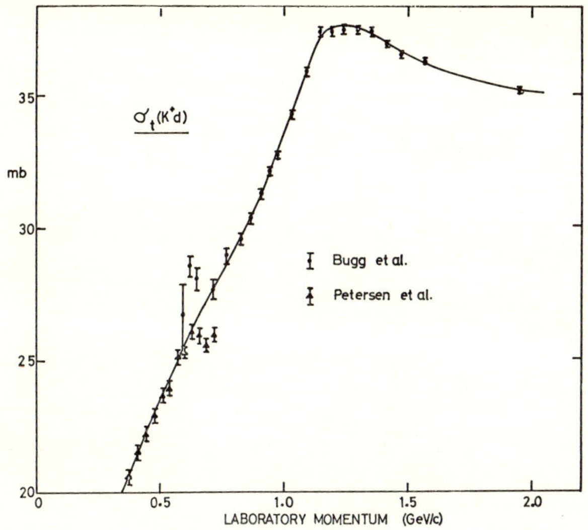

Fig. 1: K^+d total cross section data used by Dowell for the analysis of the I=0 scattering.

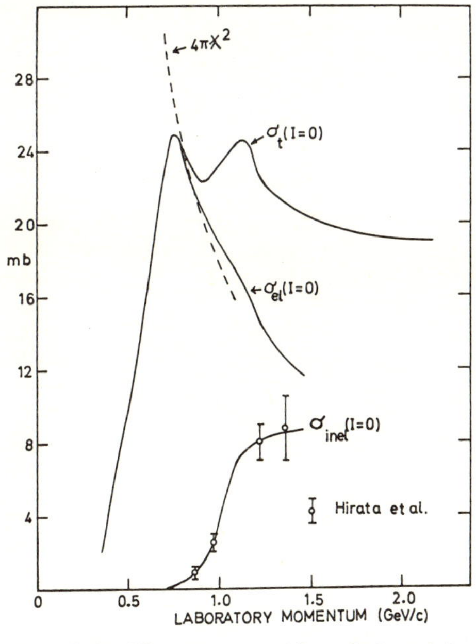

Fig. 2: The I=0, K^+ cross section deduced by Dowell.

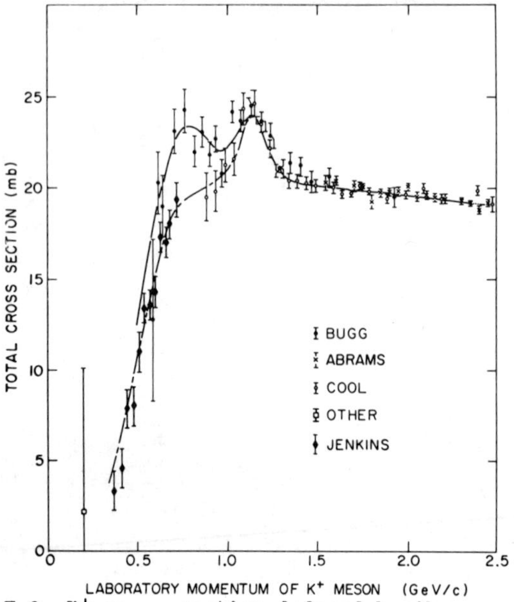

Fig. 3: The I=0, K^+ cross section deduced by Abrams, et al. from all of the available data.

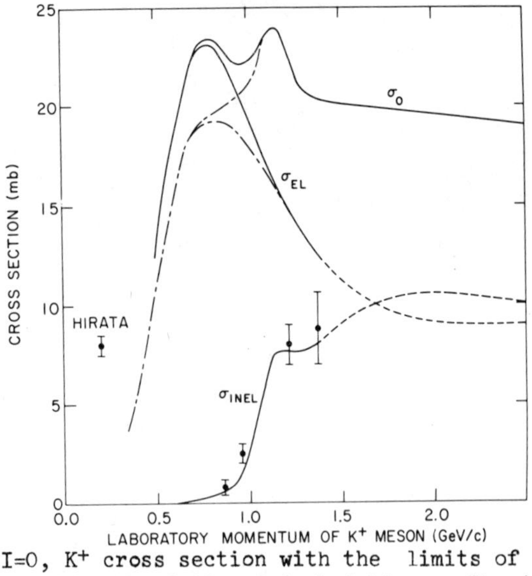

Fig. 4: The I=0, K^+ cross section with the limits of uncertainty indicated by the full and dashed lines. The inelastic and elastic contributions are indicated separately.

DISCUSSION

Hemingway: Is not Dowell's smooth curve more reasonable than yours?

Lynch: No I don't think so. I think that it is more likely that one experiment has a systematic error over a sizable momentum interval than that both experiments have considerable systematic errors in their overlap region and slight systematic errors in the regions where they do not overlap.

Cool: I think I'll open the session for the general discussion, but first I know that we have a paper submitted to the conference by John Dowell. I think the slide should be shown for an interpretation of the results of the Rutherford group for the isospin zero K^+ state. I don't know whether Damerell has any remarks to make --.

Damerell: As has been shown, uncertainties in the data from different groups, and uncertainties in the unfolding of the effect of Fermi momentum, makes the existence or otherwise of a double peaking in the I=0 total cross section rather uncertain. However, the behavior of the measured I=0 inelastic cross section indicates that in some sense there are really two peaks present. There is firstly a peak in the elastic scattering cross section, followed at higher momentum by the rising inelastic cross section. Whether or not they are resolved in the total cross section is possibly irrelevant.

Kalbfleisch: Do you have a slide showing all the Z^*, Λ and Σ states from the Particle Data Group or others like Levi-Setti (Lund, 1969) as vertical lines on your σ (total cross section) versus p (momentum) curves?

Lynch: I do not have a graph like that. I took the point of view that the determination of resonance parameters from total cross section data alone is such an underdetermined problem that it is not worth pursuing.

Moneti: How did you parameterize your smooth curve?

Lynch: By drawing a curve by hand through the data points on graph paper, reading points off of this graph at convenient intervals and doing quadratic interpolations between these points. The "smooth curves" drawn on the figures were made by connecting points at 10 MeV/c intervals with straight lines.

Cool: I would like to remark that several smooth curves can reproduce the data after unfolding for deuteron motion and other corrections. What really matters is the choice between experiments which disagree. This disagreement mostly comes from normalization problems in deuterium.

Tuan: Can you tell me where are the positions of the two $K^+n(I=0)$ peaks and their tentative widths.

Cool: Yes, we find two bumps with the following masses and widths:
1st bump: $M = 1868$ MeV, $\Gamma = 160$ MeV
2nd bump: $M = 1780$ MeV, $\Gamma = 565$ MeV.
The first bump has been there for a long time. For details see R. J. Abrams et al., Physics Letters 30B, 564, 1969.

Tripp: Is it possible to renormalize one experiment to get overall agreement?

Cool: Yes, you get very good agreement when you move the Jenkins data up by 8 percent.

Jenkins: The difference between our data and the Rutherford data lies just at that shoulder. There the data differ by 8 percent.

Cool: Yes, in this shoulder lies the difference.

Goldhaber: We have analyzed the inelastic data and I will show the angular distribution data on K^+N, I=0, to K^*N. The inelasticity which is observed is not in a single wave, it is spread over many waves, it is sharply forward peaked.

Rosenfeld: I did not understand your remark concerning the shiftings of the Jenkins data by eight percent.

Cool: I am talking about the I=0 state you see, including the Jenkins data brings the original data (based on Rutherford work) down. There you can draw a completely different curve. Which one is the right one, I do not know.

Tripp: Have you done the same for the K^- data?

Cool: I have not done this.

Damerell: The Rutherford data and the Arizona data agree in K^-.

Hemingway: How long do we have to wait for the situation to be resolved?

Cool: There is this problem of unfolding the deuterium data. There is a disagreement between different groups concerning the normalization. This has to be resolved first.

Barnett: Were there any significant differences between the experimental methods of Jenkins and Bugg in the overlapping regions?

Jenkins: I am not familiar enough with the Bugg method to be able to say, but our points agree for hydrogen.

ISOTOPIC SPIN ANALYSIS OF K^+p AND K^+d TOTAL CROSS SECTIONS

J. D. Dowell

Department of Physics, University of Birmingham

Recently new preliminary total cross section results for K^+p and K^+d scattering were reported by Petersen et al.[1] These extend the range of precise counter measurements down to 0.36 GeV/c; the lowest momentum prior to these results was 0.59 GeV/c from the work of Bugg et al.[2] The new K^+p results agree well with those of Bugg et al. (Fig. 1) but those for K^+d (Fig. 2) show a discrepancy of about 8% in the region of overlap. The important new qualitative information is however that whereas the K^+p cross section is approximately constant at low momenta the K^+d continues to fall with decreasing momentum implying that the $I = 0$ part of the cross section is falling quite rapidly. In order to investigate this in detail we have carried out an isotopic spin analysis along the lines described in reference 2, where we take into account the effect of smearing by the Fermi motion in the deuteron as well as the Glauber-Wilkin correction[3]. The smooth curves drawn through the data in figs. 1 and 2 were used in the unfolding procedure. Only the date of Petersen et al.[1] and Bugg et al.[2] have been used. The data of Cool et al.[7] are in good agreement with those of Bugg et al.[6] above 0.9 GeV/c, their lowest momentum, and are omitted for simplicity.

The kinematics used assumed the struck particle to be off the mass shell and the spectator on the mass shell. The Hulthén wave function was used to calculate the smearing effect. This was applied to the K^+p cross sections, read from the smooth curve, which were then subtracted from the K^+d results at the corresponding momenta

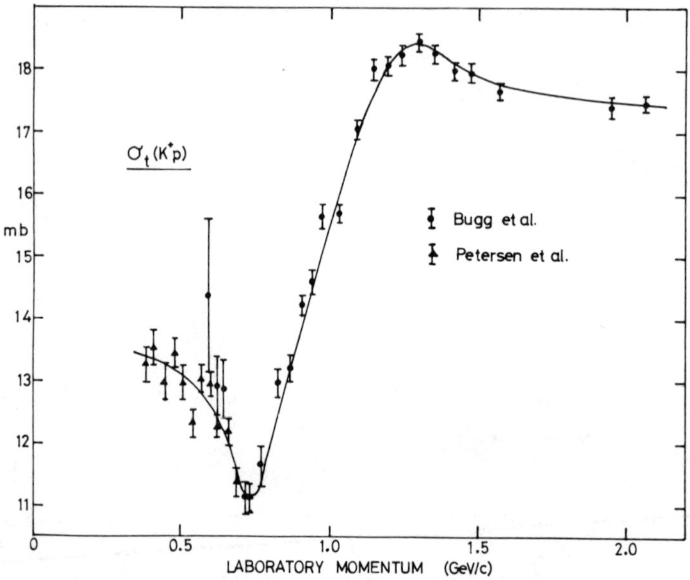

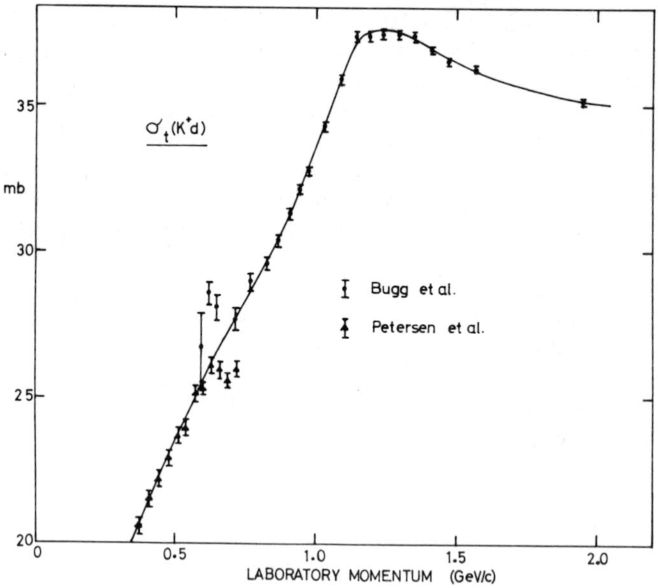

Figures 1 and 2. K^+p and K^+d total cross sections.

also taken from the smooth curve. The screening correction[3] was calculated with $\langle r^{-2} \rangle = 0.029\ \text{mb}^{-1}$ and account was taken in this correction of the real parts of the forward scattering amplitudes using the dispersion relation calculations of Carter[4].

Fig. 3 shows the result of the calculation for the I = 0 total cross section. Plotted on the same figure is the inelastic part of the I = 0 total cross section taken from the work of Hirata et al.[5]. Subtracting this from $\sigma_t(I = 0)$ yields the elastic (and charge exchange) I = 0 cross section. The striking feature of the results is the large peak in this cross section which reaches the unitarity limit $4\pi \lambda\!\!\!^{-2}$ for a $J = \frac{1}{2}$ resonance at 0.8GeV/c. This feature has already been noted by Hirata et al.[5], and in an analysis carried out by Abrams et al.[6] where the peak was fitted with a resonance formula (M = 1780 MeV, Γ = 565 MeV). These studies relied only on the data of Bugg et al.[2] at low momenta.

An appreciable contribution to the I = 0 peak at 0.75 GeV/c results from the valley in the K^+p total cross section which has been confirmed by the work of Petersen et al. The discrepancy between the two sets of K^+d measurements does not seriously affect the general conclusions. If we abitrarily renormalize the results of Petersen et al.[1] by a factor 1.08 then the effect is simply to broaden the I = 0 peak on the low momentum side by about 60 MeV/c, as indicated by the dashed curve on figure 3.

We thank Dr. D. V. Petersen of the Arizona group for kindly supplying us with numerical values of the data presented at Chicago. The errors shown on these results are typical values.

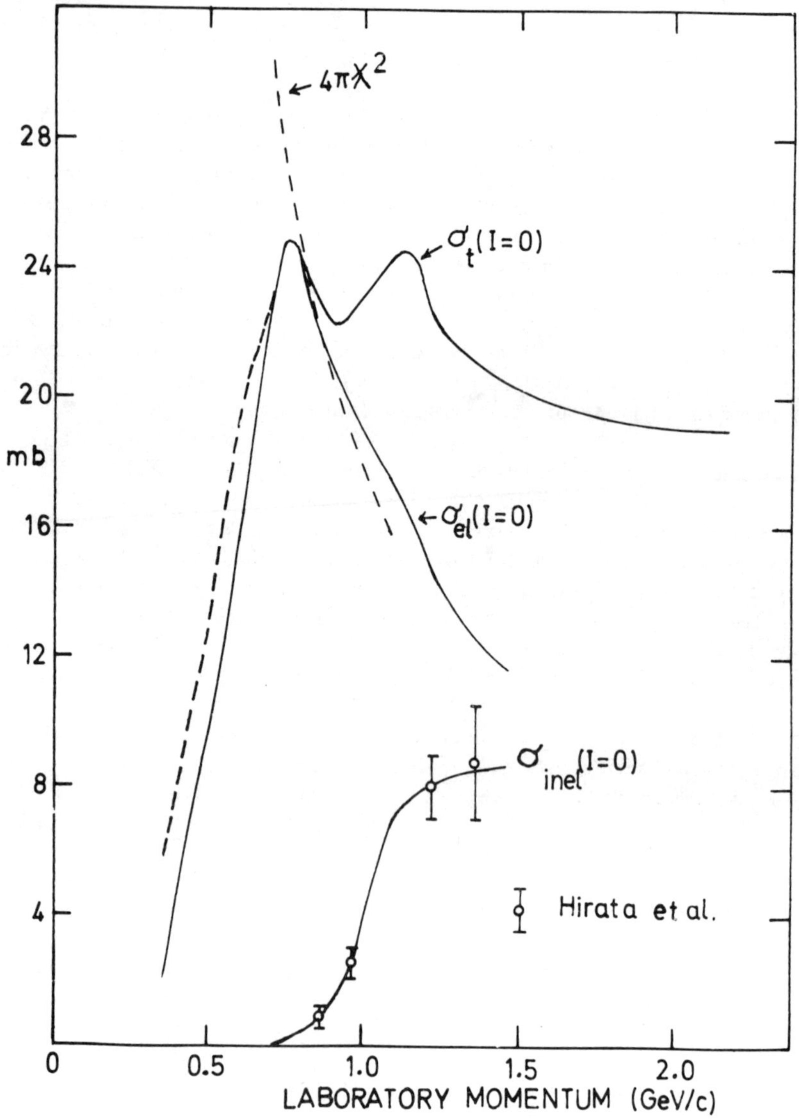

Figure 3. Unfolded I = 0 cross sections.

REFERENCES

1. D. V. Petersen, T. Bowen, P. K. Caldwell, F. Ned Dikmen, E. W. Jenkins, R. M. Kalbach, A. E. Pifer. Contribution to 1970 Chicago Meeting of the American Physical Society and private communication.

2. D. V. Bugg, R. S. Gilmore, K. M. Knight, D. C. Salter, G. H. Stafford, E. J. N. Wilson, J. D. Davies, J. D. Dowell, P. M. Hatterseley, R. J. Homer, A. W. Odell, A. A. Carter, R. J. Tapper and K. F. Riley. Phys. Rev. 168, 1466 (1968).

3. R. J. Glauber, Phys. Rev. 99, 630 (1955) 100, 242 (1955), C. Wilkin, Phys. Rev. Letters 17, 561 (1966).

4. A. A. Carter, Cavendish Laboratory Report HEP 68-10.

5. A. A. Hirata, C. G. Wohl, G. Goldhaber, and G. H. Trilling. Phys. Rev. Letters 21, 1485 (1968)

6. R. J. Abrams, R. L. Cool, G. Giacomelli, T. F. Kycia, K.K. Li and D. N. Michael, Phys. Letters 30B, 564 (1969).

7. R. L. Cool, G. Giacomelli, T. F. Kycia, B. A. Leontic, K. K. Li, A. Lundby and J. Teiger, Phys. Rev. Letters 17, 102 (1966).

SESSION II

LOWER ENERGY FORMATION

Discussion Leader: David Cline

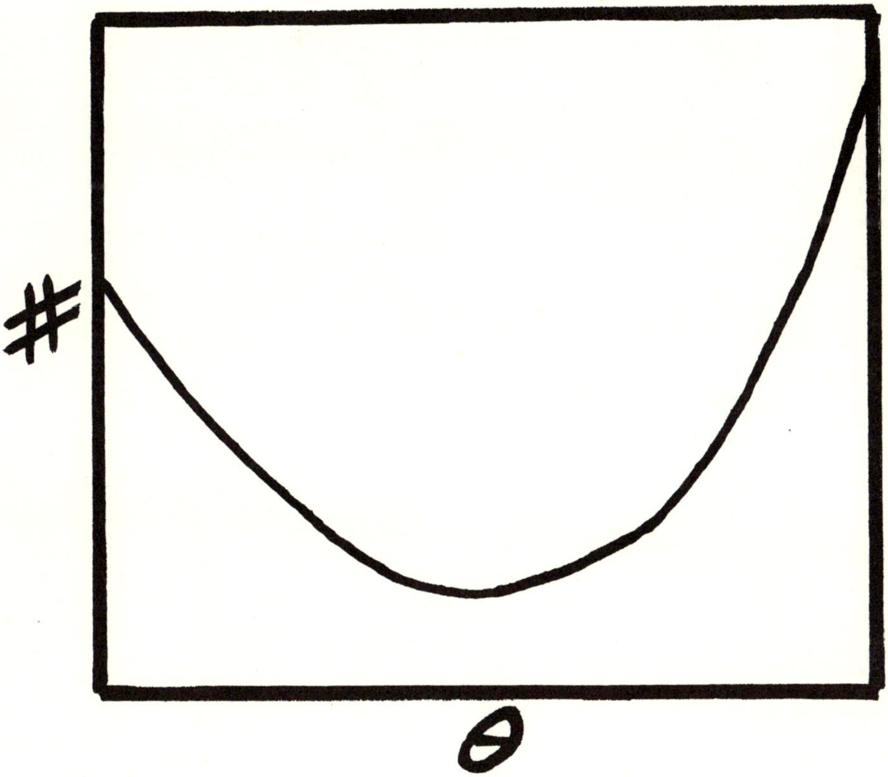

CONSTANT K-MATRIX FIT TO KN DATA UP TO 430 MEV/C*

J. Thompson

Brookhaven National Laboratory, Upton, New York 11973

As Dalitz and Tuan have pointed out,[1] in the K^-p system at low energies, because of the strong inelastic processes ($K^-p \rightarrow$ pion-hyperon), the conditions of unitarity (probability conservation) strongly affect the reaction amplitudes. The Dalitz-Tuan constant K-matrix formalism explicitly includes the unitarity conditions and has fewer parameters--because it allows less complicated energy dependence--than a K-matrix with effective ranges.[2] These attractive features, together with recently available data in the range 200-430 MeV/c[3,4] challenge one to see how wide an energy range of available data can be fit acceptably with the basic constant K-matrix parameterization.

Previously, S-wave fits[5,6] have been made to the data below 300 MeV/c;[5,7] fits have been made to the data around 400 MeV/c[8,4] region, and one fit using a K-matrix with effective ranges has been made using data from 0-530 MeV/c.[9] Also, an energy-dependent fit using constant K-matrix S-waves and constant scattering length isospin 1 P-waves has been made over the range 0-430 MeV/c[10] using only isospin 1 data above 300 MeV/c.[3,11]

The analysis of Ref. 4 indicates that, <u>except for the D_{03} wave associated with the Λ(1520)</u>, D-waves appear to be negligible below about 400 MeV/c, although their effects rise rapidly at higher momenta. Also, as the momentum increases, the three-body channels ($\Lambda\pi\pi$, $\Sigma\pi\pi$) are more important. The question is, how high in momentum can one go before the three-body channels, the presence of higher partial waves, and the more rapid energy variation of the amplitudes preclude a fit with the simple constant K-matrix parameterization.

---

*Work performed under the auspices of the U. S. Atomic Energy Commission.

This paper reports that several similar fits with R = (χ^2/degrees of freedom) ranging from 1.52-1.63 have been found for data[3,4,5,7,8,11] in the momentum range 0-430 MeV/c whereas no fit with R $<$ 1.86 was found when data at 450 MeV/c[12] were included. Previous fits, without the recent high statistics data near 400 MeV/c, found R $\sim$ 1.0-1.1. As statistical errors decrease, systematic errors become more important. The larger R's reported here reflect possible underestimation of these systematic errors, or possible parameterization inadequacies. The parameterization used was:

1) constant scattering length D_{13} waves in KN channels; these waves produce visible effects only above about 400 MeV/c,

2) Λ(1520) Breit-Wigner D_{03} wave in KN, $\Sigma\pi$ channels,

3) constant K-matrix for P_{01}, P_{11}, P_{03}, and P_{13} waves in KN, $\Sigma\pi$, and $\Lambda\pi$ channels,

4) constant K-matrix for S_{01} and S_{11} waves with amplitudes modified for the K^o-K^- mass difference and Coulomb effects in the K^-p system,[1]

5) for the K^op interactions, the strangeness +1 amplitudes are taken from recent analyses of K^+p and K^+d interactions. The S_{11} wave is described as a scattering length with an effective range,[13] while the other waves are extrapolated from 600 MeV/c,[14] assuming constant scattering lengths; the amplitudes thus found are consistent with the lower momentum work.[15] Ignoring all but the dominant S_{11} wave increases the χ^2 by about 3% and does not change the character of the K^o predictions.

The minimizing was done with the program MINFUN, with and without the data of Ref. 3, since these authors are still studying possible systematic errors. The solution of Ref. 4 was used as input. Varying the starting parameters and the step size for the various parameters yielded a family of solutions with generally similar χ^2, parameters, and amplitudes, but with some parameters differing by three to six times the quoted errors (χ^2 change of one). The best solution obtained with the Kadyk data had a χ^2 of 585 for 360 degrees of freedom. The best solution without the Kadyk data had a χ^2 of 466 for 306 degrees of freedom. The Tables give the parameters and amplitudes for this solution; the Figures compare the fit predictions with the data.

I would like to acknowledge suggestions from M. Sakitt, N. Samios, and W. Willis, and the encouragement and support of N. Samios.

Table I

	S	$P_{1/2}$	$P_{3/2}$	$D_{3/2}$	
Isospin 0				Isospin 0	
K^{o}_{KK}	-1.684 ± .020	.181 ± .010	.053 ± .007	E_R	1520.6 ± .4
$K^{o}_{K\Sigma}$	-.862 ± .010	-.154 ± .010	.023 ± .005	Γ_R	24.1 ± .4
$K^{o}_{\Sigma\Sigma}$	-.086 ± .025	-.039 ± .026	-.011 ± .073	Γ_K	6.3 ± .2
				Γ_Σ	16.2 ± .5
				X^2	.96 ± .7
Isospin 1				Isospin 1	
K^{1}_{KK}	.662 ± .020	-.051 ± .008	-.066 ± .006	a	-.014 ± .004
$K^{1}_{K\Sigma}$	-.993 ± .010	.059 ± .008	-.032 ± .004	b	0.0 ± .000002
$K^{1}_{K\Lambda}$	-.617 ± .007	-.052 ± .005	-.050 ± .002		
$K^{1}_{\Sigma\Sigma}$	1.265 ± .010	-.095 ± .072	-.207 ± .052		
$K^{1}_{\Lambda\Sigma}$	.039 ± .014	-.072 ± .020	-.044 ± .016		
$K^{1}_{\Lambda\Lambda}$	.882 ± .023	.051 ± .010	.102 ± .016		

Parameters of constant K-matrix fit. K^{I}_{AB} is the K-matrix element (dimensionless) for the isospin I transition from channel A to channel B. a and b (in fm^{-5}) are the real and imaginary parts of the D_{13} scattering length; since b = 0, the parameters describing the $\Sigma\pi$, $\Lambda\pi$ channels are omitted. The widths, Γ's, and center-of-mass energy, E_R, of the resonance are in MeV, and X^2 (in fm^{-2}) is the square of the inverse of the interaction radius. The errors quoted are the variable changes necessary to change the X^2 by 1. Only diagonal errors are given, although strong correlations certainly exist among some of the parameters, particularly those describing a single partial wave. More realistic estimates of the uncertainty in the Λ(1520) parameters are 1-3 MeV, to allow for systematic momentum errors in Ref. 4. The preliminary high statistics LRL data on charged channels near 400 MeV/c, reported at this Conference[16] indicate a larger elasticity and smaller width for the Λ(1520) than the previous (1963) results[8] which were available for this analysis. A larger elasticity and smaller width would also agree with recent publication experiments as reported by Levi-Setti.[8]

Table II

Amplitudes for Constant K-Matrix Solution with χ^2/D.F. = 466/306

	P_{K^-} (MeV/c)	S_{01}	S_{11}	P_{01}	P_{11}	P_{03}	P_{13}	D_{03}	D_{13}
KN	5	-.15 + i.10	0. + i.03	0. + i0.	0. + i0.	0. + i0.	0. + i0.	0. + i0.	0. + i0.
	70	-.37 + i.29	.02 + i.11	0. + i0.	0. + i0.	0. + i0.	0. + i0.	0. + i0.	0. + i0.
	170	-.35 + i.53	-.01 + i.25	.03 + i.01	-.01 + i.00	.01 + i0.	-.01 + i0.	0. + i0.	0. + i0.
	270	-.30 + i.67	-.05 + i.33	.11 + i.04	-.04 + i.01	.04 + i0.	-.04 + i.01	.03 + i0.	-.01 + i0.
	350	-.25 + i.73	-.07 + i.38	.21 + i.11	-.07 + i.03	.07 + i.01	-.09 + i.02	.11 + i.06	-.02 + i0.
	370							.13 + i.14	-.03 + i0.
	390	← S L O W L Y A N D S M O O T H L Y V A R Y I N G →						.05 + i.25	-.04 + i0.
	410							-.09 + i.22	-.05 + i0.
	430	-.22 + i.77	-.10 + i.42	.29 + i.24	-.13 + i.08	.12 + i.02	-.15 + i.04	-.13 + i.14	-.06 + i0.
Σπ	5	-.29 + i.09	-.08 - i.09	0. + i0.	0. + i0.	0. + i0.	0. + i0.	0. + i0.	0. + i0.
	70	-.40 + i.14	-.17 - i.14	-.02 + i0.	.01 + i0.	0. + i0.	0. + i0.	0. + i0.	
	170	-.32 + i.24	-.23 - i.26	-.06 + i0.	.02 + i0.	.01 + i0.	-.01 + i0.	.01 + i0.	
	270	-.28 + i.27	-.23 - i.29	-.15 - i.01	.05 + i.01	.02 + i0.	-.03 + i.01	.04 + i0.	
	350	-.26 + i.28	-.24 - i.30	-.23 - i.04	.09 + i.01	.04 + i0.	-.04 + i.03	.17 + i.09	
	370							.21 + i.22	
	390	← S L O W L Y A N D S M O O T H L Y V A R Y I N G →						.08 + i.40	
	410							-.15 + i.36	
	430	-.25 + i.28	-.24 - i.30	-.31 - i.08	.12 + i.01	.06 + i.01	-.05 + i.06	-.21 + i.22	↓
Λπ	5		-.06 - i.07		0. + i0.		0. + i0.		
	70		-.13 - i.11		-.09 + i0.		-.01 + i0.		
	170		-.17 - i.19		-.03 + i0.		-.03 - i.01		
	270		-.18 - i.21		-.07 - i.02		-.07 - i.01		
	350		-.19 - i.21		-.11 - i.04		-.10 - i.01		
	430		-.19 - i.22		-.15 - i.07		-.13 - i.01		

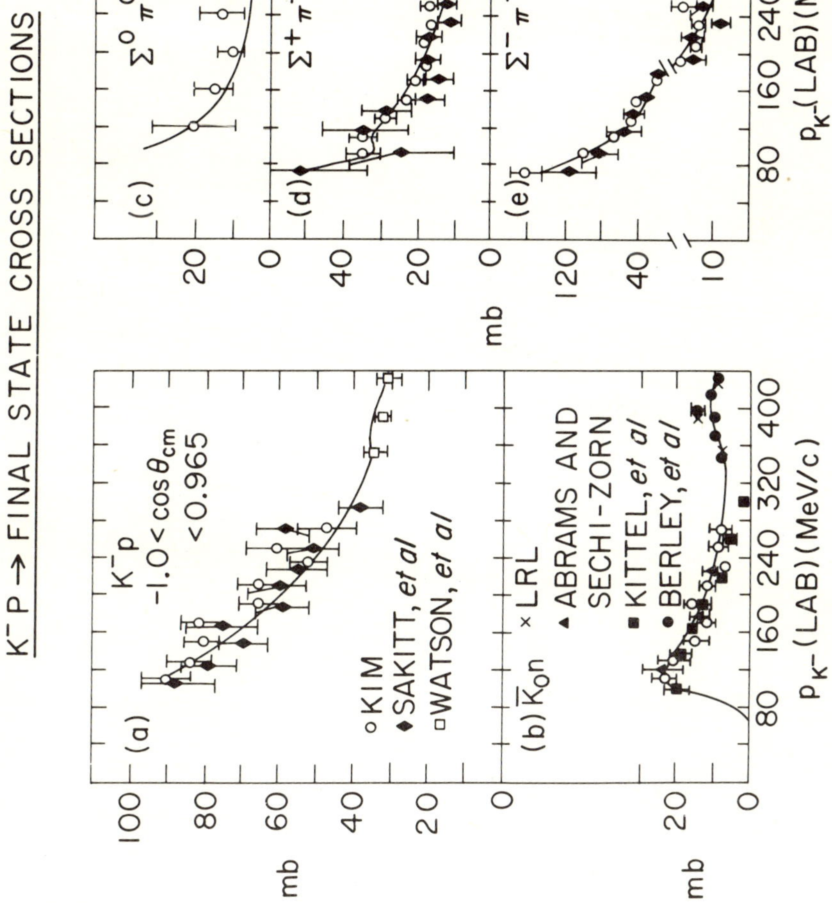
K⁻P → FINAL STATE CROSS SECTIONS
(a)
K⁻p
-1.0 < cos θcm < 0.965
○ KIM
◆ SAKITT, et al
□ WATSON, et al
(b) K̄₀n
× LRL
▲ ABRAMS AND SECHI-ZORN
■ KITTEL, et al
● BERLEY, et al
(c) Σ⁰π⁰
(d) Σ⁺π⁻
(e) Σ⁻π⁺
mb
p_K⁻ (LAB) (MeV/c)
80
160
240
320
400
100
80
60
40
20
0
120
10

ISOSPIN 1 DATA

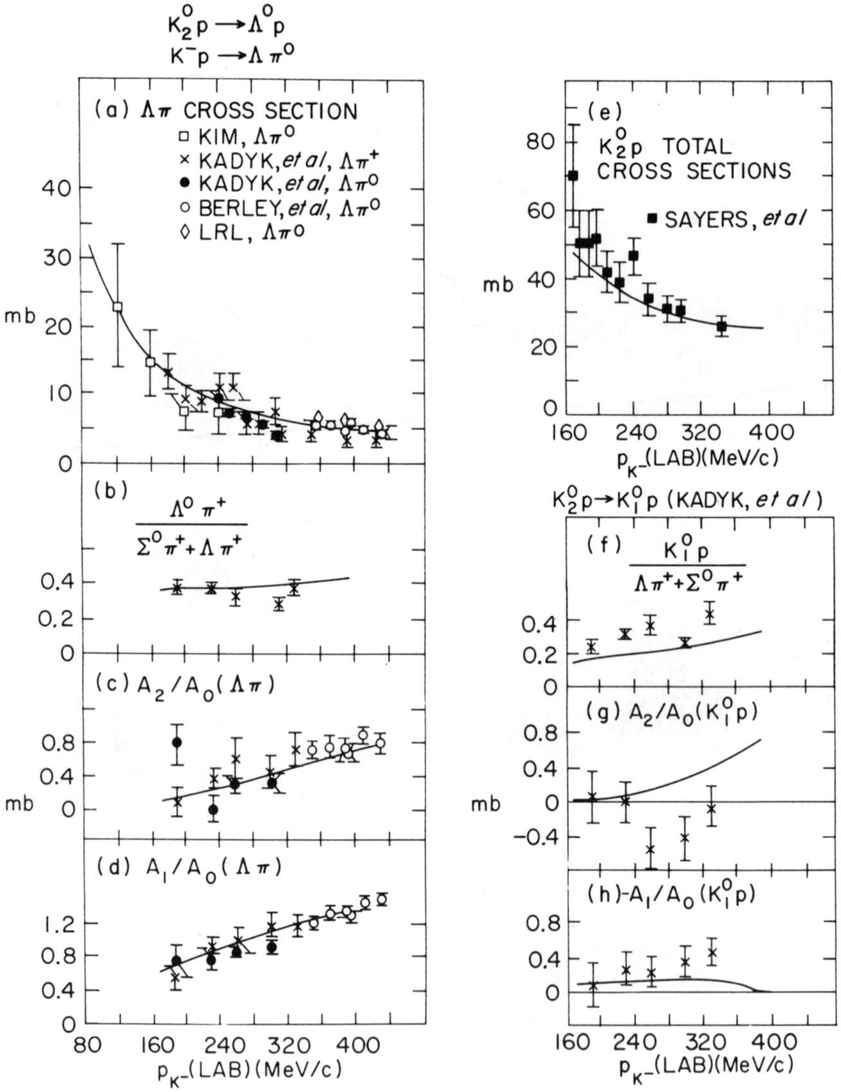

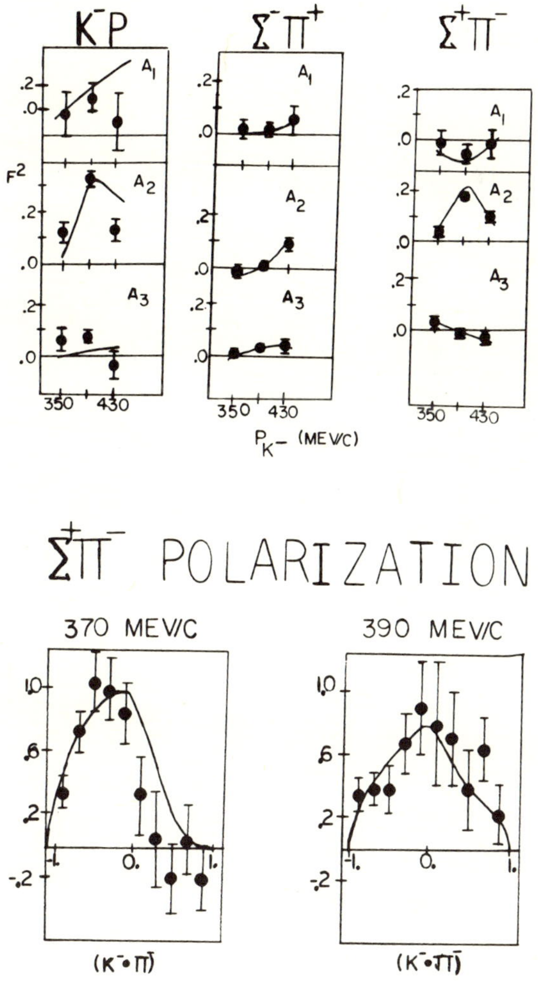

Data on $K^-p \rightarrow$ charged final states near 400 MeV/c. The Legendre coefficients for K^-p, $\Sigma^-\pi^+$, and $\Sigma^+\pi^-$ angular distributions are from Ref. 1. Polarization data for the $\Sigma^+\pi^-$ channel are from R. Bangerter, A. Barbaro-Galtieri, J. P. Berge, J. J. Murray, F. T. Solmitz, M. L. Stevenson, and R. D. Tripp, Phys. Rev. Letters 17, 495 (1966).

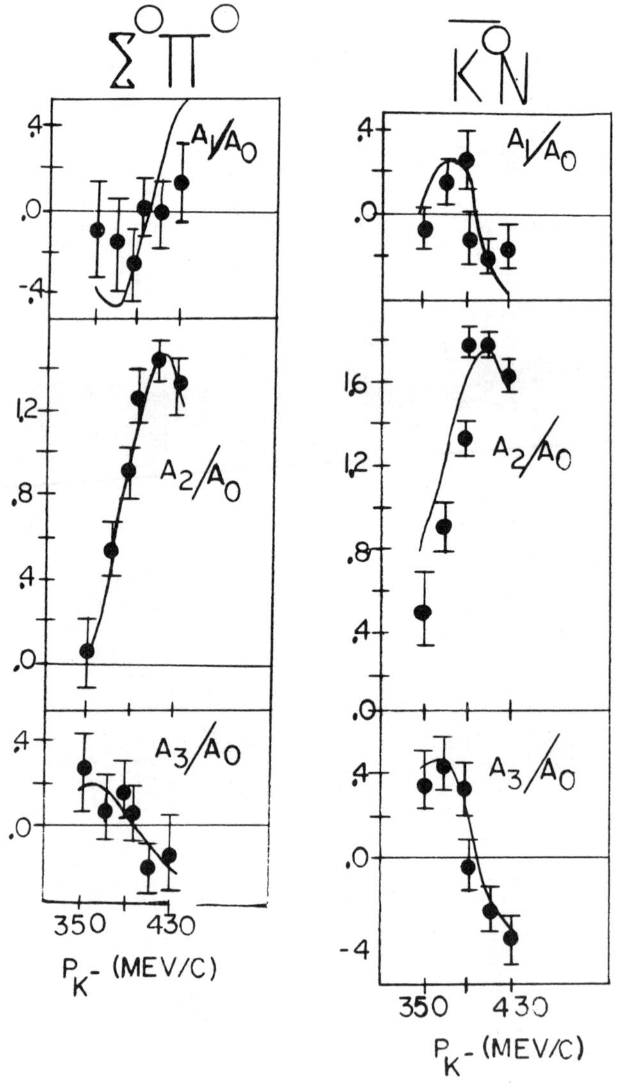

Data on $K^- p \rightarrow \Sigma^0 \pi^0$, $\bar{K}^0 n$ final states near 400 MeV/c. The Legendre coefficient ratios for the angular distributions are from Ref. 4.

REFERENCES

1. R. H. Dalitz and S. F. Tuan, Ann. Phys. 3, 307 (1960).

2. M. H. Ross and G. L. Shaw, Ann. Phys. 13, 147 (1961).

3. J. A. Kadyk, J. H. Chan, G. Goldhaber, and G. H. Trilling, UCRL 18325 (1968).

4. D. Berley, P. Yamin, R. Kofler, A. Mann, G. Meisner, S. Yamamoto, J. Thompson, and W. Willis (to be published in Phys. Rev.).

5. W. E. Humphrey and R. R. Ross, Phys. Rev. 127, 1305 (1962); M. Sakitt, T. B. Day, R. G. Glasser, N. Seeman, J. Friedman, W. E. Humphrey, and R. R. Ross, Phys. Rev. 139, B719 (1965); J. K. Kim, Phys. Rev. Letters 14, 29 (1965) and Nevis Report 149 (May 1966).

6. B. R. Martin and M. Sakitt, Phys. Rev. 183, 1345 (1969).

7. G. S. Abrams and B. Sechi-Zorn, Phys. Rev. 139, B454 (1965); W. Kittel, G. Otter, and I. Wacek, Phys. Letters 21, 349 (1966).

8. M. B. Watson, M. Ferro-Luzzi, and R. D. Tripp, Phys. Rev. 131, 2248 (1963); also, more recent work by the LRL group, as reported by R. Levi-Setti at the Lund Conference.

9. J. K. Kim, Phys. Rev. Letters 19, 1074 (1967).

10. A. D. Martin and G. G. Ross, An S and P Wave Multichannel Analysis of the Low Energy K^0_2p and K^-p Data, preprint (Aug. 1969).

11. G. A. Sayer, E. F. Beall, T. J. Devlin, P. Shepard, and J. Solomon, Phys. Rev. 169, 1045 (1968).

12. N. S. Wong, Thesis, Yale University (1969).

13. A. D. Martin and R. Perrin, Nucl. Phys. B10, 125 (1969).

14. A. K. Ray, R. W. Burris, H. E. Fisk, R. W. Kraemer, D. G. Hill, and M. Sakitt, Phys. Rev. 183, 1183 (1969).

15. S. Goldhaber, W. Chinowsky, G. Goldhaber, W. Lee, T. O'Halloran, T. F. Stubbs, G. M. Pjerrou, D. H. Stork, and H. K. Ticho, Phys. Rev. Letters 9, 135 (1962); V. J. Stenger, W. E. Slater, D. H. Stork, H. K. Ticho, G. Goldhaber, and S. Goldhaber, Phys. Rev. 134, B1111 (1964).

16. R. Tripp, Proceedings 1970 Duke Conference on Hyperon Resonances.

AN INVESTIGATION OF THE $Y\pi$ PARTIAL WAVE AMPLITUDES

BELOW $\bar{K}N$ THRESHOLD USING $\bar{K}D$ REACTIONS

D. Cline[†], R. Laumann and J. Mapp

University of Wisconsin

Madison, Wisconsin

We have studied $Y\pi$ angular distributions for the reaction $K^-D \rightarrow Y\pi N$ at a K^- momentum of 400 MeV/c. Events with $Y\pi$ effective mass below 1500 MeV are assumed to correspond to K^- scattering off virtual nucleons into open $Y\pi$ channels. This analysis confirms previous spin and parity assignments for $Y_o^*(1405)$, $Y_o^*(1520)$ and $Y_o^*(1385)$ and agrees with SU(3) multiplet assignments. A preliminary partial wave analysis suggests that the S_{01} channel has appreciable repulsive background in addition to the $Y_o^*(1405)$.

The low energy behavior of two body channels such as $\pi\ \Lambda$ and $\pi\ \Sigma$ is of the interest in the systematic study of hadron interactions. Unfortunately the unavailability of hyperon targets precludes a straight forward study of these interactions. An alternate technique using the reactions $\bar{K}N \rightarrow Y\pi$ has been widely used starting with the analysis devised by Dalitz et al. This method involves the parameterization of $\bar{K}N$ scattering data in the physical region followed by a lengthy extrapolation into the unphysical region. Several specific techniques for making the extrapolation have been devised including the zero effective range, and K matrix approaches.[1, 2] Generally speaking the results of such analysis will depend on the parameterizations used. In addition the study of higher partial waves below threshold seems difficult using this method. In such analysis it is of great importance to obtain internal consistency checks. It is an open question to what extent such consistency checks have not been made.

In contrast to the below threshold extrapolation procedure, a possible alternate technique is to utilize the virtual nucleons bound in deuterons to obtain $Y\pi$ mass values below $\bar{K}N$ threshold.

Provided sufficient internal consistency checks can be obtained, this technique may be useful in the study of the region below threshold. Furthermore the virtual nature of the struck nucleon and large relative $\bar{K}'N'$ momentum may allow the possibility of studying P and D wave contributions in the below $\bar{K}N$ threshold region.

In this paper we present the results of a study of the angular distribution and $Y\pi$ mass spectrum for the reactions

$$K^-d \to K^-(n)P \to \Lambda\pi^-p \qquad (1)$$
$$\to K^-(n)P \to \Sigma^o\pi^-p \qquad (2)$$
$$\to K^-(p)n \to \Sigma^\pm\pi^\pm n \qquad (3)$$

for a K^- beam momentum of 400 MeV/c. The data comes from an exposure of the LRL 25' bubble chamber to the Murray Septum separated K^- beam at the Bevatron.

A previous and published study of reaction (1) within the framework of the impulse approximation gave good agreement with the $\bar{K}NY_1^*(1385)$ coupling constant expected from SU(3) considerations.[3] This result was in considerable disagreement with the K Matrix analysis of Kim.[4] This analysis also showed that the impulse approximation calculation gives a good description of the $Y\pi$ mass spectrum. In the present paper we assume the validity of the impulse approximation but do not perform calculations with it; we prefer instead to consider the collision of the $\bar{K}$ with the virtual nucleon in the same way as is done with free nucleon targets. The most severe limitation to this approach is in obtaining the flux factor for these collisions and therefore obtaining the s dependence of the processes

$$\bar{K}\ \text{"n"} \to Y\pi$$

where "n" denotes a virtual nucleon. Our approach in this paper is meant to be as a model independent as possible.

In Fig. 1 we show the $\Lambda\pi$ mass spectrum for reaction 1 with appropriate cuts to remove the observed double scattering effects.[5] A strong $Y_1^*(1385)$ is observed. The peaking at the higher mass values comes from experimentally imposed cut off's in the data. In Fig. 2 the A_n coeficients of a Legendre polynomial fit to the data in the $\Lambda\pi$ center of mass are shown. In this figure all available hydrogen bubble chamber data is included for the purpose of comparison.[6] It is evident that there is good agreement between the coefficients obtained from scattering on virtual nucleons in the deuteron and the free proton data. This gives us confidence that virtual nucleons act like real nucleons and that deuteron corrections to the angular distributions are small. A partial wave anal-

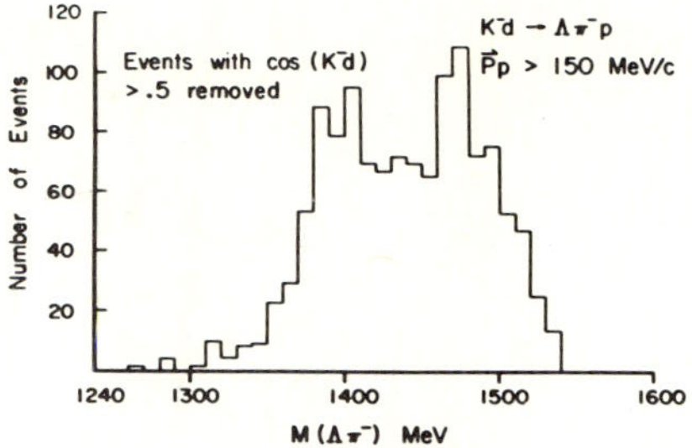

Fig. 1. Λπ Mass Spectrum

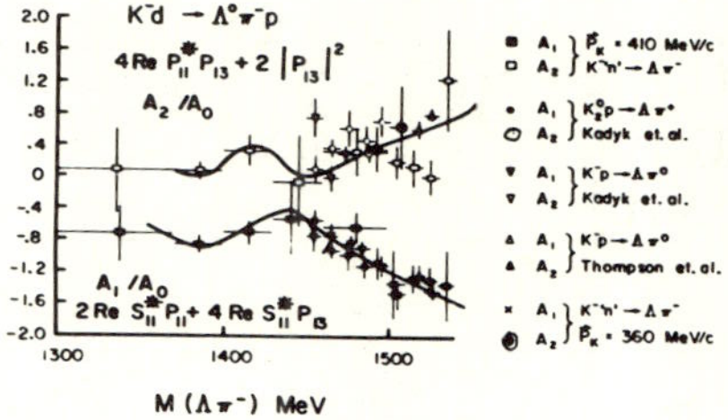

Fig. 2. Λπ Legendre Coeficients

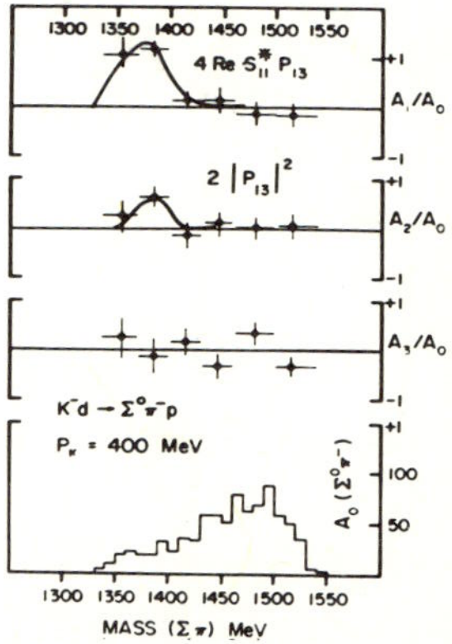

Fig. 3. $\Sigma^o\pi^-$ Legendre Coeficients and Mass Spectrum

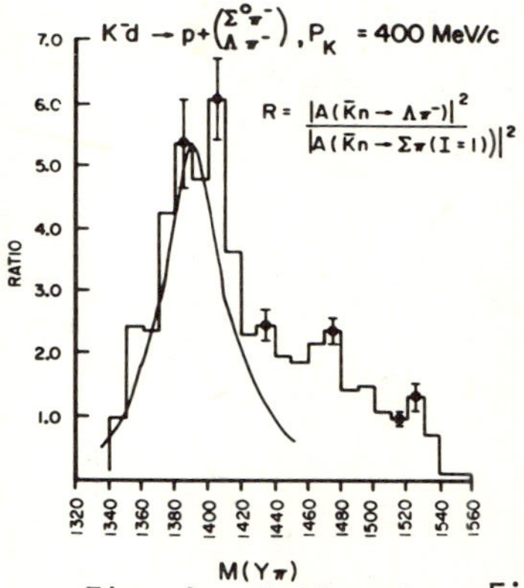

Fig. 4. Ratio of Λπ to $\Sigma^o\pi^-$ Mass cross sections

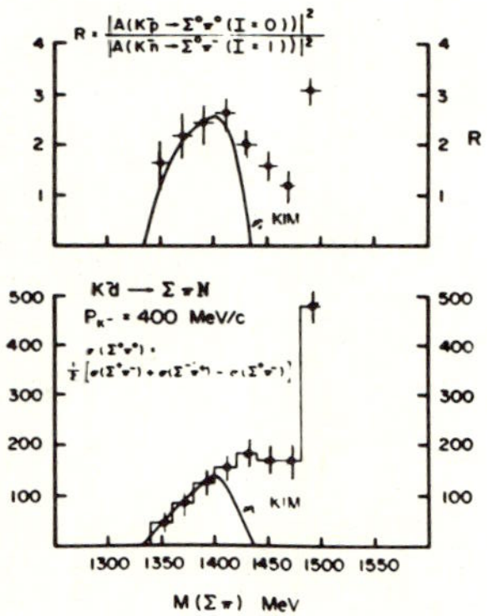

Fig. 5. $\Sigma^o\pi^o$ Mass Spectra and Ratio of $\Sigma^o\pi^o/\Sigma^o\pi^-$ cross sections

ysis of the $\Lambda\pi$ channel has been done. Assuming the existence of the $Y_1^*(1385)$ it is possible to obtain crude phase shift information. The only striking results of this analysis is the suggested importance of the P_{11} amplitude.

We now turn to the I = 1 $\Sigma\pi$ channel. Fig. 3 shows the Legendre coeficients and the mass distribution for the $\Sigma^\circ\pi^-$ system (reaction 2). Although the mass distribution does not show evidence for the $Y_1^*(1385)$ the number of events in the 1385 mass region are more than required to be consistent with the expected branching ratio of $Y_1^*(1385) \rightarrow \Sigma^\circ\pi^-$. The Legendre coefficients show striking evidence for the $Y_1^*(1385)$ interfering with S_{11} background. This provides an independent check on the spin-parity assignment of P_{13} for this state.

In order to obtain the cross section for the process $\bar{K}n \rightarrow Y\pi$ it is necessary to estimate the flux factor for $\bar{K}$'s colliding with virtual nucleons. We attempt to approximate the flux factor for virtual nucleons well off the mass shell using the measured cross section for reaction 2. It is expected that the $\Sigma^\circ\pi^-$ matrix element will be dominated by the S_{11} partial wave and there will be little structure in the amplitude. A consistency check on this view point is provided by dividing the cross section for $\Lambda\pi$ production by that for $\Sigma^\circ\pi^-$ at each $Y\pi$ invariant mass. The result of this analysis which is shown in Fig. 4 preserves the $Y_1^*(1385)$ structure in the $\Lambda\pi$ channel. The ratio of $\Lambda\pi$ to $\Sigma^\circ\pi^-$ at 1385 MeV is found to be ~ 5 compared to a value of ~ 20 if the $\Sigma^\circ\pi^-$ channel were dominated by the $Y_1^*(1385)$ contribution at low energy, which also suggests S_{11} dominance of the $\Sigma^\circ\pi^-$ channel.

Assuming the impulse approximation, a combination of cross sections (for a given $\Sigma\pi$ mass) of $\Sigma^\pm\pi^\mp n$ and $\Sigma^\circ\pi^-p$ can be used to deduce the $\Sigma^\circ\pi^\circ n$ cross sections. The resulting mass plot is shown in Fig. 4b. There is apparently a broad enhancement at low mass. This data is not consistent with a narrow $Y_0^*(1405)$ suggested in recent constant scattering length analyses of low energy data.[7] Fig. 4a shows the ratio of the $\Sigma^\circ\pi^\circ$ and $\Sigma^\circ\pi^-$ cross sections. This ratio which is expected to have the same $Y\pi$ mass structure as the $|\text{amplitude}|^2$ for the $\bar{K}'n' \rightarrow \Sigma^\circ\pi^\circ$ process. A striking peak is observed in the mass range expected for the $Y_0^*(1405)$. For comparison the shape of the $Y_0^*(1405)$ obtained by Kim in an effective range K matrix analysis is also shown.[4] The shape of the $Y_0(1405)$ appears to be consistent with the analysis of Kim if some background is assumed. The width of the $Y_0^*(1405)$ appears to be greater than 50 MeV again being inconsistent with the narrow width's suggested recently.[7]

A Legendre polynomial fit was made to the angular distributions and polarizations for reactions 3. The B_N coefficients for the Σ^+

polarization are shown in Fig. 6 as a function of $\Sigma^+\pi^-$ mass. Appreciable Y_o*(1405), Y_1*(1385) interference can be seen in the B_1 coefficient. In Fig. 7 the $\Sigma^\circ\pi^\circ$ combination of the Legendre coefficients for the final states $\Sigma^+\pi^-$, $\Sigma^-\pi^+$ and $\Sigma^\circ\pi^-$ are shown. Appreciable interference between the Y_0*(1405) and Y_o*(1520) can be observed in the A_2/A_0 coefficients. In addition, an appreciable P_{01} amplitudes appears to be required in the vicinity of the Y_0*(1405) and the Y_0*(1520). Subtracting the coefficients for $\Sigma^+\pi^-$ and $\Sigma^-\pi^+$ isolates the interference terms between the I = 1 and I = 0 amplitudes. The resulting distributions are shown in Fig. 8. The A_1/A_0 coefficients indicate the presence of appreciable Y_o*(1405) - Y_1*(1385) and Y_o*(1520) - Y_1*(1385) interference. The A_3/A_0 term shows a striking interference between the Y_1*(1385) and Y_o*(1520).

Using the information in Figs. 3, 5, 6 and 7 and assuming that only S, P and D waves are important up to 1530 MeV, a crude partial wave analysis was carried out. The Y_0*(1520) amplitude was taken as the reference amplitude which was assumed to execute a background free circle on the argand diagram as shown in Fig. 9. The other amplitudes which have been deduced relative to the Y_0*(1520) are also shown in this figure. The most striking features of the partial wave amplitudes are:

1. The Y_1*(1385) is in the lower half of the argand plot relative to the Y_o*(1405) and Y_o*(1520) as it should be from SU(3) considerations, if the I = 0 states are predominately singlet states.

2. The Y_0*(1405) appears to ride on an appreciable repulsive background. This appears to be at odds with the K matrix analysis. Of course unknown deuteron effects could be partially responsible for this background.

In conclusion it appears that the existing data on reactions 1, 2 and 3 in the Y*(1385), Y_o*(1405) and Y_o*(1520) regions are consistent with general expectations. In addition the data compare favorably with free nucleon data in all cases so far studied. We conclude that it is likely that these reactions can be used to probe the below $\bar{K}N$ threshold region in a more or less direct way. It is also possible that this technique could provide a method of measuring the $\Sigma\pi$, $\Lambda\pi$ scattering lengths. Such analyses may provide interesting contrasts to the usual K matrix type analyses.

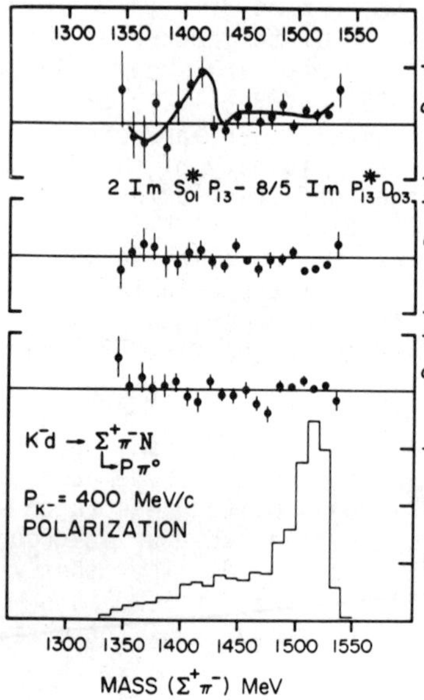

Fig. 6. Polarization and Mass Spectra

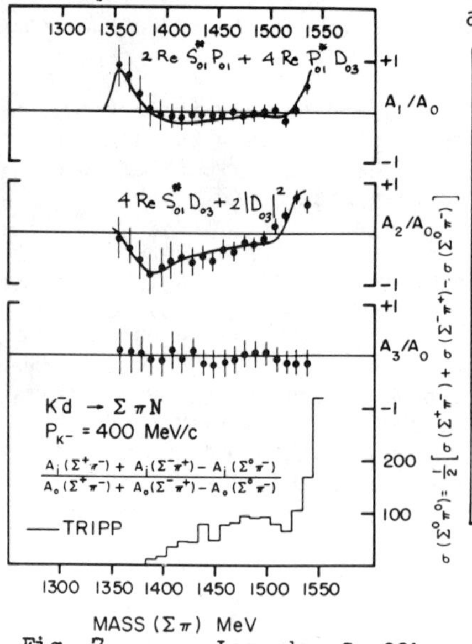

Fig. 7. Legendre Coefficients and Mass Spectra

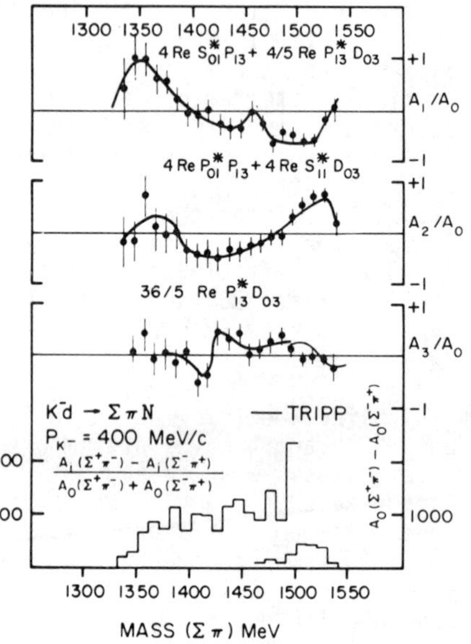

Fig. 8. Difference between $\Sigma^+\pi^-$, $\Sigma^-\pi^+$ Legendre Coeficients and Mass Spectra

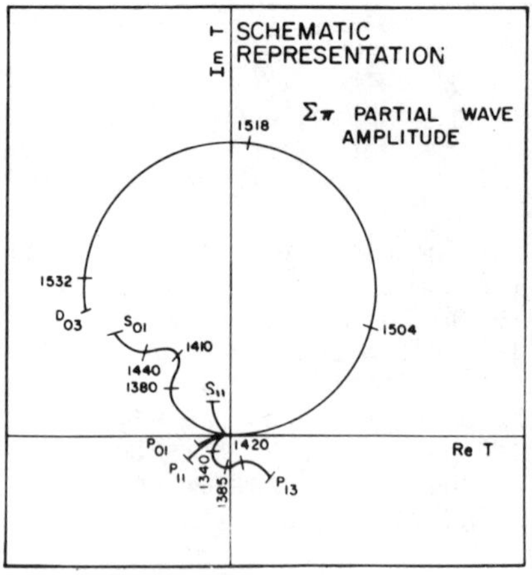

Fig. 9. Schematic of Partial Wave Analysis of Below Threshold Region

References

1. R. H. Dalitz and S. F. Tuan, Am. Phys. 10, 307 (1960).
2. M. H. Ross and G. Shaw, Am. Phys. 13, 147 (1961).
3. K. O. Bunnell, D. Cline, R. Laumann, J. Mapp and J. Uretsky, Letter al Nuovo Cimento 3, 224 (1970).
4. J. Kim, Phys. Rev. Letters 19, 1074 (1967).
5. D. Cline et al. Phys. Rev. Letters 21, 1372 (1968), G. Alexander et al. Phys. Rev. Letters 22, 483 (1969).
6. J. Kadyk, (private communication), J. Thompson et al. Yale preprint.
7. B. R. Martin and M. Sakitt, Phys. Rev. 183, 1345 (1969), A. D. Martin and G. G. Ross, Mucl. Phys. (to be published).

STUDY OF $Y_1^*(1385)\pi$ PRODUCTION NEAR THE $Y_0^*(1520)$ RESONANCE

Shu-bon Chan, J. Button-Shafer, S.S. Hertzbach,
R. R. Kofler, G. W. Meisner, and S. S. Yamamoto,
University of Massachusetts, Amherst;
M. Schiff, J. Thompson, and W. J. Willis,
Yale University; and
D. Berley and P. Yamin, Brookhaven National
Laboratory

The purpose of the experiment reported here was to determine to what extent the production of $Y_1^*(1385)$-π derives from the $Y_0^*(1520)$ resonance. Other observations of this decay mode have been reported by Cline et al. and Levi Setti et al.[1] The decay is a sensitive test of SU_3 mixing, as it is allowed for only the octet component of the $Y^*(1520)$.

The data were obtained from 180,000 pictures taken with the BNL 30-in. bubble chamber at K^- settings of 375 and 415 MeV/c. Confidence-level cuts as well as a "missing-mass" technique separated $\Sigma^o\pi^+\pi^-$ events rather cleanly and yielded 2590 $\Lambda\pi^+\pi^-$ events. Cut-offs on Λ length and opening angle were applied and corrected for. With incident flux determined by tau decays, it was found that the $Y^*(1520)$ resonance caused a 2.55 mb enhancement over a 0.70 mb background. The $M^2(\Lambda\pi^+)$ spectrum is presented in Fig. 1.

Partial-wave formalisms of Peierls and Olsson-Yodh were studied;[2] and a modified version of the Olsson-Yodh program was used to fit the data. Results are shown in Fig. 2. The dominant amplitude is that for $(L_iL_f)_{I,2J} = (DS)_{0,3}$. Attempts to fit the data with $\Lambda\pi$ systems nonresonant were unsuccessful.

Further investigation of $Y^*(1385)$-π production was made with the "moment method," an extension of the Byers-Fenster technique.[3] Spin-parity information on the parent state was obtained from the intensity $I(\theta)$ and multipole parameters $I\langle T_{\ell m}\rangle(\theta)$ for the $Y^*(1385)$, θ being the Y^* production angle. Odd-ℓ (polarization) parameters were found zero within statistical errors and thus indicated that

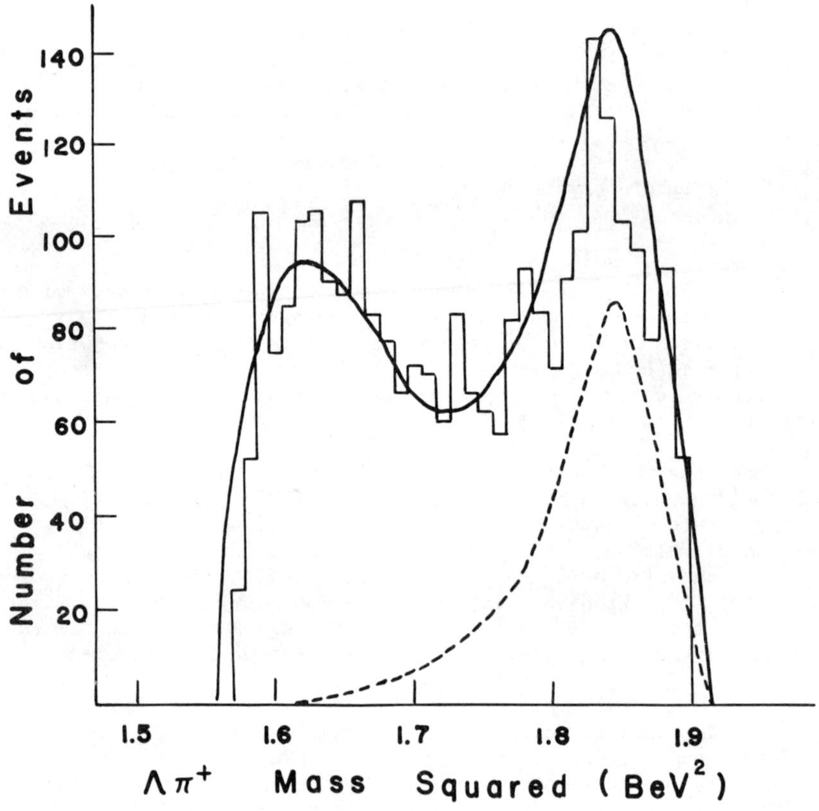

Fig. 1. $M^2(\Lambda\pi^+)$ spectrum for the reaction $K^- + p \rightarrow \Lambda + \pi^+ + \pi^-$, with the best fit obtained by partial-wave analysis. The dashed curve represents production of the $Y^*(1385)^+$ alone.

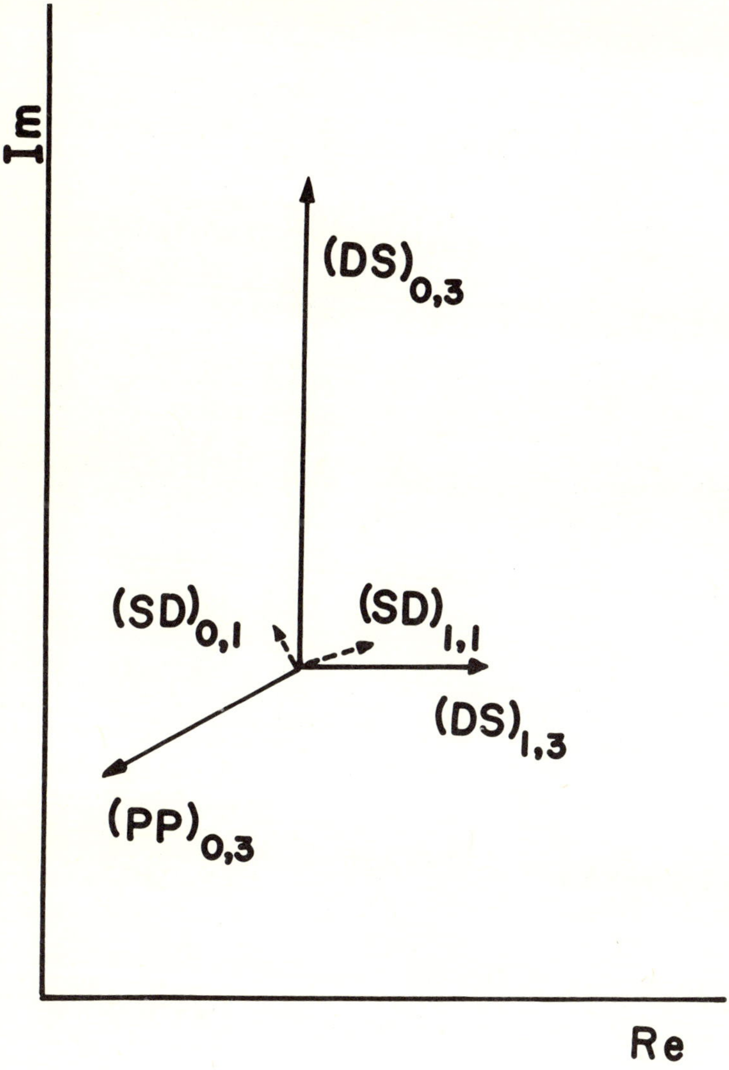

Fig. 2. Partial-wave amplitudes obtained with the modified Olsson-Yodh program. Solid lines are resonant amplitudes; dashed lines, nonresonant.

only the lower ℓ wave was operative in the Y*(1520) (plus background) decay, as expected from energy considerations.

The even-ℓ (alignment) parameters as well as $I(\theta)$ were found to fit calculated distributions very well for the assignment of $3/2^-$ to the parent state, when $Y^*(1385)^+$ and $Y^*(1385)^-$ data were added together. Other assignments for spins 1/2, 3/2, and 5/2 fitted data very poorly. Moments evaluated separately for $Y^*(1385)^+$ or $Y^*(1385)^-$ data gave information on the interference between I = 0 and I = 1 amplitudes.

An example of one of the multipole parameters, $I\langle T_{22}\rangle(\theta)$, is presented in Fig. 3. (We believe this has not been utilized before in treatment of formation-resonance decay into spin 3/2.) A χ^2 for fitting all intensity and multipole-parameter data was evaluated for each of the parent-state assumptions, with a value of 14.5 (10 degrees of freedom) obtained for the $3/2^-$ hypothesis. This best-fit hypothesis of $3/2^- \rightarrow 3/2^+ + 0^-$ improved markedly as mass restrictions were tightened for the Y*(1385), and thus again supported the resonant interpretation of the $\Lambda\pi$ systems.

To study SU_3 mixing, amplitudes from the partial-wave analysis were used to generate the $M^2(\Lambda\pi^+)$ spectrum for the production of $Y^*(1385)^+$ alone (and for $Y^*(1385)^-$ alone) without interference and background. Fig. 1 shows the $Y^*(1385)^+$ production (dashed curve). With the total cross section, as a function of $p_{incident}$ (or E*), weighted to account for $Y^*(1385)^+$ production alone, a fit was made to the amplitude (background plus 1520 resonance) $T = (a+bp_{inc})\, e^{i(c+dp_{inc})} + \sqrt{\Gamma_e\Gamma_r}\,/(E^*-E_0^* + i\Gamma/2)$ Then the coupling constant for 1520 decay into 1385+π was extracted according to $\Gamma_r\,(p_{inc}) = G^2_{1520}\, M_N\, \overline{p^*}(p_{inc})/\, E^*(p_{inc})$, where $\overline{p^*}$ is the average momentum for the Y*(1385)-π in the C.M. frame. With phase-space and barrier factors included for all modes, the fitting of these expressions yielded, as an average for $Y^*(1385)^+$ and $Y^*(1385)^-$ data, $G^2_{1520} = 0.014 \pm 0.003$.

Two techniques were used to evaluate the SU_3 mixing angle: the calculation of $\sin^2\theta$ from the ratio of G^2_{1520} to $G^2_{(8)}$ and the calculation of $\tan^2\theta$ from the ratio of G^2_{1520} to G^2_{1690} (where the (8) subscript stands for the unmixed I=0 octet member). With data from $\Xi^*(1815)$ and N*(1515) decays providing the value of $G^2_{(8)}$, the former technique yielded a mixing angle of $16 \pm 3°$; the latter yielded an angle of $\gtrsim 23 \pm 3°$.

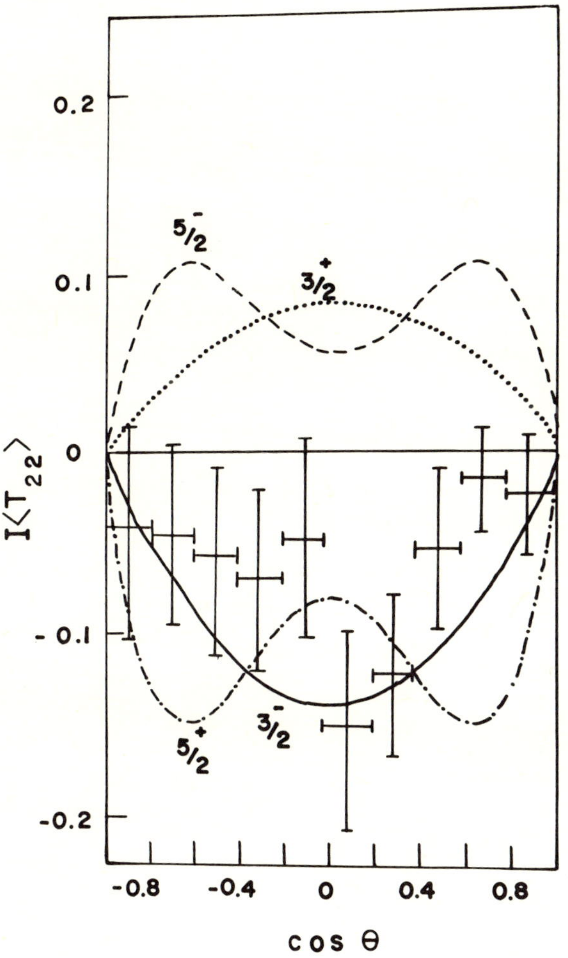

Fig. 3. Example of a multipole parameter characterizing the $Y^*(1385)$ spin state as a function of cos θ for $Y^*(1385)$ production.

Footnotes

1. D. Cline, R. Laumann, and J. Mapp, paper 922 of the International Conference on High-Energy Physics, Vienna, 1968; R. Levi Setti, rapporteur talk at Lund International Conference on Elementary Particles, Lund, Sweden, June, 1969.

2. R. F. Peierls, Brookhaven National Laboratory Report 9621 (1965); M. Olsson and G. B. Yodh, Phys. Rev. Letters 10, 353 (1963).

3. J. Button-Shafer, Phys. Rev. 150, 1308 (1966).

4. Data on $\Xi^*(1815)$ and $N^*(1515)$ decays were from Rosenfeld et al., UCRL-8030, January, 1970. The coupling constant for $Y_0^*(1690) \to Y_1^*(1385)+\pi$ came from the Levi Setti report of Ref. 1.

DISCUSSION

Goldberg: You indicated that you put barrier factors in to determine the SU(3) couplings and mixing angles in the $1520 \to 1385 + \pi$. I wonder how this was done, since you were right at threshold, and it seems to me that everything will be very sensitive to exactly what form of barrier factor you put in.

Button-Shafer: There is no barrier factor; that is, there is only s-wave for the 1385- π decay mode of the 1520. However, there is Γ -total in the denominator, and it is made up largely of contributions from $\bar{K}N$ and $\Sigma\pi$ decay modes. And, unlike the situation in the Watson-Ferro-Luzzi-Tripp article where for various reasons they decided to approximate barrier factors, since we had a fair number of events, we did put in the d-wave barriers for each mode. But we can talk about details later if you like.

(Note added in proof) The phase-space factor is of course troublesome for the decay $1520 \to 1385 + \pi$. To obtain $\Gamma(E)$ the factor p/E was averaged over p; and to obtain a branching ratio (not really necessary) this factor would be averaged over both p and E, that is, over the mass range of both the daughter and the parent resonances. Our branching ratio so defined agreed very nicely with Levi-Setti's Lund Conference value, ours being 3.9 ± 1.0 %.

DETERMINATION OF THE $\bar{K}YP$ COUPLING CONSTANTS: THE IMPORTANCE OF THE LOW ENERGY $\bar{K}N$ DATA*

T. A. Lasinski

Enrico Fermi Institute

University of Chicago

INTRODUCTION

In this article an attempt is made to summarize the many determinations[1] of the $\bar{K}Yp$ coupling constants. In particular, the importance of the low energy $\bar{K}N$ data to these calculations is discussed. Present estimates of the $K\Lambda p$ coupling constant seem to cluster at two values, $g^2_{\bar{K}\Lambda p}/(4\pi) \sim 5$ or ~ 13. Experimentally, this discrepancy reflects the lack of a precise knowledge of the S and P - waves from the elastic threshold to $P_{LAB} \sim 600$ MeV/c. It is to be stressed that small differences in these partial waves above threshold can lead to much larger differences below threshold and, consequently, to the above discrepancy.

SUM RULE FOR $g^2_{\bar{K}\Lambda p}/(4\pi)$

In what follows the discussion is restricted to the $K\Lambda p$ coupling constant since the greatest discrepancy seems to be in its determination. In order to evaluate $g^2_{\bar{K}\Lambda p}/(4\pi)$, the following sum rule[2] must be evaluated:

$$\frac{g^2_{K\Lambda p}}{4\pi} = \frac{1}{X_\Lambda}\left\{\frac{m_p + m_k}{4m_p}(2\bar{a}_o + a_o - 3a_1) - \frac{m_K}{\pi}\int_{E_{\pi\Sigma}}^{m_K} dE' \, \frac{\mathrm{Im}\bar{f}_o(E')}{E'^2 - m_K^2} + \frac{m_K}{4\pi^2}\int_{m_K}^{\infty} \frac{dE'}{P'}\left[\sigma_-^n(E') - \sigma_+^n(E') - 2\sigma_-^P(E') + 2\sigma_+^P(E')\right]\right\}. \quad (1)$$

In this expression $\bar{f}_o$ is the $\bar{K}N$ laboratory amplitude and the σ's refer to $K^{\pm}$ total cross sections from neutrons or protons. $\bar{a}_I$ and a_I are the K^-N (real part) and K^+N scattering lengths for isospin I = 0, 1. E and P refer to the laboratory energy and momentum of the kaon. X_Λ is a constant[3] depending only upon m_K, m_Λ, and m_p.

Table I[2] summarizes the relative sizes of the various contributions to the sum rule. At first glance it appears as though there are large cancellations which in themselves might lead to difficulties. However, since the values of rows 1, 3-6 depend on fairly well measured quantities, such as scattering lengths and total cross sections, they are not a serious source of discrepancy.

The high energy contribution (row 7) rests upon the assumption that $\sigma_+ - \sigma_- = c/p^\alpha$ where $\alpha > 0$ and c is a constant. If α were equal to zero, the integral would diverge logrithmically and Eq. (1) would no longer be valid. It should be noted that most calculations of $g^2_{\bar{K}\Lambda p}/(4\pi)$ have essentially determined c and α from data below ~20 GeV/c. In view of the recent Serpukhov[4] data which show σ_- (above 20 GeV/c) flattening out some 3 mb above σ_+ (at ~20 GeV/c), the value of row 7 must necessarily change. Indeed, if α turns out to be zero, not only will the sum rule have to be modified but the present intuitive picture of high energy diffraction scattering (Imf >> Ref) will also have to be abandoned.[5]

It is, however, the so-called below threshold contribution (row 2) which is most relevant to the work of the low energy $\bar{K}N$ experimentalist. To appreciate the problems associated with this contribution,it is necessary to review the K - matrix formulation.

K - MATRIX FITS AND BELOW THRESHOLD CONTRIBUTION

In order to compute the contribution from the $\pi\Sigma$ threshold to the K^-p threshold, it is first necessary to obtain an analytic expression for the scattering amplitude above threshold. This is best done by determining through a fit to the data the parameters of the K - matrix, since this formalism correctly treats the effect of opening thresholds upon a scattering amplitude.

For S - wave only, the cm scattering amplitude is given by

$$f = K \frac{1}{1 - iqK} \tag{2}$$

	Contributions to $g^2_{\bar{K}\Lambda p}/(4\pi)$	Value
1)	$2\,\bar{a}_o + a_o - 3a_1$	-18.2
2)	$-\frac{m_K}{\pi X_\Lambda}\int_{E_{\pi\Sigma}}^{m_K} dE' \frac{Im\bar{f}_o(E')}{E'^2 - m_K^2} + \frac{m_K}{4\pi^2 X_\Lambda}\int_{m_K}^{0.3\ GeV/c} \frac{dE'}{P'}\,[\sigma_-^n - 2\sigma^P]$	+35.1?
3)	$+\int_{0.3\ GeV/c}^{5.0\ GeV/c} \frac{dE'}{P'}\,\sigma_-^n(E')$	+ 8.6
4)	$-\int_{0.3\ GeV/c}^{5.0\ GeV/c} \frac{dE'}{P'}\,2\sigma_-^P(E')$	-23.5
5)	$+\int_{m_K}^{5.0\ GeV/c} \frac{dE'}{P'}\,2\sigma_+^P(E')$	+10.4
6)	$-\int_{m_K}^{5.0\ GeV/c} \frac{dE'}{P'}\,\sigma_+^n(E')$	- 5.7
7)	$\frac{m_K}{4\pi^2 X_\Lambda}\int_{5.0\ GeV/c}^{\infty} \frac{dE'}{P'}\,[\sigma_-^n - \sigma_+^n - 2\sigma_-^P + 2\sigma_+^P]$	- 2.7?

TABLE I. Typical (see Ref. 2) contributions to the various terms of Eq. (1).

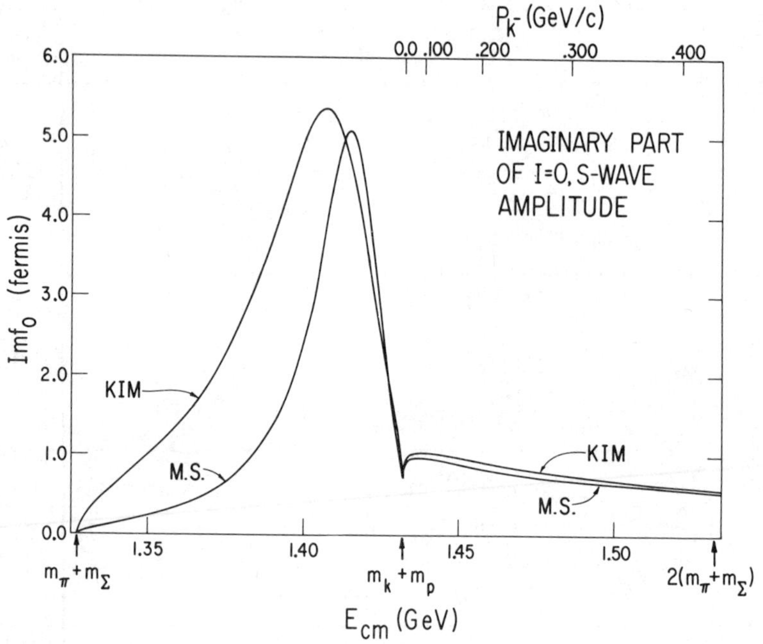

Fig. 1. Comparison of "zero range" (Ref. 7) and "effective range" (Ref. 8) amplitudes.

Consider the inverse K - matrix

$$K^{-1}(E_{cm}) = K^{-1}(m_K + m_P) + \frac{1}{2}\begin{pmatrix} r_K\, q_K^2 & 0 \\ 0 & r_\Sigma (q_\Sigma^2 - q_{\Sigma o}^2) \end{pmatrix} \qquad (3)$$

Two approaches have been taken in fitting the low energy data. In one[(7)] only the first term is used ("zero range", constant K - matrix) and the data are fit only up to ~550 MeV/c.** In order to extrapolate below threshold, one simply replaces the q in Eq. (2) by $i\,|q|$ for $E_{cm} < m_K + m_p$.

To illustrate the vast difference two such fits can give below threshold, Fig. 1 compares $\mathrm{Im}\bar{f}_o$ in the cm frame from the fits of Martin and Sakitt[(7)] and of Kim[(8)]. Although the two curves agree fairly well above threshold, there is a significant difference below threshold. Not only are the mass and width of the $\Lambda(1405)$ affected, but the value of $g^2_{\bar{K}\Lambda p}/(4\pi)$ from Kim's

** See correction on page 93.

amplitude would be considerably larger than that from Martin and Sakitt's (see Table I., row 2).

So far as the S - wave is concerned, the essential difference between the fits is Kim's use of the "effective range" inverse K - matrix. Part of the motivation for "zero range" fits is that Kim determined P and D_3 waves as well. Although these waves are necessary in a fit up to 550 MeV/c, it may be that the large number of additional parameters required are not well enough determined. In addition, the $\pi\pi\Lambda$ channel, which is difficult to handle in the K - matrix formalism, becomes more significant at higher momenta. Thus it was necessary to see whether a fit in the region where only S - wave is significant (below $\sim$300 MeV/c) would yield the same result.

The difficulty with the "zero range" fits on the other hand, is what might be called their "reliability".[9] To compute $g^2_{\overline{K}\Lambda p}/(4\pi)$ it is necessary to extrapolate the fitted amplitude down to the $\pi\Sigma$ threshold. As is indicated in Fig. 1, this corresponds to extrapolating over an energy range twice as great as that over which the amplitude above threshold was determined ($\sim$300 MeV/c).

How, then does one decide between these solutions? Clearly the correct solution should approximately agree with partial wave analyses of data above $\sim$500 MeV/c. In Fig. 2 are shown the S_{01} and S_{11} partial waves of Refs. (7) and (8) with solutions from other experiments.[10] The experimental difficulty is now evident. The "effective range" (Kim) fit gives better agreement with S_{01}, whereas the "zero range" (M.S., M.R.) fit seems better for S_{11}. In fact, these discrepancies persist when the inelastic partial waves are considered.[10] It is clear that anyone who attempts to compute the $\overline{K}YP$ coupling constants had best check the consistency of the amplitude he extrapolates with the higher energy solutions. In addition, experimentalists who perform partial wave analyses should also keep such consistency considerations in mind.

OTHER METHODS FOR DETERMINING $g^2_{\overline{K}\Lambda p}/(4\pi)$

Chan and Meiere have written a "continuous moment" sum rule for $\overline{f}(E)/(E - m_K)^{1-\beta}(E - E_{\pi\Sigma})^{\beta}$, where $0 < \beta < 1$. It is identical to Eq. (1) when $\beta = 0$ and introduces $\text{Re}\, f_o$ into the integral below threshold when $\beta \neq 0$. Varying β, they were thus able to perform a consistency check on Kim's amplitude below threshold. Although Kim's solution was consistent, an earlier calculation by Lusignoli et al., which used a constant scattering length for $\overline{f}$, failed this check. Martin, Queen, and Violini[1], however,

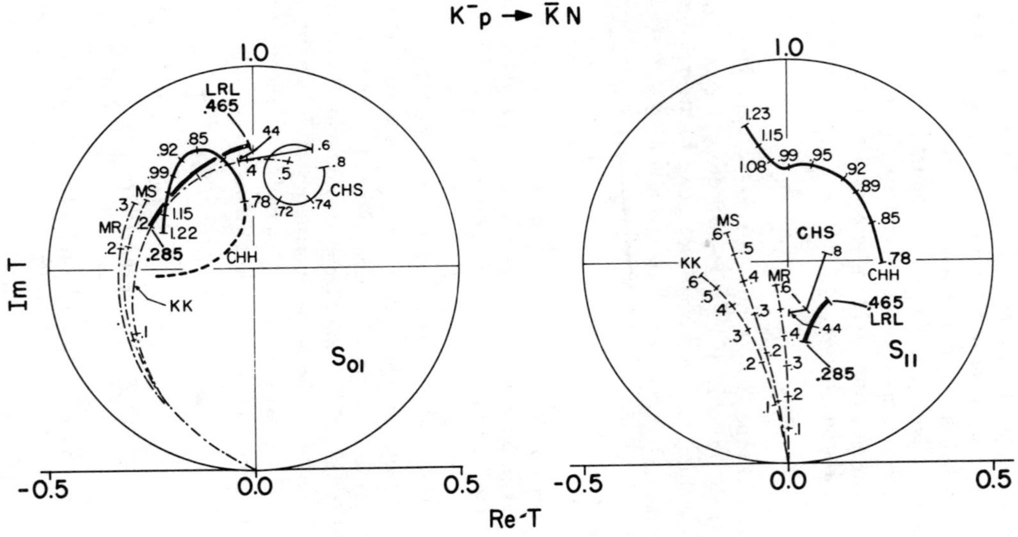

Fig. 2. Comparison of low energy K - matrix fits to higher energy solutions.

pointed out that the real difficulty is that the constant scattering length formalism does not correctly treat threshold effects, thus stressing the need for the K - matrix formalism.

The importance of such sum rules is that in principle they could use the experimentally determined real parts of the amplitude (above threshold) to constrain further the determination of the $\bar{K}$YP coupling constants. To illustrate to what sensitivity this is possible, the ratio $Re\bar{f}/Im\bar{f}$ is compared with dispersion calculations (which, of course, require the knowledge of $g^2_{\bar{K}Yp}$) in Fig. 3. It should be stressed[11] that the experimental values indicated in Fig. 3 are rather sensitive to both the order of Legendre fit to $d\sigma/d\Omega$ and to the total cross sections. In particular, the discrepancy below ~700 MeV/c may reflect imprecise total cross section data in that region.

There are other methods of calculating $g^2_{\bar{K}\Lambda p}/(4\pi)$ (see, e.g., Martin and Perrin[1], Deo and Cutkosky[1]) which in principle are believed to be relatively insensitive to the behavior of the low energy $\bar{K}$N data. These methods have, therefore, not been discussed in detail since they are not as instructive as to the problems associated with the low energy data. Furthermore, it must be noted that the values obtained for $g^2_{\bar{K}\Lambda p}/(4\pi)$ are not consistent (compare the values of the above authors).

CONCLUDING REMARKS

The present status of the determination of $g^2_{\bar{K}\Lambda p}/(4\pi)$ is best summarized by Fig. 4. Most authors seem to admit that systematic errors--which are generally not indicated--could shift their values quite a bit. This undoubtedly means that experimentally the low energy $\bar{K}$N data are still not adequate to permit a reliable determination of the $\bar{K}$YP coupling constants. In addition, many of these calculations are not independent in that they are often reanalyses of earlier estimates.

Finally, it is tempting to speculate that the ultimate value of $g^2_{\bar{K}\Lambda p}/(4\pi)$ will be ~9. Such a value would be within 3 standard deviations of the various determinations. Furthermore, since SU(3) seems to favor a value of ~13, something on the order of 9 would imply a 30% breaking effect.

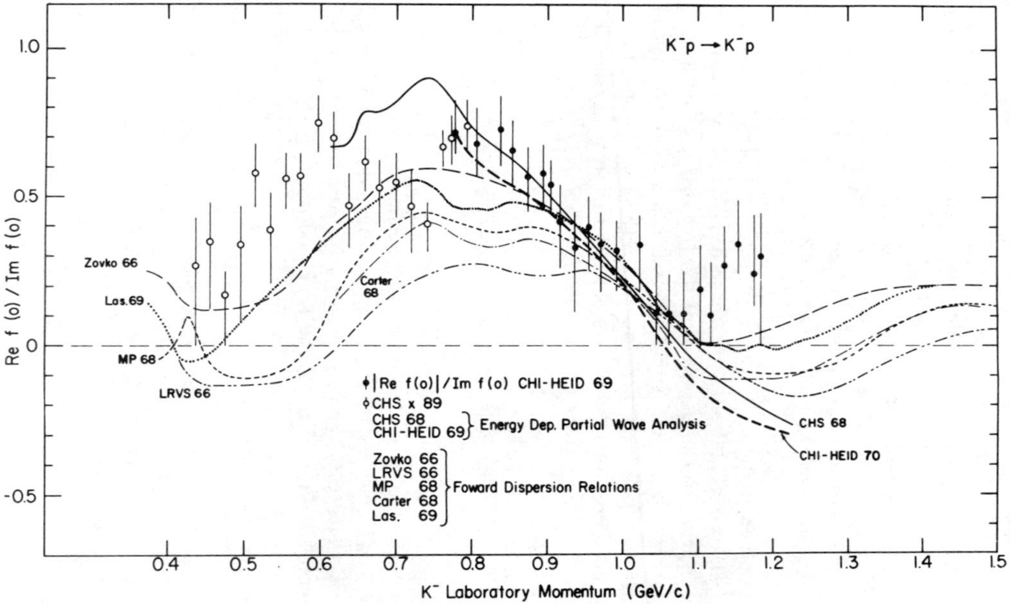

Fig. 3. Disperation calculations compared to data in the 1 GeV/c region. Note that data can only give |Re f|.

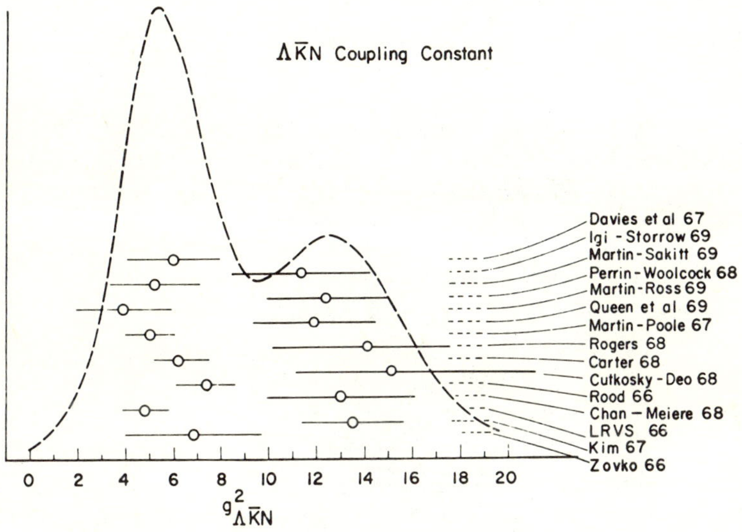

Fig. 4. Ideogram of determinations of the rationalize $\Lambda \bar{K} p$ coupling constant.

ACKNOWLEDGEMENT

I would like to thank Riccardo Levi Setti for his always constructive criticism and for the loan of several of his LUND figures.

** Correction to sentence following Eq. 3:
"In one[7] only the first term is used("zero range", constant K-matrix) and the data are fit only up to 300 MeV/c. The other[8] approach includes the second term ("effective range") and fits the data up to 550 MeV/c."

REFERENCES

*Work supported by the Air Force Office of Scientific Research, Office of Aerospace Research, U.S. Air Force, Contract No. F-44620-69-C-0107.

1) Relevant papers in roughly chronological order: Zovko, Zeit. für Phys. 192, 346, Phys. Letts. 23, 143; Lusignoli et. al., Phys. Letts. 21, 229, Nuovo Cimento 45A, 792; Rood, CERN, TH-733, Nuovo Cimento; Carter, Phys. Rev. Letts. 18, 801; Martin and Poole, Phys. Letts. 25B, 343, Nucl. Phys. B4, 467; Kim, Phys. Rev. Letts, 19, 1074, 1079; Chan and Meiere, Phys. Rev. Letts. 20, 568; Cutkosky and Deo, Phys. Rev. Letts. 20, 1272; Rogers, UCRL-18481; Lasinski, EFI BC Group #1; Martin, Queen, and Violini, Phys. Letts. 29B, 311; Igi and Storrow, CERN, TH-993; Martin and Ross, University of Durham (England) preprint; Martin and Sakitt, Phys. Rev. 183. Notice that the combined output is approximately equal to that of CHS.

2) The sum rule is well-known. See, e.g., Martin and Sakitt, Ref. 1.

3) $X_\Lambda = m_K [(m_\Lambda - m_p)^2 - m_K^2]/[2 m_p^2 (E_\Lambda^2 - m_K^2)]$.

4) IHEP - CERN Collaboration, Phys. Letts. 30B, 500 (1969).

5) None of the authors who calculated sum rules similar to Eq. (1) was aware of the Serpukhov results. Since Re $\bar{f}$ Im $\bar{f}$ (given $g^2_{K\Lambda p}$) agrees qualitatively with the data (see Fig. 4), there is weak evidence for the validity of Eq.(1). On the other hand, it is possible that the effect of $\sigma_+(\infty) \neq \sigma_-(\infty)$ does not really affect this low energy comparison. It thus becomes of great importance to measure the real parts of the K^+p amplitudes either through Coulomb interference at small $|t|$ or through the potentially more subtle technique of K^o regeneration from protons.

6) Dalitz and Tuan, Ann. Phys. 10, 307; Shaw and Ross, Phys. Rev. 126, 814.

7) Martin and Sakitt (R.S.), Martin and Ross (M.R.), Ref. 1.

8) Kim, Ref. 1.

9) Both classes of fits give "good χ^2". Some of the data used in Ref. 7 was not available to Kim.

10) See R. Levi Setti's Lund Report for references.

11) For further discussion of the experimental difficulties in determining |Ref| see Conforto et. al., Chicago-Heidelberg preprint, to be submitted to Nucl. Phys. B.

Λ (1520) REVISITED

Robert D. Tripp

Lawrence Radiation Laboratory

Berkeley, California 94720

K^--proton interactions in the region below 500 MeV/c have been under investigation by bubble chambers for many years. The general behavior of the dominant S-wave amplitudes in both I = 0 and 1 and resonant D_{03} amplitude now seem fairly well understood. However, considerable uncertainty remains concerning the smaller P- and D-wave amplitudes. Large scale bubble chamber experiments now in their final stages of analysis should soon clarify the situation.

Figure 1 is a compilation of a decade of K^-p path length from 0 to 500 MeV/c. The seven experiments[1-7] displayed are described in the figure caption. Notice that the ordinate is logarithmic so that the K-65 experiment for example has about 100 times the path length of the earlier WFT experiment in the region of Λ(1520). In this report I shall review the experimental situation in the vicinity of Λ(1520), the region covered by the old experiment of Watson, Ferro-Luzzi, and Tripp[4] (WFT), and the recent experiments of Brookhaven, Massachusetts, and Yale[5] (BMY) and of the Berkeley group[6] (K-65).

Following the discovery and establishment of the quantum numbers of Λ(1520) by WFT, this resonance was seen over the years in a number of production experiments. These production experiments[8] gave a branching fraction for Λ(1520) into $\bar{K}N$ of about 0.45 whereas the elasticity found by WFT was only 0.30. One would normally be inclined to favor the formation experiment because of the intrinsically more detailed analysis to which the data are amenable. However, in the production experiments Λ(1520) was often produced comparatively free from background, and it is difficult to fault their straightforward results. K-65 now finds that an elasticity of 0.45

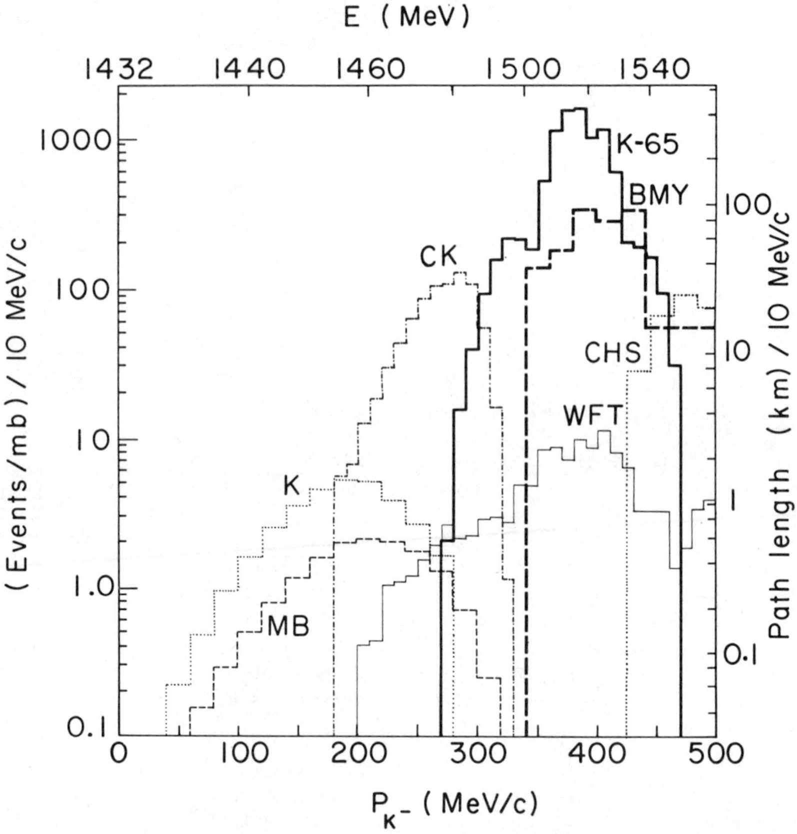

FIG. 1. Path lengths for the seven major hydrogen bubble chamber experiments on K^-p interactions from 0 to 500 MeV/c. The abbreviations are as follows: K(Kim[1]), MB(Maryland-Berkeley[2]), CK (Chan and Kadyk[3]), WFT(Watson, Ferro-Luzzi, and Tripp[4]), BMY (Brookhaven, Massachusetts, and Yale[5]), K-65(unpublished Berkeley experiment[6]), and CHS(CERN, Heidelberg, and Saclay[7]). The BMY path length shown corresponds only to that part which has been analyzed for certain strong interactions; their total path length in the region of $\Lambda(1520)$ is comparable with K-65.

is indeed correct and it is instructive to see how the low elasticity found by WFT came about.

Figure 2 shows the angular distribution of K^-p elastic scattering in the vicinity of $\Lambda(1520)$ to have a very strong $\cos^2\theta$ dependence in the data of K-65. The corresponding data of WFT appear in Fig. 3,

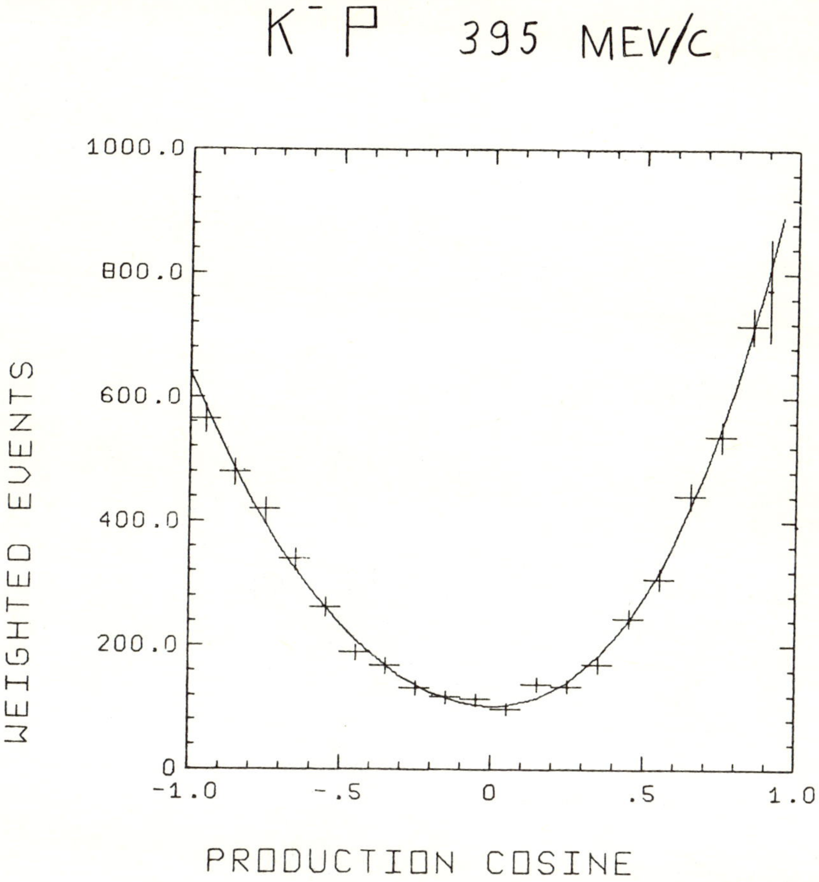

FIG. 2. K^-p elastic angular distribution in the momentum interval 390 to 400 MeV/c as measured by K-65.

where at 390 MeV/c the $\cos^2\theta$ term is seen to be less pronounced and to rapidly disappear at momenta above and below the resonance. Fig. 4 shows this effect more quantitatively. Here the ratio A_2/A_0 in the Legendre polynomial expansion of the elastic angular distribution (which is a measure of the amount of $\cos^2\theta$) is exhibited as a function of the K^- laboratory momentum for both K-65 and WFT. There is a clear disagreement in the vicinity of the resonance. This we now understand to be mainly a result of the poor momentum resolution in the latter experiment. The experimentally measured momentum resolution of WFT is shown in Fig. 5. The peculiar double-peaked structure is due to the fact that the forward and backward scatters result in a stopping particle whose momentum is well measured by range, whereas the group of events with a resolution of

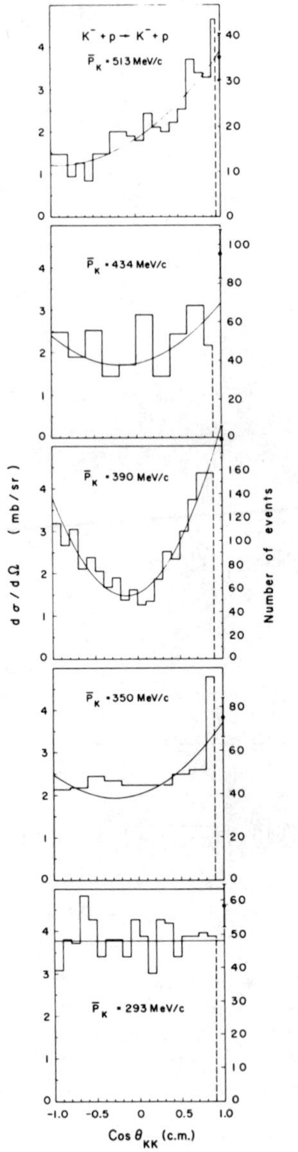

FIG. 3. K^-p elastic angular distributions in five different momentum regions as measured by WFT.

about 12 MeV/c are those particles scattered near 90 deg, where both particles leave the chamber. However, not even a resolution function as large as 12 MeV/c when folded into the new data is sufficient to reconcile the two experiments; at least 15 MeV/c is required. The poor momentum resolution of WFT compared to K-65 arises from three causes:

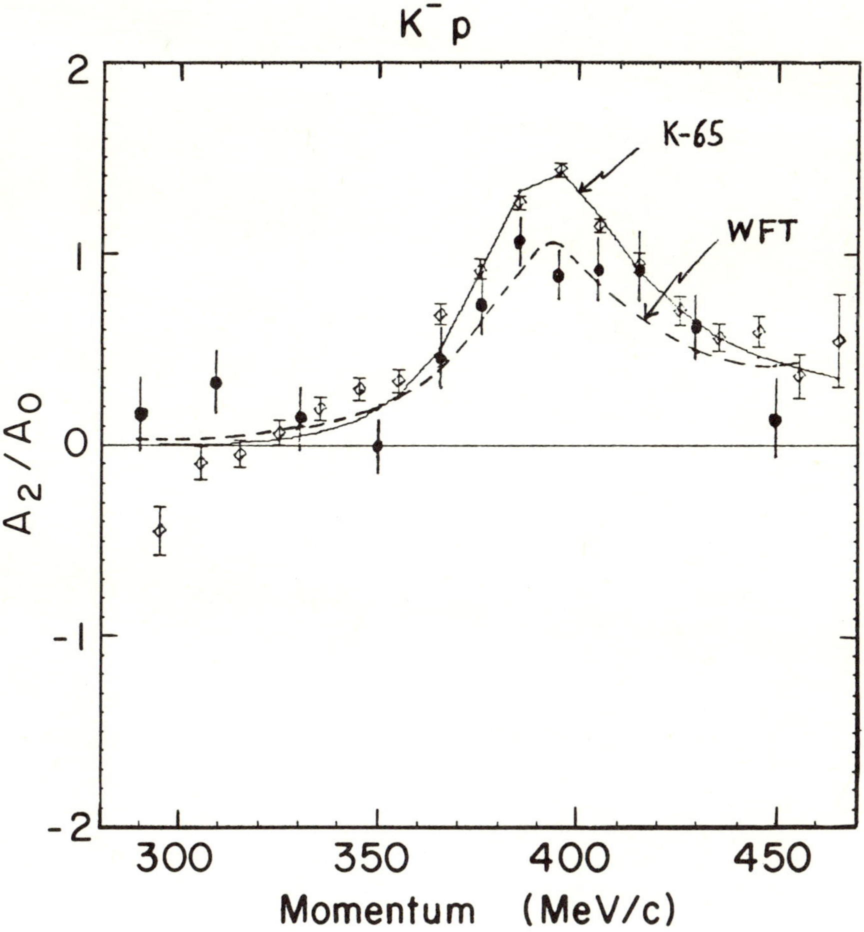

FIG. 4. The ratio A_2/A_0 in the Legendre polynomial expansion of the K^-p angular distribution vs momentum. Open diamonds are data from K-65; solid circles are data from WFT. The solid line and dashed lines are multichannel scattering length fits to the corresponding data.

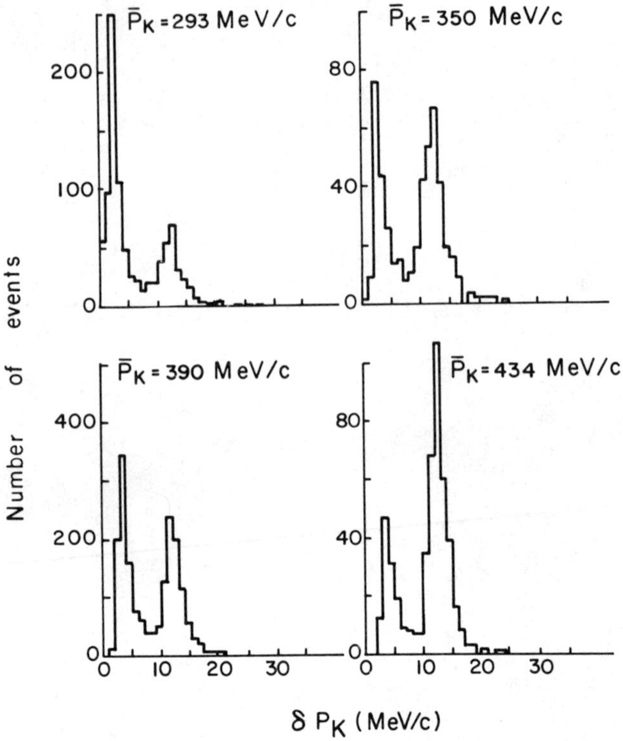

FIG. 5. Experimentally measured momentum resolution of WFT.

(1) The beam was degraded from 760 MeV/c so that the momentum uncertainty on an incoming track was about 30 MeV/c, whereas in K-65 degraders were not used and the momentum spread was less than 5 MeV/c.

(2) The chamber magnetic field was about 11 kG vs 18.7 kG for K-65.

(3) The 15-inch chamber used by WFT had perhaps the world's worst turbulence and gave χ^2 distributions for fitted events that were considerably too large, leading to error assignments that were underestimated by about 20%. The momentum scale in Fig. 5 should therefore be increased by this factor.

Because the K^-p elastic cross section is so much larger than reactions with better momentum resolution, such as $K^-p \rightarrow \bar{K}^0 n$, it dominated in the χ^2 fitting of the coupled-channel analysis. Hence the solution of WFT followed the trend of elastic data whereas it did not reproduce the bump observed in the charge-exchange cross section, as can be seen in Fig. 6. Here an elasticity 50% greater would increase the resonance bump by 2.25, resulting in a better fit.

This dramatizes some of the problems of a detailed high-statistics analysis, serving as an apology for why, after a number of years, the newer experiments have not as yet produced final results. Later I shall return to a discussion of the branching fractions for Λ(1520) as now measured in K-65.

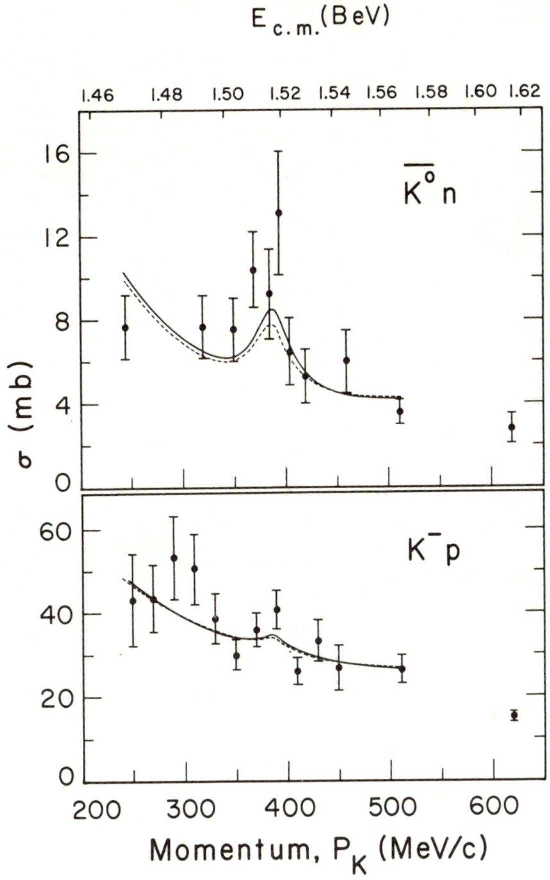

FIG. 6. Cross sections for K^-p charge exchange and elastic scattering as a function of momentum as measured by WFT. The solid lines represent the best fit to all the data in the coupled-channel analysis of WFT.

The BMY group has recently submitted for publication[5] a coupled-channel partial-wave analysis of the K^-p system from 350 to 430 MeV/c. Since they have analyzed only the "0 prong V^0 topology" (i.e. $\bar{K}^0n$, $\Lambda\pi^0$, and $\Sigma^0\pi^0$), they have included in their analysis, data for other channels as reported by WFT as well as some published data on Σ polarization from K-65.[6] That they have residual experimental problems is evident in Fig. 7 where their charge-exchange coefficients are shown as a function of momentum. Considerable disagreement between two runs at different momenta is apparent in the overlap region at 390 MeV/c. Their best-fit solution predicts only a very small bump in the charge-exchange cross section. This, as in the analysis of WFT, is probably a result of poor momentum resolution. Thus they report an elasticity for $\Lambda(1520)$ of only 0.26, which is even lower than that of WFT, while the resonant width comes out to be 26 MeV, some 50% too large. Figure 8 summarizes their elastic partial-wave amplitudes for the S and P waves parametrized either with effective ranges or constant K matrices. Also shown are the amplitudes of Kim[9] and Martin and Sakitt[10] for the S waves.

Turning now to a discussion of the preliminary data of K-65,[6] covering the momentum range from 280 to 470 MeV/c, Table I lists the seven channels involved in the coupled-channel analysis with the number of measured events in each channel. Apart from the elastic scatters, nearly all events on the film were measured so that these represent essentially the number of events for each channel within the fiducial volume. Only about 10% of the two prongs (elastic scatters) were measured in the momentum region from 350 to 420 MeV/c where, as can be seen in Fig. 1, some 83% of the K-65 path length is contained. From 280 to 350 MeV/c (10% of the path length) and from 420 to 470 MeV/c (7% of the path length), nearly all the elastic scatters were measured. For the entire experiment the total number of measurements was about 1/2 million. The immense job of analyzing all this data is in the able hands of our graduate student Terry Mast and research associate Roger Bangerter.

We have used two different energy-dependent expressions for the nonresonant partial-wave amplitudes: 1) a constant scattering length parametrization in a manner essentially identical to that of WFT, and 2) parametrization of the S_{01}, S_{11}, and P_{13} amplitudes by means of the Ross and Shaw[11] effective range. The latter, with a number of extra parameters, produces only a slightly improved χ^2 despite the fact that, barring anomalies, it should give a more exact description of the true energy dependence of these amplitudes. This illustrates the point that when fitting over a limited energy region, extraction of the energy dependence of the amplitudes is extremely difficult. However, once we incorporate into the analysis the published lower-energy data and perhaps some at higher energy from other experiments, meaningful effective range parameters probably can be deduced.

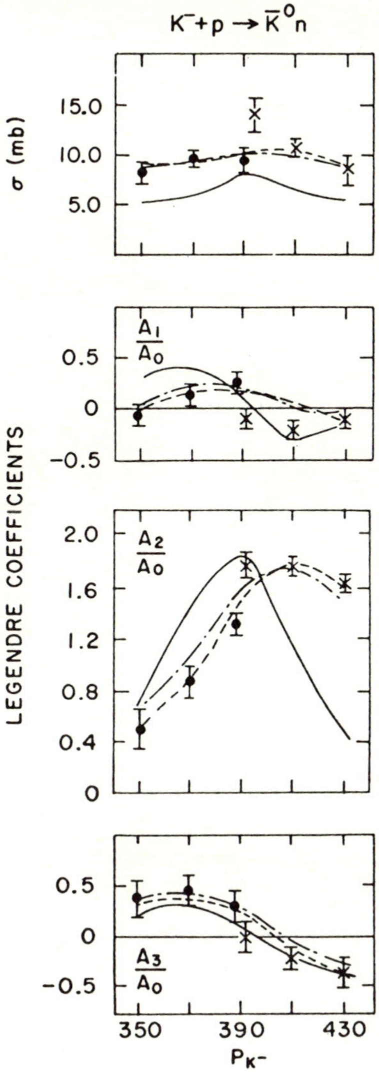

FIG. 7. Charge-exchange angular distribution coefficients from the BMY experiment as a function of momentum. Solid circles and crosses correspond to two different momentum settings. The solid curves are predictions from J. Kim[9] based on older data; the other curves are K matrix fits of BMY.

$K^- + p \rightarrow \bar{K}N$
PARTIAL WAVE AMPLITUDES

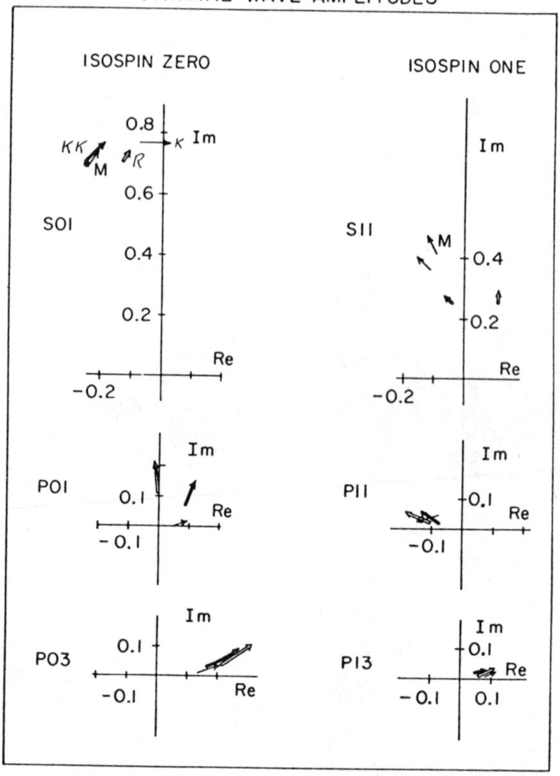

FIG. 8. Elastic partial-wave amplitudes of BMY. The beginning of the arrow is at 350 MeV/c and the tip at 430 MeV/c. Shown are the constant K-matrix (KK) and effective-range (R) analysis of BMY, as well as the solution of Kim (K) and Martin and Sakitt (M) for the S wave.

Table I summarizes the χ^2 obtained from the constant scattering length fit broken down according to channel and according to whether it arises from the partial cross section, differential cross-section polynomial coefficients, or polarization polynomial coefficients. The polarization coefficients fit best; this is reasonable since they should be least affected by the expected biases. The cross sections fit least well, owing mainly to continuing problems associated with the different momentum resolution of the tau decays from which the path length is determined, compared to that from various other topologies. The elastic scatters and $Y^0\pi^0$ channels are still in quite preliminary form, while for the other channels we

Table I. The ratio of χ^2 to expected χ^2 according to channel and type of measurement for K-65.

		$\chi^2/\langle\chi^2\rangle$		
Channel	Measured events	σ	A_n/A_0	B_n/A_0
K^-p	46,000		2.23	
$\bar{K}^0N$	30,000	1.34	1.80	
$\Sigma^+\pi^-$	85,000	1.85	1.61	0.87
$\Sigma^-\pi^+$	61,000	1.74	0.98	1.25
$\Sigma^0\pi^0$ }	40,000	1.89	2.71	1.27
$\Lambda\pi^0$ }		2.18	1.09	0.99
$\Lambda\pi^+\pi^-$	10,000	1.77		
Total	272,000	1.86	1.71	1.07

Overall $\chi^2/\langle\chi^2\rangle = 1.45 \pm 0.05$

do not anticipate substantial changes in the final data. The overall ratio of χ^2 to expected χ^2 is 1.45 ± 0.05, so in a chi-squared sense the fit is poor. However, the general trend of the data and nearly all structure is fit rather well. In Fig. 9 the ratio $\chi^2/\langle\chi^2\rangle$ is plotted as a function of momentum for this solution. If the assumed momentum dependence of the amplitudes is responsible for the high χ^2, then, since most of the path length is from 350 to 420 MeV/c, one would expect that this region would fix the amplitudes and that both above and below this region the χ^2 would rise. Such a behavior is not observed. Conversely, systematic effects will show up most visibly in the region containing most data. Since χ^2 is not enhanced here, this cannot be the primary reason for high χ^2. Perhaps a combination of both effects gives rise to the momentum independence observed in Fig. 9.

In order to explore the possibility of unexpected structure in the momentum dependence of the amplitudes, we have broken the data into three momentum intervals centered at 325, 395, and 425 MeV/c. All amplitudes were free to vary in each interval [except for a fixed mass and width of Λ(1520)]. The χ^2 did not improve, suggesting again that systematic effects rather than improper momentum dependence is the major source of high χ^2. Figures 10, 11, and 12 show the Argand plots for these amplitudes. Figures 13 and 14 show the corresponding $\bar{K}N$ S-wave amplitudes and S- and P-wave amplitudes for other low-energy experiments as compiled by Levi-Setti for the 1969 Lund Conference.[8] The clockwise behavior of the

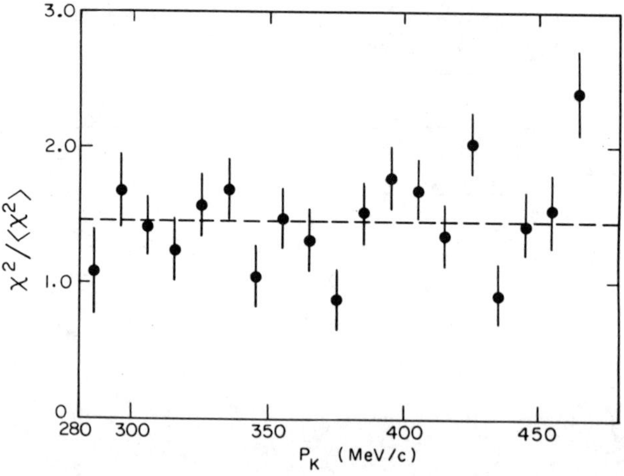

FIG. 9. The ratio of χ^2 to expected χ^2 as a function of momentum for a solution of an energy-dependent coupled-channel analysis of K^-p reactions in the K-65 experiment.

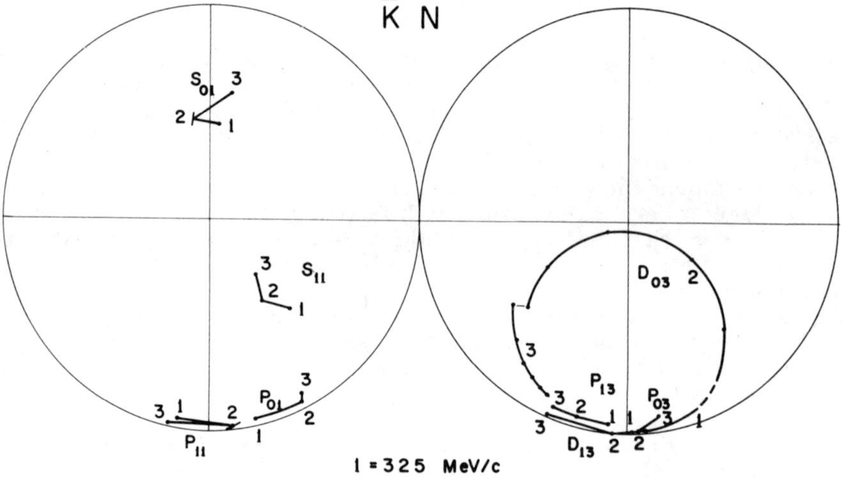

FIG. 10. $\bar{K}N$ amplitudes from the K-65 experiment with data broken into three momentum intervals centered at (1) 325 MeV/c, (2) 395 MeV/c, and (3) 425 MeV/c. The coupled-channel analysis was done in the constant scattering-length approximation except for the resonant D_{03} amplitude. Typical variations over fitted momentum intervals are indicated by several thin lines.

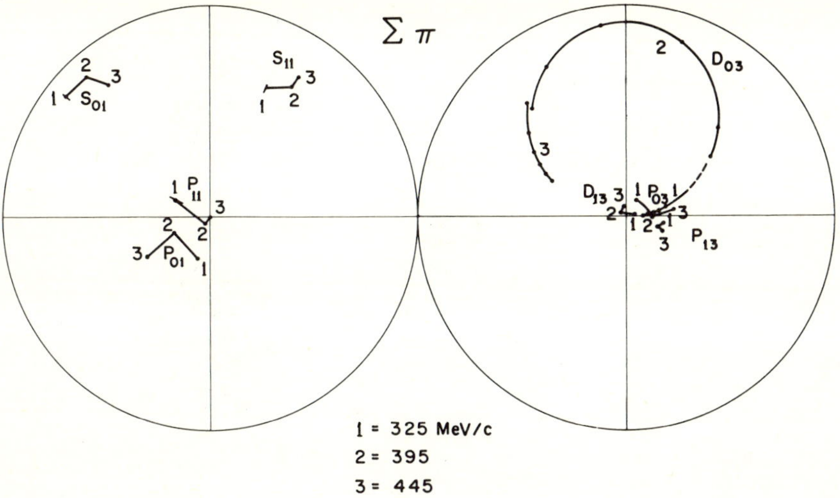

FIG. 11. Same as Fig. 10, but for the $K^-p \rightarrow \Sigma\pi$ amplitudes.

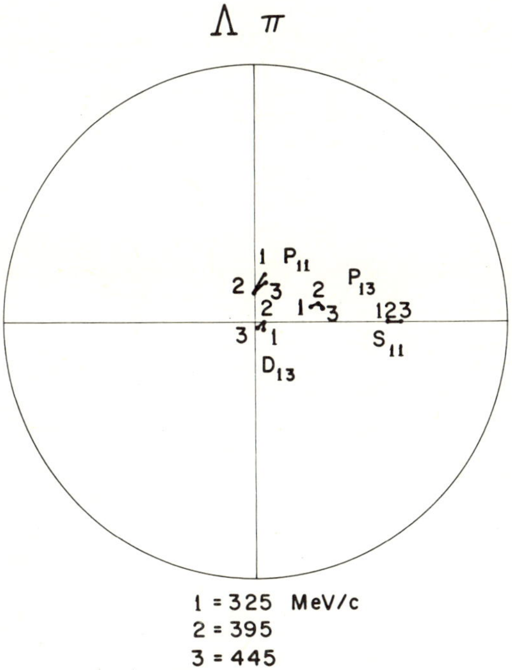

FIG. 12. Same as Fig. 10, but for the $K^-p \rightarrow \Lambda\pi$ amplitudes.

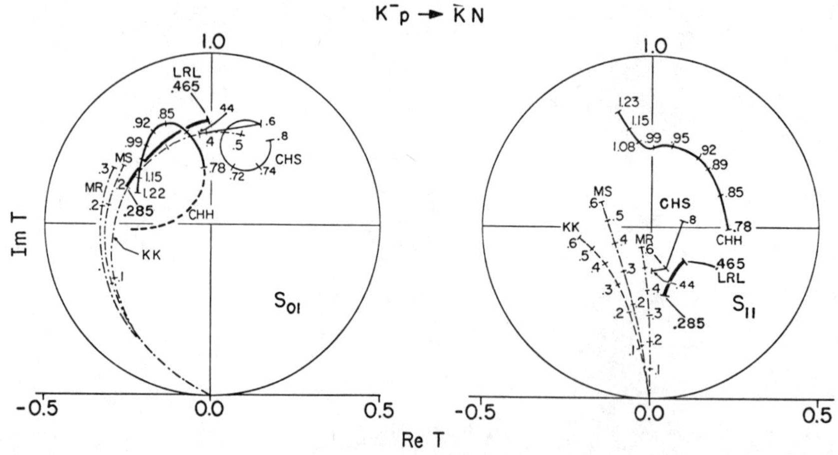

FIG. 13. A compilation[8] of the low energy $K^-p \rightarrow \Sigma\pi$ amplitudes.

$\bar{K}N$ S_{01} amplitude indicated by the low-energy K-matrix solutions in Fig. 13 is not strongly suggested by our data in Fig. 10, but probably there is sufficient uncertainty in our lowest-energy point to accommodate such a behavior. In the $\Sigma\pi$ channel the agreement is better. Perhaps the only anomalous structure suggested by Fig. 10 [apart from the D_{03} amplitude for $\Lambda(1520)$ which is parametrized as a resonance] is some possible structure in the $\bar{K}N$ P_{11} state. However, it must be emphasized that both the data and analysis are preliminary.

New branching fractions for $\Lambda(1520)$ were reported by K-65 at the Lund Conference. We have updated them with some minor changes for this conference with the percentages listed in Fig. 15. The numbers are still not final, but they should not change outside of the indicated statistical uncertainty. The figure also shows some other mass-dependent effects associated with $\Lambda(1520)$ branching fractions. Starting on the left are the branching fractions predicted for a pure SU_3 singlet. Next the $\bar{K}N$-$\Sigma\pi$ mass difference splits them as indicated; however, the experimental shift is in the opposite sense as shown in the middle column. This can be reconciled by introducing a mixing angle of 18 deg between the $3/2^-$ singlet and octet, thereby also greatly improving the agreement of the octet $\Lambda(1690)$ decay rates with decay rates for other members of that octet. We have further introduced different centrifugal barriers for decays into different charge states. This reduced χ^2 slightly (by about 1%) and gave the branching fractions depicted in the next column. Finally, in order to see if the electromagnetic mass splittings (which are comparable to the half-width of the resonance) could lead to further violations of isospin than those embodied in the centrifugal barrier

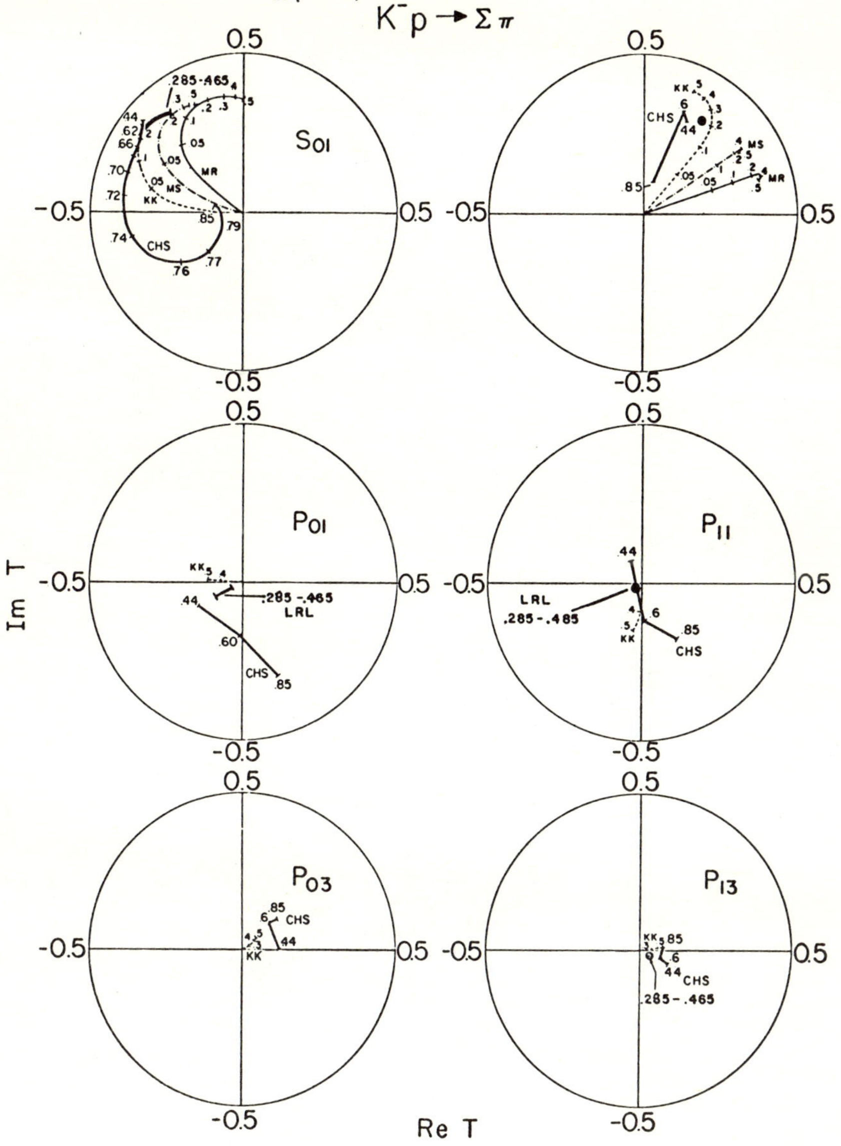

FIG. 14. A compilation[8] of the low energy $K^-p \rightarrow \Sigma\pi$ amplitudes.

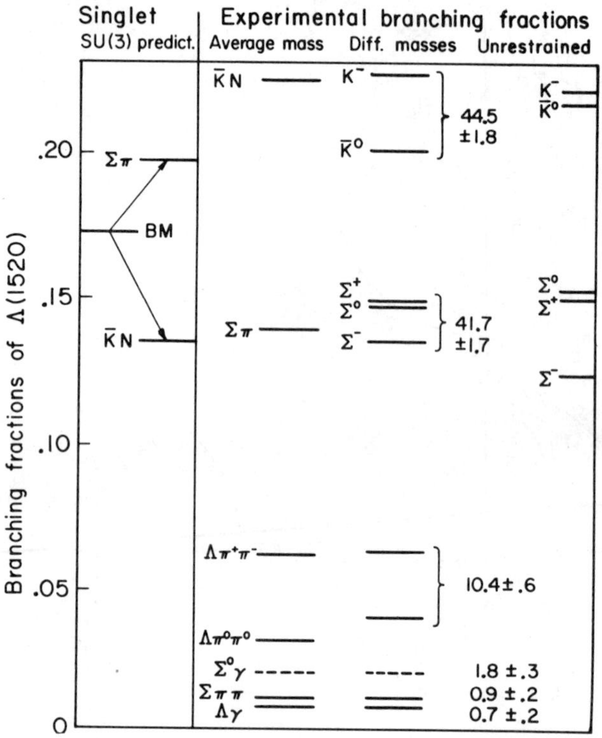

FIG. 15. Branching fractions for Λ(1520).

corrections, we have left all $\bar{K}N$ and $\Sigma\pi$ branching fractions unrestrained, thereby introducing three new parameters into the analysis. The χ^2 again lowered by only about 1% and gave the branching fractions illustrated in the last column. They are sufficiently close to those of the previous column that within errors we have no indication of isospin violation in the decay of Λ(1520).

As a further test of isospin conservation we have studied the charge-independence triangle formed by the $\Sigma^+\pi^-$, $\Sigma^-\pi^+$, $\Sigma^0\pi^0$ reaction amplitudes at various momenta and angles. Specifically, charge independence requires that $\vec{T}_+ + \vec{T}_- + 2\vec{T}_0 = 0$, where T represents the reaction amplitude at a given angle and the subscript indicates the Σ charge state. Violation of charge independence is indicated by nonclosure of the triange, which can be expressed by the condition that the area of the triangle be ≤ 0. This leads to the requirement for charge independence

$$2\sigma_+\sigma_- + 8\sigma_+\sigma_0 + 8\sigma_-\sigma_0 - \sigma_+^2 - \sigma_-^2 - 16\sigma_0^2 \geqslant 0,$$

where σ denote the differential cross section in the charge state indicated by the subscript. A preliminary test at 20 angles and 17 momenta using quite primitive $\Sigma^0\pi^0$ data yielded no statistically significant indication of charge-independence violation at any of the 340 points. Final data will be subjected to this test in the near future. Since the reaction amplitude at any angle is composed of many partial waves, this is a more gross test of charge independence, although less model-dependent than the one specific to Λ(1520).

Future plans for the K-65 experiment call for the introduction of the full correlation matrix into the coupled-channel analysis of the polynomial coefficients. This will certainly give an improved solution and may reduce the overall χ^2. It is also planned to introduce the effective range matrix parametrization into all partial waves. Although this will bring in many new parameters, it is hoped that an improved χ^2 will result. Finally, there are some hints of structure in the data that we cannot accommodate at the present time. The most prominent is shown in the A_1/A_0 coefficient for K^-p elastic scattering in the vicinity of 425 MeV/c as seen in Fig. 16. If such structure persists in the final data it is difficult to see how it can be accommodated in the present parametrization with only one resonant amplitude.

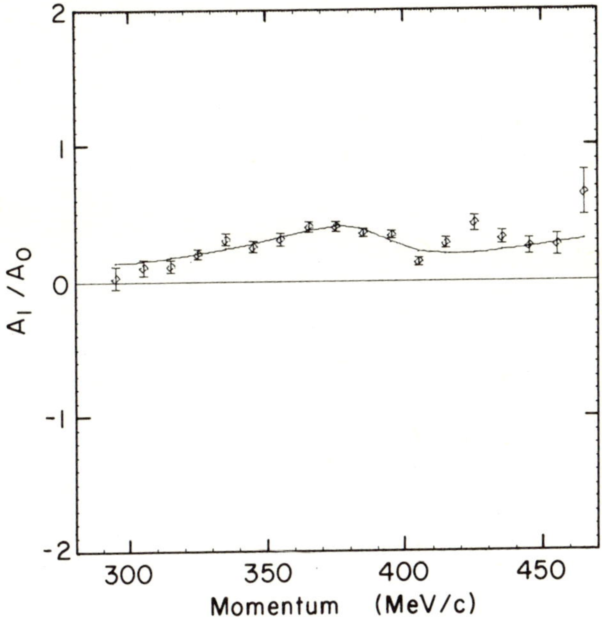

FIG. 16. The A_1/A_0 coefficient vs momentum for K^-p elastic scattering from K-65.

REFERENCES

1. J. K. Kim, Phys. Rev. Letters 14, 29 (1965).
2. M. Sakitt et al., Phys. Rev. 139, B719 (1965).
3. J. Chan and J. Kadyk, private communication of an unpublished Berkeley experiment.
4. M. B. Watson, M. Ferro-Luzzi, and R. D. Tripp, Phys. Rev. 131, 2248 (1963).
5. D. Berley et al., Phys. Rev. (to be published).
6. Some of the $\Sigma^{\pm}$ data appears in R. Bangerter et al., Phys. Rev. Letters 17, 495 (1966) and R. Bangerter, Lawrence Radiation Laboratory Report UCRL-19244, July 1969.
7. R. Armenteros et al., CERN/D. Ph. II/Phys 70-7; submitted to Nuclear Physics B.
8. R. Levi-Setti, in Proceedings of the Lund International Conference on Elementary Particles, 1969 (Berlingska Boxboktryckeriet, Lund, Sweden, 1969).

DISCUSSION

Kim: It looks like your poor χ^2 ratio tends to come from specific regions. For instance, in the elastic channel it comes mainly from the 420 region, and in say the $\Lambda\pi$ channel it looks like it comes from the 320 region. Have you tried to improve these particular regions, especially?

Tripp: Yes, it's true that the $\Lambda\pi$ doesn't fit very well. We have done a number of things, like trying to put in s-wave and p-wave resonances to patch up these little anomalies. The program always spits them out.

Kim: There's no way that the anomalies can be fit by putting in arbitrary resonances or something like that?

Tripp: Well, we have to work on it more. We tried to do it, as I said, and we haven't succeeded in our first efforts; but, you know, you have to be clever about what wave you put in and so on. Maybe we haven't been clever enough so far.

Kim: I'm asking because when I use the effective range parameterization, the poor χ^2 ratio you talk about doesn't show up that much; the ratio is not big at all.

Tripp: You are not using as much data as we are. Remember, we have a lot of events now.

Kim: I'm not using your data; however, other people's data show the same thing. But the way I analyze it, χ^2 really goes down. So poor χ^2 means, I think, that a special region fits poorly. Once you do your parameterization such that you take care of that, probably χ^2 goes down. It's not a systematic problem for the general region, but some specific region that gives poor χ^2.

Tripp: Well it's clear, I think, that the $\Lambda\pi^0$ could benefit by putting in the effects of the 1385. I haven't really tried to tack on a resonance as well as an effective range.

The analysis is continuing, and I expect we should have our final results within a year or so. The plans for the future are to remove most of the systematic effects of the momentum resolution in the experiment. Another thing we've always neglected is the correlation matrix; we've only put in the diagonal terms of the error matrix. Now, the off-diagonal elements are quite large, even though we make a polynomial expansion. I think that this is going to reduce χ^2, although people who understand these things better than I do say that statistically it shouldn't change the χ^2. Also it will be essentially an effective range analysis.

Kim: When you do these regions separately, do you use constant scattering length analysis?

Tripp: Yes, just out of convenience. But we should probably not because, as you saw, the s-wave amplitudes tend to be going almost perpendicular to the longitudes in this region, so that at least in the s-wave it is not a good approximation.

SESSION III

MODERATE ENERGY FORMATION

Discussion Leader: Riccardo Levi-Setti

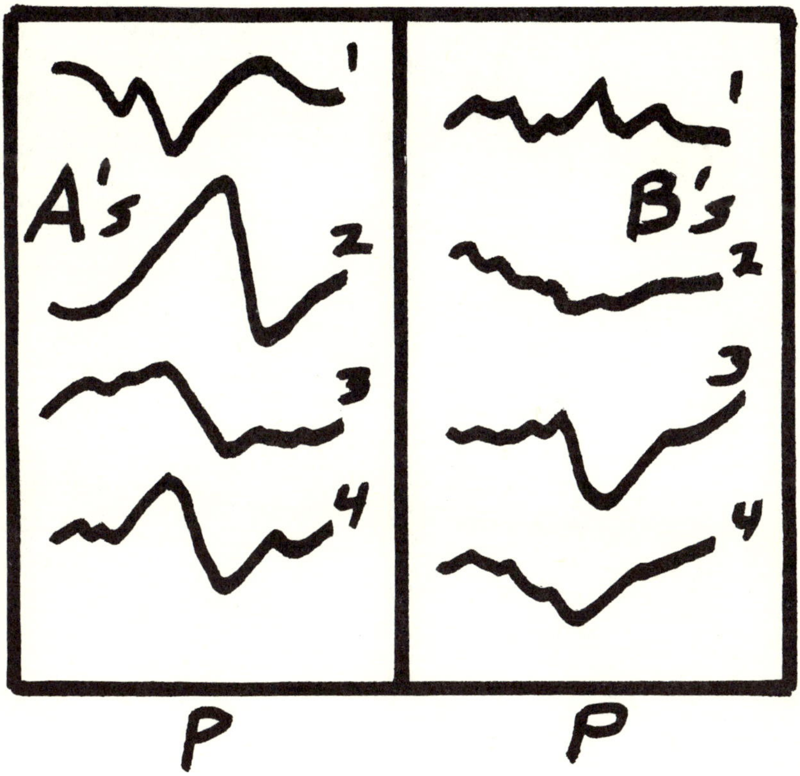

$\overline{K}N$ PARTIAL-WAVE AMPLITUDES

A COMPILATION

R. Levi Setti, D. W. Merrill;* D. J. Herndon**

* Enrico Fermi Institute, Univ. of Chicago, Chicago, Ill.

**Lawrence Radiation Laboratory, Univ. of California, Berkeley, Cal.

One of the recent undertaking of the LRL Particle Data Group is the compilation of partial wave amplitudes for various elastic scattering and reaction processes. In addition to a tabulation of the amplitudes, this review will contain computer generated Argand diagrams, speed plots and total cross section contributions for each partial wave.

The success of this task will depend upon the willingness and cooperation of all those groups contributing to the study of the $\overline{K}N$ interaction in supplying the relevant information. A preliminary and incomplete version of the $\overline{K}N$ compilation (UCRL-20030 $\overline{K}N$, April 1970) was available at the time of this Conference based partly on data collected from the literature, partly on detailed data contributed by various authors. It will be most helpful for the completion of this compilation if the results of recent partial wave analyses, not yet included in the preliminary version, could be made available in tabular form. The information which we wish to obtain is the following:

a) Channel and type of data included.
b) Resonant and non-resonant partial waves included in the fit and their parametrization if energy dependent.
c) Real and imaginary part of each partial wave amplitude, at each of the momenta considered. If a given partial wave is described solely by a Breit-Wigner form, we could, of course, compute real and imaginary parts as a function of momentum, provided the exact B-W-formulation is indicated, together with the relevant resonant parameters.
d) Unitarity constraints imposed.
e) χ^2 of the fit versus expected χ^2.

f) For the reaction amplitudes, please indicate the convention adopted as to the sign of resonant amplitudes for I = 0 and I = 1.
g) If the phase shift analysis is energy independent, some details as to the method used.
h) Appropriate reference, which will also be incorporated in the figure headings.

Even a xerox copy of original graphs would be of great help, if a tabulation is too much to ask. We would like to limit to a minimum the reading of data from published graphs in view of the inherent loss of precision. If this should become necessary, a qualifying statement will accompany such data.

A FIT OF BARYON DECAY RATES WITHIN SU(3)[‡]

D.E. Plane, P. Baillon, C. Bricman[*], M. Ferro-Luzzi, J. Meyer, E. Pagiola, N. Schmitz[**]

CERN, Geneva

E. Burkhardt, H. Filthuth, E. Kluge[+], H. Oberlack

Institut für Hochenergiephysik, Heidelberg

R. Barloutaud, P. Granet, J. P. Porte, J. Prevost

C.E.N. Saclay

ABSTRACT

A test of the consistency between observed resonant amplitudes or decay rates for baryon resonances with SU(3) predictions has been made. Coupling constants describing the decays are determined and predictions are given for decays into stable baryons and mesons.

---

‡ The full text of this paper has been submitted to Nuclear Physics

* Visitor from I.I.S.N. Brussels

** Present Address, Max Planck Institut fur Physik und Astrophysik, Munich

\+ Now at SLAC, Stanford.

DISCUSSION+

Galtieri: May I ask a question about the amplitudes of F_{15}. You said that everything seems to agree. What size of the amplitude did you use? What was the value for it?

Plane : For the resonant amplitudes in the $\Lambda\pi$ and $\Sigma\pi$ final states we took the results of the CRS collaboration. They give resonant amplitudes of -0.09 and -0.08 for the $\Lambda\pi$ and $\Sigma\pi$ channels respectively. The predictions from the analysis are in good agreement with these values.

Butterworth: I will be changing those numbers slightly later on but not significantly.

Plane: We have taken errors of $\pm$ 0.05 so unless those numbers will change by more than that the results will still be valid.

Butterworth: Your χ^2's generally are much too low by a factor of 3. Have you been increasing the errors quoted by authors?

Plane: Yes we have taken an error of ± 0.05 on nearly all resonant amplitudes. You may compare this with the statistical error quoted, for example, on the $\Sigma(1660)$ elasticity which is ± 0.01. We feel that the statistical errors from formation experiments do not take into account the possible presence of background and consequent departure from the assumed Breit-Wigner energy dependence.

Feld: I am not quite sure that I understand the rationale behind that analysis. You take into account singlet-octet mixing for the Λ 's, but you assume that there is no octet-decimet mixing for the Σ 's and the Ξ 's. Is there any rationale for that or is it just completely ad hoc?

Plane: Somewhat ad hoc. As far as I know there is not yet any evidence for octet-decuplet mixing, and by taking only singlet octet mixing we get satisfactory fits for the $1/2^-$ and $3/2^-$ multiplets. There may be further mixing but we have tried only the simplest possible interpretation.

Feld: The main reason that that interpretation worries me is that if you take the simplest other check, which would be the mass formulae, you run into trouble if you assume that there is no mixing except between the singlet and the octet, so that obviously that must be an over simplification. Maybe the errors are large enough

---

+ This discussion concerns the paper "A Fit of Baryon Decay Rates Within SU(3)" presented at this conference by the C.H.S. collaboration and to be published in Nuclear Physics.

to cover it now, but in the final analysis, when the errors become small and all the data are in, I suspect that you are also going to have to take into account octet-decimet mixing as well as singlet-octet.

Tuan: I am still a little bit confused about your statement about the J^P $5/2^+$ octet because I thought at least from the latest Particle Data Group tables that the Σ (1910) decay into $\Sigma\pi$ is still very small and that poses the major problem.

Plane: Since then there has been new data from the CRS collaboration. They start around 1915 MeV and use a few points of CHS data below that energy.

Butterworth: We got four energy points from CHS.

Plane: Particularly to look for the 1915.

Tuan: And you have a large $\Sigma\pi$ decay mode now?

Plane: They have an amplitude at resonance of -.08. (Updated below)

Galtieri: I am going to be discussing this in my talk later.

Goldberg: In regard to the χ^2 being too low, we found that if you assume mixing and if you leave the mixing angle open, in most of these octets you can actually fit anything. You can fit decay rates that are an order of magnitude different. I notice in the results of this fit, there were no errors quoted on those mixing angles. You'll find that the mixing angle for example is typically 20° with an error of about 50°. So in fact the determinations that are made of the mixing angle when you leave that open and don't fit it from the mass formula are almost meaningless because the data just are not of good enough quality to fit a singlet amplitude, an octet amplitude, and a mixing angle at the same time.

Plane: For the $3/2^-$ multiplet we have a mixing angle of -25° with an error of 6°. For the $1/2^-$ nonet we have a mixing angle of 18° with an error of 17°. Furthermore how can you calculate a mixing angle from the mass formula if the spins and parities of the cascades have not been measured, so you do not know which multiplet to put them in.

Goldberg: But in addition, at the beginning you made some indication that this was a better procedure because you had good errors on t, but from the presentation of your last data you hardly have any errors on t for most of the resonances that you considered.

Plane: The resonances considered in our SU(3) fits all came from energy dependent solutions and are all well established. Thus we did not include the new P wave resonance which I presented, nor did we include the cascades, apart from Ξ (1530), since their J^P values have not yet been determined.

Goldberg: You presented another item also that puzzles us. You had a Ξ^* $5/2^+$ and $5/2^-$ at masses of 1950 and 1990. Where did you get those from.

Plane: We predict cascades at 1950 and 1985 for the $5/2^-$ and $5/2^+$ multiplets. The latter should decay mainly into ΣK.

Goldberg: Did you say that there was evidence for it?

Plane: The data on Ξ^* come directly from the Particle Data Group. There are cascade resonances with masses of 1930 and 2030 which we tentatively take as candidates for the $5/2^-$ and $5/2^+$ octets.

Butterworth: Could I just give the 1915 branching ratios from the latest CRS fit. For the Σπ modes it is -.13 ± .03 and for the Λπ mode .07 ± .015.

THE $\bar{K}N$ INTERACTION BETWEEN 400 AND 1200 MeV/c : FIRST ATTEMPT TO AN ENERGY INDEPENDENT PARTIAL WAVE ANALYSIS IN THE $\bar{K}N$, $\Sigma\pi$ AND $\Lambda\pi$ CHANNELS

R. Armenteros, P. Baillon, C. Bricman,*
M. Ferro-Luzzi, E. Pagiola, J. O. Petersen,
D. E. Plane
CERN, Geneva
E. Burkhardt, H. Filthuth, E. Kluge,+
H. Oberlack
Institut für Hochenergiephysik, Heidelberg

Abstract:

The reactions $\bar{K}N \to \bar{K}N$, $\bar{K}N \to \Sigma\pi$ and $\bar{K}N \to \Lambda\pi$ have been analyzed in terms of partial waves. The data employed come from a series of bubble chamber exposures to K^- with momenta ranging between $\sim$400 and $\sim$1200 MeV/c. The method adopts a parametrization for the energy dependence of a restricted number of partial waves and uses the continuity of the data over the entire region to obtain the remaining waves.

* Visitor from IISN, Bruxelles.

+ Present address: SLAC, Stanford, California.

1. INTRODUCTION

The $\bar{K}N$ interaction has been fruitfully analyzed in terms of partial wave amplitudes from threshold to momenta near 2 GeV/c. Depending on the momentum region and the preference of the analyst, these analyses have variously introduced the constant scattering-length approximation, the effective-range parametrization and other more arbitrary forms for the energy dependence of the amplitudes. These approaches, although quite adequate in providing information on the gross features of the interaction, reach very quickly the limit of their usefulness when the behaviour of the amplitudes is investigated in detail. The inadequacy of the methods is particularly evident in cases where resonant phenomena occur in the presence of sizeable backgrounds. There is, at the moment, a real need for new methods of analysis which depend as little as possible on the constraining effects of the parametrization. The energy-independent analyses of the type successfully performed for the πN interaction are unfortunately not yet applicable here due to the insufficiency of the experimental data.

As an intermediate stage between the fully parametrized and the unparametrized methods, we present an analysis of the data between ~400 and ~1200 MeV/c which borrows features from both approaches. The reactions studied are $\bar{K}N \to \bar{K}N$, $\bar{K}N \to \Sigma\pi$ and $\bar{K}N \to \Lambda\pi$. The general trend of the partial wave amplitudes in these channels has already been suggested by previous analyses(*). On the basis of these, we select a certain number of amplitudes where the stronger and less controversial effects have been established. These effects are usually ascribed to resonances; we have limited our selection to those resonances whose existence has been verified by several independent experiments, either of the formation or the production type. We assume, further, that their description in terms of the conventional Breit-Wigner formalism is sufficient for our purposes. By considering these amplitudes as known, the problem is greatly simplified and we can derive the remaining amplitudes at each momentum without further assumptions.

(*) See, for example, ref. 1 for a review of the most recent results.

The results suggest a considerable amount of structure in most of the amplitudes. The interpretation of the Argand diagrams is not straightforward; several new resonances can be inferred, some with more confidence than others. A preliminary version of this analysis has been presented at the Lund Conference[1].

2. METHOD

The procedure was described in detail in connection with an earlier analysis of the $\bar{K}N \rightarrow \Lambda\pi$ reaction[2]. Apart from minor modifications, the method is substantially the same. We recall the basic steps:

a) the data are fitted to a power series of the incident K^- momentum (P) over the region under study. The results of the fit are referred to as "smooth data" and used to replace the original discrete set of data.

b) a random choice is made for the initial values of the partial wave amplitudes at a starting momentum P_o. The values are then allowed to vary so as to best predict the "smooth data" at the momentum P_o.

c) the final values of the amplitudes at P_o are taken as initial values at the adjacent momentum $P_o + \Delta P$ and then fitted to the data. The interval ΔP can be made as small as convenient, the "smooth data" being defined over the whole momentum region. With ΔP sufficiently small, the final values of the amplitudes at $P_o + \Delta P$ are generally close to those at P_o, thus insuring local continuity.

Step (c) is continued on both sides of P_o until the boundaries of the momentum region are reached. The requirement for an acceptable solution is that all the amplitudes are continuous from one end to the other over the complete momentum interval. Step (b) is repeated at different starting momenta to search for alternative solutions.

The experimental conditions are considerably different for the three reactions under study. We have introduced different assumptions for each case and have dealt with the data according to the specific situation. The details of the method and the results obtained are discussed separately for each reaction. The definition of the partial waves and the notations are as in ref. 3.

3. RESULTS

3.1 $\bar{K}N \rightarrow \Lambda\pi$

Of the three reactions considered this is the

simplest to analyse. An analysis of this reaction along the lines described above has already been performed between 600 and 1200 MeV/c[2]. We have extended the analysis towards the lower momenta, using the new data between 436 and 600 MeV/c[4]. The data above 600 MeV/c are those already used in ref. 2 and consist of differential cross sections ($d\sigma / d\Omega$) and polarizations ($\vec{P}$) for the reactions $K^-p \rightarrow \Lambda\pi^o$ [5] and $K^-n \rightarrow \Lambda\pi^-$ [6] at intervals of ~20 MeV/c. Below 600 MeV/c the only reaction available is $K^-p \rightarrow \Lambda\pi^o$. In all cases, the cross sections and polarizations at each momentum have been replaced by the coefficients A_n and B_n of the following expansions:

$$d\sigma / d\Omega = \lambda\!\!\!^{-2} \sum_{n=o}^{N} A_n P_n (\cos\theta) \tag{1}$$

$$\vec{P} (d\sigma / d\Omega) = \lambda\!\!\!^{-2} \hat{n} \sum_{n=1}^{N} B_n P_n^1 (\cos\theta) \tag{2}$$

The angles θ refer to the direction of the meson in the c.m. system, the unit vector $\hat{n} = \vec{K} \times \vec{\pi} / |\vec{K} \times \vec{\pi}|$ is the normal to the production plane, $\lambda\!\!\!^{-}$ is the wavelength of the incident channel, P_n and P_n^1 are the Legendre and first-associated Legendre polynomials of order n, respectively. The coefficients A_n and B_n, with their errors, have then been used to generate the "smooth coefficients" according to the prescriptions above (Section 2). Fig. 1 shows the original coefficients for $K^-p \rightarrow \Lambda\pi^o$ and the curves corresponding to the preferred solution. These curves reproduce closely the general behaviour of the smooth data.

As in ref. 2, the D_{15} amplitude ($I = 1$, $J^P = 5/2^-$) has been taken as known, being completely specified by its Breit-Wigner resonant parametrization corresponding to $\Sigma(1760)$. This assumption, together with a limit on the amplitudes to those having $J \leqslant 5/2$, is sufficient to eliminate the phase ambiguity inherent to the inelastic channels. Below 600 MeV/c, however, the effects due to $\Sigma(1760)$ are negligible and the problem would become indeterminate without the introduction of another constraint. Thus, below this momentum, we have arbitrarily fixed the phase of the S_{11} amplitude. This amplitude has a magnitude sufficiently large to provide a meaningful effect in the analysis and a phase which hopefully does not vary so quickly as to make the hypothesis unrealistic. However, it should be kept in mind that

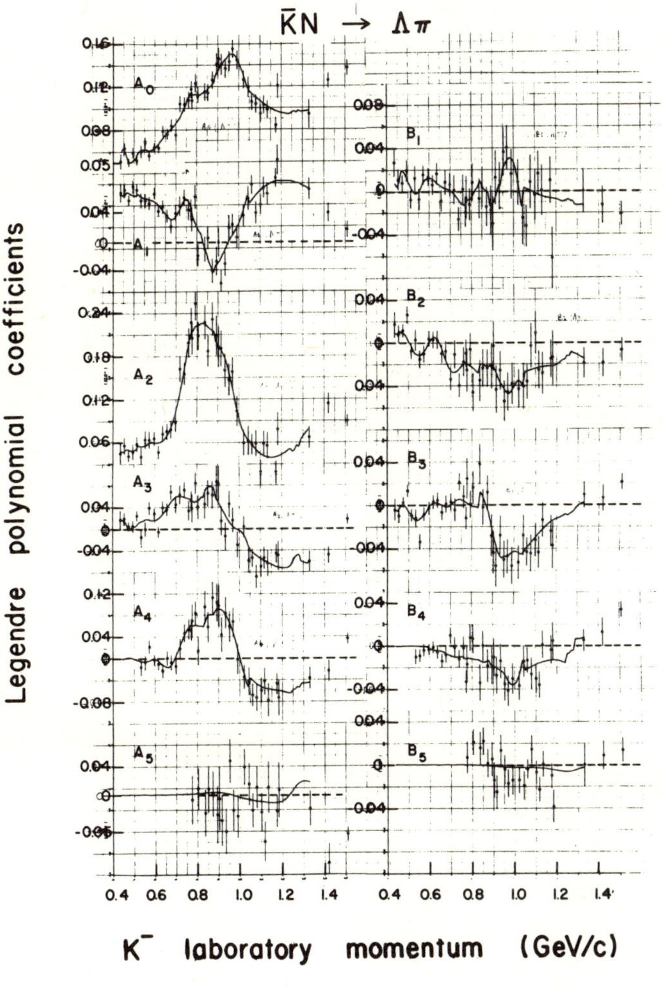

Figure 1. Legendre polynomial coefficients for the reaction $K^-p \rightarrow \Lambda\pi^0$. The curves are the predictions of the partial wave amplitudes.

the possibility of a variation in the S_{11} phase between 436 and 600 MeV/c implies that all the other amplitudes are also indetermined within an overall phase.

Table 1 summarizes the situation concerning the unknowns and the useful data at each momentum. Table 2 gives the parameters introduced in this and in the subsequent analyses of the other channels*.

With the additional data below 600 MeV/c we still find two solutions similar to those given in ref. 2. However, only one of them, shown in Fig. 2, joins reasonably well the solution below 400 MeV/c.[7] The salient features are:

i) A large loop in S_{11} between 700 and 1000 MeV/c. It suggests the existence of a resonance with $J^P = 1/2^-$, $M \simeq 1730$ MeV, $\Gamma \simeq 80$ MeV and $t_{\Lambda\pi} \simeq 0.25$.

ii) A small resonant structure in D_{13}, at the correct momentum to account for $\Sigma(1660)$. An estimate of the resonant parameters suggests $M \simeq 1660$ MeV, $\Gamma \simeq 50$ MeV and $t_{\Lambda\pi} \simeq -0.10$. It is worth remarking that for this amplitude the alternative solution gives almost identical results.

iii) Other structures of a more uncertain nature in P_{11} between 600 and 800 MeV/c and in P_{13} between 700 and 900 MeV/c.

3.2 $\bar{K}N \rightarrow \Sigma\pi$

Both $I = 0$ and $I = 1$ amplitudes are present in this reaction. This brings to 12 the number of complex amplitudes with $J \leqslant 5/2$ to be determined. The available

---

* In all cases, the Breit-Wigner expression $T(E) = t / (\epsilon - i)$ has been used for the amplitude T of the resonance at the energy E; t is the amplitude at resonance and $\epsilon = 2(E_R - E)/\Gamma$ with E_R being the resonant mass and Γ the width. With the exception of $\Lambda(1520)$, for which a constant Γ was used, the energy dependence of the width has been taken as $\Gamma(E) = \Gamma(E_R) \cdot k\, B_\ell(k)$, where k is the c.m. momentum and $B_\ell(k)$ is the non-relativistic centrifugal barrier of ref. 3 with the radius of interaction equal to 1 fermi. For the inelastic channels we use the convention of a positive (negative) sign in front of t to denote that the phase of the amplitude at resonance is 0° (180°) with respect to the imaginary axis.

TABLE 1

Available data and partial waves in $\bar{K}N \to \Lambda\pi$ at each momentum P_K.

	$P_K < 600$ MeV/c	$P_K > 600$ MeV/c
Unknown amplitudes	$\|S_{11}\|$, P_{11}, P_{13}, D_{13}	S_{11}, P_{11}, P_{13}, D_{13}
Known amplitudes	Phase of S_{11}, $D_{15} = \Sigma(1760)$	$D_{15} = \Sigma(1760)$ $F_{15} = \Sigma(1910)$
Data $\begin{cases} K^-p \to \Lambda\pi^0 \\ K^-n \to \Lambda\pi^- \end{cases}$	A_0 to A_5 ; B_1 to B_5	

TABLE 2

Parameters of the resonant amplitudes used as fixed input. The parameters are those determined by the latest energy-dependent analysis of each channel and no attempt has been made to introduce the same values for mass and width in the different channels. The amplitudes at resonance (t_i, $i = \Lambda\pi, \Sigma\pi, \bar{K}N$) are related to the respective branching fractions (x_i) by $t_i = (x_{\bar{K}N}\, x_i)^{1/2}$.

Amplitude	Resonance	Mass (MeV)	Width (MeV)	Amplitude at Resonance
(a) $\bar{K}N \to \Lambda\pi$				
D_{15}	Σ (1760)	1767	108	0.26
F_{15}	Σ (1910)	1915	65	0.05
(b) $\bar{K}N \to \Sigma\pi$				
D_{03}	Λ (1520)	1518	16	-0.43
D_{03}	Λ (1690)	1681	82	0.34
F_{05}	Λ (1820)	1819	70	0.27
G_{07}	Λ (2100)	2108	140	-0.16
D_{13}	Σ (1660)	1661	44	0.21
D_{15}	Σ (1760)	1766	114	0.09
F_{15}	Σ (1910)	1915	60	-0.08
(c) $\bar{K}N \to \bar{K}N$				
D_{03}	Λ (1520)	1518	16	0.46
D_{03}	Λ (1690)	1690	33	0.18
D_{05}	Λ (1830)	1807	123	0.09
F_{05}	Λ (1820)	1817	71	0.62
F_{07}	Λ (1870)	1864	39	0.12
G_{07}	Λ (2100)	2100	160	0.25
D_{13}	Σ (1660)	1662	49	0.08
D_{15}	Σ (1760)	1768	128	0.45
F_{15}	Σ (1910)	1910	50	0.08
F_{17}	Σ (2030)	2040	150	0.25

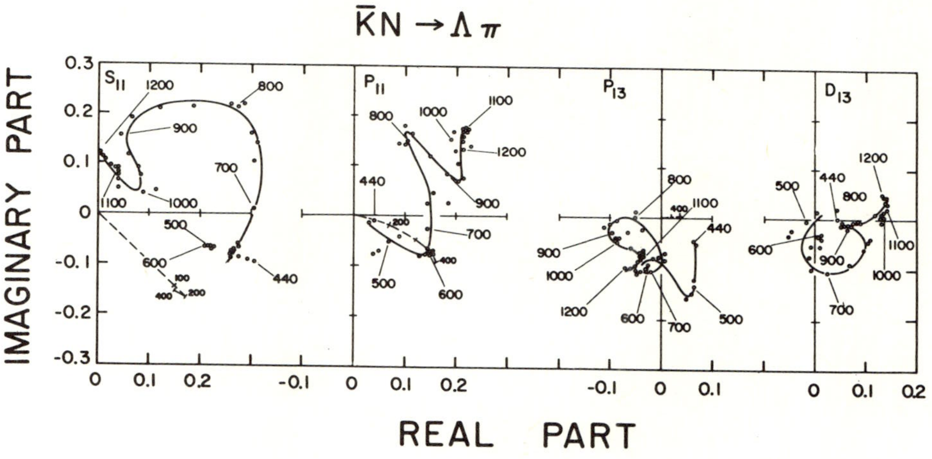

Figure 2. Argand diagrams of the partial waves in $\overline{K}N \rightarrow \Lambda\pi$ from 440 to 1200 MeV/c. Also given are the results of a previous analysis at lower momenta[7]. The amplitudes should be inverted to agree with the proposed convention of reference 1. The curves are hand-drawn and indicate the general trend of the amplitudes.

experimental information[4-6] consists of: a) differential cross sections for the mixed isospin reactions $K^-p \to \Sigma^\pm \pi^\mp$ over the whole region; b) differential cross sections for the I = 0 reaction $K^-p \to \Sigma^o \pi^o$ between 436 and 800 MeV/c; c) differential cross sections for the I = 1 reactions $K^-n \to \Sigma^{\bar{o}} \pi^{\underline{o}}$ between 700 and 1200 MeV/c; d) polarization for $K^-p \to \Sigma^+ \pi^-$ measured via the asymmetry of the $\Sigma^+ \to p\, \pi^o$ decay, over the whole region; e) polarization for $K^-p \to \Sigma^o \pi^o$ through the decay $\Sigma^o \to \Lambda\, \gamma$ and the asymmetry of the subsequent $\Lambda \to p\, \pi^-$ decay, between 436 and 800 MeV/c.

The above data have been reduced to coefficients of expansions (1) and (2) and then fitted to a power series of P (fig. 3a and b). The order of the Legendre expansions is not uniform for all reactions; particularly in the case of $K^-p \to \Sigma^o \pi^o$ and $K^-n \to \Sigma^{\bar{o}} \pi^{\underline{o}}$, the significance of the higher order coefficients is severely limited by the poor statistics. This fact, together with the lack of information on either one or the other of the above reactions over part of the momentum range under study, has the consequence of underconstraining the system of equations which relate the unknown amplitudes with the data. The complexity of the analysis is further increased by the overall phase indetermination common to the inelastic channels. As a result, a large number of artificial constraints had to be introduced. Table 3 lists the amplitudes chosen as unknown and those parametrized as resonant. This choice is the result of several attempts in different conditions and represents an empirical compromise. The parameters of the fixed resonances are given in Table 2(b). It has been verified that the results of our analysis do not depend critically on these precise values.

Only one reasonably smooth solution was obtained; this is shown in Fig. 4. The following features are worth commenting upon:

i) The large loop in S_{01} between 650 and 800 MeV/c confirms the resonant behaviour already noticed in refs. 8 and 9 and attributed to the decay $\Lambda(1670) \to \Sigma\pi$. The parameters of the resonance, estimated from visual inspection of Fig. 4, are in good agreement with those previously quoted.

ii) The situation concerning P_{01} is somewhat confused. The distribution of the points along the lines shown in Fig. 4 suggests the existence of a resonant state with $M \sim 1750$ MeV, $\Gamma \sim 70$ MeV and $t \sim -0.20$. Other resonances in the same amplitude can not be excluded but their existence and resonant parameters are much more doubtful.

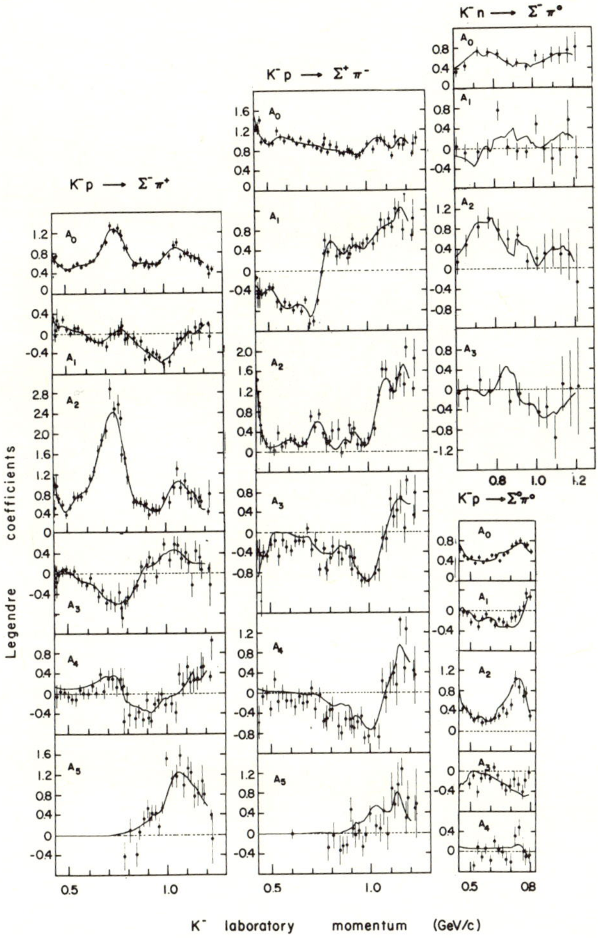

Figure 3a. Legendre polynomial coefficients for the reactions $K^-p \rightarrow \Sigma^+\pi^-$, $\Sigma^-\pi^+$, $\Sigma^o\pi^o$ and $K^-n \rightarrow \Sigma^-\pi^o$. The curves are the predictions of the partial wave amplitudes.

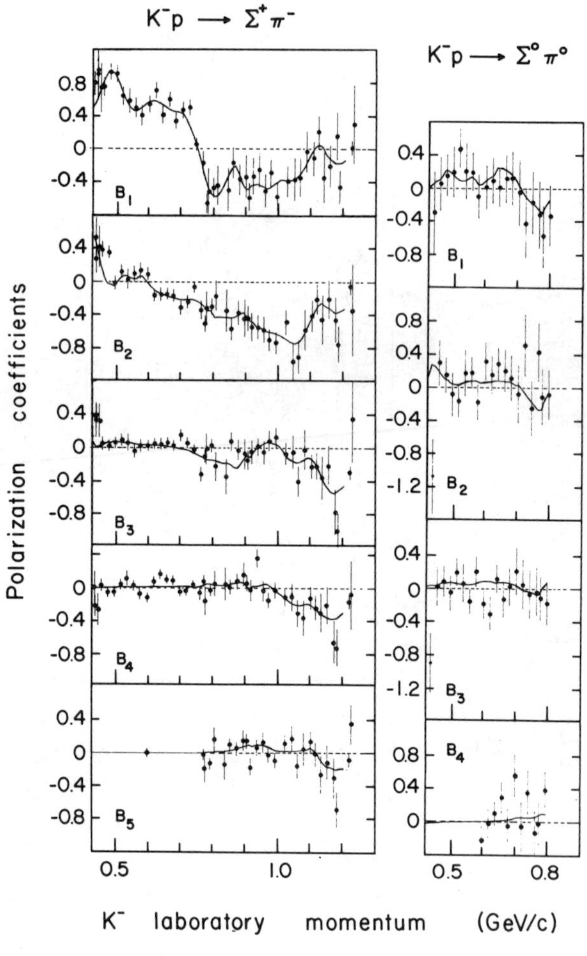

Figure 3b. Polarization coefficients for the reactions $K^-p \rightarrow \Sigma^+\pi^-$ and $\Sigma^o\pi^o$.

TABLE 3

Available data and partial waves in $\bar{K}N \rightarrow \Sigma\pi$ at each momentum P_K.

	$P_K < 700$ MeV/c	$P_K > 700$ MeV/c
Unknown amplitudes	S_{01}, S_{11}, P_{01}, P_{11}, P_{03}, P_{13}	S_{01}, S_{11}, P_{01}, P_{11}, P_{03}, P_{13}, D_{05}
Known amplitudes	$D_{03} = \Lambda(1520)$ $D_{03} = \Lambda(1690)$ $D_{13} = \Sigma(1660)$	$D_{03} = \Lambda(1690)$ for $P_K < 1100$ $F_{05} = \Lambda(1820)$ $G_{07} = \Lambda(2100)$ $D_{13} = \Sigma(1660)$ for $P_K < 1100$ $D_{15} = \Sigma(1760)$ $F_{15} = \Sigma(1910)$ for $P_K < 1100$
Data: $K^-p \rightarrow \Sigma^+ \pi^-$	A_0 to A_4, B_1 to B_4	A_0 to A_5, B_1 to B_5
Data: $K^-p \rightarrow \Sigma^0 \pi^0$	A_0 to A_4, B_1 to B_4	none
Data: $K^-p \rightarrow \Sigma^- \pi^+$	A_0 to A_4	A_0 to A_5
Data: $K^-n \rightarrow \Sigma^- \pi^0$	none	A_0 to A_3

iii) The downward loop noticeable in P_{11} does not exhibit a resonant pattern in the momentum dependence of the points above 600 MeV/c. Below this momentum the rapid variation of the amplitude, although suggestive of a resonant behaviour, coincides with a large increase in the size of the errors (typically ± 0.05). Thus no definite statement can be made concerning the existence of a resonance in this amplitude.

iv) The D_{05} amplitude, as studied in earlier energy-dependent approaches of these[10] and other data [11,12] has given significant indications of being resonant. The parameters of this resonance, called $\Lambda(1830)$, were determined as $M \simeq 1827$ MeV, $\Gamma \simeq 75$ MeV and $t_{\Sigma\pi} \simeq 0.15$.[10] Although the solution presented here is not at variance with this result, this resonance occurs too near the end of the momentum range to allow a more precise determination of its parameters.

As for the $\Lambda\pi$ analysis, the coefficients predicted by the solution (i.e. the curves in Fig. 3) are practically identical to the smoothed input data.

3.3 $\bar{K}N \to \bar{K}N$

The conditions for the analysis are notably worse for this than for the previous two reactions, mainly because of the lack of polarization data in our energy range. The number of unknown amplitudes are again 12 (if we stop at values of $J = 5/2$) and the available data[4-6,13-15] are limited to: a) differential cross sections for the reactions $K^-p \to K^-p$ and $K^-p \to \bar{K}^0n$ over the whole momentum region; b) total K^-p cross sections over the complete interval and separate $I = 0$ and $I = 1$ cross sections between 600 and 1200 MeV/c; c) differential cross sections for the $K^-n \to K^-n$ reaction between 800 and 1200 MeV/c; d) polarization of $K^-p \to K^-p$ at a few momenta above 1 GeV/c. It should be stressed that item c) above does not cover the complete angular region and has a very limited statistical significance; item d) is only available at 3 well separated momenta. Both c) and d) have only been employed as a check.

In view of the inadequate experimental situation, particularly below 700 MeV/c, the method described above is unable to produce meaningful results, even assuming as known all the amplitudes corresponding to the established resonances. We have, therefore, modified the procedure outlined in section 2 and substituted the random selection of the starting values of the amplitudes with a biased choice of these amplitudes as given by the solution of the earlier energy-dependent analyses[9,16] in

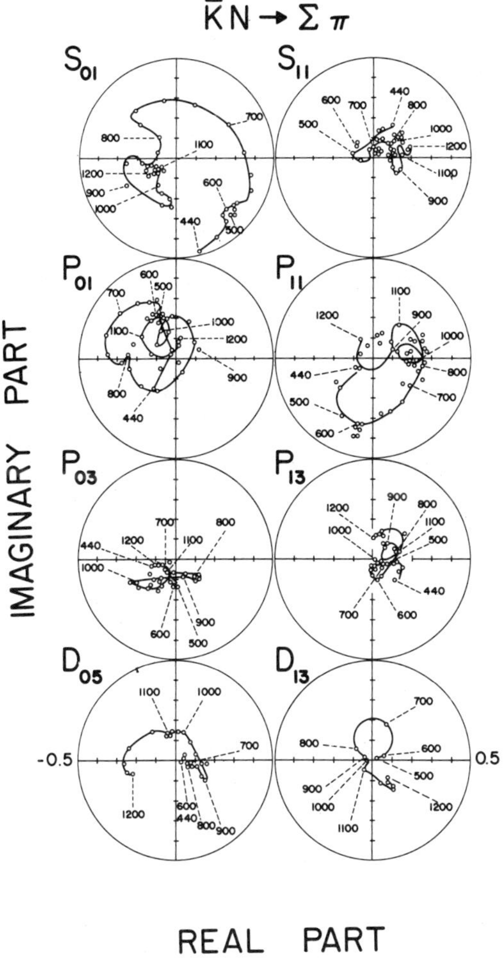

Figure 4. Argand diagrams of the partial waves in K̄N → Σπ from 440 to 1200 MeV/c. The amplitudes are indicated by the points every 20 MeV/c. The I = 0 amplitudes should be inverted to agree with the proposed convention of reference 1. The curves are hand-drawn and indicate the general trend of the amplitudes.

the neighbourhood of 900 MeV/c. The amplitudes are then propagated to each end of the momentum region as for the $\Lambda\pi$ and $\Sigma\pi$ analyses. The solutions obtained are somewhat dependent on the starting momenta.

The method was also modified with regards to the way of employing the experimental data. As mentioned above, several differential cross sections are not measured over the complete angular region; as a consequence, the expansions (1) and (2) are not always reliable. Therefore we have chosen to use directly the available angular distributions. In order to ensure continuity of the partial waves through continuity of the input data, we still need to employ values of $d\sigma / d\Omega$ which vary smoothly as a function of momentum and angle. For this purpose, we fit the measured values of $d\sigma / d\Omega$ over the whole range of momenta and angles considered. The method yields interpolated values of $d\sigma / d\Omega$ at any momentum and angle. This amounts to a substitution of the smooth coefficients A_n and B_n discussed in section 2 by smooth differential cross sections. The latter are then directly compared with the predictions implied by the initial amplitudes. A fit is then performed by allowing the amplitudes to vary so as to best agree with these data.

Figs. 5 show examples of the data employed, together with the curves predicted by the fitted partial wave amplitudes. Also shown are the polarizations predicted for the same reactions.

Table 4 lists the amplitudes chosen as variable in the fit and those parametrized as resonant. The parameters of the resonances can be found in Table 2. The results of such an analysis, it should be stressed, are somewhat dependent on the starting values taken from the energy-dependent solution. However, the partial waves at each momentum are free to drift away from their original position independently of the requirements from the other momenta; they are only bound by the continuity of the data. Thus, any structure which may have escaped detection in the energy-dependent analyses has now, in principle, the opportunity to manifest itself.

The solution obtained is that shown in Fig. 6. The overall behaviour is quite similar to that of the parent energy dependent solutions. The downward loop in S_{01} at about 700 MeV/c indicates the presence of $\Lambda(1670)$, which was introduced as a resonant circle superposed to a linear background in refs. 9 and 16. In addition, there are a few new features worth noticing:

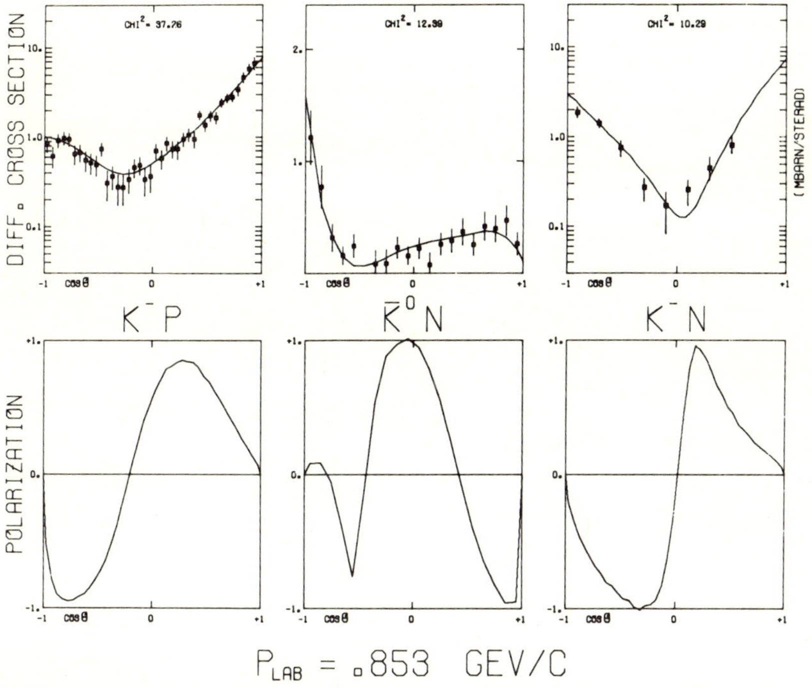

Figure 5a. Example of the data used for the $\overline{K}N$ analysis. The incident K^- momentum is 853 MeV/c. The graphs refer to the reactions $K^-p \to K^-p$, $K^-p \to \overline{K}^0n$ and $K^-n \to K^-n$. The curves are the differential cross sections (in mb/sr) and the polarizations predicted by the partial wave amplitudes obtained in the analysis. The angle θ refers to the c.m. direction of the outgoing meson with respect to the incident K^-.

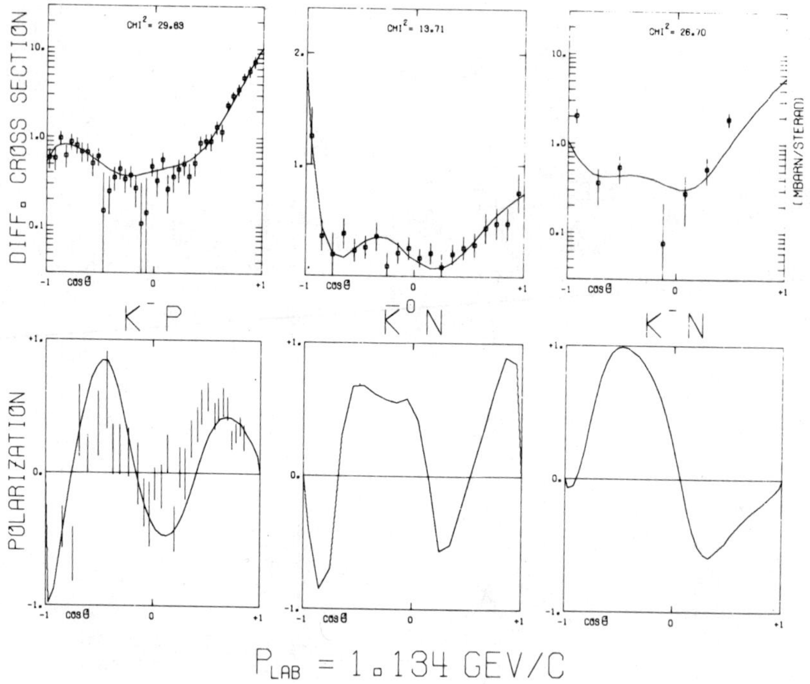

Figure 5b. Example of the data used for the $\bar{K}N$ analysis. The incident K^- momentum is 1134 MeV/c. The total number of momenta used in the analysis is 42.

TABLE 4

Available data and partial waves in $\bar{K}N \to \bar{K}N$ at each momentum P_K.

Unknown amplitudes	S_{01}, S_{11}, P_{01}, P_{11}, P_{03}, P_{13}
Known amplitudes	$D_{03} = \Lambda$ (1520), $D_{03} = \Lambda$ (1690), $D_{13} = \Sigma$ (1660), $D_{05} = \Lambda$ (1830), $D_{15} = \Sigma$ (1760), $F_{05} = \Lambda$ (1820), $F_{15} = \Sigma$ (1910), $F_{17} = \Sigma$ (2030),
Data	$K^-p \to K^-p$, $K^-p \to \bar{K}^o n$ differential cross-sections $K^-n \to K^-n$ differential cross-sections (used only as a consistency check) $\sigma_{tot}(K^-p)$, σ_1^a

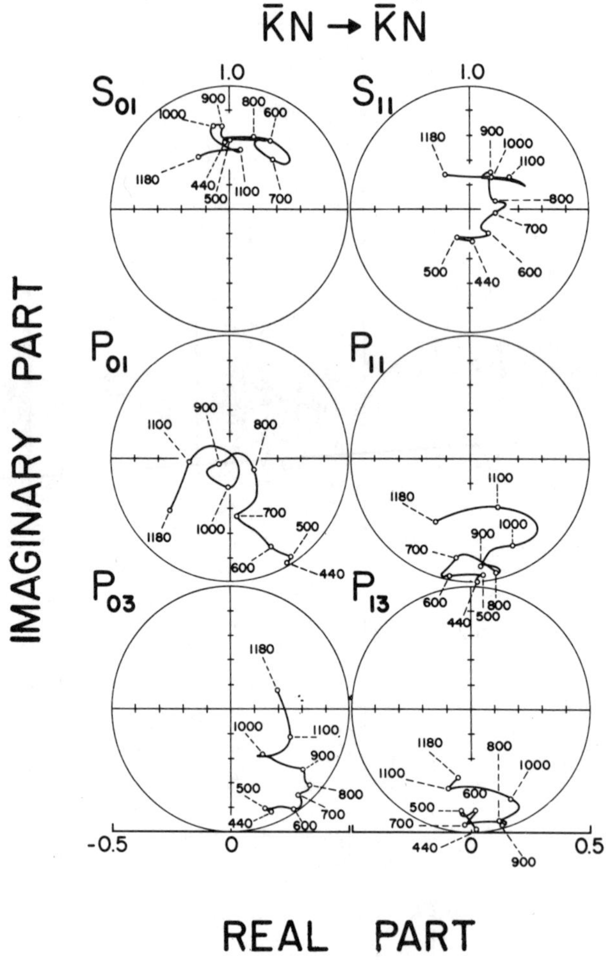

Figure 6. Argand diagrams of the partial waves in $\bar{K}N \rightarrow \bar{K}N$ from 440 to 1180 MeV/c. The momenta are marked every 100 MeV/c. The curves describe the end points of the amplitudes.

i) A loop appears in P_{01} between 900 and 1100 MeV/c. If due to a resonance, the parameters are: $M \simeq 1800$ MeV, $\Gamma \simeq 30$ MeV and $t_{\bar{K}N} \simeq 0.15$.

ii) P_{11} could well be resonant near the end of the range, at ~ 1100 MeV/c. A visual estimate of the resonant parameters gives $M \simeq 1850$ MeV, $\Gamma \simeq 30$ MeV and $t_{\bar{K}N} \simeq 0.20$.

iii) A rapidly changing structure appears in P_{13} but we note that the maximum speed occurs in the neighborhood of the clockwise turn at 1100 MeV/c.

A remark about the effect of choosing a different starting momentum is in order. The only appreciable effects are observed in the P_{01} and P_{11} amplitudes, which then loose the resonant structures quoted above. However, the solution which we present comes from a starting point which is reliable in the framework of the method used.

4. CONCLUSIONS

Table 5 summarises the parameters of the possible resonances resulting from the three analyses. It has to be remarked that the requirements of unitarity were only imposed on the separate channels

$$|T_{\Lambda\pi}|^2 \leqslant \tfrac{1}{4}, \quad |T_{\Sigma\pi}|^2 \leqslant \tfrac{1}{4}, \quad |T_{\bar{K}N}|^2 \leqslant \mathrm{Im}\, T_{\bar{K}N}$$

Overall unitarity, on the other hand, requires:

$$|T_{\Lambda\pi}|^2 + |T_{\Sigma\pi}|^2 + |T_{\bar{K}N}|^2 \leqslant \mathrm{Im}\, T_{\bar{K}N}$$

It has been verified that the solutions obtained satisfy the overall unitarity condition.

A wave by wave comparison with the results of our earlier analyses[2,8,9,16] may be summarized as follows:

a) S_{01} · The new results agree with the old and confirm the existence of $\Lambda(1670)$.

b) P_{01} · Previous results[16] indicated a possible resonant behaviour for this wave in the $\bar{K}N$ channel with $M \simeq 1745$ MeV, $\Gamma \simeq 147$ MeV and $t_{KN} \simeq 0.40$. The new results, although still suggestive of a resonant behaviour in the $\bar{K}N$ and in the $\Sigma\pi$ channel, are not in very good agreement with the old. It has to be noticed, however, that a reliable determination of the parameters is very difficult, given the fairly large uncertainties in the experimental data.

TABLE 5

Summary of the parameters of the new resonances suggested by the analysis. The values of the parameters are the result of a visual estimation and should be considered as tentative. Taking into account only the statistical uncertainty, we estimate approximate errors of ± 10 MeV on masses and widths and ± 0.02 on amplitudes at resonance.

Amplitude	Channel	Mass (MeV)	Width (MeV)	Amplitude at resonance* (t_i)
P_{01}	$\bar{K}N$	1800	30	0.15
	$\Sigma\pi$	1750	70	-0.20
S_{11}	$\Lambda\pi$	1730	80	0.25
P_{11}	$\bar{K}N$	1850	30	0.20
	$\Sigma\pi$	1500-1600	50	0.20
D_{13}	$\Lambda\pi$	1660	50	-0.10

*In order to agree with the convention proposed in reference 1, the $\Lambda\pi$ amplitudes and the I = 0 $\Sigma\pi$ amplitudes should be inverted.

c) P_{03} · As previously, no evidence is found for any significant structure.

d) D_{05} · This wave has been fixed as resonant in the $\bar{K}N$ channel but left free in the $\Sigma\pi$. The resonant behaviour is confirmed and the parameters agree reasonably well with those found in ref. 10.

e) S_{11} · A resonant behaviour was noticed for this amplitude at $\sim$ 1650 MeV in the earlier analysis of the $\Lambda\pi$ channel[2]. The present $\Lambda\pi$ solution, although different from the earlier, still exhibits resonant features. These occur now in a different mass region ($\sim$ 1730 MeV) and may be connected to the $\Sigma\eta$ threshold enhancement and other effects reported by various sources in favour of an $I = 1$ $J^P = \frac{1}{2}^-$ resonance in this mass region. On the other hand, no corresponding resonant behaviour is apparent in the amplitudes of the $\bar{K}N$ and $\Sigma\pi$ channels.

f) P_{11} · Two uncorrelated resonant-like loops are seen in the $\bar{K}N$ ($M \sim$ 1850 MeV) and $\Sigma\pi$ ($M \sim$ 1550-1600 MeV) channels. As mentioned earlier, the behaviour of this wave in the $\bar{K}N$ channel depends critically on the starting point for the analysis. The momentum at which the possible resonance occurs is furthermore too near the end of the range to consider the evidence too seriously.

g) P_{13} · As for the P_{03} wave, no obvious suggestions for resonant behaviour are observed.

h) D_{13} · This amplitude has been assumed to be purely resonant and well described by Σ (1660) in the $\bar{K}N$ and $\Sigma\pi$ channels. Left free in the $\Lambda\pi$ channel, a small but well defined resonant structure is observed. The mass and width agree well with those expected; the amplitude at resonance, $t_{\Lambda\pi}$, is smaller than that of the earlier solution[2].

We would like to emphasize that, although the new results have been summarized in a single table, this does not imply that the evidence is always equally strong or weak. It is clear that data both new (in particular polarization in elastic scattering and differential cross sections for K^-n) and more statistically precise are needed to reach final conclusions.

REFERENCES

1. R. Levi-Setti, Rapporteur's talk on strange baryon resonances at the 1969 Lund Conference (Berlingska Boktryckeriet, Lund).

2. R. Armenteros et al, Nucl. Physics B8 (1968) 183.

3. R. D. Tripp, Baryon resonances, in Proc. of the Intern. School of Physics, Varenna 1964 (Academic Press, New York).

4. R. Armenteros et al., The $\bar{K}N$ interaction from 440 to 800 MeV/c: K^-p cross sections. CERN TC Report D.Ph.II/Phys. 70-7 sub. to Nuclear Physics (1970).

5. R. Armenteros et al., Nucl. Phys. B8 (1968) 233.

6. R. Armenteros et al., The $\bar{K}N$ interaction between 800 and 1200 MeV/c (IV): Cross sections of K^-n reactions. CERN TC Report D.Ph.II/Phys. 69-33. Submitted to Nucl. Phys. (1970).

7. J. K. Kim, Phys. Rev. Letters 17 (1966) 832; A. Barbaro-Galtieri et al., reported at the 1969 Lund Conference (Berlingska Boktryckeriet, Lund).

8. R. Armenteros et al., Nucl. Phys. B8 (1968) 223.

9. R. Armenteros et al., Nucl. Phys. B14 (1969) 91.

10. R. Armenteros et al., Phys. Letters 24B (1967) 198.

11. R. Armenteros et al., Nucl. Phys. B3 (1967) 592.

12. R. B. Bell, Phys. Rev. Letters 19 (1967) 936.

13. D. V. Bugg et al., Phys. Rev. 168 (1968) 1466.

14. R. L. Cool et al., Phys. Rev. Letters 16 (1966) 1228.

15. C. R. Cox et al., Phys. Rev. 184 (1969) 1443.

16. R. Armenteros et al., Nucl. Phys. B8 (1968) 195.

DISCUSSION

Steinberg: Could you comment briefly on the lack of ambiguity in your solutions. The situation here is different from our experience with the K^+p analysis in which several competing solutions appear. You seem to always have a unique solution. For example in your energy independent method, what happens if you start at different random points? Do you generate different solutions that you have?

Plane: In the $\Lambda\pi$ channel where we seem to have most success with this energy-independent approach, we tried over 200 starting values, taking a set of random amplitudes at different momenta throughout the entire energy region. We find only one solution which is continuous throughout the whole region. We also attempted the same approach in the $\overline{K}N$ analysis where we found many solutions. For this reason we had to use a set of starting amplitudes coming from a previous energy dependent analysis of the $\overline{K}N$ channel. This was done at one momentum only. Naturally we tried taking these biased starting points at several different momenta and found that the only significant differences between the solutions were in the P_{01} and P_{11} waves, which sometimes lose their resonant like structure which I showed. However the S_{01} wave always seems to resonate whatever starting momentum is chosen.

Hemingway: In most of the fits you have used a certain amount of fixed data and you have used an amplitude which describes a perfect circle in the Argand plot. None of your phase shifts show perfect circles so is it not a possibility that the new ones you have found are a direct result of that assumption that they are perfect circles?

Plane: I'm sorry you're saying . . . . . . ?

Hemingway: That many of the resonances you have fixed. You have assumed that they are Breit-Wigner, and that therefore they describe circles in the Argand plot. Very few resonances do in fact describe circles, so the fact that you have fixed quite a lot of them may in fact produce structure in the ones that you are varying, which are really faked.

Plane: The analysis to which we refer was presented as an intermediate step between an energy dependent approach and a fully energy independent analysis, which I think is the best which can be done with the data presently available. Clearly our results depend on the resonances we have parameterized as pure Breit-Wigner and if you were to find that, for instance, the Σ (1760) was not after all coupled to the $\Lambda\pi$ channel, a very unlikely possibility, then the whole $\Lambda\pi$ analysis would be invalidated. You will recall that I initially compared resonance parameters from different energy dependent partial wave analyses, and showed that they agreed rather well.

Hemingway: Yes, however, you never gave any errors on all those parameters. I was somewhat worried. Looking at your plots it seemed very difficult to assume that the errors would be small.

Plane: Well if you look into a detailed plot where you can see the speed more accurately then of course this helps you to find the maximum width. Clearly from your position in the room you can not possibly see how the speed goes along those lines.

Hemingway: Perhaps you could tell us what the typical errors are on the masses and widths and elasticities.

Plane: The results of determining resonance parameters from Argand diagrams can be rather subjective. The width is more difficult to find than the mass. In the $\Lambda\pi$ analysis we claim a new resonance in the S_{11} partial wave amplitude, which we have also confirmed in an energy dependent analysis. The resonance parameters which we find independently in the separate analyses agree very well. As far as errors are concerned the problems are far more difficult. We quote errors on masses and widths of ± 10 MeV and on resonant amplitudes of ± 0.05, but these errors should not be taken too literally.

Tripp: I have two questions. One is that in your $\bar{K}N$ analysis it seems to me you disregarded completely the total cross section measurements in K^-p and K^-d which are very precisely determined; if I understand the analysis correctly.

Plane: No, those were included. Total cross sections and the cross sections in the I=0 and I=1 states were included.

Tripp: In the 1.0 to 1.3 GeV/c analysis that you discussed? Where you found the S_{01} resonance?

Plane: Yes in both energy dependent and the energy independent approaches these were included.

Tripp: Fine. And the second question; this new S_{01} resonance that you notice at 1850. It is just above threshold for Ξ K. Maybe Dick Dalitz might have some statement about the possibility that it might be related to a threshold effect there. Certainly Ξ K doesn't have much of a bump in that region but could there be some threshold effec

Dalitz: For the $S_{01}(1850)$ Y^* resonance which lies close to the Ξ K threshold, I think that the proper point of view is to ask what possible effect this resonance could have on the threshold Ξ K behavior, rather than to ask whether the resonance is a consequence of a strong Ξ K attraction. If the s-wave Ξ K channel is strongly coupled with this S_{11} resonance, then this resonance will have a strong effect on the Ξ K production reaction at threshold. But there is no necessity that this coupling should be strong.

Since we are speaking about this S_{01} (1850) resonance, it is interesting to remark that, for the nucleonic members of the (70,L=1-) supermultiplet, the ^4P states are approximately degenerate in mass, consisting of $D_{15}(1670)$, $D_{13}(1680)$, seen in the CERN-EXPT phases but not included in the tables, and $S_{11}(1700)$. For the negative-parity Y^* states, we notice a similar mass degeneracy in the states discussed today, which we would assign to the ^4P configuration by comparison with the corresponding N^* states. For I=0, we have the states $D_{05}(1830)$ and $S_{01}(1850)$. We are still lacking the further state D_{03} to go with these. For I=1, we have the states $D_{15}(1765)$ and $S_{11}(1755)$. We are still lacking a further state D_{13} to go with these. But at least this situation, taken together with the corresponding N^* situation, is rather suggestive that the ^4P states for each (I,Y) should be approximately mass degenerate.

Kim: In your 1620 resonance in the Lund Conference your value for the mass was something like 1550 for the $\Sigma\pi$ decay mode. Now you are giving 1620. What is the difference between the two results?

Plane: Well since the Lund Conference we have added more data at lower and higher energies.

Kim: You mean you have more data between 450 and 600.

Plane: No, we used lower energy data.

Kim: But it seems that for instance the amplitide doesn't agree with Tripp's amplitude at low energy, at 400 for your . . . .

Plane: The results of the energy dependent analyses tie on very well; those of the energy independent analyses do not agree very well in some cases with Tripp's.

Kim: But this change in the mass is interesting, for instance from 1550 to 1620. You know originally the $\Lambda\pi$ had a 1620 mass and it is gone. At the Lund Conference you had a 1550 in $\Sigma\pi$, now it has become 1620. I am interested in these differences and so on.

Plane: We have essentially a new solution as we have added more data. You might expect to modify previous solutions if you add more data.

Kim: And also your D_{13} and D_{03} waves which you put in come originally from energy dependent studies, right? Do you think that energy dependent results and energy independent results would give the same D_{13} and D_{03} parameters? I guess the F waves or D_{15} waves are way up there but you know the D_{13} and D_{03} resonances are right at the 1660 region. How much do you think that affects the mass of these other things you got.

Plane: What we have done is to take the energy dependent approach and then to make it slightly energy independent by freeing the S and P waves. Within the framework we take the results of our previous energy dependent analyses to fix, for instance, the parameters of the $\Sigma(1660)$. We have checked, for the $\Lambda\pi$ analysis, that small variations in the fixed resonant parameters do not produce solutions with different resonances.

Hart: It is not a question exactly but just so we won't run out of resonances and to correct your introductory slide, 240,000 K^-p pictures were taken at the BNL 30" the summer of 1967, eight momenta 680-840 MeV/c. We have got about 150,000 events going into SUMX.

Hemingway: How many resonances can you throw out without ruining your fit?

Plane: The energy independent approach just does not work that way. Your question is only applicable to energy-dependent analyses. In the results I gave, the S and P waves were free to move at each momentum; they were not parameterized in any way.

Hemingway: So presumably if you had gone to an infinite number of waves you would get an infinite number of resonances. Is that right? The more parameters you have, the more variations you are going to get. So there is a limit presumably. I am trying to get a feeling for how significant some of these effects are.

?: But you will not get resonances.

Hemingway: It must not always be said that there are resonances.

Plane: That's why the Argand plots are presented, so you can judge for yourself.

Hemingway: Well, it seems to me that in certain cases you are saying that there is not really much evidence because there is a little cusp at the end where in other cases you didn't.

Plane: If you observe a large resonance-like loop in the middle of the energy region analysed, then you can have more confidence that you have detected a resonance than if you only see a little cusp at the end of the energy region, since in the latter case you do not have any amplitudes to tie on with and so you cannot be certain that the amplitudes are still continuous.

Levi-Setti: I would only mention that several ambiguities might be removed by a multi-channel analysis. For this I am quite curious to hear what Kim will have to say a little later.

Kalmus: I really have a couple of very short comments and a question. The first comment is that is seems to me that this kind of analysis uses a different philosophy to that in the past. In the past it has been a philosophy of maximum simplicity: how few resonances can one use to fit the data. Here there is a different approach which I think is interesting and seems to me to be guided somewhat by SU(3) in that the search for loops or kinks in the Argand plot is rather strongly correlated to ones theoretical prejudice. In particular the ascribing of some of these to be resonances and of others to be not significant. I presume that since the data has been fitted satisfactorily with an energy dependent fit that the data really doesn't need all these loops to be resonances, of course they might still be there. Then there is another comment, or a question I guess now, and that is, is there some theorem that links the order of the polynomial you fitted your A coefficients to, with the number of resonances one gets?

Plane: We have made a smoothing of the $\Lambda\pi$ data with an eighth order polynomial and we do not find eight new resonances, only two.

Kalmus: Well it might not be nine but it might be connected to the order you use. And finally an interesting point perhaps. There is some K^-n elastic scattering data at just about the energy that you showed on one of your earlier slides from Nathan Jew at Berkeley with a good deal more data than I think you showed. By looking at it quickly it seems to fit your curve rather better than your own data did.

Plane: These were only used as a check in our analysis.

Kim: May I make one comment on Kalmus' comment. You said that the energy dependent study does not require the resonances but in fact the χ^2 value you get on the energy dependent solution isn't that good. And by using energy independent study actually you are making the χ^2 much lower so energy dependent study really didn't rule out the resonances because χ^2 improved a lot. For instance, looking at $Kp \rightarrow \Sigma\pi$ in their data when they did the energy dependent analysis the χ^2 was quite high.

AN IMPROVED PARTIAL WAVE ANALYSIS OF $\overline{K}N \to \overline{K}N$ FROM 800 TO 1200 MeV/c

B. Conforto, D.M.Harmsen***, T. Lasinski, R. Levi Setti, M. Raymund†; and E. Burkhardt, H. Filthuth, S. Klein, H. Oberlack, H. Schleich‡

†Enrico Fermi Institute and Department of Physics, University of Chicago, Chicago, Illinois*
‡Institut für Hochenergiephysik der Univ. Heidelberg**

Abstract

An extensive search for resonant states has been performed through an energy dependent partial wave analysis of $\overline{K}N$ data in the momentum region 777 - 1226 MeV/c. The $K^-p \to K^-p$, $\overline{K}^\circ n$ differential cross sections, polarization angular distributions and total K^-p cross sections were included in the fit. In addition to the well established hyperons, evidence is reported for a resonant state $\Sigma(1760)$ of $J^P = 1/2^-$ and for the assignment $J^P = 3/2^+$ for a $\Lambda(1870)$ state previously assigned to $J^P = 7/2^+$. From the $K^-p \to K^-p$ differential and total cross sections between 436 and 1226 MeV/c the ratio of the real to the imaginary part of the forward scattering amplitude has been derived as a function of K^- incident momentum. These results are found to be consistent with the predictions of several dispersion relation calculations.

The full text of this paper is to appear in Nuclear Physics.

---

* Research sponsored by the Air Force Office of Scientific Research Office of Aerospace Research. United States Air Force. Contract No. AF 49 (638)-1652 and The National Science Foundation.

** Work supported by the Bundesministerium für Bildung und Wissenschaft.

*** Now at CERN, Geneva.

A STUDY OF THE I=1 REACTION $K^-n \to \Sigma^{\pm}\pi^{\mp}\pi^-$ FROM 1585 TO 1735 MeV*

E. B. Brucker[†], W. C. Harrison[‡], W. H. Sims[#],

J. R. Albright, J. P. Chandler, J. E. Lannutti

and B. G. Reynolds[††]

Department of Physics, The Florida State University, Tallahassee, Florida 32306

From an exposure of the BNL 30" deuterium filled bubble chamber to separated K^- beams of momenta 678, 727, 777 and 835 MeV/c we have extracted all events that are samples of the reactions $K^-d \to \Sigma^{\pm}\pi^{\mp}\pi^- p$. Assuming that an impulse approximation is valid and that the Fermi momentum of the neutron is described by the Fourier transform of the Hulthén wave function, we have derived the cross sections for the pure I=1 reactions:

$$K^-n \to \Sigma^-\pi^+\pi^- \quad (1a)$$

$$\Sigma^+\pi^-\pi^- \quad (1b)$$

Because of the order of magnitude increase in statistics of this experiment over other experiments[1] in the energy region 1585-1735 MeV we are able to present these cross sections in 10 MeV steps across this region. These cross sections are displayed in Fig. 1. In addition to the expected $\Sigma(1765)$, there is evidence for an enhancement in the summed cross section, Fig. 1c, with an energy E_o = 1681±3 MeV, width Γ = 30±10 MeV and a cross section at resonance of 340±65 μb above background. We show that this enhancement may be interpreted as resonance structure in the $\Lambda(1405)\pi$ final state.

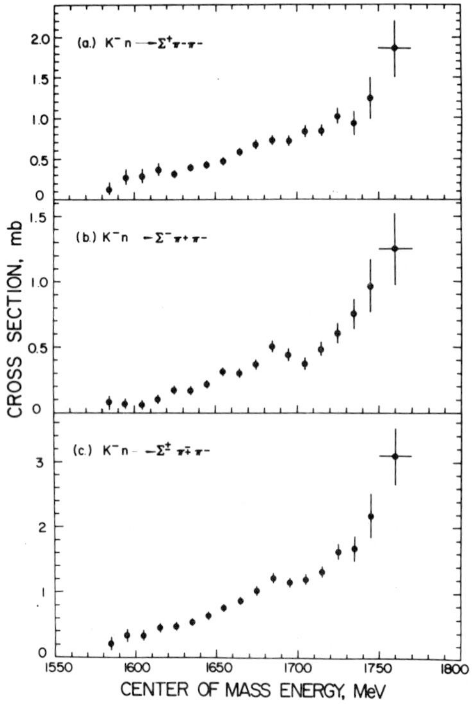

FIG. 1. The total cross section as a function of the center of mass energy of the $\Sigma\pi\pi$ system for the reactions (a) $K^-n\rightarrow\Sigma^+\pi^-\pi^-$, (b) $K^-n\rightarrow\Sigma^-\pi^+\pi^-$, and (c) the sum of (a) and (b). The errors are statistical only and do not include an overall systematic uncertainty of $\leq$ 10%.

In Fig. 2 we display the differential cross section in $\Sigma^-\pi^+$ effective mass from reaction (1a). There is a clear separation between the $\Sigma(1385)$-$\Lambda(1405)$ and the $\Lambda(1520)$. The contributions of the $\Lambda(1520)$ are consistent with the observation of the CHS collaboration[2] that this final state dominates the I=1, T=0 cross section in the region of the $\Sigma(1765)$. The numerical values agree with the values obtained by Delaney, et al[3]. The comparable distribution for the $\Sigma^+\pi^-\pi^-$ final state (not shown) is complicated by the two possible ways of coupling the Σ^+ with a π^- to produce a Y^*.

We have derived the I=1 cross sections for $\Sigma\pi$ combinations in the T=0 state (σ_{10}) by using the expression

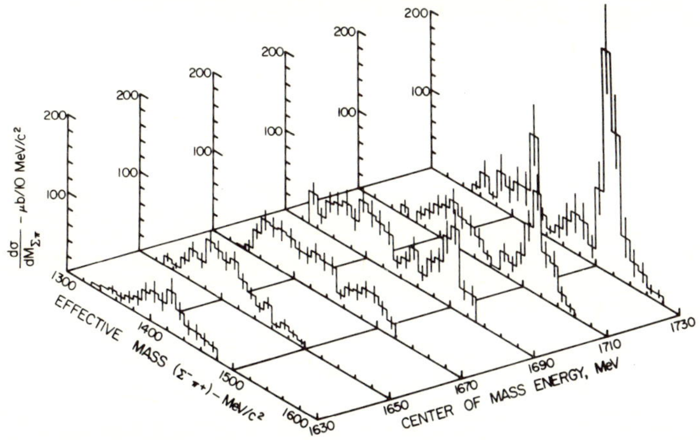

FIG. 2. Differential cross section in $\Sigma^-\pi^+$ effective mass as a function of center of mass energy. These cross sections are normalized to give the total cross section for a 20 MeV center of mass energy interval centered about the value shown.

$$\sigma(K^-n \to \Sigma^{\pm}\pi^{\mp}\pi^-) = \frac{2}{3}\,\sigma_{10} + \frac{1}{2}\,\sigma_{11} \quad .$$

The σ_{11} cross sections have been determined from the experiment of Sims, et al[4] who showed that the $\Lambda\pi\pi$ final state is dominated by $\Sigma(1385)\pi$. Subtracting this contribution to the $\Sigma\pi\pi$ data we derive σ_{10} as displayed in Fig. 3. The CHS collaboration[5] has derived the same cross section in a different manner. The two results disagree appreciably in the energy range 1640-80 MeV where the overlap of crossing resonance bands is maximal. The "1680" peak remains but now is limited to be either $\Lambda(1405)\pi$ and/or $\Lambda(1520)\pi$. Using the values for the mass and width found above we obtain a cross section at resonance of 400±100 μb.

In order to abstract the $\Lambda(1405)\pi$ contribution we have used only reaction (1a) and limited the $\Sigma^-\pi^+$ effective mass to be below 1465 MeV/c². Since we are not using the summed cross sections an interference term between the T=0 and T=1 amplitudes has been estimated from the difference between reactions (1a) and (1b). The $\Sigma(1385)\pi$ contribution was removed as before. The results are displayed in Fig. 4. These data have been fit to a Breit-Wigner resonance shape plus an incoherent background. Fixing the mass and width at the previously determined

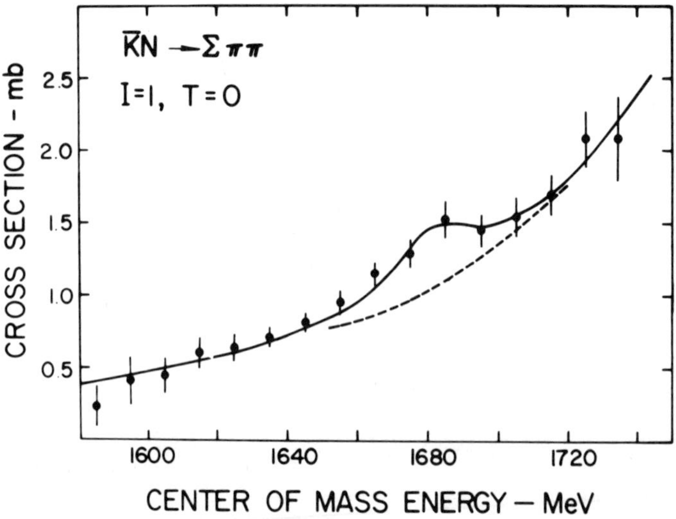

FIG. 3. Cross sections for $\overline{K}N$ interactions leading to the $\Sigma\pi\pi$ final state. I is the isospin of the $\overline{K}N$ system with the $\Sigma\pi$ subsystem having isospin T. The solid curve is the best fit to an incoherent sum of $\Sigma(1765)$, $\Lambda(1405)\pi$ phase space and a Breit-Wigner resonance. The dashed curve is the estimated background under the resonance.

values, the cross section at resonance is now 390±100 μb. This fit is displayed as the solid curve in Fig. 4. If we assume that this resonance is due to only one partial wave ($J^P=\frac{3}{2}^-$) the corresponding elasticity is then

$$X_{\overline{K}N}X_{Y^*\pi} = 0.007 \pm 0.002 \quad .$$

With the previously established rate for $\Sigma(1660)\rightarrow\Sigma(1385)\pi$[4] we find the branching ratio:

$$\text{B.R.}\left(\frac{\Sigma(1660)\rightarrow\Lambda(1405)\pi}{\Sigma(1660)\rightarrow\Sigma(1385)\pi}\right) = 0.23 \pm 0.08 \quad .$$

These results are relatively insensitive to either the amount of interference assumed to be present or the width used for the resonance. In addition, we get essentially the same branching ratio by fitting the $\Lambda\pi\pi$ cross section of Ref. 4 to a resonance plus a background and ignoring the partial wave determinations.

Finally, our branching fraction for $\Sigma(1660)\rightarrow\Lambda(1405)\pi$ is larger than the upper limit determined previously[5]. The mass and width we have determined for the $\Lambda(1405)\pi$ enhancement are the same, within errors, as those deter-

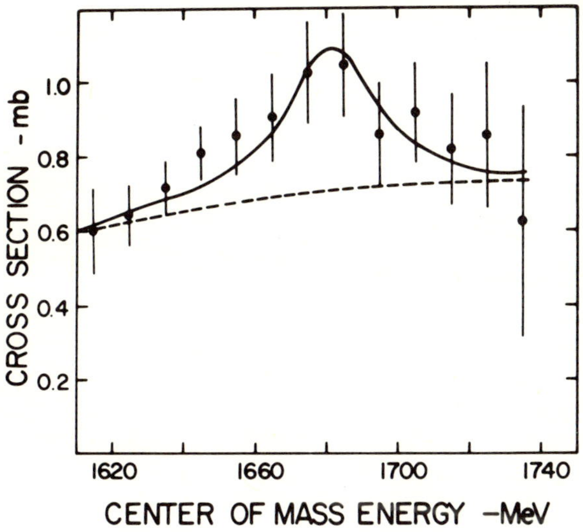

FIG. 4. The cross section for $\Lambda(1405)\pi$ as a function of center of mass energy. The solid curve is the best fit to the data by a Breit-Wigner plus incoherent background. The dashed curve is the estimated background contribution to the fit.

mined by R. Armenteros, et al[6] for the reaction $K^-n \rightarrow \Sigma^{\circ}\pi^{\bar{o}}$ (M=1680 MeV, Γ=47 MeV). Using their value for the rate $\Sigma(1660)\rightarrow\Sigma\pi$ we find a branching ratio for $\Sigma(1660)\rightarrow\Lambda(1405)\pi$ to $\Sigma(1660)\rightarrow\Sigma\pi$ of ~0.2 which would agree with the ratio found by Eberhard, et al[7] for their "less peripheral" $\Sigma(1660)$.

FOOTNOTES AND REFERENCES

*Work supported in part by the U. S. Atomic Energy Comm.
†Present address: Rutgers University, New Brunswick, N. J.
‡Present address: Harvard University, Cambridge, Mass.
#Present address: Brookhaven Nat'l Lab., Upton, L. I.,N.Y.
††Present address: Ohio University, Athens, Ohio

1. R. Armenteros, P. Baillon, P. Lexa, A. Minten, K. H. Nguyen, E. Pagiola, V. Pelosi, R. Barloutaud, F. Bigata, M. Crozon, C. Louedec, J. L. Narjoux, F. Pierre, CERN/D.Ph. II/Phys 69-33 (unpublished).

2. R. Armenteros, et al, Phys. Letters 19, 338 (1965).
3. W. C. Delaney, P. Guidoni, G. W. London, G. Martelotti Nuovo Cim. Letts. 1, 497 (1969).
4. W. H. Sims, J. R. Albright, E. B. Brucker, J. T. Dockery, J. E. Lannutti, J. S. O'Neall, B. G. Reynolds J. H. Bartley, R. M. Dowd, A. F. Greene, J. Schneps, M. Meer, J. Mueller, M. Schneeberger, S. Wolf, Phys. Rev. Letts. 21, 1413 (1968).
5. R. Armenteros, et al, Nuc. Phys. B8, 216 (1968).
6. R. Armenteros, P. Baillon, A. Minten, N. H. Khanh, V. Pelosi, M. Crozon, C. Louedec, J. L. Narjoux, F. Pierre, Nuc. Phys. B10, 459 (1969).
7. P. Eberhard, J. H. Friedman, M. Pripstein and R. R. Ross, Phys. Rev. Letts. 22, 200 (1969).

MULTICHANNEL K MATRIX ANALYSIS OF $\bar{K}N$ INTERACTION IN THE REGION 0 TO 1200 MeV/c

Jae Kwan Kim

Department of Physics, Harvard University

Cambridge, Massachusetts

Multichannel phase shift analysis has been performed on the available experimental data for all the two stable particle channels of $\bar{K}N$ interactions in the momentum region from 0 to 1200 MeV/c. In this analysis, the K matrix parametrization has been used. The preliminary results of the argand diagrams of the reaction amplitudes and the estimated resonance parameters within the energy region are presented. The results indicate some new resonances. The branching ratios obtained from this multichannel analysis for the resonances lying in the confused 1660 to 1700 MeV mass region are appreciably different from the results of previous single channel analyses. The final results of this analysis will be published soon.

The experimental data for the reactions $\bar{K}N \rightarrow \bar{K}N$, $\bar{K}N \rightarrow \Sigma\pi$,and $\bar{K}N \rightarrow \Lambda\pi$ below 1200 MeV/c have been accumulating very rapidly during last three years. More data is still needed for a better analysis, but it has become possible to make a meaningful multichannel phase shift analysis incorporating simultaneously above three reactions with $\bar{K}N$ total cross sections. For this analysis, the following data have been used. The most recent and extensive series of experiments have been carried out by CERN-Heidelberg-Saclay collaboration for the momentum region between 440 to 1200 MeV/c. They have analyzed the above three reactions seperately by using the energy dependent method, where they have expanded the partial wave amplitudes linearly in the incident beam momentum for the background and the Breit-Wigner amplitude has been used for the resonant partial wave.[1] They also have analyzed their data using the semi-energy independent method, where the resonance parameters obtained from the previous energy dependent fit have been fixed as the Breit-Wigner amplitudes for those waves and have searched for the undetected resonances mainly in the S and P waves energy independently.[2] The earlier experiment in the momentum region from 600 to 820 MeV/c was performed by Brookhaven group and the results were analyzed by using the energy dependent method similar as the CHS parametrization for the above three reactions seperately.[3] Recently Yale group finished an experiment for the region from 450 to 670 MeV/c and have analyzed their results using a multichannel analysis.[4] Yale-Brookhaven-Massachusetts collaboration finished a high statistical experiment covering the narrow momentum region from 350 to 430 MeV/c and also a multichannel analysis was performed on the data.[5] These new data from above four experiments cover momentum region from 350 to 1200 MeV/c. The region below 350 MeV/c is taken care of by the existing data from Berkeley and Columbia.[6]

Using the above data, a first multichannel phase shift analysis has been performed for the momentum region from 0 to 1200 MeV/c. The parametrization of the amplitude used in this analysis is similar to the previous multichannel effective range analysis done for the region from 0 to 550 MeV/c except that non-diagonal ranges are also included in this work whereas previously only the diagonal ranges were included.[7] Since the linear expansion of the inverse of K-matrix is not expected to be valid for such a wide energy region, it is expanded linearly from 0 to 500 MeV/c and successively linearly expanded again at the 500 MeV/c for the additional 100 MeV/c momentum interval at a time until the highest momentum interval is reached. In doing this successive expansions, only the range parameters were varied in order to continue the adjacent momentum intervals smoothly. In this preliminary result, the D_5 and F_5 waves have been parametrized using the known Breit-Wigner amplitudes and those waves were fixed. All S_1, P_1, P_3 , and D_3 waves were parametrized by using the K matrix expansion and those were searched for any resonant like behaviors. Thus altogether, 64 parameters were used for the expansion energy K matrix elements and the additional 64 parameters were used for the ranges. These 64 matrix elements also contain extra channels to allow for the three particle productions which are approximately 10 to 20 percent for the momentum region from 400 to 1200 MeV/c. However, in this result, the total three body cross sections have not been used to constrain these extra parameters but these are left free to vary. In the final result, these extra parameters will be tightly constrained by using the total three body cross sections and also D_5 and F_5 partial waves will be parametrized by K matrix.

The experimental informations on the differential cross sections and the polarizations from hyperon decays were used in the form of Legendre series expansion coefficients. These experimental Legendre coefficients were compared with the calculated values from the parameters and using the experimental errors on the coefficients the chisquare function was formed. This chisquare function was searched and minimized by using the program Minfun and also the local minimizing program Stepit. The results of this analysis are summarized in Fig. 1 and Fig. 2 as the argand diagrams of the reaction amplitudes. The resonance parameters estimated are summarized in Table 1 and Table 2.

Fig. 1 shows the argand diagrams of the reaction amplitudes in the total isospin zero channel. The S_{01} amplitude resonates once at 740 MeV/c and again at 1000 MeV/c. The resonance at 740 MeV/c with the mass 1670 MeV is strongly seen in all the channels $\bar{K}N$, $\Sigma\pi$, and $\Lambda\eta$. The elasticity of this resonance is a factor of two larger than the presently accepted value. This reduces the branching ratio into the $\Lambda\eta$ channel considerably. The resonance at 1000 MeV/c with the mass 1790 MeV is seen more clearly in the $\Sigma\pi$ channel but it is also seen in the $\bar{K}N$ channel as a small half loop. The P_{01} amplitude does not show the strong resonant behavior around the 1750 MeV region contrary to the conclusions of CHS result. However, $\Sigma\pi$ channel shows the resonant behavior around the 530 MeV/c with a sizable loop. This can be interpreted as a new resonance with the mass 1575 MeV. The P_{03} amplitude shows small resonant loop around the 820 MeV/c. This can be interpreted as a new resonance with mass 1710 MeV. The well known D_{03} amplitude resonates twice at both 1520 and 1690 MeV. The branching ratio of this well known 1690 MeV resonance into the $\Sigma\pi$ channel is larger than the result presently accepted and also there is not much three particle decay mode from this multichannel analysis.

Fig. 2 shows the argand diagrams of the reaction amplitudes in the total isospin one channel. The S_{11} amplitude seems to resonate twice. It resonates mainly into the $\Lambda\pi$ channel at 760 MeV/c with the mass 1680 MeV and then it seems to resonate again at 1000 MeV/c with the mass 1790 mainly this time into the $\Sigma\pi$ channel. The P_{11} amplitude has very strong resonant effect in the $\Sigma\pi$ channel and also has sizable half loop in the $\bar{K}N$ channel both at the momentum of 740 MeV/c with the resonance mass 1670 MeV. The P_{13} amplitude is in general very small but it has a small resonant like behavior at around the 1040 MeV/c with the mass 1820 for both the channels $\Sigma\pi$ and $\Lambda\pi$. The D_{13} amplitude resonates at the mass value of 1665 MeV, mainly into the $\Sigma\pi$ channel but its branching into the $\Lambda\pi$ channel is almost negligible.

The resonance parameters of relatively sure hyperon resonances are summarized in Table 1. This table includes the resonance mass, the total decay width, branching fractions into two particles, and the estimate of branching fractions into three particles. The resonance parameters of relatively weakly indicated hyperon resonances are summarized in Table 2. The resonances belonging to Table 2 need further investigations. For instance, the fitting of the total three particle cross sections may strongly affect some of these states as some of these are strongly coupled to these channels. Also the removal of the constraints on D_5 and F_5 waves by the known Breit-Wigner amplitudes and the replacements of these waves by freely varying K matrix parametrization may affect these states strongly.

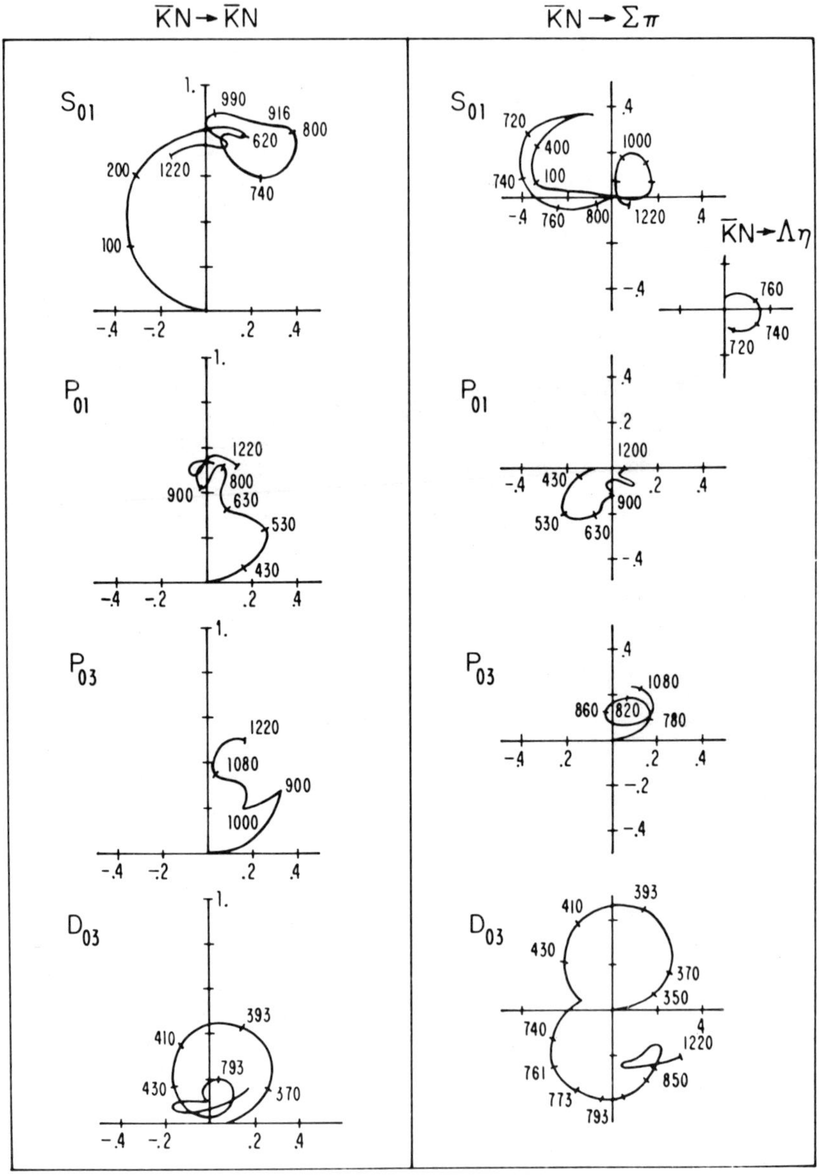

FIG. 1 ARGAND DIAGRAMS OF THE I = O AMPLITUDES

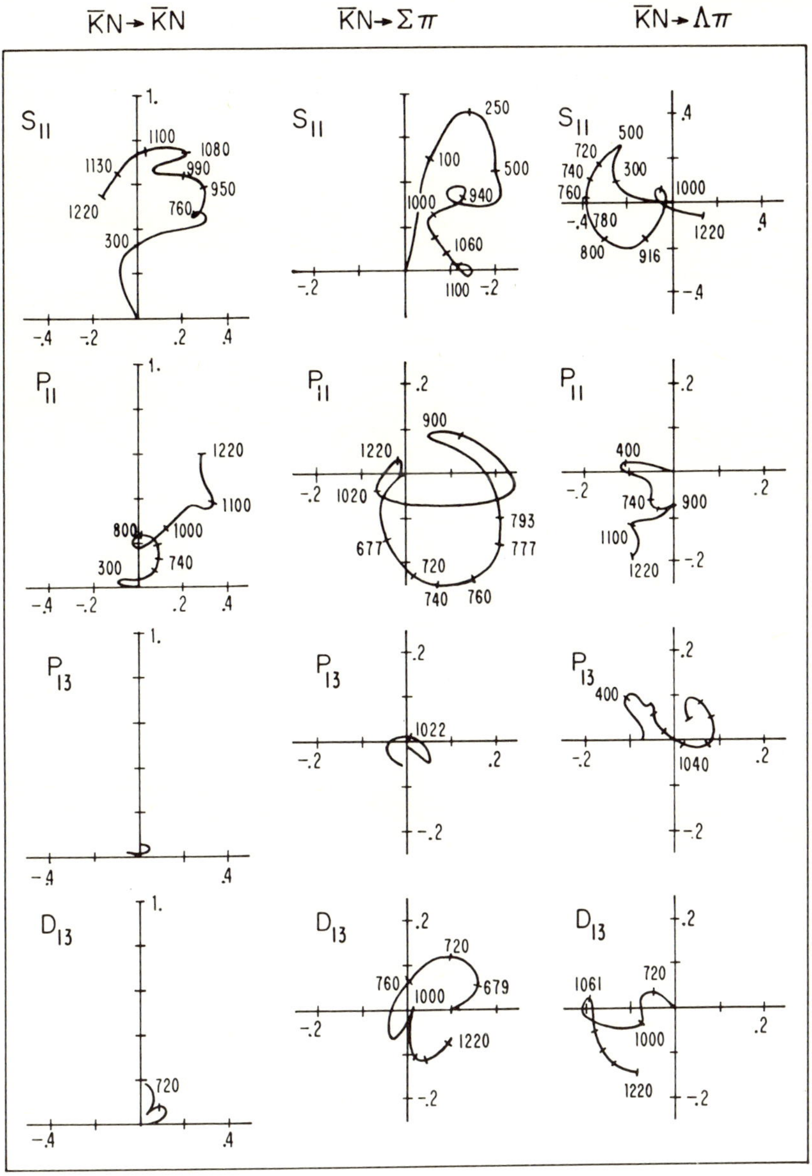

FIG. 2 ARGAND DIAGRAMS OF THE I=1 AMPLITUDES

Table 1

The resonance parameters of relatively sure baryon states.

Amplitude	Mass(MeV)	Width(MeV)	$x_{\bar{K}N}$	$x_{\Sigma\pi}$	$x_{\Lambda\pi}$	$x_{\Lambda\eta}$	$x_{\Sigma\pi\pi,\Lambda\pi\pi}$
S_{01}	1670	38	.30	.59	-	.11	-
S_{01}	1790	50	.10	.40	-		
P_{11}	1670	60	.20	.40	.03	-	.37
D_{03}	1690	60	.20	.75	-		.05
D_{13}	1665	45	.08	.36	.06	-	.50

Table 2

The resonance parameters of relatively poorly known states.

Amplitude	Mass(MeV)	Width(MeV)
S_{11}	1680	70
S_{11}	1790	60
P_{01}	1575	50
P_{03}	1710	40
P_{13}	1820	80

Therefore, only the mass and the width parameters have been given for the Table 2. These values are the preliminary results of the disentangling of these complicated branching ratios through the first multichannel phase shift analysis. The final results of this analysis will be soon published after more careful study.

References

(1) Armenteros et. al., Nuclear Physics B8,195(1968); B8,223(1968); B8,183(1968); B8,233(1968); B14,91(1969).
Armenteros et. al., The $\bar{K}N$ interaction from 436 to 793 MeV/c: K^- p cross sections. CERN TC Report D.Ph. II/Phys.
(2) R. Levi-Setti, Rapporteur's talk on strange baryon resonances at the 1969 Lund Conference(Berlingska Boktryckeriet, Lund).
(3) L. Bertanza et. al., Phys. Rev. 177 ,2036(1969).
(4) J. Wong and W. Willis, APS Bulletin 14 , 592(1969).
J. Wong's thesis, Yale University (1969).
(5) J. Thompson's thesis, Yale University (1969).
(6) M. B. Watson, M. Ferro-Luzzi, and R.D. Tripp , Phys.Rev. 131, 2248(1963).
J. K. Kim, Phys. Rev. Letters 14,29(1965); Columbia University Report No. NEVIS-149,1969.
(7) J. K. Kim, Phys. Rev. Letters 19 , 1074(1967).

DISCUSSION

Rosenfeld (to the Chairman): I really ask it as a question to you because I'm scared to make a comment which would make everyone else in the room mad. Kim's coupled-channel approach seems to work beautifully and I don't have to look first at a slide of K^-p then of $\bar{K}^0 n$ then of $\Sigma\pi$ and so on. What's wrong? Why doesn't everyone do it this way?

Levi-Setti: Computer limitations.

Rosenfeld: But Kim did it.

Ferro-Luzzi: Plotting things all crowded into one graph doesn't make them much more clear, right? Just separate all those graphs into separate graphs and you get the CHS solutions.

Rosenfeld: But presumably there is some constraint built into this which is harder to handle in separate channels.

Ferro-Luzzi: The amplitudes, from what I can see just by looking at that crowded slide, seem to me very similar to what Plane has presented before. They were just on separate slides.

Kim: Can I also say that the Argand diagrams which we obtain from such a method are in fact quite similar to the energy independent study of the CERN people, except that here I have analyzed all the reactions, 2000 pieces of data simultaneously, and also unitarity was used and so on. So I think that things are better constrained and not only that, the study of the K-matrix gives better criteria on the resonances.

Levi-Setti: How many hours of 6600 did you use for this analysis?

Kim: Well, it is hard to say really it has been going on quite a while part-time, also the Sigma-7 has been used.

Thompson: Well just roughly, within a factor of two, how many hours?

Kim: I think it would be maybe a couple of hundred of hours of 6600.

Lasinski: How many parameters have you used?

Kim: In the linear expansions it is 64 parameters for the ranges but the zero ranges are another 64 parameters. The idea is that once you find the parameters in one region, you only vary the ranges for adjacent energy ranges to have continuity. So in successive analyses only 64 parameters vary.

Lynch: Did you use the total cross sections.

Kim: Yes, I used the K^-p total cross sections and the I=1, I=0 separate total cross sections.

Ross: If you take into account unitarity, how do you take account of the fact that some of the channels are not two-body channels?

Kim: When I mentioned unitarity it means for instance that when there are inelastic reactions like $Kp \rightarrow \Lambda\pi\pi$ or $\Sigma\pi\pi$, I allow additional channels for that and then of course unitarity there is whatever is left over. Eventually in fact I may try to constrain these additional channels by fitting to the rest of the inelastic reactions. I haven't done that. So that is essentially left free. But one more thing that I want to comment on is that the D_{03} resonance at 1680 doesn't decay much into $\Sigma\pi\pi$ or $\Lambda\pi\pi$. It mainly decays into $\overline{K}N$ and $\Sigma\pi$ channels. The branching ratio to a 3 body state is much smaller than given in the Rosenfeld table. And in fact at that location it is the P_{01} that decays into $\Lambda\pi\pi$ or $\Sigma\pi\pi$ not the D_{03}.

Roper: Did you start from different initial input sets of these parameters to see if you could get different solutions?

Kim: The idea is that I have tried to search parameters at different regions, certain finite energy regions like 100 MeV and try to see what kind of solution I obtain and then try to continue on from there so these have been tried in various combinations but it is hard to say these solutions are unique or anything like that.

ASYMMETRY AND DIFFERENTIAL CROSS SECTION FOR ELASTIC SCATTERING OF K^- MESONS BY POLARIZED PROTONS*

M. Zeller, R.D. Ehrlich, A. Etkin,** P. Glodis, V.W. Hughes, K. Kondo,+ D.C. Lu, R. Patton, S. Mori, G.A. Rebka, Jr., J.E. Rothberg,‡ and P. Thompson

Gibbs Laboratory, Yale University

In an experiment at the Brookhaven National Laboratory[1] we have recently completed measurements of the asymmetry and differential cross section for K^- p elastic scattering from polarized protons at five incident K^- momenta of 1.8, 1.9, 2.0, 2.1 and 2.2 GeV/c corresponding to an invariant mass range from 2150 to 2300 MeV. The K^- mesons were obtained in a partially separated beam with a momentum spread $\Delta p/p$ of ±2% and were further identified by a differential gas Cerenkov counter. A polarized proton target of the L.M.N. type was used. The detection of the recoil particles was made with magnetostrictive wire chambers, permitting good spatial resolution and high signal to background ratios. The acceptance of the detection apparatus was typically from $\cos\theta^* = 0.9$ to $\cos\theta^* = -0.7$, and a total of approximately 12,000 elastic scattering events was collected at each momentum point. Analysis of the data is beginning, and very preliminary results indicate asymmetry values consistent with the results of a CERN experiment.[2]

---

* Research supported in part by U.S. Atomic Energy Commission.

** NSF Predoctoral Fellow.

+ On leave from University of Tokyo, Tokyo, Japan.

‡ Present address: University of Washington, Seattle, Wash.

1. G.A. Rebka, Jr., et al., Phys. Rev. Letters 24, 160 (1970).
2. C. Daum, et al., Nuclear Physics B6, 273 (1968).

SESSION IV

HIGHER ENERGY FORMATION

Discussion Leader: Riccardo Levi-Setti

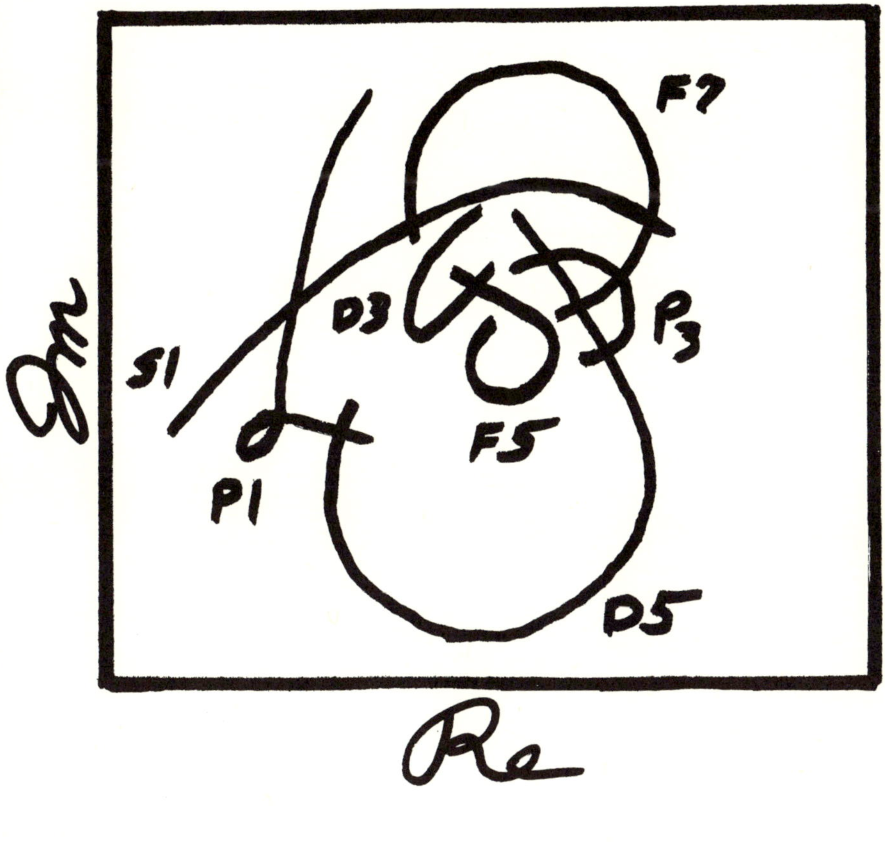

REVIEW OF PARTIAL-WAVE ANALYSES OF THE $\bar{K}N$ SYSTEM ABOVE 1.1-GeV/c K^- BEAM MOMENTUM*

Angela Barbaro-Galtieri

Lawrence Radiation Laboratory

Berkeley, California 94720

ABSTRACT

A review of all partial-wave analyses done so far in the mass region 1850 to 2250 MeV is presented. Only the $\Lambda\pi$ and $\Sigma\pi$ channels have been analyzed in detail for hyperon states in all partial waves. Some information is also available for the K^-p and $\bar{K}^0n$ systems as well as the ΞK system. A total of four $I = 0$ states and seven $I = 1$ states have been claimed in this energy region, of which only one and five respectively have been seen in at least two channels. The uniqueness of solutions obtained in the analysis of a single channel is discussed.

1. INTRODUCTION

The energy region discussed in this review starts at the upper end of the large bump seen in total cross section at 1 GeV/c incident K^- momentum. The isospin 0 and isospin 1 components of the total $\bar{K}N$ cross section are shown in Fig. 1. These are taken from Lynch's review of total cross-section measurements presented at this conference.[1]

The partial-wave analyses performed in the 450- to 1100-MeV/c region have been reviewed by Plane.[2] This paper is a review of all the analyses done at higher energy. The region in question is shown in Fig. 1. The $I = 0$ cross section shows only one bump, whereas the $I = 1$ shows two bumps. The analyses discussed in this paper claim 4 states in the $I = 0$ and 7 in the $I = 1$ states as listed in Table I. Of these 11 states only 6 are present in more than one channel.

This energy region has not been as extensively studied as the lower one. The analysis is more complicated because many partial

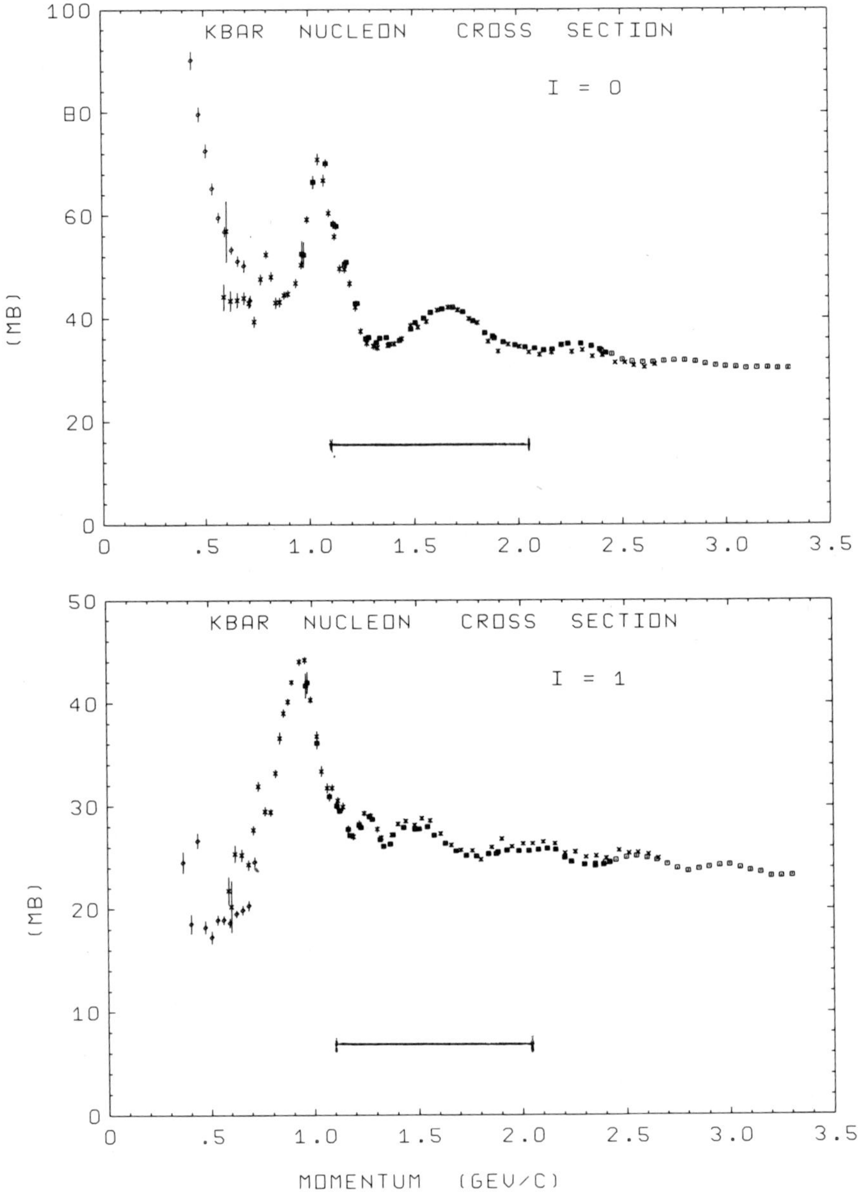

FIG. 1. The I = 0 and I = 1 part of the total $\overline{K}N$ cross section taken from Lynch.[1] Horizontal bars indicate the region discussed in this paper.

Table I. Proposed states in the M = 1850-2250 MeV region. Arrows indicate new states and states not yet confirmed. Crosses indicate the channels where each state has been claimed. Po means possible.

	State			$K \to \bar{K}N$	$\Lambda\pi$	$\Sigma\pi$	ΞK
I = 0	$\Lambda(1870)$	F_{07} or P_{03}	←	×			
	$\Lambda(2015)$	F_{07}	←			×	
	$\Lambda(2040)$	D_{03}	←			×	Po
	$\Lambda(2100)$	G_{07}		×		×	×
I = 1	$\Sigma(1900)$	P_{11}	←		×		
	$\Sigma(1915)$	F_{15}		×	×	×	
	$\Sigma(1940)$	D_{13}	←		×	×	Po
	$\Sigma(2020)$	F_{17}		×	×	×	×
	$\Sigma(2070)$	P_{13}	←		×		
	$\Sigma(2120)$	G_{17}	←		×	×	
	[a]$\Sigma(2250)$	G_{19} or F_{17}		×	Po	×	

[a]This resonance is really outside any of the partial-wave analyses performed, except for the K^-p, as it can be seen in Fig. 2. Some of the analyses, however, have included it as a G_{19} amplitude, of which only the lower tail is used.

waves are involved and the uniqueness of a solution is very questionable. Many channels are open: in addition to the two-body channels ($\bar{K}N$, $\Lambda\pi$, $\Sigma\pi$, ΞK), many quasi-two-body channels are certainly possible [$\Lambda(1405)\pi$, $\Lambda(1520)\pi$, $\Sigma(1385)\pi$, $\Sigma(1765)\pi$, etc.]. The elasticities tend to be small; in fact, as seen in Fig. 1, none of the bumps seen in total cross section is very large, although the spins tend to be larger than at lower energy. [The contribution of a resonance to the total cross section at resonance is $\sigma_T = 4\pi \lambda\!\!\!^{-2}(J + 1/2)x_e$].

Figure 2 shows what data have been used for partial-wave analyses and which channels have been analyzed. For the elastic channel no complete analysis exists as yet; that is, a simultaneous fit of the K^-p and $\bar{K}^0n$ data.

Throughout this paper the definitions used for the differential cross section $d\sigma/d\Omega$ and polarization ($\vec{P}$) expansions, unless otherwise stated, are as follows:

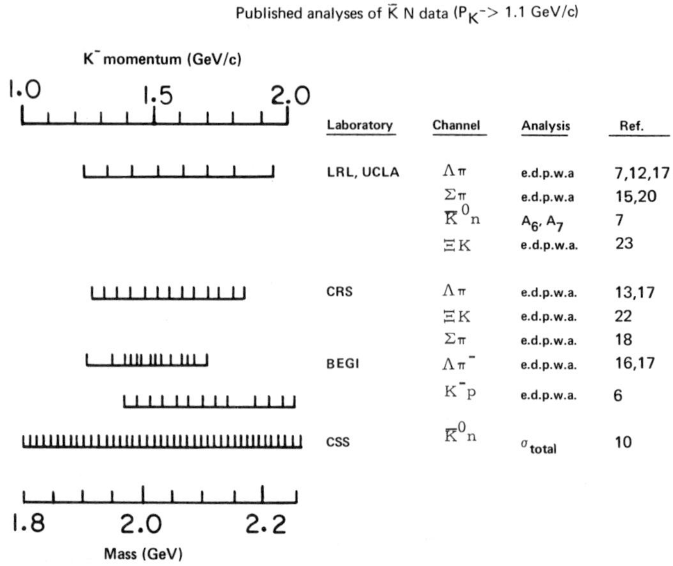

FIG. 2. Summary of published analyses above 1.1-GeV/c incident K^- momentum. Each mark on the lines at left represents a momentum at which data are available. In the column marked analysis, the symbol e. d. p. w. a. stands for "energy-dependent partial-wave analysis." References are given to papers where the data have been analyzed.

$$\frac{d\sigma}{d\Omega} = \lambda\!\!\!^{-2} \sum_i A_i P_i (\cos\theta), \tag{1}$$

$$\vec{P}\,\frac{d\sigma}{d\Omega} = \vec{n}\,\lambda\!\!\!^{-2} \sum_i B_i P_i (\cos\theta), \tag{2}$$

where $\lambda\!\!\!^{-} = \hbar/p$, p being the momentum of the K^- in the K^-N c. m. s.; P_i are Legendre polynomials; and P_i^1 are the first associated Legendre functions. The coefficients A_i and B_i are expressed in terms of the amplitudes squared or product of amplitudes.[3]

All of the partial-wave analyses reported here are energy-dependent analyses; that is, the amplitudes have been given an energy dependence assumed valid over a wide range of incident momentum. For each partial wave the assumption is:

$$T = T_b + e^{i\Phi} T_R, \tag{3}$$

where T_b is a background amplitude, Φ is an arbitrary phase, and T_R is the Breit-Wigner resonant form. The background has been parameterized as

$$T_b = (A + Bk + Ck^2 + Dk^3) + i\,(E + Fk + Gk^2 + Hk^3) \tag{4}$$

or as

$$T_b = (a + b\,k + ck^2)e^{i(d+ek+fk^2)}, \qquad (5)$$

where k is the incident K^- momentum (or a simple function of it like $k = p_K - p_0$) and the coefficients are real constants. For T_R the expression used is

$$T_R = \frac{\pm\sqrt{x_e x}}{\epsilon - i} = \frac{t}{\epsilon - i}, \qquad (6)$$

where x_e is the elasticity of the resonance, $x = \Gamma_c/\Gamma$ is the branching ratio into the channel c in question, and ϵ is $(E-E_R)/(\Gamma/2)$ or a more sophisticated expression which includes centrifugal barrier factors etc.[3] The sign ± is related to the SU(3) multiplet to which the resonance belongs. It is always positive for the $\overline{K}N$ channel and it can have either sign for the other channels.

The sign convention used here agrees with the one used by Levi-Setti[4] and used in the Particle Data Group Tables of SU(2) and SU(3) coefficients.[5] Whenever necessary the sign of the amplitude has been changed to agree with this convention. In order to avoid confusion Levi-Setti's summary figure is reproduced here in Fig. 3.

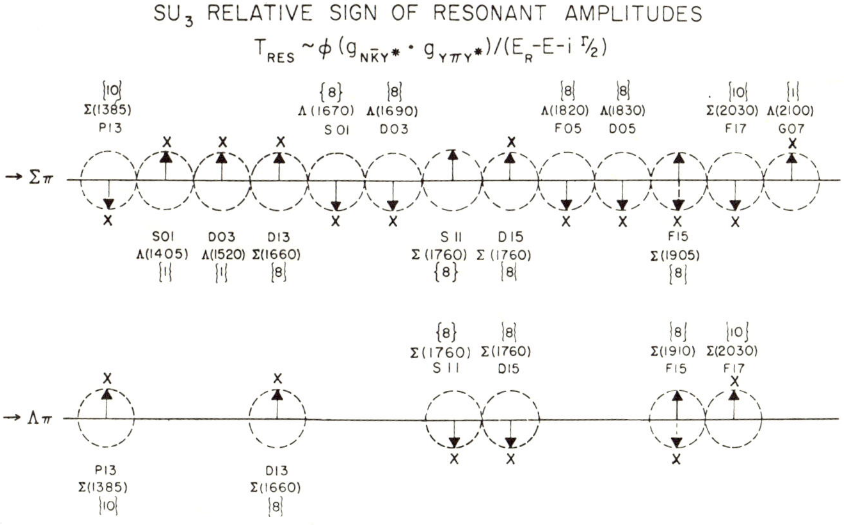

FIG. 3. Relative signs of the resonant $\Sigma\pi$ and $\Lambda\pi$ amplitudes as predicted by SU(3) using the currently accepted multiplet assignment for these resonances.[4] The arrows indicate the predicted and the X marks the observed phases. This plot is taken from Levi Setti,[4] who uses the same sign convention adopted here.

2. THE ELASTIC CHANNEL

As mentioned earlier, a complete partial-wave analysis of the elastic channel has not been performed yet. By "complete" analysis is meant an analysis of K^-p, $\bar{K}^0n$ differential cross sections, polarizations (available for K^-p)[6], and total-cross-section data.

The difficulty of this analysis is due to the fact that the major contribution to the K^-p differential cross section comes from diffraction scattering. Even in the backward direction the contribution from the background is comparable with that from the resonances, as illustrated in Fig. 2 of the paper by Daum et al.[6] Table II summarizes what analysis have been done in the elastic channels.

The discovery of two $J = 7/2$ states in the 2000-MeV mass region goes back to 1966 when Wohl et al.[7] reported the evidence for these two states in the $\bar{K}^0n$ and $\Lambda\pi$ channels. The elasticities of each were found to be equal, as seen in Table II. In this paper, however, only the A_6 and A_7 coefficients of the Legendre polynomial expansion were fitted and only these two states were included in the analysis, so the presence of other states was not tested.

Total-cross-section measurements in this energy region by Cool et al. and Bugg et al.,[8] confirmed the presence of two bumps in the total cross sections (Fig. 1). The two experiments are in very good agreement[1] and the elasticities for the two $J = 7/2$ states are reported in Table II. These measurements showed an

Table II. Results of the analyses of the elastic channel data. States listed here are only the ones that have been included in the analyses. The last four columns list elasticities quoted by the various authors.

States		M	Γ	Wohl[7]	σ_{tot}[8]	Daum[6]	Bricman[10]
[a]Λ(1870)	$\frac{7}{2}^+$?	1870	~ 40		.10		.07
Λ(2100)	G_{07}	2120	~145	.25	.33	.33	.24
Σ(1915)	F_{15}	1910	~ 50		.10	.12	.07
Σ(2020)	F_{17}	2040	~125	.25	.10	.11	.27
[b]Σ(2250)	$\frac{7}{2}$?, $\frac{9}{2}$?	2240	~180		.08	.08	.04

[a]The partial-wave analysis of Armenteros et al.[11] suggests this state; the elasticity was found to be .12; Bugg et al.[8] quote x = .10; however, Conforto et al.[25] suggest $J^P = 3/2$. See Section 7.

[b]Elasticity obtained by assuming $J = 7/2$.

additional small bump in the I = 1 cross section at M = 1915 MeV. It has been argued for some time that this bump may be due to the unfolding technique used, since it does not show in the measured cross sections (see Fig. 4 for the K^-p data). A resonance is now seen in the partial-wave analyses of $\Sigma\pi$ and $\Lambda\pi$ channels, as discussed later, as well as in a production experiment,[9] therefore it is fairly well established.

Daum et al.[6] have measured the proton polarization of the K^-p elastic scattering. They have done a partial-wave analysis that

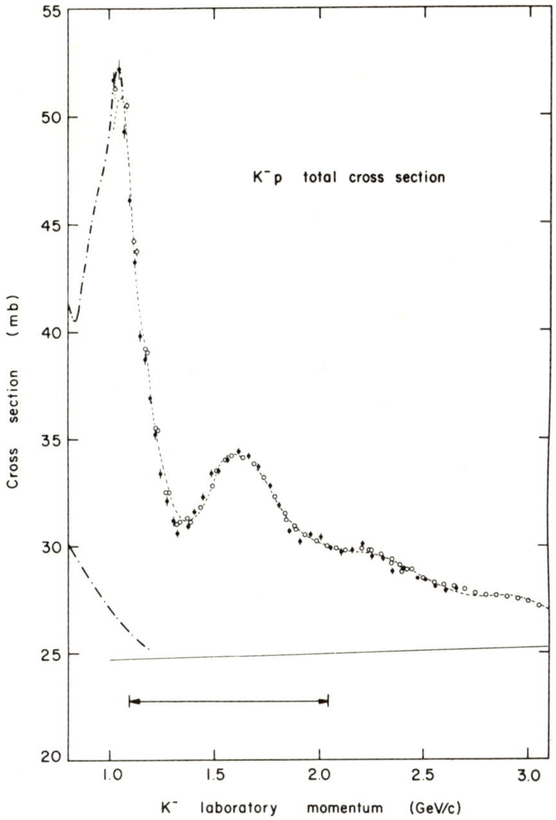

FIG. 4. Compilation of the total K^-p cross section data,[8] taken from Bricman et al.[10] The dashed line is their fit to the data, the solid line is the contribution from the assumed linear background. The horizontal line shows the region of interest in this paper.

included the differential cross sections and polarizations for the K^-p channel as well as the total-cross-section data.[8] The background amplitudes were parameterized in three different ways, one being our expression (4) carried only to the linear term. The results of the three analyses are in agreement. The J^P assignments of Wohl et al.[6] for $\Sigma(2020)$ and $\Lambda(2100)$ were checked, but no new states were added in the analysis. The results are shown in Table II. As for $\Sigma(2250)$ the analysis could not find a definite J^P assignment; everything from 5/2 to 11/2 fits the data equally well.

Finally Fig. 5a shows the total cross section for the charge exchange reaction $K^-p \to \bar{K}^0 n$ as measured by Bricman et al.[10] At resonance the contribution of a resonance to this cross section is

$$\sigma = \pi \lambda\!\!\!^{-2} (J + 1/2)\, x^2,$$

where x is the elasticity and the I - spin factors have been included. The contribution of the same resonance to the total cross section is

$$\sigma_T = 2\pi \lambda\!\!\!^{-2} (J + 1/2)\, x.$$

Bricman et al. have fitted the K^-p cross section of Fig. 4 and the $\bar{K}^0 n$ cross section of Fig. 5a to the sum of a background term [Eq. (4) quadratic in P_{K^-}] plus resonances. In the region of interest they have included the resonances listed in Table II and obtained the elasticities quoted. Figure 5b shows the effect of $\Sigma(1915)$: the contribution of each resonance to σ is shown as a solid line, whereas the interference terms are shown as dashed lines. They find that the large $5/2^+$ interference term is necessary to fit the data and substantiates the presence of $\Sigma(1915)$. Fits to the data without this resonance give considerably worse χ^2 (400 compared with 145 for the best fit, where 133 is expected).

Table II gives a summary of the situation. The evidence for the F_{07} state listed as $\Lambda(1870)$ comes from partial-wave analysis of the lower-energy data[11] and it is discussed in Plane's review.[2] Masses and widths listed are estimated values consistent with the ones used by the various authors. Finally, the comparison with Table I shows that six more states have been claimed in other channels and are not included in the elastic channel. As usual, a more detailed analysis is likely to show more structure, therefore we will have to wait.

3. THE $\Lambda\pi$ CHANNEL

The partial-wave analysis of the differential cross-section and polarization coefficients has been done by five different groups. Table III is a summary of the results. In this table the signs of t [the amplitudes at resonance defined by Eq. (6)], have been changed whenever necessary to agree with our sign convention.

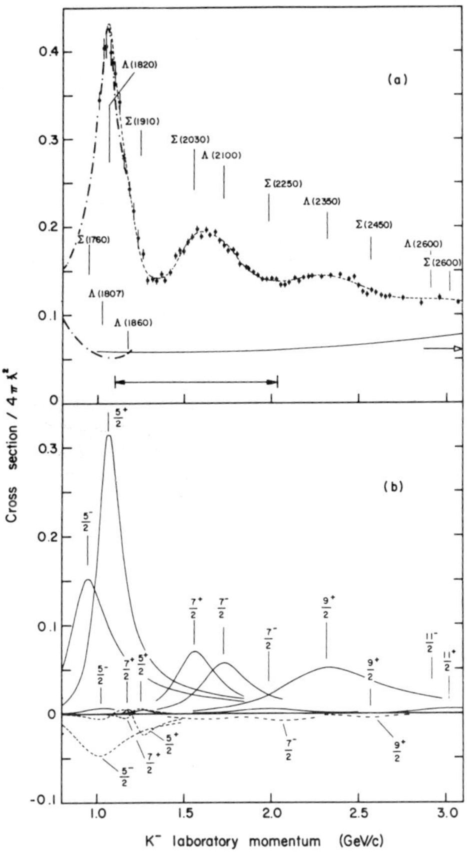

FIG. 5(a). Cross section divided by $4\pi \lambda\!\!\!^{-2}$ versus incident K^- momentum for the reaction $K^- p \to \overline{K}^0 n$ as measured by Bricman et al.[10] The dashed curve is the result of their fit, the solid line is the contribution of the background, and the vertical lines indicate the position of the resonances included in the fit. (b) The solid lines represent the contribution of the resonances to the fitted curve, the dashed lines represent the interference terms for each partial wave.

1) Smart[12] made an analysis over a wide region, including the data of Wohl et al.[7] and Dauber et al.[20] The background parameterization used is given by Eq. (5), with the expansion up to different order for different partial waves. Each resonance had four parameters, one being a free phase. A P_{11} state was first reported by this analysis as shown in Table III. Figure 6 shows the amplitudes corresponding to the best fit obtained. The P_{13} and

Table III. Results of partial-wave analyses in the $\Lambda\pi$ channel. In the second column N = 50 means that 50 parameters were used in the overall fit. In the first row to the right, N_{par} indicates how many parameters were used for that partial wave. R(4) means that the resonant amplitude included four parameters (M, Γ, |t|, Φ).

		N_{par}	P_{11}	P_{13}	D_{13}	F_{15}	F_{17}	G_{17}	[d]G_{19}
[a]Smart[12]	$\Lambda\pi^-$, $\Lambda\pi^0$	N_{par}	4+R(4)	6	6	4+R(4)	R(4)	3	3
	.75-1.9 GeV/c	M	1882±40			1902±11	2032±6		
	N=50	Γ	222±150			52±25	160±16		
	χ^2/DF =418/390	t	-.11±.03			-.08±.02	.21±.01		
CRS[13]	$\Lambda\pi^0$	N_{par}	4	8	4	8+R(4)	R(3)	4	---
	1.134-1.84 GeV/c	M				1910±20	2030±10		
	N=39	Γ				60±20	165^{+30}_{-15}		
	χ^2/DF =319/295	t				-.10±.02	.20±.02		
[b]Cox et al.[16]	$\Lambda\pi^-$	N_{par}	6	3+R(3)	3	3+R(3)	R(3)	3	3
	1.18 -1.71 Gev/c	M		[b]2100±20		1903±10	2027±6		
	N=40	Γ		87±20		77±27	158±16		
	χ^2/DF =258/199	t		-.16±.03		-.09±.02	.19±.01		
[c]Litchfield[17]	$\Lambda\pi^-$, $\Lambda\pi^0$	N_{par}	5+R(3)	4+R(3)	4+R(3)	5+R(3)	R	7	---
	1.0 -1.85 GeV/c	M	1920±30	2070±30	1940±30	1895±10	2022±4		
	N=54	Γ	170±40	250±40	280±40	70±15	170±15		
	χ^2/DF =705/636	t	-.14±.03	-.09±.03	-.14±.03	-.070±.015	.200±.008		
[e]This paper	$\Lambda\pi^0$	N_{par}	4+R(4)	4	R(4)	4+R(4)	R(3)	R(3)	R(1)
	1.0 -1.85 GeV/c	M	1950±50		1940±50	1905±30	2010±15	2060±20	[2250]
	N=38	Γ	200±50		200±50	70±20	115±15	70±30	[140]
	χ^2/DF =299/245	t	-.09±.04		-.12±.04	-.11±.03	.16±.03	-.07±.02	-.18

[a]The F_{15} state Σ(1765) was also included in the analysis; the following parameters were found: M = 1775±7, Γ = 146±9, t = -.266±.017.

[b]Parameters reported here are the ones of fit 13, except for the P_{13} where mass and errors have been estimated by us, examining the various fits reported by the authors. The other errors are only statistical.

[c]The results quoted are the ones from fit A.

[d]Σ(2250), G_{19} is really outside the range of most analysis, but its lower tail can be included instead of a G_{19} background.

[e]Quantities in square brackets have been kept fixed. See text for a fit from .61 to 1.85 GeV/c.

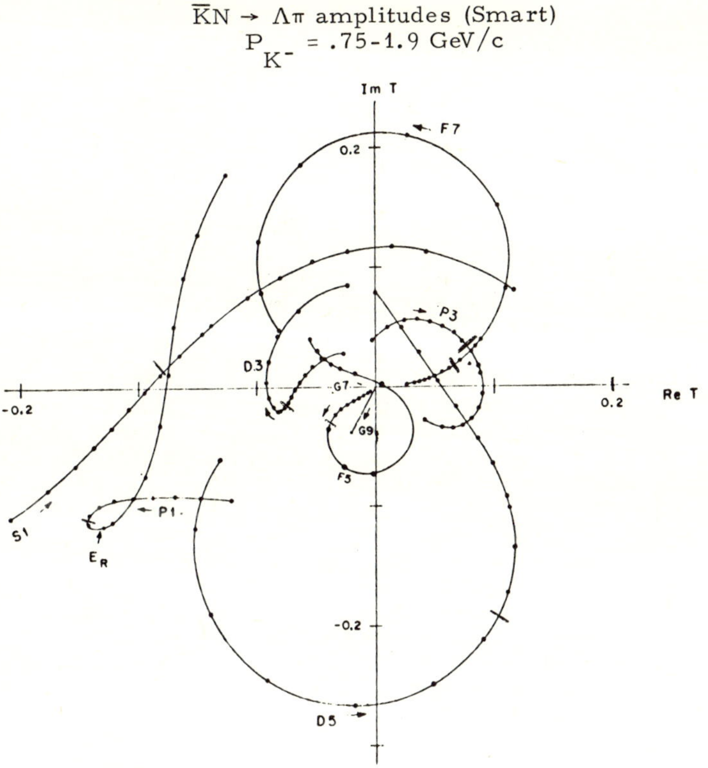

FIG. 6. Partial-wave amplitudes for the reaction $\overline{K}N \rightarrow \Lambda\pi$ obtained by Smart[12] (rotated by 180 deg so as to agree with our sign convention). Arrows represent the direction of increasing energy. Dots represent the various momenta used. P_{K^-} = 1.1 GeV/c is between the eighth and ninth points (where a line is drawn).

D_{13} amplitudes were not parameterized as resonances, but had a quadratic background dependence on P_{K^-}. The errors quoted in Table III are twice the statistical errors.

2) Berthon et al.[13] called CRS in the table, have included some data of the CHS collaboration[14] in order to determine the F_{15} amplitude in the 1915-MeV region. Their parameterization of the background differs from Eqs. (4) and (5) in that, for each partial wave, they use:

$$T_b = \sum_n f_n P_n(k) \tag{7}$$

where P_n are Legendre polynomials, k is the incident momentum, and f_n are complex constants. This parameterization does not differ from Eq. (4), but the authors prefer it because it saves time in their fitting procedure. The order of the expansion for each partial wave was experimentally determined to be the one that fits the data better.

Only the F_{15} and F_{17} resonances were included in this analysis. The parameters reported in Table III were obtained by examination of the results of many different fits. As for the D_{13} which had been reported earlier in the $\Sigma\pi$ channel,[15] the authors state that "the D_{13} is adequately parameterized by a linear background." We will discuss the D_{13} in more detail later. Figure 7 shows the partial-wave-amplitude results of their fit A, excluding the F_{17} resonant amplitude.

3) Cox et al.[16] have used the reaction $K^- n \to \Lambda\pi^-$. The background parameterization was the same as the previous analysis (Eq. 7). These authors suggest a new state in the P_{13} partial wave at a mass of ~ 2100 MeV. Figure 8 shows the partial-wave amplitudes corresponding to their fit 13; the D_{13} is again a straight line.

4) Litchfield[17] has made a detailed analysis of all the available data is the 1.0- to 1.85-GeV/c momentum interval.[7, 12, 13, 14, 16] The background parameterization used is given by Eq. (7) and the method used is the one described in the CRS analysis above. He has in addition used a second method; that is, instead of parameterizing ReT_b and ImT_b as in Eq. (7), he uses η and δ expanded as in Eq. (7) and then writes $T_b = (\eta e^{2i\delta} - 1)/2i$. The two methods give similar results. In Fig. 9 the amplitudes obtained with method A are shown. The D_{13}, which was not present in any of the previous analyses, is now included and seems to be required by the data. The P_{11} and P_{13} are also included as resonances added to background.

5) This analysis. The analysis of Litchfield was not available to me before this conference, therefore I made a fit to the data of Berthon et al.[13] as well as data of the CHS collaboration[19] from .617 GeV/c, in order to test the possibility of the $D_{13}(1940)$ resonance[15] being present in the $\Lambda\pi$ system.

The data were fitted in two separate parts: (a) from .617 to 1.2 GeV/c, (b) from 1.0 to 1.85 GeV/c, and finally all together from .617 to 1.85 GeV/c. In each fit partial waves up to G_{19} were used, parameterized as linear background (S_{11}, P_{13}), linear background plus resonances (P_{11}, D_{15}, F_{15}), as resonances alone (F_{17}, G_{17}, G_{19}), and as a sum of two resonances with a free phase (D_{13}). Various fits were obtained.

A) The best fit for region (a) had a $\chi^2 = 179$ for 218 degrees of freedom (35 variables were used, including t values for all the resonances above 1.2 GeV/c). For the D_{13} and D_{15} resonances the following parameters were obtained:

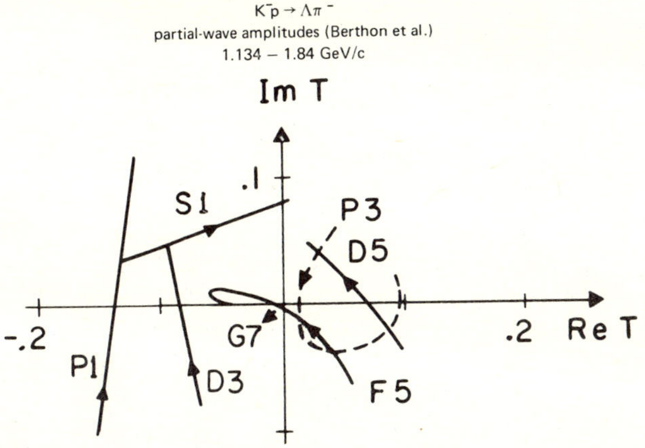

FIG. 7. Partial-wave amplitudes for the $K^-p \rightarrow \Lambda\pi^0$ reaction obtained by Berthon et al.[13] (rotated by 180 deg so as to agree with our sign convention). Fit shown here is fit A. The F_{17} circle is not shown.

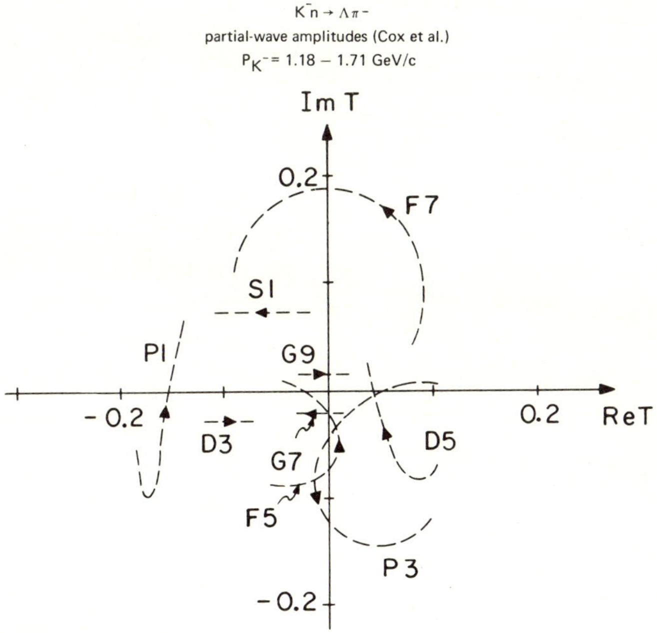

FIG. 8. Partial-wave amplitudes for the reaction $K^-n \rightarrow \Lambda\pi^-$ obtained by Cox et al.[16] (rotated by 180 deg so as to agree with our sign convention). Arrows indicate the direction of increasing energy.

D_{13}	D_{15}	
M = 1665 ± 10 MeV	M = 1765 ± 10 MeV	
Γ = 50 ± 10 MeV	Γ = 115 ± 10 MeV	(8)
t = .09 ± .02	t = -.22 ± .03	

The errors quoted are estimated by inspection of different fits obtained with different background conditions.

B) The best fit to region (b) is the one shown in Table III. The values for M, Γ, t and their errors quoted in this table are obtained by inspection of all the various fits done (including C and D below).

C) A first fit to all the data (from .617 to 1.85 GeV/c) was attempted by alternatively fixing the background parameters in one of the regions and letting the background of the other region and the resonance parameters readjust over the entire energy region. The best fit was obtained for χ^2/DF = 435/452. The resonant parameters changed within the errors quoted in (8) and Table III. Figures 10 and 11 show the A_i and B_i coefficients as functions of incident K^- momentum; the curves represent the fit to the data just described.

D) A final fit was done over the whole region, using the same method as C above, with the additional condition that the amplitudes at 1 GeV/c be the same for the lower and the higher region. The χ^2 got worse as expected (χ^2/DF = 582/450), but the parameters of the resonances changed within the errors quoted in (8) above and Table III. Figure 12 shows the amplitudes obtained in this fit, which have the same features as the ones obtained for the other three fits, but are different enough to increase the χ^2 to 582 (from 435). Obviously some of these partial waves are not adequately represented by a linear momentum dependence.

4. CONCLUSIONS ON $\Lambda\pi$ ANALYSES

An inspection of Figs. 6-9 and 12 and Table III tells us about the agreement of the five partial-wave analyses discussed so far. All five groups claim to have found no alternative solution to the one shown and all claim a good χ^2. Different sets of data points were used by the different groups, but the various sets of data are in very good agreement as testified by the good χ^2 obtained by Litchfield,[17] who has used all available data.

> Are there any differences between the five analyses?
>
> Out of eight partial waves, three agree and five do not agree.

a) The D_{15}, F_{15}, and F_{17} are in good agreement in all five analyses. It is reassuring to note that the F_{15}, $\Sigma(1915)$, turns out to be the same in all five analyses, although different background was used.

b) The S_{11}, P_{11}, P_{13}, D_{13}, G_{17} amplitudes are somewhat in diagreement in the five analyses, although, except for the G_{17} they are not any smaller than the D_{15} and F_{15}.

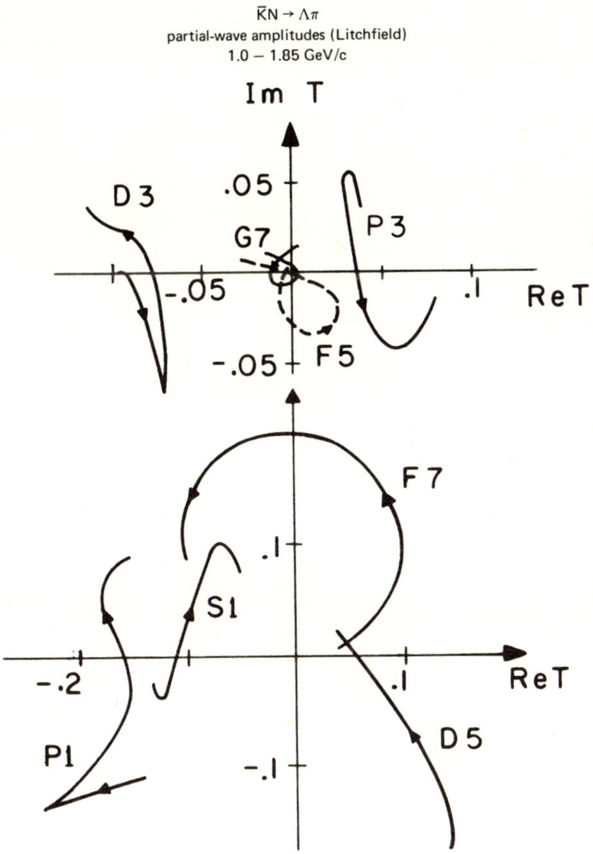

FIG. 9. Partial-wave amplitudes for the $\Lambda\pi$ channel, result of the analysis of Litchfield.[17] Only fit A is shown, fit B being very similar. Litchfield used all the data included in the three previous analyses.

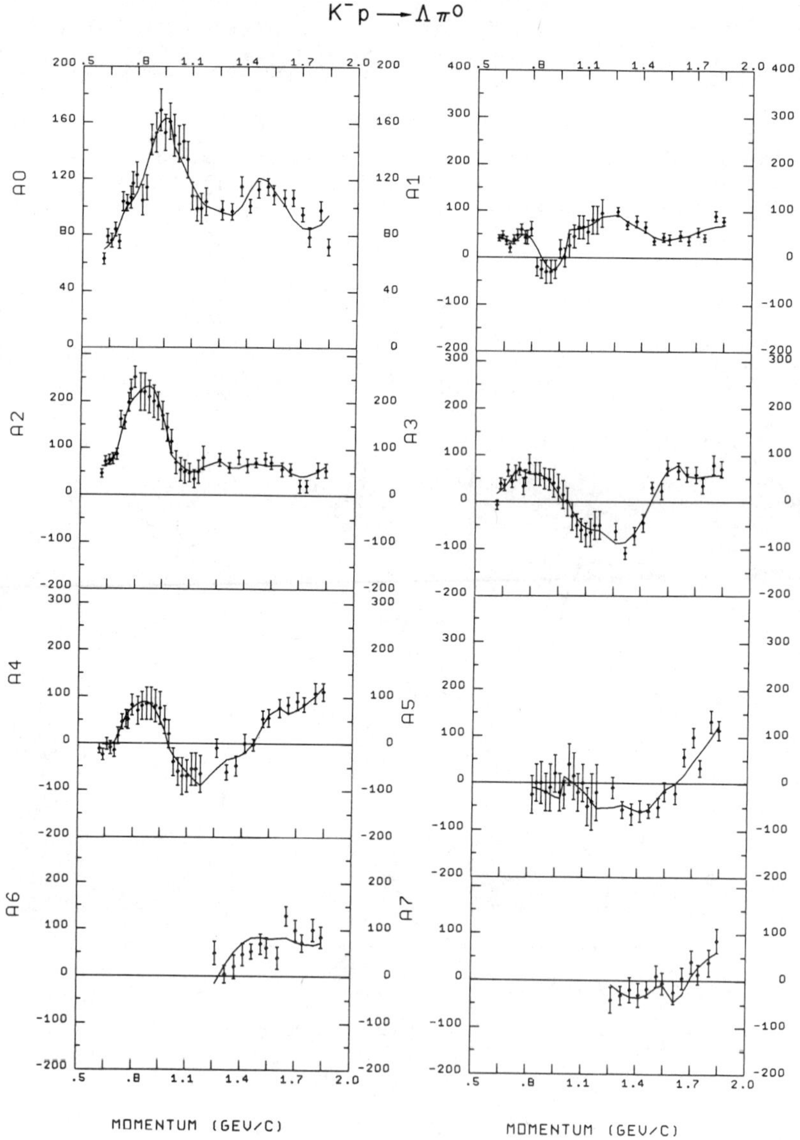

FIG. 10. The coefficients of the Legendre polynomial expansion of the differential cross section for the reaction $K^-p \to \Lambda\pi^0$ from .617 to 1.85 GeV/c. Data are from Berthon et al.[13] and Armenteros et al.[14, 19]. Curves are the best fit of the partial-wave analysis reported here, resulting from fit C.

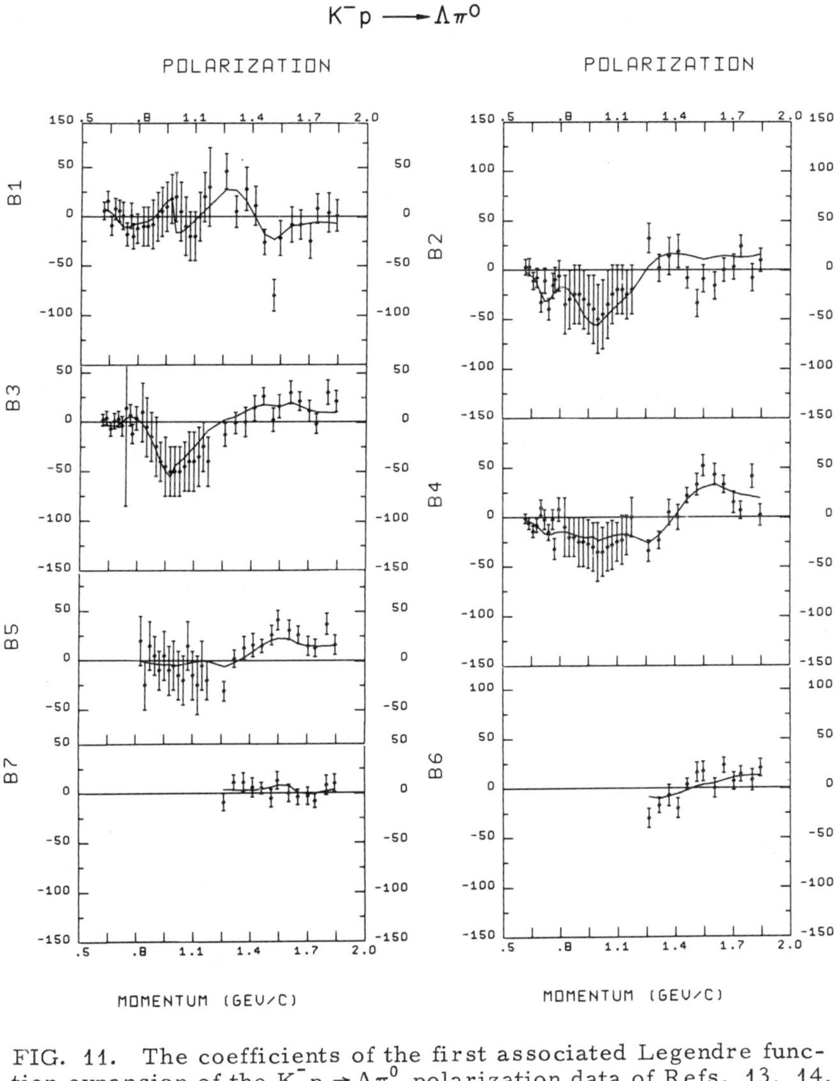

FIG. 11. The coefficients of the first associated Legendre function expansion of the $K^-p \rightarrow \Lambda\pi^0$ polarization data of Refs. 13, 14, 19. Curves are the best fit obtained by the analysis reported here, resulting from fit C.

1) The S_{11} of Cox et al.[16] and of the analysis reported here do not agree with the other three analyses and neither agree with each other.

2) The P_{11} of this analysis goes in the opposite direction of the other four. Notice, however, that a resonance with a negative t is included in the analyses of Smart,[12] Litchfield,[17] and this paper and the amplitudes of Figs. 6, 9, and 12 are not very inspiring. No doubt this state has to wait until it is found in some other channel, possibly with less background.

3) The P_{13} of the three British analyses agree with each other, while Smart and this paper do not agree with the other three analyses and neither agree with each other. Is there a resonance in this partial wave? In view of the disagreement I would suggest waiting for confirmation from some other channel.

4) The D_{13} of the five analyses have in common only the average value in that they lie in the same quadrant, but the orienta tion and variation with energy are in bad disagreement. Litchfield[17] and the analysis reported here have included a resonance whose parameters are in good agreement although different background was used. So what should we do with this state? We will come back to this question in Conclusions.

5) The G_{17} is the smallest amplitude present in these analyses and it again is different in the five analyses. At least three possibilities are present: II quadrant, III quadrant, resonance across III and IV quadrant. We will again discuss this amplitude in Conclusions.

In conclusion, the differences between the analyses reported here show that, assuming that none of the analyses is wrong, the uniqueness of the solution is a myth. Obviously we have to be very careful in accepting resonances found in only one channel. It would be desirable to accept a state only after it appears very clearly, i.e., practically background-free, in at least one channel. Unfortunately, if the elasticity is very small, conclusions can be drawn only by looking at many channels simultaneously.

5. THE $\Sigma\pi$ CHANNEL

In the $\Sigma\pi$ channel both I-spin zero and I-spin one are present, which makes the analysis more complicated. Since the $\Sigma^0\pi^0$ channel is very difficult to measure, only $\Sigma^{\pm}\pi^{\mp}$ data are used, and only Σ^+ provide polarization information. Therefore, compared with $\Lambda\pi$ channel, we have twice as many amplitudes, but only one additional piece of information (A_i^+, B_i^+, A_i^-).

So far only one complete analysis is available for this review;[15]

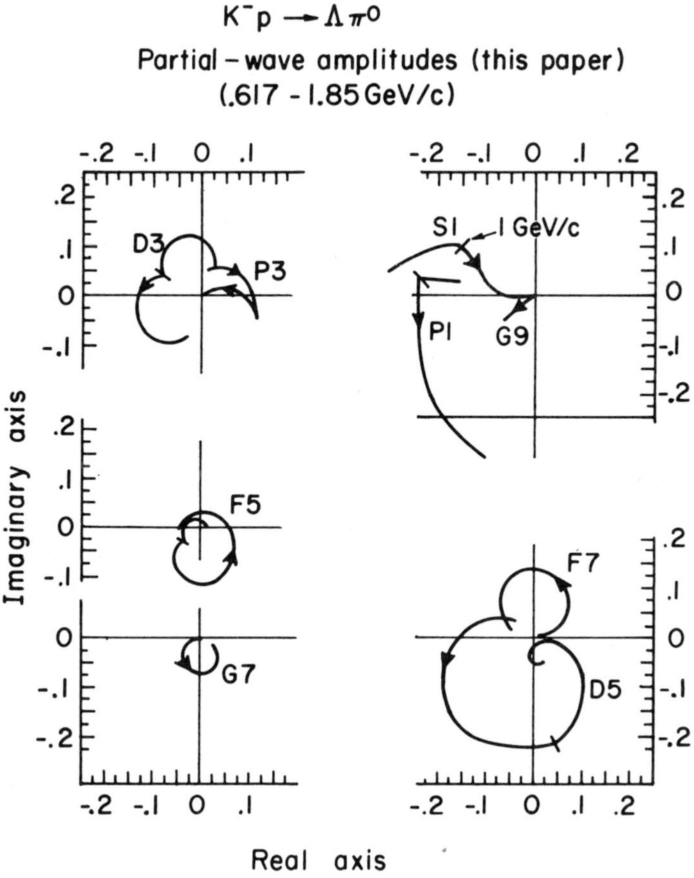

FIG. 12. Partial-wave amplitude result, for the reaction $K^- p \to \Lambda\pi^0$, from the analysis reported in this paper (fit D). The analysis extends from .617 to 1.85 GeV/c. The 1.0 GeV/c point is where a sharp change occurs. Arrows indicate the direction of increasing energy.

in addition, at this conference new data and a new analysis from the CRS collaboration have been reported by Butterworth[18] in a preliminary form, but not yet available for inclusion here. Dauber et al.[20] have published an analysis of the $\Sigma^-\pi^+$ data around 2.2 GeV center-of-mass energy, which shows the presence of a $J = 9/2$ partial wave, which can probably be associated with the $\Sigma(2250)$ seen in total cross section,[8] although the width quoted by Dauber et al. (80 MeV) is smaller than the one seen in total cross section (about 200 MeV). It is possible that more structure will appear at this mass when more data become available.

The analysis performed by this author was done as described for the $\Lambda\pi$ channel. The overall analysis goes from .617 to 1.95 GeV/c. Below 1.2 GeV/c the data used were taken from the CHS collaboration,[21] the 1.35 GeV/c point was taken from Trower,[24] above 1.7 GeV/c the data from UCLA were added,[20] and in the 1.2 to 1.7 GeV/c a few A_0 coefficients from the CRS collaboration[18] were included. The analysis was performed first in region (a), that is, .617 to 1.2 GeV/c, then in region (b) (1.0 to 1.95 GeV/c), and finally in the whole region. The partial waves included in the fit were S through G_9 waves. They were parameterized as follows:

Linear background	S_{11}, P_{11}, P_{03}, P_{13}
Linear back. + reson + phase	S_{01}, P_{01}, D_{15}, F_{15}
One resonance	D_{05}, F_{05}, F_{07}, F_{17}, G_{07}, G_{17}, G_{19}
Two resonances + a phase	D_{03}, D_{13}.

The following four solutions were found:

A) The best fit for the lower-energy region had $\chi^2/DF = 567/400$ (58 parameters were used, including the t values of some of the higher-energy resonances).

B) The best fit for the higher-energy data (above 1.0 GeV/c gave $\chi^2/DF = 428/267$ (321 data points were included and 54 parameters were used).

C) The lower- and higher-energy data were fitted at the same time by fixing the background amplitudes in one of the two regions and letting all the other parameters readjust. The best fit had $\chi^2/DF = 898/608$ (a total of 673 data points were used and 65 variables were varied). The curves shown in Fig. 13 were obtained with this fit.

D) Finally the background amplitudes for the lower- and higher-energy part were required to be the same at 1.0 GeV/c. The best fit found for this case had $\chi^2/DF = 1022/609$ (the fit got worse, as expected), but the main features of the amplitudes and the resonance parameters remained the same. Figure 14 shows the partial-wave amplitudes obtained in this fit.

Notice that the χ^2/DF obtained in the four fits above are considerably worse than for the $\Lambda\pi$ case. The fits of the CHS collaboration at lower energy also show the same trend.[21]

The results of the above fits are shown in Table IV. Here again the parameters quoted have been obtained by inspection of the various fits done. Only results at the higher energy ($M > 1850$ MeV) will be discussed here; see Ref. 15 for discussion of the results at lower energy.

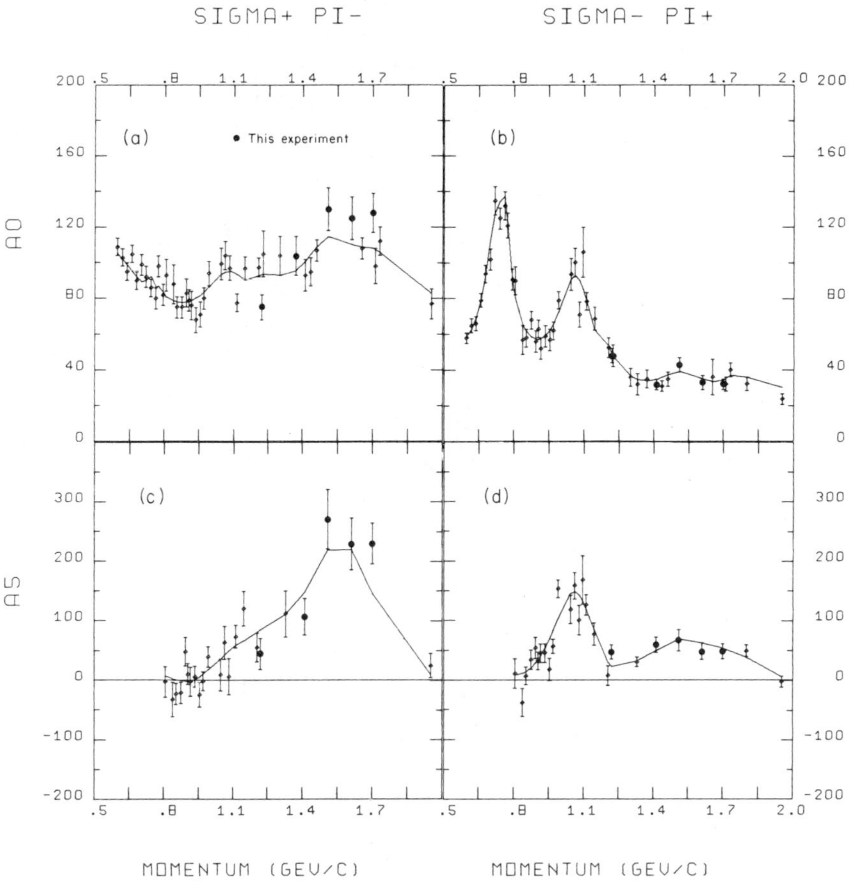

FIG. 13. Some data for the reaction $K^-p \to \Sigma\pi$. (a) and (b) The $A_0 = \sigma/4\pi\lambda\!\!\!^{-2}$ coefficients for the $\Sigma^+\pi^-$ and $\Sigma^-\pi^+$ channel respectively. (c) and (d) The A_5 coefficients for the same two channels. Data are from Refs. 21, 15, and 20. The curves are the result of fit C discussed in the text.

(a) I = 1 states. This analysis, as previously reported,[15] suggests two new states, the $D_{13}(1940)$ and the $G_{17}(2120)$, out of the four indicated with an arrow in Table I. The preliminary analysis of the CRS collaboration presented at this conference[18] does not include either of these two, whereas a D_{15} at 2070 MeV is suggested. The results of the CRS analysis in final form are not available to me, but we have compared the Legendre polynomial coefficients (A_i^+, A_i^-, B_i^+) found in the two experiments and there is no difference in the behavior of the coefficients except that the

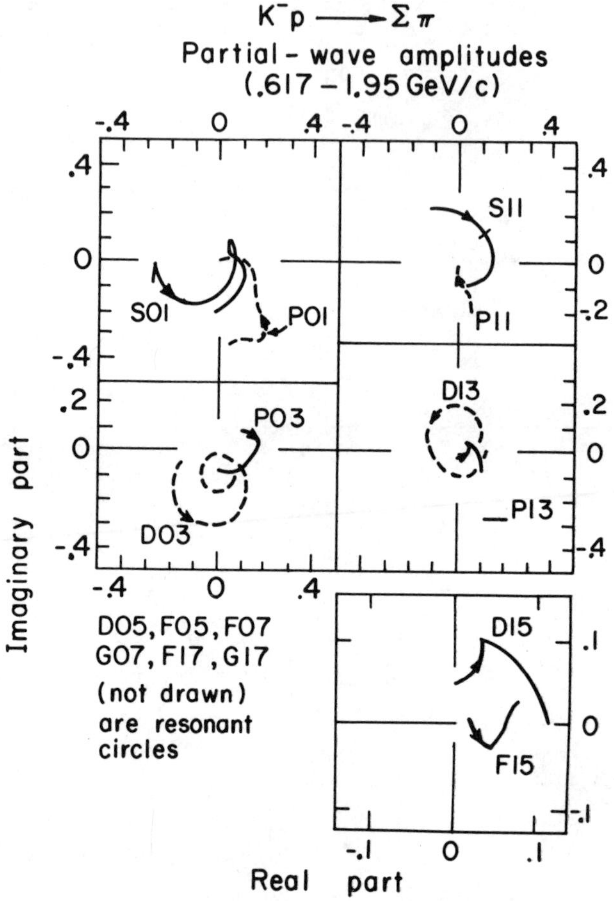

FIG. 14. Partial-wave amplitudes for the $K^-p \rightarrow \Sigma\pi$ channel result of the analysis reported in this paper (and Ref. 15). The 1.0-GeV/c point is where a sharp change occurs. Arrows indicate the direction of increasing energy.

CRS data is more numerous, the new data points fitting smoothly in with my old ones. Here again the question arises:

<u>Are these two different solutions?</u>

The results on the $F_{15}(1915)$, however, are in agreement in the two experiments, the values of t being compatible within errors. (CRS quotes t = -.13 ± .03).

Table IV. Results of the partial-wave analysis for the $\Sigma\pi$ channel.[15] Quantities in square brackets have been kept constant. N_{par} is the number of parameters used for that particular partial wave.

	N_{par}	M(MeV)	Γ(MeV)	t
S_{01}	4+R(4)	1683 ± 5	25 ± 5	-.29 ± .03
P_{01}	4+R(4)	1690 ± 10	22 ± 5	-.13 ± .03
D_{03}	R(3)	1680 ± 5	85 ± 10	-.31 ± .03
D_{05}	R(3)	1840 ± 15	150 ± 30	-.16 ± .03
F_{05}	R(3)	1820 ± 10	100 ± 20	-.26 ± .03
D_{03}	R(4)	2010 ± 30	130 ± 50	-.20 ± .04
F_{07}	R(3)	2020 ± 20	160 ± 30	-.15 ± .02
G_{07}	R(3)	2110 ± 20	60 ± 25	.06 ± .03
D_{13}	R(3)	1662 ± 5	48 ± 5	.18 ± .06
D_{15}	4+R(2)	[1765]	[115]	.06 ± .03
F_{15}	4+R(2)	[1900]	[70]	-.06 ± .03
D_{13}	R(4)	1940 ± 40	200 ± 50	-.12 ± .03
F_{17}	R(3)	2000 ± 20	100 ± 40	-.052 ± .010
G_{17}	R(3)	2120 ± 30	135 ± 30	.13 ± .02
G_{19}	R(1)	[2250]	[140]	.07

(b) I = 0 states. This analysis[15] suggests again two new states, $F_{07}(2015)$ and $D_{03}(2040)$. Here again the preliminary analysis of CRS disagrees, their best solution requires a D_{05} at 2110 MeV, therefore the same question as above applies.

We will discuss this point farther in Conclusions.

6. THE ΞK CHANNEL

Data in the Ξ^-K^+ and Ξ^0K^0 are very scarce because the cross sections for these two reactions are very small. For Ξ^-K^+ the cross section in the region of interest is less than 180 μb, for the Ξ^0K^0 channel, it is less than 80 μb. The Ξ^0K^0 presents the additional problem that only 1/3 of the K^0 have visible decays, therefore the available statistics are very small. For these reasons only two groups have done partial wave analyses so far.

Burgun et al[22] have done a partial wave analysis in the 1.26-1.84 GeV/c region. The background amplitudes were parameterized as linear functions of the K^- momentum (Eq. 4). Figure 15 shows their A_0 coefficients, Fig. 16 shows the coefficients A_0 through A_7 as function of momentum for the $\Xi^- K^+$ reaction. The curves shown correspond to the best fits obtained: (A) with constant background and the $\Lambda(2100)$ and $\Sigma(2020)$ resonances included, (B) with linear background and the same two resonances, (C) as B with the addition of a new state at 2070 MeV, $\Gamma \sim 120$ MeV, $J = 3/2^{\pm}$ and x_e x = .015 ±.003. The parameters obtained for $\Sigma(2020)$ and $\Lambda(2100)$ tion B are shown in Table V.

Table V. Results of partial-wave analyses in the ΞK channel. Quantities in square brackets were kept constant.

	Burgun et al.[22]			Muller[23]	
	M(MeV)	Γ(MeV)	t[a]	M(MeV) Γ(MeV)	t
$\Sigma(2020)$ F_{17}	[2030]	[160]	<.05	[2030, 120]	.023
$\Lambda(2100)$ G_{07}	2080 ± 10	80 ± 10	.09 ± .01	[2100, 140]	.003

[a]These authors quote only x', in order to calculate $t = \sqrt{x_e x'}$ we assume that they used x_e from the Particle Data Group[5] compilation of August 1968. Values quoted are from solution B.

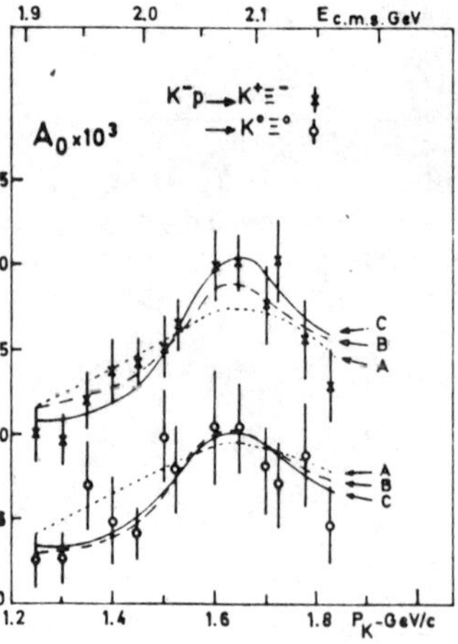

FIG. 15. The $A_0 = \sigma/4\pi\lambda\!\!\!^{-2}$ coefficient for the $K^- p \to \Xi^- K^+$ and $\Xi^0 K^0$ channels. Plot taken from Burgun et al.[22]

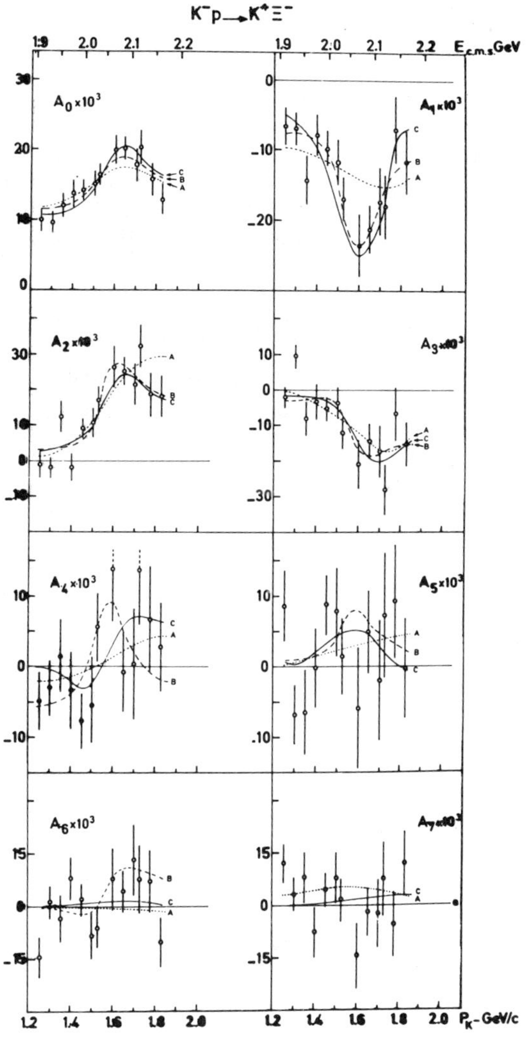

FIG. 16. The coefficients of the differential cross section for the reaction $K^-p \rightarrow \Xi^-K^+$ reported by Burgun et al.[22] The curves correspond to the three fits discussed in the text.

Muller[23] has analyzed the LRL and UCLA data (see Fig. 2) in the momentum region 1.2 to 2.7 GeV/c. In this analysis the partial waves up to D_5 were parameterized as baryon-exchange amplitudes; the higher partial waves included only resonances. The results obtained for the F_{17} and G_{07} resonances are shown in Table V. As for the lower partial waves, they look very similar to the ones shown for the $\Lambda\pi$ or $\Sigma\pi$ channels. Figure 17 shows three of these partial waves. Whether they include resonances or not is hard to say at the moment, but certainly the possibility is not excluded. The D_{13} has also the characteristic "loop" behavior.

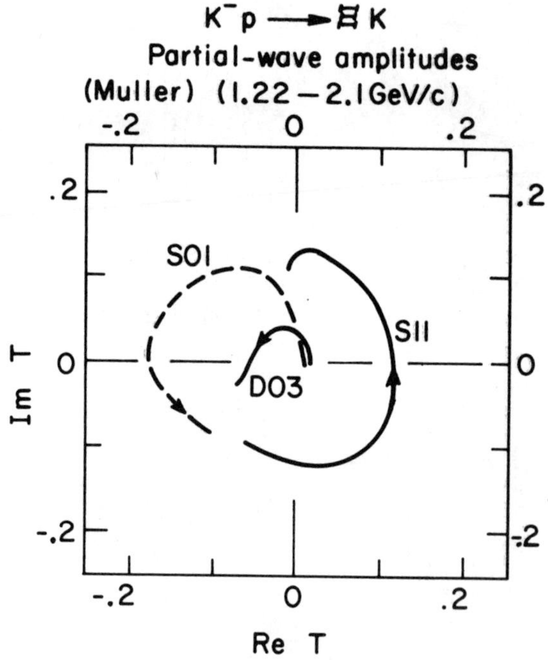

FIG. 17. Some partial-wave amplitudes for the reaction $K^-N \rightarrow \Xi K$ as found by Muller[23], using a u-channel exchange parameterization. The energy range shown here is from 1.22 to 2.1 GeV/c.

Both analyses show a surprisingly small amplitude for the $\Lambda(2100)$, which indicates that the enhancement at this mass in the cross sections of Fig. 15 has a different origin. The CRS fit C suggests the possibility of some other resonance in this region. Muller's fit does not completely account for the peaks in A_0, and shows a contribution from $\Lambda(2100)$ even smaller than the CRS value. It is clear that more data and more fits are necessary.

The $\Sigma(2020)$ also has a small coupling to this channel, as seen from Table V. Muller included in his analysis all the Y^* resonances with $J \geqslant 5/2$ listed in the Table of Particles of the Particle Data Group,[5] and found that, except for $\Sigma(2020)$, all the amplitudes for these resonances are less than 0.01 [$\Sigma(1915)$ has an amplitude $t = 0.008$].

7. CONCLUSIONS

The main conclusion to be drawn from this review is that the uniqueness of a fit, when so many partial waves are involved, is quite in doubt. In the $\Lambda\pi$ channel, with five analyses performed, only three out of eight partial-wave amplitudes are in agreement, as discussed in Sections 4 and 5. At lower energy ($P_{K^-} < 1.1$ GeV/c) the problem seems to be less serious. The analyses by Armenteros et al.[11] and Conforto et al.[25] for the $\bar{K}N$ channel differ only in that the $\Lambda(1870)$ resonance appears as a P_{03} resonance in the latter and an F_{07} in the former; the rest of the partial waves are in agreement.

All the analyses discussed here were energy-dependent analyses that used similar types of background (Eqs. 4 and 5 differ only in that the first one tends to find straight lines, the second one prefers curved lines). How many solutions will be found when an energy-independent analysis is attempted is for anybody to guess.

My conclusions are the following:

(a) We should not accept a resonance claimed in only one channel, unless the resonant amplitude is large (for example, $|t| > 0.2$) and the partial wave involved is background-free.

(b) A resonance which is suggested by more than one channel, but is always accompanied by a large background so that its Argand plot does not look like a resonance at all, is very suspicious. We should wait and accept it only when it looks like a circle or part of a circle in at least one channel.

Following these rules, let us now go through the 11 states discussed in this paper. Table VI is a summary of the results discussed earlier.

$F_{07}(1870)$. The status of this resonance is probably discussed in Plane's review. It is included here because, if its J^P assignment is $7/2^+$, the partial wave analyses done above 1.1 GeV/c should include it along with the other $J = 7/2$ states. This state is not required by the $\Sigma\pi$ or ΞK analyses, so there is no way at present to confirm it. The fact that its width is narrow makes it difficult with

Table VI. Summary of the results obtained for the 11 states discussed in this paper. M and Γ are expressed in MeV. For the $\overline{K}N$, x, the elasticity, is listed. For the other channels the quantity listed is t, the amplitude at resonance (Eq. 6). The last column is this author's rating for the resonance.

		M	Γ	$\overline{K}N$ x	$\Lambda\pi$ t	$\Sigma\pi$ t	ΞK t	Status
Λ(1870)	F_{07}?	1870	40	.10				Poor
Λ(2015)	F_{07}	2020	160			-.15		Wait
Λ(2040)	D_{03}	2010	130			-.20	(?)[a]	Wait
Λ(2100)	G_{07}	2100	60-145	.29		.06	seen[b]	Good
Σ(1900)	P_{11}	~1900	200		-.11			Poor
Σ(1915)	F_{15}	1905	60	.10	-.09	-.09	.008	Fair
Σ(1940)	D_{13}	1940	200		-.13	-.12	(?)[a]	Wait
Σ(2020)	F_{17}	2020	100-170	.1-.25	.19	-.05	.023	Good
Σ(2070)	P_{13}	~2100	87-250		-.12			Poor
Σ(2120)	G_{17}	~2100	~100		-.07	.13		Wait
[c]Σ(2250)	G_{19}?	2040	~160	.05	(-.18)[c]	(.07)[c]		Fair

a. The analysis of Burgun et al.[22] suggests a J = 3/2 state with either I -spin in this mass region.
b. The two analysis of the ΞK channel[22, 23] do not agree on this value, see Table V.
c. This state is really outside the region where the analysis have been done. Here it has been assumed to be a G_{19} state, as suggested by the analysis of Dauber et al.[20]

the present data to draw any conclusions. As mentioned earlier, the analysis by Conforto et al., who used the same data as CHS, has yielded a different J^P assignment for this effect ($J^P = 3/2^+$) with parameters M = 1873 ± 10 MeV, Γ = 70 ± 20 MeV, x = 0.21 ± 0.03. So the situation is more confused than ever.

F_{07} (2015). The analysis of the $\Sigma\pi$ channel[15] requires this amplitude to explain the large difference between the $\Sigma^+\pi^-$ and $\Sigma^-\pi^+$ data. Figure 13 shows the A_0 and A_5 coefficients, where this difference is evident. In fact the interference of two F_7 resonances and two G_7 resonances can easily explain the data. In view of the fact that it is seen in only one channel we should probably wait for confirmation.

D_{03} (2040). This state also is seen in only one channel; the $\Sigma\pi$ data of Ref. 15 suggest this state, but a fit to the data with linear background instead of a resonance is still acceptable, although worse.[15] It is difficult to detect a 3/2 state in a region where already four 7/2 states are present. Again we should wait for confirmation from another channel.

G_{07}(2100). An I-spin zero bump at this mass is clearly seen in total cross section. The G_{07} assignment is not challenged yet; however, the width from the total cross section data seems to be larger than from partial-wave analyses, supporting the possibility that the F_{07} and D_{03} just discussed may contribute to this bump.

P_{11} (1900). As this is seen only in the $\Lambda\pi$ channel, as discussed in Section 4, the evidence is not compelling, as of now.

F_{15} (1915). For the first time since it was suggested in 1966 by the total cross section data, this state appears now to be on somewhat stable grounds. As mentioned in Section 2 it appeared in the I = 1 total cross section only after the unfolding of the K^-p and K^-d data. The $\bar{K}^0n$ cross sections of Bricman et al.[10] (Section 2) indicate its presence, so do the $\Lambda\pi$ and $\Sigma\pi$ analyses. It is not clear which bump has been seen in the production experiment by Barnes et al.,[9] since no spin analysis is available as yet. However, their quoted width (Γ = 100 ± 30 MeV) seems to exclude the other two candidates at this mass, P_{11} and D_{13}. One more reassuring fact about Σ(1915) is that now it seems to fit well in the SU(3) $5/2^+$ octet. As reported at this conference, Plane et al.,[26] using values and signs of t as given in Table VI, obtain a good fit for the $5/2^+$ octet, therefore the "flip-flop" situation described by Levi-Setti at the Lund Conference[4] seems now to be cured.

D_{13} (1940). This state has been reported in two channels, so one has more confidence than for the other states classified as "poor" or "wait" in Table VI. The situation of the D_{13} amplitude in the $\Lambda\pi$ channel, as discussed in Section 4, is not very reassuring. In addition it should be pointed out that, as shown in Fig. 2, this energy region is at the lower edge of the experiments discussed here and at the upper edge of the CHS data.[14,21] This condition may be acceptable for studying the higher-spin F_{15}(1915), but it makes studying the underlying structure difficult. My suggestion is to wait rather than to accept this state.

F_{17} (2020). There seems to be no problem in accepting this resonance, although the coupling to the $\Sigma\pi$ channel is somewhat smaller than the SU(3) decuplet assignment would require.[4,26,27]

P_{13}(2070). Seen only in $\Lambda\pi$, where the various analyses do not agree (see Section 4), this needs confirmation from some other channel.

G_{17}(2120). The reason for waiting before accepting this state is that it is at the upper edge of the experiments done so far, therefore the data are scarce. From the analyses of $\Lambda\pi$ and $\Sigma\pi$ done by this author, it appears that a resonance fits the data quite well; however, caution is seldom regretted.

Σ(2250). The J^P assignment ($9/2^-$) is only tentative, and the values of t for $\Sigma\pi$ and $\Lambda\pi$ are to be considered tentative also, because the Σ(2250) is almost completely outside the energy range used in the fits. The main evidence comes from the total cross section data.

In conclusion, the status of the analyses above 1.1 GeV/c does not look very good at the moment. Other experimental data and other analyses will shortly appear on the scene[28] and undoubtedly will help in understanding the situation. In addition to the two-body reactions just discussed, the quasi-two-body such as $\Sigma(1385)\pi$, $\Sigma(1660)\pi$, or $\Lambda(1820)\pi$ should help considerably.

The next question that may arise is: Assuming that more than the four states classified as good or fair are really present in these analyses, it there room for them in the current baryon spectroscopy? Meshkov[29] (in his paper presented at this Conference) discusses baryon spectroscopy using the quark model. There is certainly room for new D_{03} and D_{13} states as well as for new F_{07} and G_{17} states, if we believe that for each N^* found a Λ and a Σ state are needed to form an octet, and for each Δ found one has to find a Σ resonance to complete the decuplet. By an inspection of the Baryon Table of the Particle Data Group one can easily see that each of the 11 states, except F_{07}(1870), has one or more possible companions to form an SU(3) multiplet. However, as an experimentalist, I think we should apply ourselves to understanding what the data tell us before getting into the classification of every possible curve we may detect in an Argand diagram.

FOOTNOTE AND REFERENCES

*Work done under the auspices of the U. S. Atomic Energy Commission.

1. G. Lynch, Unfolding of Deuterium Data, presented at this conference.
2. D. Plane, Review of Partical Wave Analyses from 450 to 1200 MeV/c Incident K^- Momentum, presented at this conference.
3. For more details on this point or in general on partial-wave analysis see, for example, the review article "Baryon Resonances" by A. Barbaro-Galtieri, in Advances in Particle Physics, ed. by Cool and Marshak (Interscience Publishers, New York, 1968), Vol. II.
4. R. Levi Setti, in Proceedings of the Lund International Conference on Elementary Particles, Lund, Sweden, June 1969 (Berlinska Boktrycheriet, Lund, Sweden, 1969), p. 385.
5. Particle Data Group, Review of Particle Properties, Rev. Mod. Phys. 42, 87 (1970).
6. C. Daum, F. C. Erné, J. P. Lagnaux, J. C. Sens, M. Steuer, and F. Udo, Nucl. Phys. B7, 19 (1968).
7. C. G. Wohl, F. T. Solmitz, and M. L. Stevenson, Phys. Rev. Letters 17, 107 (1966).
8. R. L. Cool, G. Giacomelli, T. F. Kycia, B. A. Leontic, K. K. Li, A. Lundby, and J. Teiger, Phys. Rev. Letters 17, 102 (1966); R. J. Abrams et al., Phys. Rev. Letters 19, 259 (1967); D. V. Bugg et al., Phys. Rev. 168, 1466 (1968).
9. V. E. Barnes et al., Phys. Rev. Letters 22, 479 (1969), see an enhancement at this mass in the $\Sigma\pi$ channel. No information on the quantum numbers of this state is available. It is not clear if it is the F_{15} state or any of the other ones listed in Table I.
10. C. Bricman, M. Ferro-Luzzi, J. M. Perreau, G. Bizard, Y. Declais, J. Duchon, J. Seguinot, Phys. Letters 31B, 148 (1970).
11. R. Armenteros et al., Nucl. Phys. B8, 195 (1968) and B8, 223 (1968).
12. W. S. Smart, Phys. Rev. 169, 1330 (1968); and W. S. Smart, A. Kernan, G. E. Kalmus, and R. P. Ely, Phys. Rev. Letters 17, 556 (1966).
13. A. Berthon et al., College de France-Rutherford Laboratory-Saclay collaboration, Nucl. Phys. (to be published).
14. R. Armenteros et al., CERN-Heidelberg-Saclay collaboration, Nucl. Phys. B8, 183 (1968).
15. A. Barbaro-Galtieri, Abstract No. 90 in Proceedings of the Lund International Conference on Elementary Particles, Lund, Sweden, June 1969 (Berlinska Boktryckeriet, Lund, Sweden, 1969).
16. Cox et al., Birmingham-Edinburgh-Glasgow-Imperial College collaboration, Nucl. Phys. (to be published).

17. P. J. Litchfield, Partial Wave Analysis of $\bar{K}N \rightarrow \Lambda\pi$ between 1.0 and 1.85 GeV/c, Rutherford Laboratory Preprint RPP/H/66(1970).
18. I. Butterworth, reported at this conference.
19. Only part of the coefficients with their errors were available to me in table form (up to .793 GeV/c). In the region .825 to 1.175 GeV/c the A_i and B_i were read off the graphs of Ref. 14 as the smooth curve that CHS have actually used in their fit. These values appear on Figs. 10 and 11 of this paper.
20. P. M. Dauber et al., Phys. Letters 23, 154 (1966); only $\Sigma\pi$ data discussed. Also, P. M. Dauber, Ph. D. Thesis, UCLA (1966), which discusses both $\Lambda\pi$ and $\Sigma\pi$ data.
21. R. Armenteros et al., CHS collaboration, Nucl. Phys. B8, 233 (1968); R. Armenteros et al., The $\bar{K}N$ Interaction Between 458 and 793 MeV/c: Cross Sections; CERN TC Report D. Ph. II/Phys., to be published (1970). I thank Dr. D. Plane for making these data available to me prior to publication.
22. G. Burgun et al., CRS collaboration, Nucl. Physics B8, 447 (1968).
23. R. A. Muller, Ph. D. Thesis, Lawrence Radiation Laboratory Report UCRL-19372 (1969).
24. W. P. Trower, Ph. D. Thesis, University of Illinois Report COO-1195-54 (1966).
25. B. Conforto et al., Chicago-Heidelberg Collaboration, reported by Levi-Setti, reference 4.
26. D. Plane et al., A Fit of Baryon Decay Rates within SU(3); CERN TC Report D. Ph. II/Phys. 70-9, to be published in Nucl. Physics (1970).
27. R. D. Tripp, Proceedings of the International Conference on High Energy Physics, Vienna, Austria, August 1968 (CERN Scientific Information Service), p. 173.
28. Armenteros et al., at CERN have new data in this energy region; also R. Birge et al. at the Lawrence Radiation Laboratory, Berkeley.
29. S. Meshkov, Comments on Baryon Spectroscopy, presented at this conference.

REACTION $K^-p \rightarrow \Lambda\pi^0$ IN THE E* RANGE 1915 TO 2168 MEV

COLLEGE DE FRANCE - RUTHERFORD LABORATORY -

SACLAY COLLABORATION

The reaction $K^-p \rightarrow \Lambda\pi^0$ was studied in a total of 1.5×10^6 pictures taken at Nimrod in the Saclay 180ℓ bubble chamber at 13 momenta spaced at 50 MeV/c intervals between 1.26 and 1.84 GeV/c ($1915 < E^* < 2168$ MeV). Separation from $\Sigma^0\pi^0$ and $\Lambda\pi^0\pi^0$ used the missing mass squared to the Λ(MMS). If MMS was compatible with $m^2_{\pi^0}$ within 1 S.D. the event was retained; if MMS was higher than $m^2_{\pi^0}$ by more than 1 S.D. it was discarded (since real $\Lambda\pi^0$ events are badly contaminated for this sample); if MMS was less than $m^2_{\pi^0}$ by more than 1 S.D. the event was counted twice. This technique left the overall contamination less than 4%. The final unweighted number of $\Lambda\pi^0$ events is 4805.

Figure 1 shows the cross-section, Figures 2 and 3 the A_ℓ and B_ℓ coefficients for Legendre expansions of $\frac{d\sigma}{d\Omega}$ and $\vec{P}\frac{d\sigma}{d\Omega}$. Coefficients above A_7 and B_7 were compatible with zero at all energies.

P.W.A. was carried out. B.W's. were used for the established $\Sigma(2030)$ and $\Sigma(1765)$. Added to these resonant waves and used in other waves was $F^{J,P} = \sum_{n=0}^{N^{J,P}} f_n^{J,P} P_n(p')$ where $F^{J,P}$ is the complex amplitude of given J^P, P_n is a Legendre polynomial expressed as a function of beam momentum normalized to run between -1 at 10 MeV below the lowest momentum and +1 at 10 MeV above the highest. The complex $f_n^{J,P}$ constants are fitted for. The expansion is limited to those $N^{J,P}$ needed to give a good data fit. No assumption is made as to the behaviour of the amplitude, the expansion for each P.W. amplitude being increased until χ^2 indicates the data is adequately described. The procedure converges rapidly for non-resonant amplitudes and new resonances will hopefully be indicated provided their width is not too small.

Results (a) Σ(2030). The parity of Σ(2030) is confirmed $+^{ve}$ (G7 rather than F7 increases χ^2 by 115). Either sign of $\sqrt{xx'}$ fits the data - however if the sign were $+^{ve}$ relative to Σ(1765) this would imply a large D5 background in conflict with CHS results[1] that the D5 background is small throughout the 1765 region. Hence we conclude Σ(2030) and Σ(1765) have amplitudes of opposite sign.

$$M = 2030 \pm 10 \text{ MeV} \quad \Gamma = 165 \pm {}^{30}_{15} \text{ MeV} \quad |\sqrt{xx'}| = 0.2 \pm 0.02$$

(b) Other waves. Figure 4 shows the obtained amplitudes. Fit A is the simplest giving a good representation of the data $N^{J,P} = 3$ for F5 and P3 and $N^{JP} = 1$ for all others. (This has some probability as N = 3 for all waves for 20 fewer parameters). Fit B uses a completely different method of analysis in which data is divided into partially overlapping groups of adjacent energies. In one interval it is assumed that all amplitudes are representable by a complex constant (Σ(2030) put in as B.W.). The good agreement between the two methods is encouraging.

The fast movement of F5 at the lower energies confirms that Σ(1915) is in the F5 wave. Its sign is opposite to that of Σ(2030). Our data doesn't permit determination of parameters but adding the higher CHS points we get M = 1910 ± 20 MeV $\Gamma = 60 \pm 20$ MeV $|\sqrt{xx'}| = 0.1 \pm 0.02$.

The structure in P3 bears similarity to a resonant loop. If assumed due to a resonance $M \simeq 2060$ MeV, $\Gamma \simeq 180$ MeV.

In $K^- p \to \Sigma\pi$ Barbaro-Galtieri claimed I = 1 resonances in this region in the D3 and G7 waves[2] and Levi-Setti[3] raised the question as to whether Σ(1910) is due to D3. It is seen that in the $\Lambda\pi$ channel there is certainly an F5 Σ(1910) and with this data alone D3 is adequately parametrized by a linear background. The G7 amplitude is small even though A7 and B7 are very sensitive to G7 - F7 interference.

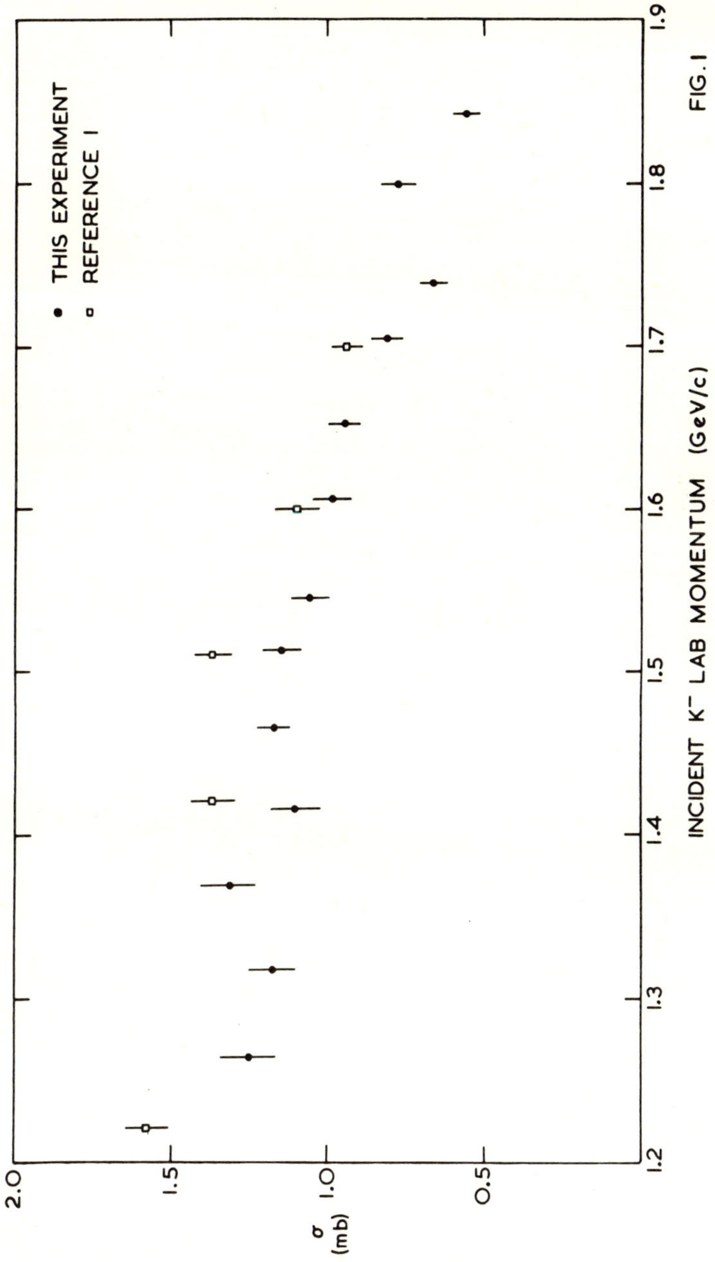
THIS EXPERIMENT
REFERENCE 1
σ (mb)
2.0
1.5
1.0
0.5
1.2
1.3
1.4
1.5
1.6
1.7
1.8
1.9
INCIDENT K⁻ LAB MOMENTUM (GeV/c)

FIG. 1

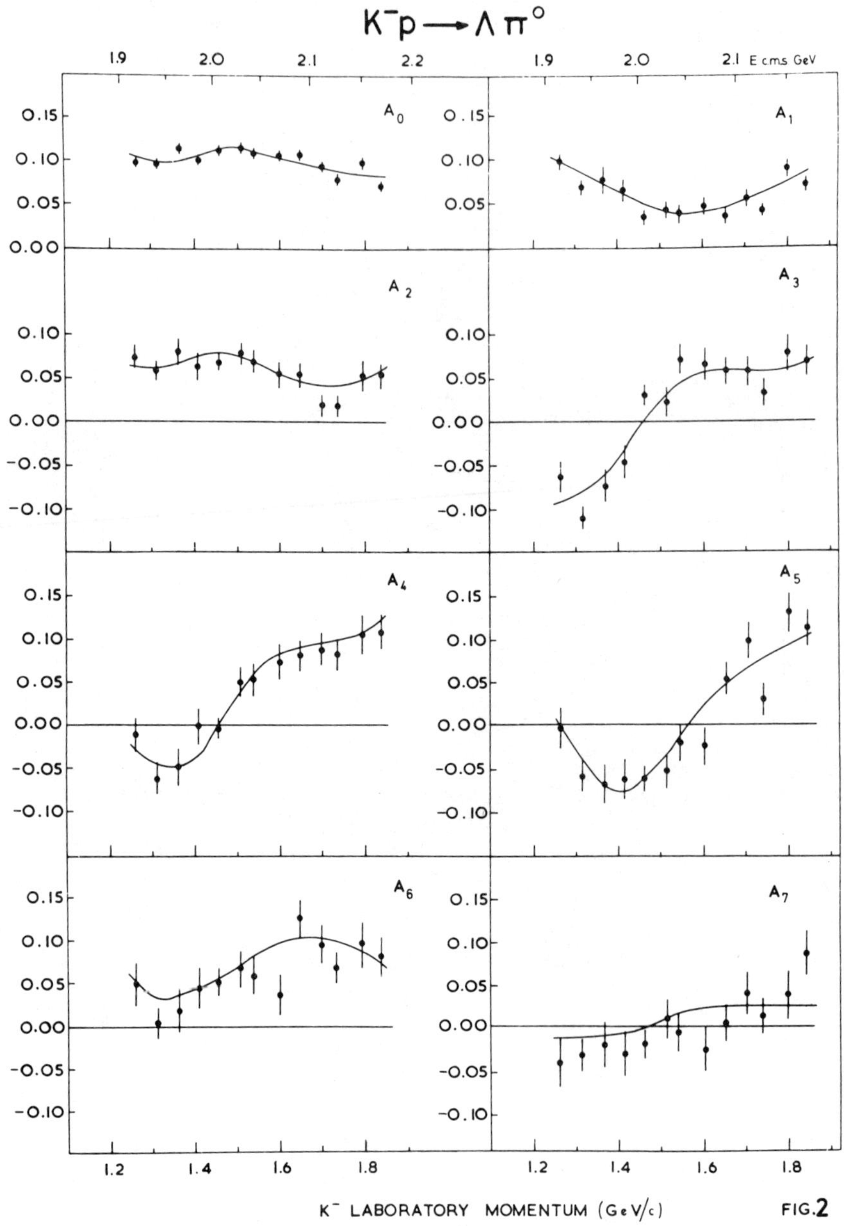
K⁻p ⟶ Λπ⁰
E cms GeV
A₀
A₁
A₂
A₃
A₄
A₅
A₆
A₇
K⁻ LABORATORY MOMENTUM (GeV/c)

FIG. 2

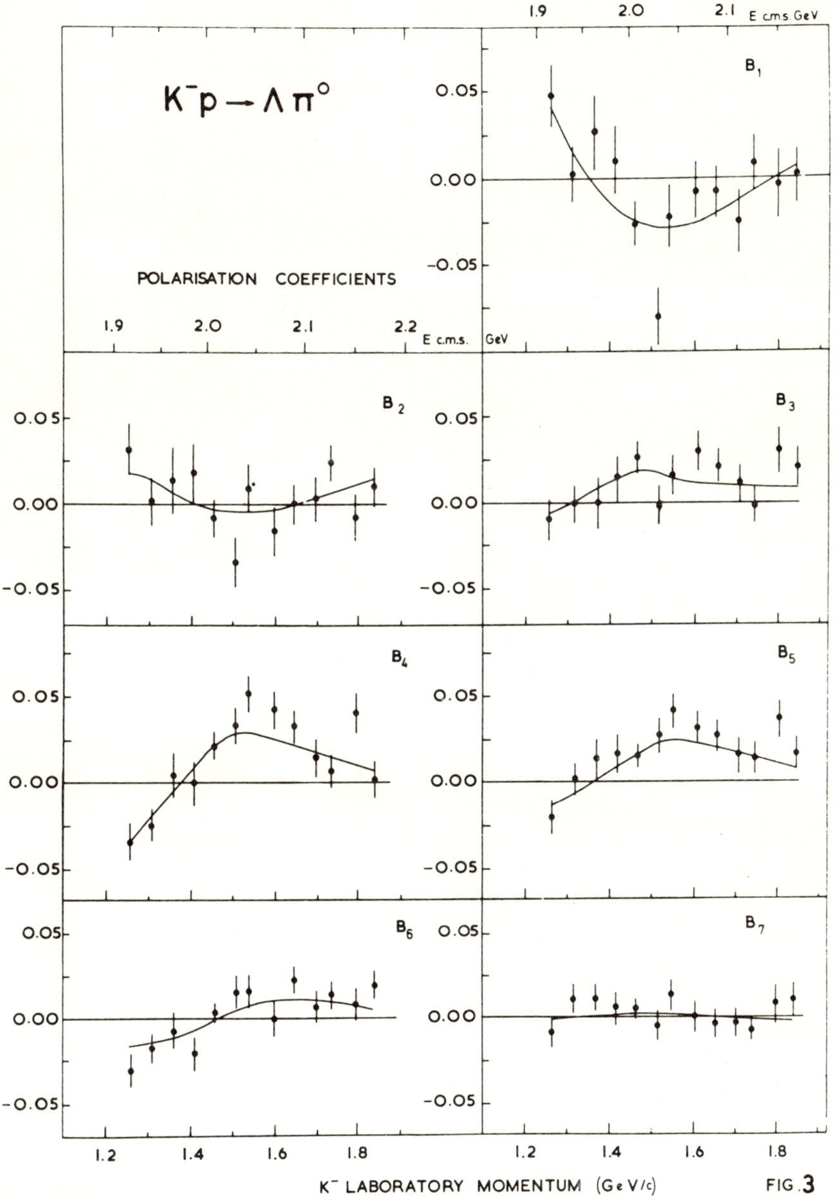

K⁻p → Λπ°
POLARISATION COEFFICIENTS
1.9
2.0
2.1
2.2
E c.m.s. GeV
E c.m.s. GeV
B₁
B₂
B₃
B₄
B₅
B₆
B₇
0.05
0.00
-0.05
1.2
1.4
1.6
1.8
K⁻ LABORATORY MOMENTUM (GeV/c)

FIG. 3

References

1. R. Armenteros, P. Baillon, C. Bricman, M. Ferro-Luzzi, H.K. Nguyen, V. Pelosi, D.E. Plane, N. Schmitz, E. Burkhardt, H. Filthuth, E. Kluge, H. Oberlack, R.R. Ross, R. Barloutaud, P. Granet, J. Meyer, J.C. Narjoux, F. Pierre, J.P. Porte and J. Prevost, Nuclear Physics, B8 (1968) 183.

2. A. Barbaro-Galtieri, Paper 90, Lund International Conference on Elementary Particles, Lund, Sweden, June 1969.

3. R. Levi Setti, Proceedings of Lund International Conference on Elementary Particles, Lund, Sweden, June 1969.

REACTIONS $K^-p \rightarrow \Sigma^{\pm}\pi^{\mp}$ IN THE E^* RANGE 1915 TO 2168 MEV

COLLEGE DE FRANCE - RUTHERFORD LABORATORY - SACLAY

COLLABORATION

The reactions $K^-p \rightarrow \Sigma^{\pm}\pi^{\mp}$ were studied in a total of 1.5 x 10^6 pictures taken at Nimrod in the Saclay 180 HBC at 13 momenta spaced evently between 1.26 and 1.84 GeV/c. Systematic bias due to bad measurement of short decay tracks or poor scan efficiency on small angle decays was corrected by eliminating events with Σ decay leading to π's and protons of momenta below 45 MeV/c and 140 MeV/c respectively, or decay angles smaller than 9°, and then weighting appropriately the remaining events. To avoid weights greater than ~ 3, $\Sigma^+ \rightarrow p\pi^0$ decays with $p_\Sigma >$ 1.35 GeV/c were rejected and replaced by $\Sigma^+ \rightarrow n\pi^+$ decays using the ratio $p\pi^0/n\pi^+ = 1.05$. (This couldn't be done to find polarizations which must use the $p\pi^0$ decay and the whole sample of such decays was used without correction for scanning bias. However the asymmetry should not be affected by this bias and of course the loss of event numbers due to bias was allowed for in computing $\vec{P}\frac{d\sigma}{d\Omega}$).

Angular distributions are very different for the two reactions. Whereas $\Sigma^-\pi^+$ events are approximately symmetric, the $\Sigma^+\pi^-$ events show sharp forward π^- peaking. However this peaking is maximal at intermediate energies and is less important above 1.7 GeV/c. At 1.6 GeV/c a fit to $\frac{d\sigma}{dt} \propto e^{-bt}$ gives $b \sim 10$ GeV^{-2} whereas at high energies the slope is only ~ 5. We thus feel it more convenient to discuss the reactions from the s-channel viewpoint when the difference between $\Sigma^+\pi^-$ and $\Sigma^-\pi^+$ is accounted for by interference between amplitudes of different i-spin adding constructively in $\Sigma^+\pi^-$ and destructively in $\Sigma^-\pi^+$.

A P.W.A. analysis has been carried out for $J \leqslant 7/2$. Since 16 amplitudes are involved only linear background terms a + bp were used where a and b are complex. In the first fit B.W's. were inserted for Σ(2030), Λ(2100) and the tails of Λ(1815) and Λ(1827). All remaining amplitudes were assumed non-resonant. The relative sign of F17 and G07 amplitudes was found to be negative. Next the Σ(1915) was sought. Since this lies at the lower energy limit, CHS points[1] at 1.134, 1.153, 1.173 and 1.183 GeV/c were included. A B.W. with free parameters added to the linear F15 background improves χ^2 by ~90. The sign of the Σ(1915) amplitude relative to that of Σ(2030) and Σ(1815) is positive. (This experiment is not sensitive to the parity of Σ(1915). A D15 assignment gives the same probability since the evidence for the Σ(1915) arises from interference with the lower 5/2 resonance but this can be either the Λ(1815) or Λ(1830)). This fit has $\chi^2/N_D = 468/326$, and though it explains most of the data is not a particularly good fit.

Barbaro-Galtieri[2] has reported a P.W.A. of this channel which claims new F_{07} G_{17} D_{03} and D_{13} states in this region. Inclusion of these new 7/2 resonances at published mass and width but re-optimized branchings produces a fit in which the well established F_{17} and G_{07} resonances go to essentially zero amplitude. The solution therefore seems implausible.

Introduction of new resonances certainly improves the fit and a number of alternative solutions are obtained. Thus for example one solution, which fits A_0 particularly well, has new F_{05} and F_{15} states. The parameters for this fit are given below. The important feature is that the Λ(2100), Σ(2030) and Σ(1915) parameters are essentially stable against inclusion of these new states.

Table

Resonance	J^P	M (MeV)	(MeV)	$\sqrt{xx'}$
Σ (1915)	$5/2^+$	1900 ± 15	75 ± 20	-0.13 ± 0.03
Σ (2030)	$7/2^+$	2035 ± 10	150 ± 20	-0.09 ± 0.02
Λ (2100)	$7/2^-$	2130 ± 20	140 ± 15	+0.16 ± 0.02
Σ(2070)	$5/2^+$	2070 ± 10	140 ± 20	+0.12 ± 0.02
Λ(2110)	$5/2^+$	2110 ± 20	185 ± 30	+0.17 ± 0.03

This fit has $\chi^2/N_D = 400/320$

References

(1) Proceedings of the Lund Conference

(2) Proceedings of the Lund Conference

DISCUSSION

Kalmus: This is a very large energy region to fit with a linear background. It seems to me that since you seem to be worried about the χ^2's, if you fit it with a background like Smart's for instance, you might do considerably better.

Butterworth: I agree with the comment and I think it is supported by comparing our $\Lambda\pi$ with our $\Sigma\pi$. The $\Lambda\pi$ has got this free parameterization which, if you like, is carrying many terms in the background, and gives very good fits. But we haven't at this stage got anything other than linear backgrounds in the $\Sigma\pi$ channel.

Kim: I am using a linear expansion of the inverse K-matrix only for a small energy region at a time and continue out with different expansions and so on. Remember, the linear expansion of the inverse of the K-matrix corresponds to much more than the linear expansion of the amplitude itself since the Breit-Wigner amplitude is actually a linear expansion of the inverse K-matrix.

PARTIAL WAVE ANALYSIS OF $K^-N \longrightarrow \Lambda\pi$ BETWEEN 1 AND 1.85 GEV/C

P.J. LITCHFIELD

RUTHERFORD HIGH ENERGY LABORATORY

A partial wave analysis is carried out on all published data on the reactions $K^-p \longrightarrow \Lambda\pi^0$ and $K^-n \longrightarrow \Lambda\pi^-$ between 1.0 and 1.85 GeV/c(1-8). The method uses the "free parametrization" of partial waves without prior assumption as to their nature as described in the study of $K^-p \longrightarrow \Lambda\pi^0$ by the College de France - Rutherford Laboratory - Saclay Collaboration.
For each partial wave amplitude a Legendre expansion was made for the real and imaginary parts and alternatively for η and δ (η was allowed to go negative so the amplitude could pass through the centre of the Argand diagram if necessary). It will be seen that though the low order structure is quite different for these two forms of parametrization, the results obtained were in each case essentially the same. The dominant F7 $\Sigma(2030)$ was explicitly included as a free parameter B.W. The D5 $\Sigma(1765)$ was included with its accepted parameters with the condition that it must dominate the D5 wave below 1.2 GeV/c. The analysis was repeated several times in its entirity using both methods of parametrization. No other solution of comparable probability was found indicating the uniqueness of the solution.

Results. The coefficients of fits to $\frac{d\sigma}{d\Omega}$ and $\vec{P}\frac{d\sigma}{d\Omega}$ are shown in Figures 1 and 2 and the partial wave amplitudes in Figures 3 and 4. The solid lines correspond to expansions in the real and imaginary parts, the dashed to expansions in η and δ. The fits agree closely. Amplitudes are shown from 1.05 to 1.8 GeV/c, although fitting was from 1.0 to 1.85 GeV/c. This is done to avoid difficulties at the ends of the momentum range due to the poorness of the constraints on the amplitudes at these points.

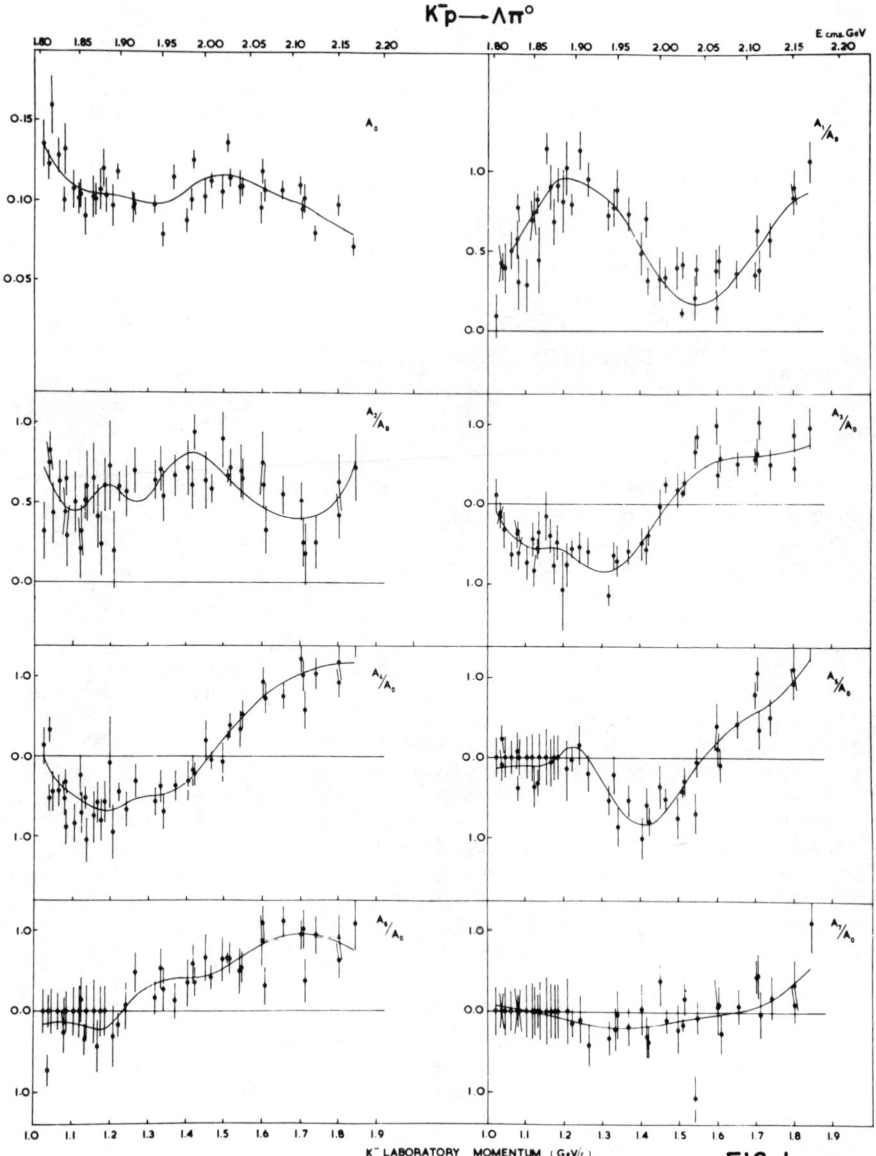

$K^-p \longrightarrow \Lambda\pi^0$
E cms GeV
A_0
A_1/A_0
A_2/A_0
A_3/A_0
A_4/A_0
A_5/A_0
A_6/A_0
A_7/A_0
K^- LABORATORY MOMENTUM (GeV/c)

FIG. 1

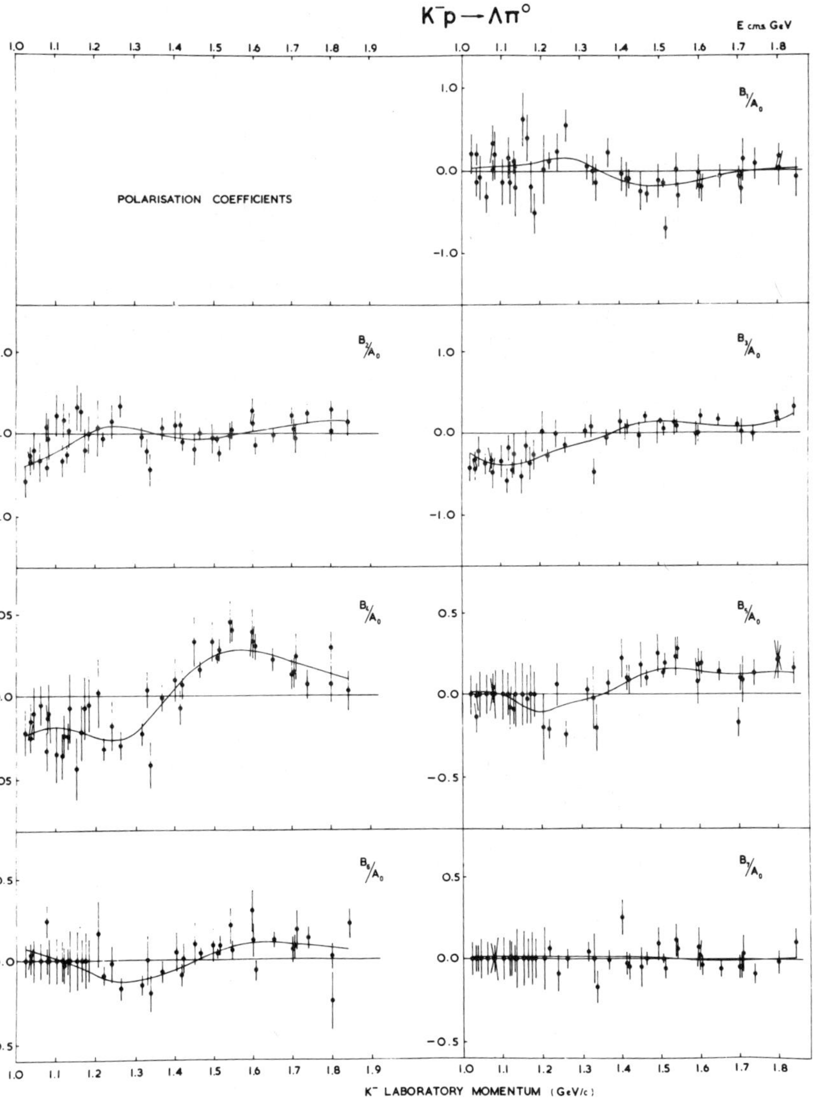
$K^-p \to \Lambda\pi^0$
E cms GeV
POLARISATION COEFFICIENTS
B_1/A_0
B_2/A_0
B_3/A_0
B_4/A_0
B_5/A_0
B_6/A_0
B_7/A_0
K⁻ LABORATORY MOMENTUM (GeV/c)

FIG. 2

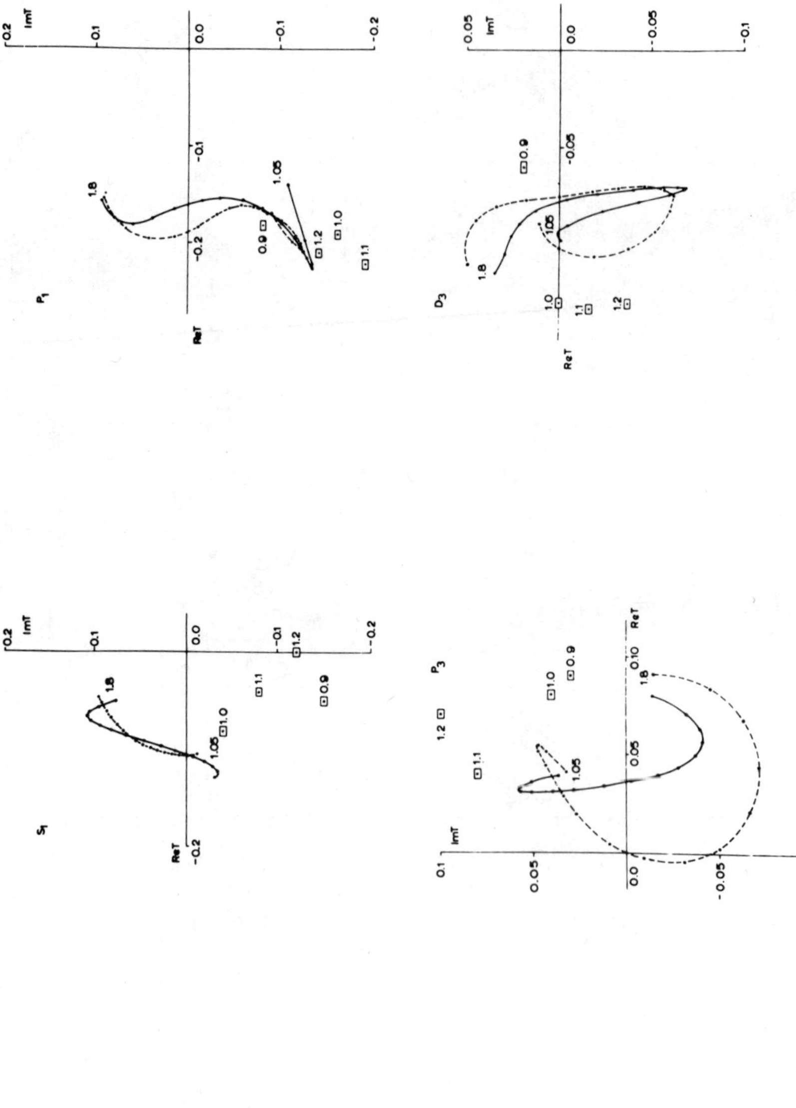

FIG. 3

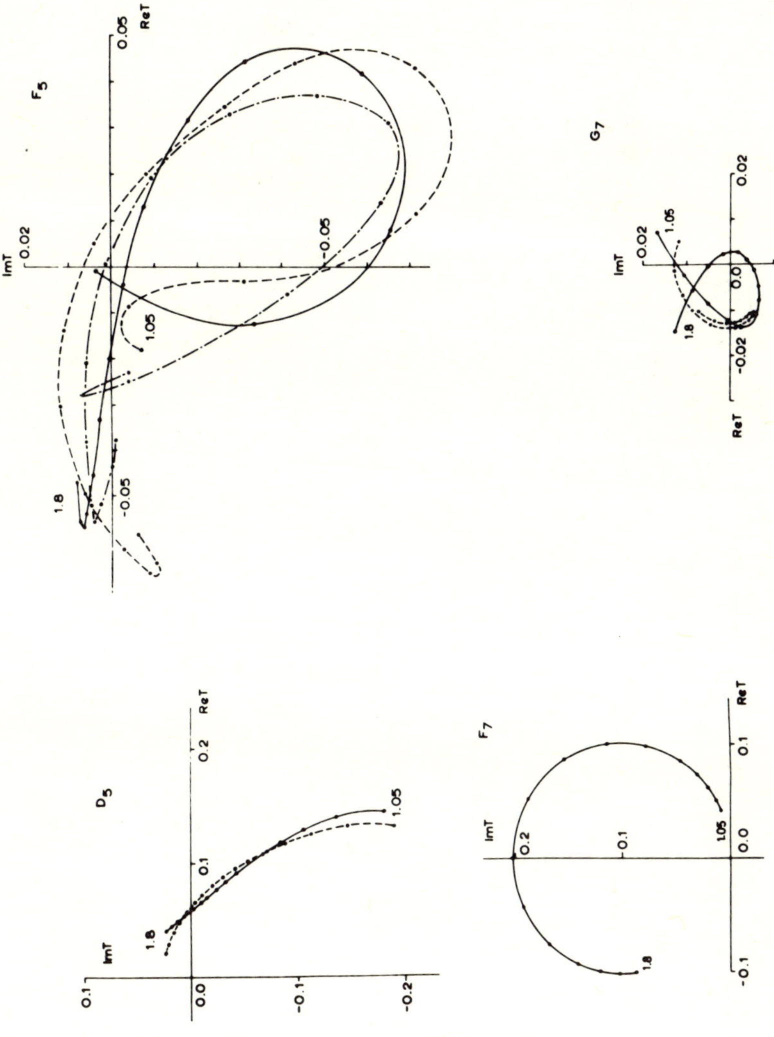
F5
ImT
ReT
0.02
0.05
-0.05
1.05
1.8
G7
1.05
1.8
0.0
-0.02
D5
0.1
0.0
-0.1
-0.2
0.2
1.05
1.8
F7
0.1
-0.1
1.05
1.8

FIG. 4

F7. No background was necessary in addition to $\Sigma(2030)$. $M = 2022 \pm 4$ MeV, $\Gamma = 170 \pm 15$ MeV, $\sqrt{xx'} = +0.200 \pm 0.008$ (Levi-Setti sign convention). It should be noted that all partial wave analyses give larger Γ values than the value of 120 MeV given in the compilation by the Particle Data Group which is dominated by measurements of σ_T.

F5. For this wave the figure shows the free fit to Re f and Im f (dot - dashed) and explicit inclusion of $\Sigma(1915)$ as a B.W. together with a free background parametrization (the other two curves). All three curves agree. Inclusion of the B.W. drops χ^2 by 69. $M = 1895 \pm 10$ MeV, $\Gamma = 70 \pm 15$ MeV, $\sqrt{xx'} = -0.070 \pm 0.015$.

G7. The data requires a small G7 at the higher momenta which could possibly be associated with the G7 $\Sigma\pi$ resonance reported by A. Barbaro-Galtieri(10), but if so $|\sqrt{xx'}| < 0.02$.

D5. No new states.

D3. This wave shows a strong reversal of direction. Replacing the amplitude by a linear form increases χ^2 by 72 for 6 parameters. If interpreted as a resonance, $M = 1940 \pm 30$, $\Gamma = 280 \pm 40$, $\sqrt{xx'} = -0.14 \pm 0.03$ (A. Barbaro-Galtieri(10) reports a D_{13} $\Sigma\pi$ state with $M = 1920$, $\Gamma \sim 160$).

P3. The CRS collaboration see structure in the P3 wave as do the BEGI collaboration(4). Addition of the world's data confirms the effect (replacement by linear form raises χ^2 by 30 for 3 parameters). $M = 2070 \pm 30$, $\Gamma = 250 \pm 40$, $\sqrt{xx'} = -0.09 \pm 0.03$.

P1. The replacement of the structure in this amplitude by a linear form raises the χ^2 by 45 for the removal of 5 parameters. The fit can be regained by a resonance with $M = 1920$ MeV, $\Gamma = 170$ MeV, $\sqrt{xx'} = -0.14$. Smart(9) reported such a resonance which was not seen in other partial wave analyses. Omission of Smart's data from the compilation, however, leaves the effect present.

S1. This amplitude is smooth with no evidence of resonance. All of the new resonances have $\sqrt{xx'}$ opposite to $\Sigma(2030)$ and are thus octet members, all can be correlated with N* states obtained in πN P.W.A.

References

1. R. Armenteros et al, Nucl. Phys., B8, 233 (1968).

2. CERN-SACLAY collaboration. Nucl. Phys. to be published.

3. W.M. Smart et al, Phys. Rev. Letters 17, 556. (1966).

4. G.F. Cox et al., Nucl. Phys. to be published.

5. C.G. Wohl, et al., Phys. Rev. letters 17, 107 (1966).

6. A. Berthon et al., Nucl. Phys. to be published.

7. W.P. Trower, Ph.D. Thesis, University of Illinois Report COO-1195-54 (1966) unpublished.

8. P.M. Dauber, Ph.D. Thesis, University of California, Los Angeles (1966) unpublished.

9. W.M. Smart, Phys. Rev. 169, 1330 (1968).

10. A. Barbaro-Galtieri, Paper 90, Lund International Conference on Elementary Particles, Lund, Sweden, June 1969.

K^-p ELASTIC DIFFERENTIAL CROSS SECTION MEASUREMENTS BETWEEN 1.7 AND 2.5 GeV/c

P.C. Barber, T.A. Broome, W. Busza*, B.G. Duff,
D.A. Garbutt**, F.F. Heymann, D.C. Imrie, G.J. Lush,
E.N. Mgbenu***, K.M. Potter, D.M. Ritson****,
L.A. Robbins, S.J. Sharrock, A.D. Smith

University College London

and

R.C. Hanna, R.A. Rosner, E.J. Sacharides

Rutherford High Energy Laboratory.

* Presently at: Massachusetts Institute of Technology
** Presently at: Imperial College London
*** Presently at: Fourah Bay College, Sierra Leone.
**** Permanent address: Stanford Linear Accelerator Center, Stanford, California 94305, USA

In an experiment currently being analysed at the Rutherford High Energy Laboratory, Chilton, England, K^-p elastic differential cross sections have been measured at 13 momenta in the range 1.1 GeV/c to 1.4 GeV/c, and at 19 momenta in the range 1.7 GeV/c to 2.5 GeV/c. The angular distributions in the lower momentum range contain approximately 5000 elastic events at each momentum. Between 1.7 GeV/c and 2.0 GeV/c the presently available data contain about 2000 elastic events per momentum. The latter number represents about half the available data. The data were taken using the same equipment as the K^+p differential cross sections presented at this conference[1)].

The angular distributions have been parametrized by means of Legendre polynomial expansions. Between 1.1 GeV/c and 1.4 GeV/c the expansion coefficients, plotted in the form An/Ao, vary smoothly with momentum. Above 1.7 GeV/c an initial normalization has been performed to total cross section data [2)], using the optical theorem assuming that the real part of the forward scattering amplitude may be neglected. The forward point required for this normalization was obtained from the Legendre expansion for the chosen order of fit. Four representative cross sections are presented in Figure 1, the plotted curves are the Legendre polynomial fits. The angular distributions are similar to previous measurements in this momentum range[3]. They exhibit strong diffraction peaks whose slope on a t scale remains roughly constant at 7 $(GeV/c)^{-2}$ up to a momentum of 2 GeV/c and then rises smoothly to a maximum of 8 $(GeV/c)^{-2}$ at 2.3 GeV/c. The diffraction peaks are followed by a dip at a constant

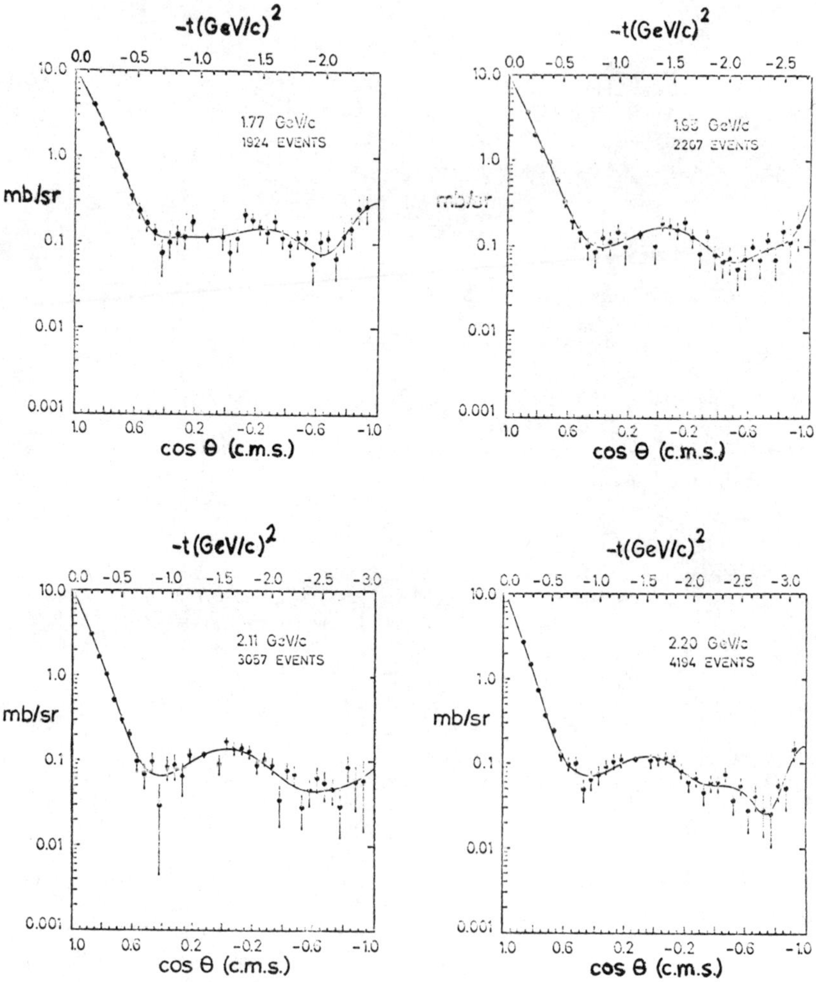

Figure 1. K^-p Elastic Differential Cross Sections.

value of $-t \approx 0.8$ $(GeV/c)^2$ and then a secondary maximum in the region of $-t \approx 1.5$ $(GeV/c)^2$. In the backward region, the dip at a constant value of $-u \approx 0.3$ $(GeV/c)^2$ reported by Daum et al [3] is clearly seen, the depth of this minimum appearing to increase with increasing momentum.

Figure 2 shows the Legendre polynomial expansion coefficients for the data in the momentum range 1.7 GeV/c to 2.5 GeV/c. The higher order coefficients tend to rise steeply with increasing momentum without showing any significant structure. In contrast the total elastic cross section decreases slowly over this momentum region. This behaviour precludes any determination of resonance parameters from an examination of the coefficients alone. A diffractive background plus resonance model is currently being used to fit the differential cross sections.

We would like to acknowledge the important contributions made by the staff of the Rutherford High Energy Laboratory to the success of this experiment. In addition we are particularly grateful to P.R.Pitts, R.H.Watson and D.B.Webb for their technical assistance and substantial aid during data acquisition.

REFERENCES

1. P.C. Barber, T.A. Broome, B.G. Duff, F.F. Heymann, D.C. Imrie, G.J. Lush, B.R. Martin, K.M. Potter, D.M. Ritson, L.A. Robbins, R.A. Rosner, S.J. Sharrock, A.D. Smith, R.C. Hanna, A.T.Lea and E.J. Sacharides.
K^+p Elastic Scattering between 1.4 and 2.3 GeV/c, contribution to this conference.

2. D.V. Bugg, R.S. Gilmore, K.M. Knight, D.C. Salter, G.H. Stafford, E.J.N. Wilson, J.D.Davies, J.D. Dowell, P.M. Hattersley, R.J. Homer, A.W.O'Dell, A.A. Carter, R.J. Tapper and K.F. Riley.
Phys.Rev., 168, 1466 (1968)

3. C. Daum, F.C. Erne, J.P. Lagnaux, J.C. Sens, M. Steuer and F. Udo.
Nucl.Phys., B6, 273 (1968)

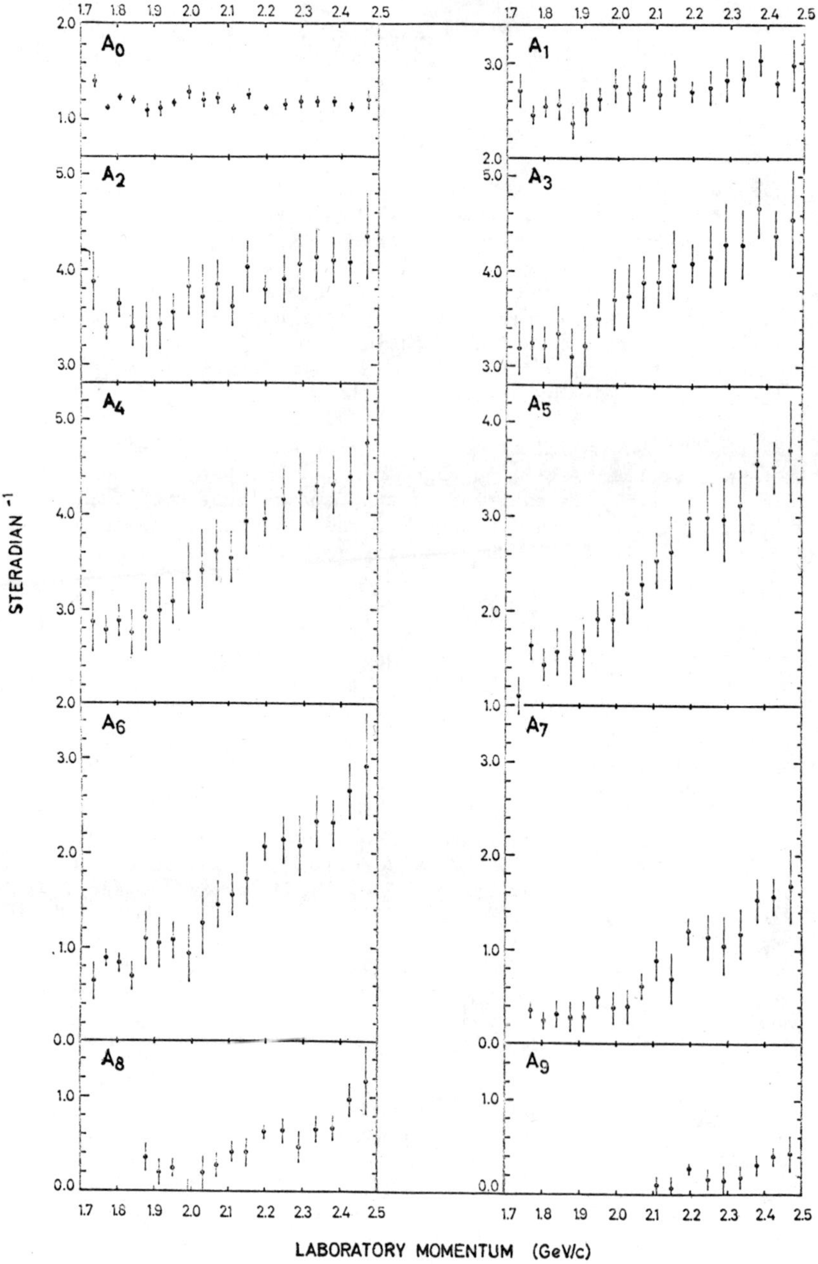

Figure 2. Legendre Polynomial Expansion Coefficients.

SESSION V

Y* PRODUCTION

Discussion Leader: Malcolm Derrick

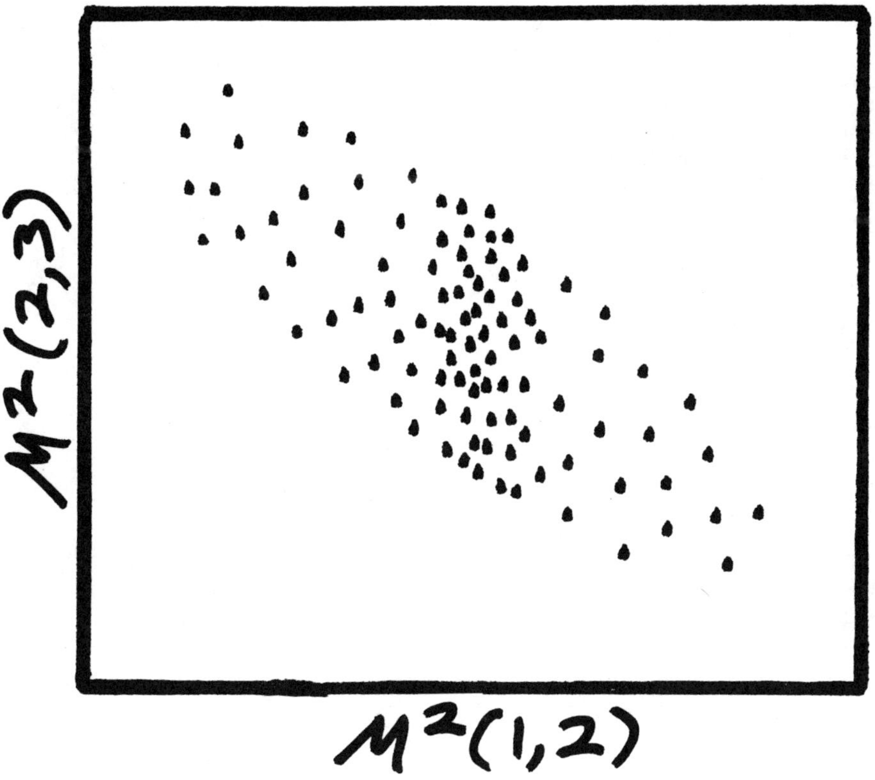

REVIEW OF S = -1 BARYON STATES AS OBSERVED IN PRODUCTION EXPERIMENTS

DAVID H. MILLER

PURDUE UNIVERSITY, WEST LAFAYETTE, INDIANA

This review will cover the published data on S = -1 baryon states seen in production experiments. It will, therefore exclude a review of phase shift analyses of low energy $\bar{K}N$ data and also bumps seen in total cross sections. This material will be covered by other speakers. Excellent reviews of Strange Baryon resonances have been given in the last year by R. Levi-Setti[1] and Robert D. Tripp[2] who, in addition to covering all the experimental data, compare it to the predictions of SU3. In addition the latest Reviews of Modern Physics[3] contains the most recent compilations of the Particle Data Group.

PHASE SHIFT ANALYSES

The phase shift analyses carried out by several groups over the last few years have revealed a large number of Y_0^* and Y_1^* states in the mass region up to and above 2 GeV. These data indicate that analysis of 'bumps' seen in mass spectra in production experiments will be extremely complicated. This is particularly true of Y_0^* final states if Y_1^* resonances exist in the same mass region and decay into the same final state. A state not fed from the Y_1^* channel is $\Sigma^0\pi^0$ which is very difficult to analyze and has not been studied in production experiments. The importance of production experiments is twofold; the first, naturally, is to study the production mechanisms and the other is to look for new states. If states exist with very small coupling to the KN channel it is possible they would not as yet have been observed in the formation experiments, whereas, they could be produced via, for example K^* exchange, in production experiments.

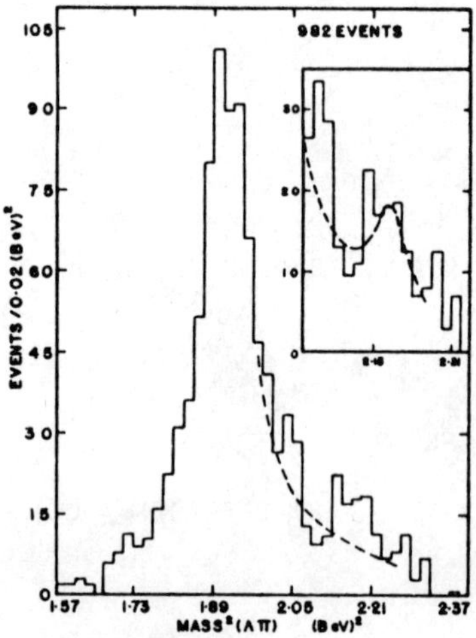

FIG. 1. Distribution of the $\Lambda\pi^+$ mass for all $\pi^+p \to K^+\pi^+\Lambda$ events. The dashed line represents the best fit obtained for $Y_1^*(1385)$ plus nonresonant background. The insert is an enlargement of the same mass distribution in the 2.17-GeV^2 region. The dashed line in the insert represents the two-resonance fit.

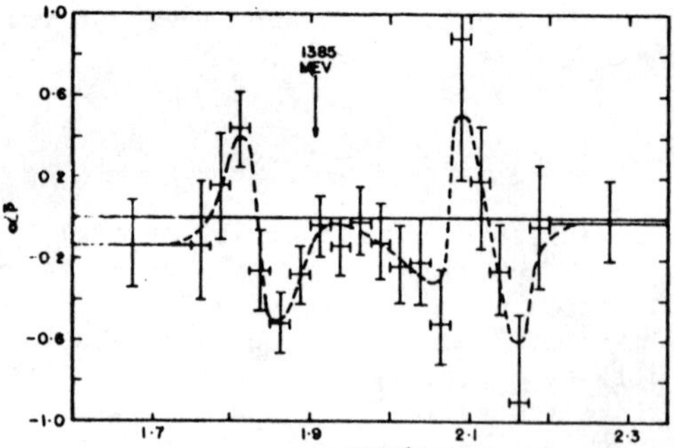

FIG. 2. The Λ polarization as a function of the $\Lambda\pi^+$ mass squared for all $\pi^+p \to K^+\pi^+\Lambda$ events. The dashed line is a free-hand curve drawn through the experimental data.

FIGURE 1

PRODUCTION EXPERIMENTS

I will review the Y^*'s seen, more or less, in order of ascending mass. I will also concentrate on the states which are less well established, and I will ignore such states as the $Y_1^*(1385)$, $Y_0^*(1405)$ and $Y_0^*(1520)$ whose properties are well known.

$Y_1^*(1480)$

This resonance was first proposed by Pan and Forman[4] who analyzed the reaction,

$$\pi^+ p \rightarrow \Lambda + K^+ + \pi^+ \text{ at } 1.7 \text{ GeV/c}$$

with an exposure of approximately 6 events/μb. In addition to a strong $Y_1^*(1385)$, they saw a peak at a mass of 1480 ± 15 MeV with a width Γ = 35 ± 20 MeV. The peak itself is not overwhelmingly convincing (Fig. 1), but in addition, the Λ polarization is observed to oscillate about 20 MeV below the $Y_1^*(1480)$. A similar oscillation is observed just below the 1385 and gives some credence to the idea that both oscillations indicate resonant behaviors. In a subsequent paper[5] they claimed evidence for $\Sigma^o\pi^+$ and $\Sigma^+\pi^o$ decays of the 1480 in the reaction $\pi^+p = \Sigma^+\pi^+K^o$ and also a possible $p\bar{K}^o$ decay. Once again the Λ and Σ^+ polarizations oscillate near the resonance mass. The $(\Sigma\pi)^+$ effective mass and the polarizations are shown in Fig. 2. The branching ratio $(\Sigma\pi)^+/\Lambda\pi^+$ = .72 ± .49. The $p\bar{K}^o$ signal is of less than 1 standard deviation significance and does not add any weight to the question of whether the state exists. As the authors state in their paper, no other comparable experiment exists on this reaction to either confirm or deny it. There are several points, however, which are not covered in the paper but which bear on the data. In the reaction $\pi^+p \rightarrow n\pi^+\pi^+$ in this energy region there is very strong $N^*(1670)$ production. Some typical data is shown in Fig. 3, together with cross sections for the $n\pi^+$ decay mode. This data comes from Benvenuti et al. at 2.26 GeV/c.[6] The total cross section at ~ 1.7 GeV/c appears to be near 1 mb. If this resonance couples to ΛK^+, then, as can be seen in Fig. 4, it could reflect at $M(\Lambda\pi)$ = 1480. Such a decay could also account for the polarization change in the 1480 region. The branching ratio into ΛK^+ would have to be ~ 1% to give a reflected peak of the magnitude seen. The angular momentum barrier and phase space would suppress the decay relative to the πN mode. The ratio of angular momentum barrier and phase space factors favor the πN by a factor of 14, so it is possible that the coupling of the N^* to ΛK^+ could be as strong as the 1% decay rate. In any case it would be nice to have an experimental limit on this quantity even if it is not the explanation. Looking at the experimental Dalitz plot it appears that the N^* decay is not prominent although one

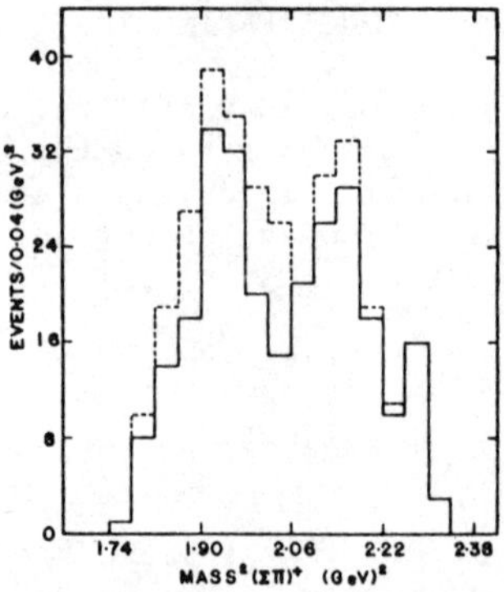

FIG. 1. The distribution of the square of the $(\Sigma\pi)^+$ invariant mass for Reactions (2) and (3). The dashed-line histogram is the total data and the solid-line histogram is the data with the cut $\Delta^2(\pi \to K) < -0.4\ \text{GeV}^2$.

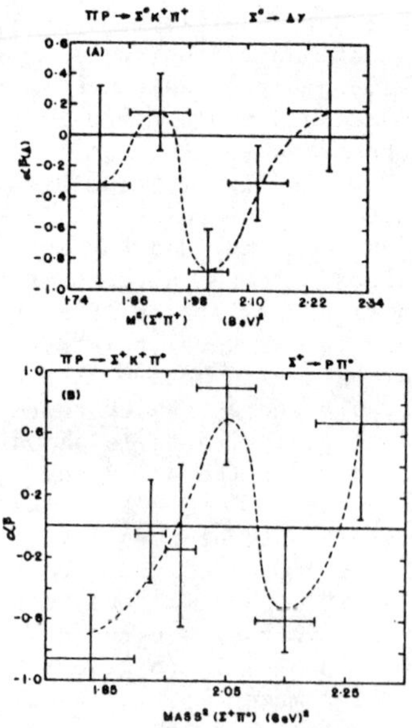

FIG. 2. (a) The Λ polarization as a function of the $\Sigma^0\pi^+$ mass squared. (b) The Σ^+ polarization as a function of the $\Sigma^+\pi^0$ mass squared. In both figures, the dashed lines are free-hand curves drawn through the experimental data.

FIGURE 2

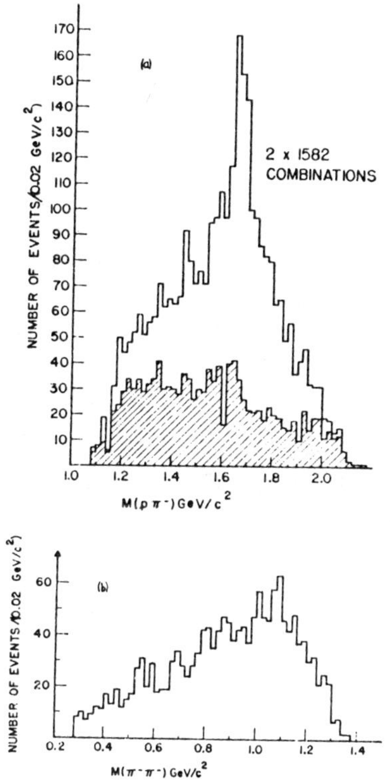

FIG. 1. (a) Effective mass of $\pi^- p$. Shaded region represents events with 4-momentum transfer from neutron to $p\pi^-$ system less than 0.6 $(\mathrm{GeV}/c)^2$. (b) Effective mass of the $\pi^-\pi^-$ system.

TABLE I. Characteristics of the 1670-MeV/c^2 peak produced in $\pi^- n$ and $\pi^+ p$ experiments.

Reaction	Beam momentum (GeV/c)	$E_{c.m.}$ (GeV)	Mass (GeV/c^2)	Width (GeV/c^2)	Cross section (μb)
(2)	1.59[a]	1.97	...	...	470 ± 50
(2)	2.0[b]	2.16	1668 ± 13	168 ± 35	775 ± 95
(1)	2.26 (this expt)	2.27	1667 ± 5	105 ± 16	930 ± 110
(2)	2.34[c]	2.30	1681 ± 14	109 ± 32	320 ± 100
(2)	2.75[d]	2.42			530[e]
(2)	...[f]				

[a] Reference 3.
[b] Reference 4.
[c] Reference 5.
[d] Reference 6.
[e] Calculated from data in Ref. 6.
[f] Indications of the peak were reported at 2.77 and 4 GeV/c but no details were given (Ref. 7).

FIGURE 3

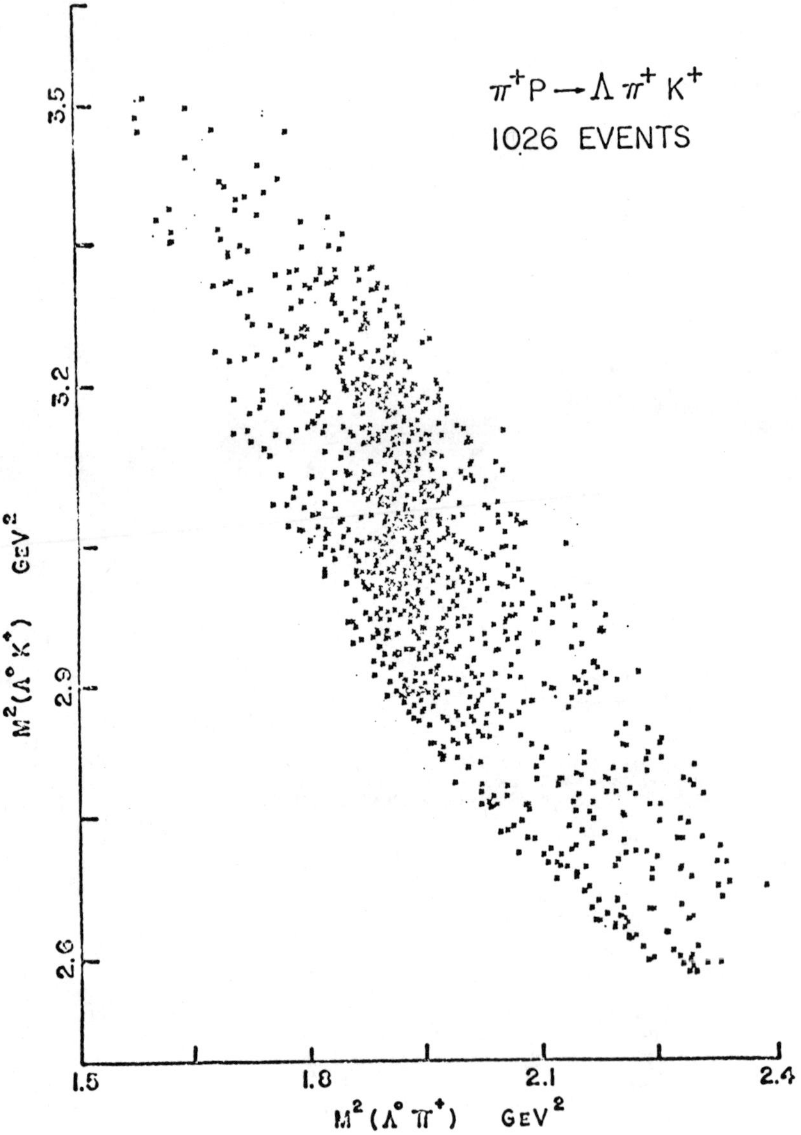

Fig. 24. Dalitz plot for $\pi^+ p \rightarrow \Lambda^o K^+ \pi^+$ events.

FIGURE 4

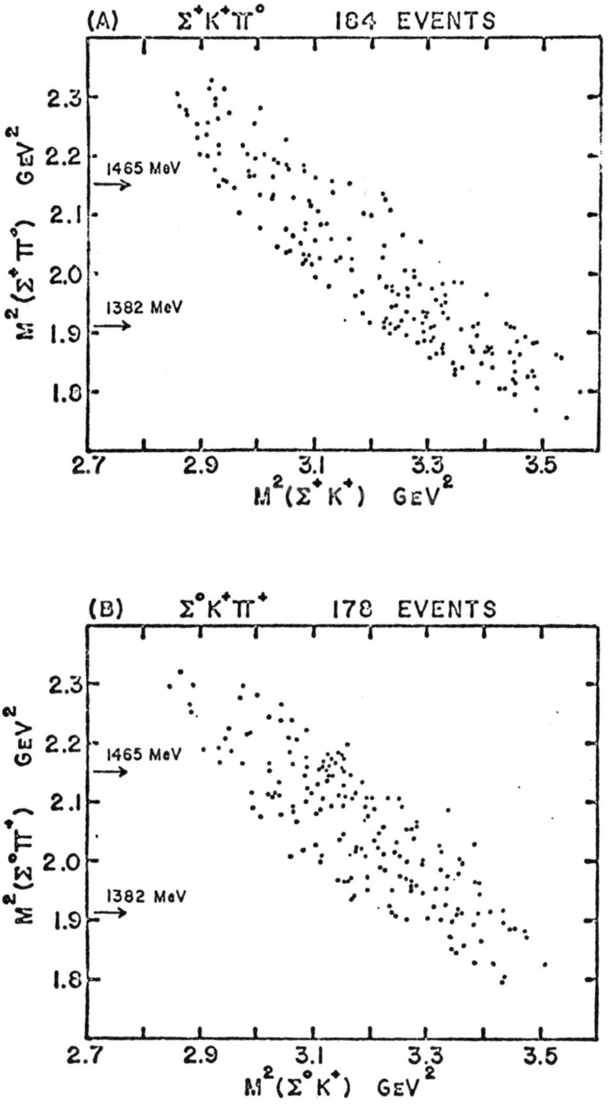

Fig. 32. Dalitz plots for a) $\pi^+p\rightarrow\Sigma^+\pi^oK^+$ and b) $\pi^+p\rightarrow\Sigma^o\pi^+K^+$.

FIGURE 5

really needs to look at a function of u to get a good branching ratio. The most striking feature apart from the 1385 is the increased population along the lower boundary. This corresponds to enhancing low $K^+\pi^+$ masses a fact which is hard to explain. The distribution of points contributing to the 1480 leads one to suspect that the effect may not be a true resonance. This is also true of the $(\Sigma\pi)^+$ data where in the case of $\Sigma^o\pi^+$ the points cluster in a group near the upper edge of the Dalitz plot (Fig. 5). It would certainly be nice to see more data on this particular state. If the resonance holds up with more data, its couplings to KN or K^*N would have to be weak not to show up in KN interactions.

$Y^*(1620)$

This resonance was first observed by Crennel et al.[7] in K^-d interactions at 3.9 GeV/c in the reaction

$$K^-d \rightarrow \Lambda\pi^-\pi^-\pi^+Ps$$

This effect was confirmed by them[8] with an increase in data of a factor of four in the same experiment. The data from this latter paper are shown in Fig. 6(a) where the $\Lambda\pi$ mass spectrum is plotted with three combinations per event. A very strong Y_1^* is observed, together with a significant enhancement near 1620. As an insert to Fig. 6(a) is a plot showing the Missing Mass Squared (MM^2) calculated from the incident beam and all the outgoing tracks except the Λ. This should peak at 1.115^2 for true $\Lambda\pi^-\pi^-\pi^+$ events and at 1.192^2 for true $\Sigma^o\pi^-\pi^-\pi^+$ events. Only events for $MM^2 <$ 1.4 GeV are plotted in Fig. 6. It is a little difficult to judge what the Σ^o contamination is for this plot, since only events fitting reaction 1 are included. It would have been more illuminating to plot this for all events to get some idea of the backgrounds involved. Figure 6(b) shows the branching ratios obtained from various final states analyzed so far. As can be seen, the $\bar{K}^op$ is quite small and, as yet, no $\Sigma\pi$ or $\Sigma\pi\pi$ ratios have been published from this experiment. In Fig. 7(a) the $\Lambda\pi^\pm$ mass plot is shown removing all events that fit $\Lambda\pi^-\pi^+\pi^-\pi^o$ down to a probability of 1%; as can be seen, the 1620 is still very significant. This rules out the possibility of low energy pi zeros contaminating the sample. There is some evidence in the data for simultaneous $Y^*(1620)\rho$ production, and possibly $Y^*(1620)\omega$ production (Fig. 7b). The overall effect is not particularly peripheral and appears in both $\Lambda\pi^-$ and $\Lambda\pi^+$ final states. This latter point is also true of the $Y^*(1385)$ where significant Y_1^{*+} is seen.

The controversy surrounding this particular effect has centered on the possibility of $\Sigma^o\pi^+\pi^-\pi^-$ final states being accepted as $\Lambda\pi^-\pi^+\pi^-$. This effect would cause a $\Sigma^o\pi$ resonance in

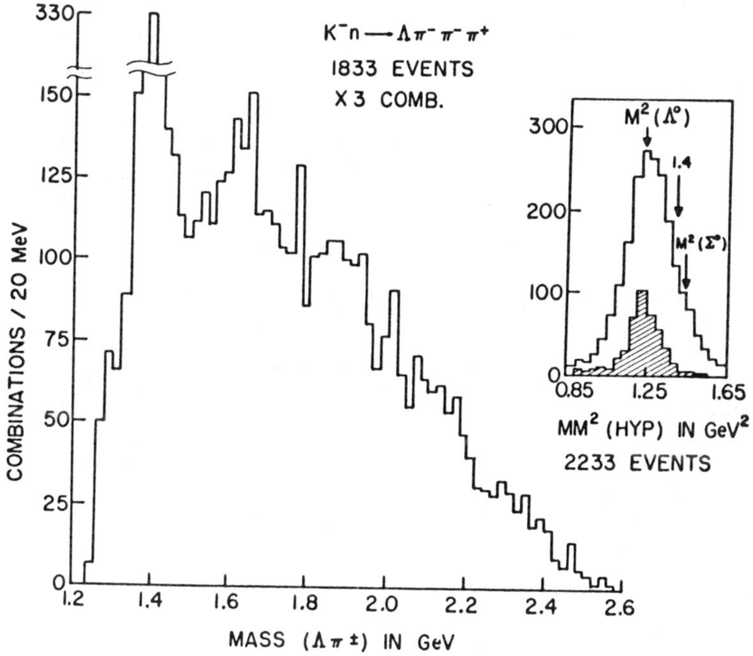

BRANCHING RATIOS FOR $Y^*(1619)$

$M = 1619 \pm 18$ MeV $\quad \Gamma = 72^{+22}_{-15}$ MeV

$$\frac{(Y^*(1385)\pi)^\pm}{\Lambda\pi^\pm} = 0.1 \pm 0.1$$

$$\frac{\bar{K}^\circ P}{\Lambda\pi^+} = 0 \pm 0.2$$

$$\frac{\bar{K}^\circ P\pi^\circ}{\Lambda\pi^+} = 0 \pm 0.02$$

$$\frac{\bar{K}^\circ n\pi^+}{\Lambda\pi^+} = 0 \pm 0.02$$

FIGURE 6

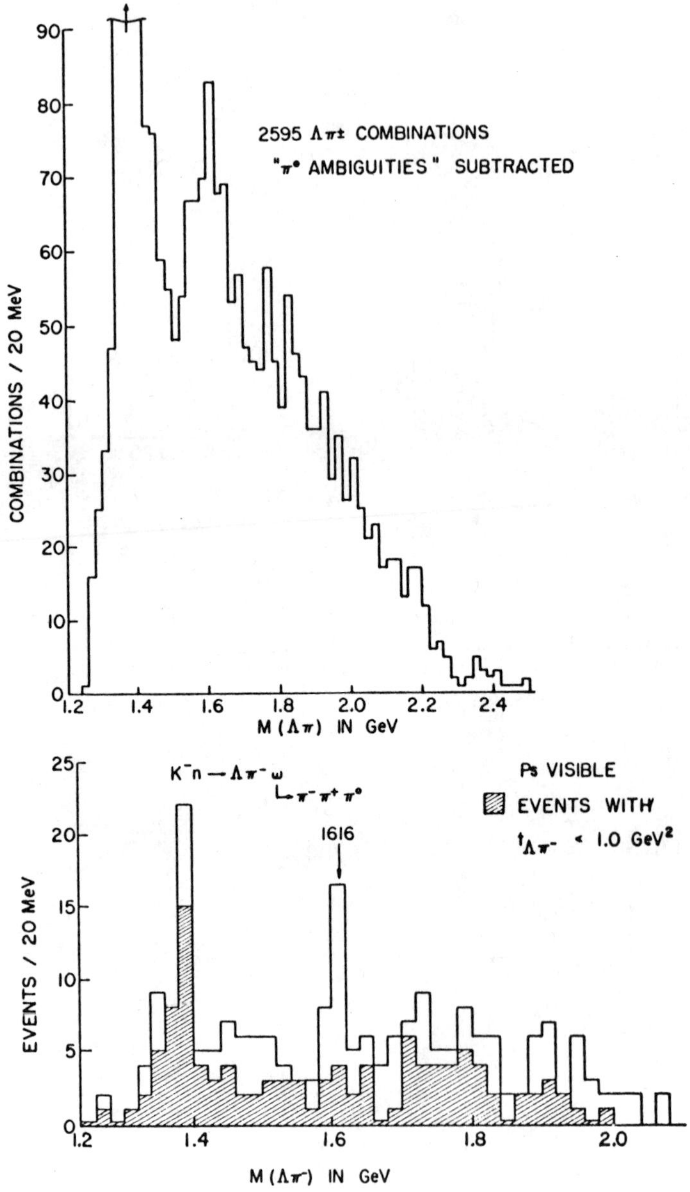
2595 Λπ± COMBINATIONS
"π° AMBIGUITIES" SUBTRACTED
COMBINATIONS / 20 MeV
M (Λπ) IN GeV
K⁻n → Λπ⁻ω
↳ π⁻π⁺π°
1616
Ps VISIBLE
EVENTS WITH
t_Λπ⁻ < 1.0 GeV²
EVENTS / 20 MeV
M (Λπ⁻) IN GeV

FIGURE 7

the 1680 MeV region to appear as a $\Lambda\pi$ resonance near 1620 MeV, or vice versa of course. This effect was pointed out very clearly by the SABRE collaboration[9] who have analyzed K^-d interactions at 3 GeV/c, which is not far from 3.9 GeV/c particularly with the center of mass energy spread due to the Fermi motion of the nucleon. Their data is shown in Fig. 8. In Fig. 8(a) is plotted the $\Lambda\pi$ combinations for events selected with the $MM^2 < 1.4$ GeV and a small signal can be seen near 1620 of 70 ± 32 events. In Fig. 8(b) are plotted events fitting $\Sigma^o\pi^+\pi^-\pi^-$ with $1.6 > MM^2 > 1.4$ to enhance Σ^o production and to cut out π^o production. A much stronger 1620 can be seen; however, there must still be a large contamination of Λ events. This is shown by the peak at 1460 in the $\Sigma^o\pi$ mass. The strong correlation between $\Sigma^o\pi^+$ and $\Lambda\pi^+$ mass for events fitting either hypothesis is shown in Fig. 8(c). In particular, the $Y_1^*(1385)$ reflects at $\Sigma\pi(1460)$, and events in the $\Lambda\pi(1620)$ region give a $(\Sigma\pi)$ peak at 1660. The very large 1385 signal, however, does show a large number of lambda events present. In Fig. 9(a,b) are plotted the $\Lambda\pi$ and $\Sigma\pi$ effective mass distributions for events only fitting either one or the other final state. It is difficult to draw any conclusions from these plots. Perhaps more convincing, is the observation of the $\Sigma(1660)$ in the $\Sigma^-\pi^o$, K^-n and $\Sigma\pi\pi$ decay modes, which then demands a $\Sigma^o\pi^-$ decay comparable to the signal observed. Their conclusions, therefore, are that they do not see any 1620 production. One thing of interest here is that there appears to be no $\Lambda\pi$ peak at 1660 in their data, whereas the $\Lambda\pi^-/\Sigma^o\pi^-$ decay is 1:1 from formation experiments,[10] and in production experiments[11] has been observed to be very high.

In any case, Crennel et al. do detailed tests to show that they are dealing with true $\Lambda\pi^+\pi^-\pi^-$ events. For example, they show that for events with missing energy < 0, (the unphysical tail of the missing energy distribution), that the 1620 is still apparent. In addition, the events thought to be $\Lambda\pi^-\pi^-\pi^+$ when fitted as Σ^o events have a γ ray which has a very anisotropic angular distribution with respect to the vector $K^- \times \Sigma^o$ in the Σ^o rest frame, whereas true events should be distributed isotropically (which their true events are). The analysis of the $\Sigma\pi$ and $\Sigma\pi\pi$ final states in Crennel's data should serve as an indication of how much 1660 is produced.

A confirmation of Crennel's data has been published by Ammann et al.[12] from K^-d interactions at 4.5 GeV/c. This data is shown in Fig. 10. Figure 10(a,b) show the $\Lambda\pi^-$ and $\Lambda\pi^+$ separately and production of a $\Lambda\pi$ state can be seen near 1640 MeV. Figure 10(c) yields a mass of 1642 ± 12 MeV and Γ = 55 ± 24 MeV. This is not inconsistent with Crennel's mass and width. The events selected as $\Sigma^o\pi^-\pi^-\pi^+$ show no obvious 1660 production (10d) and the only structure in the $p\bar{K}^o$ is above 1700 MeV. The charged Σ events were not analyzed but the decay $\Lambda(1405)\pi$ with the

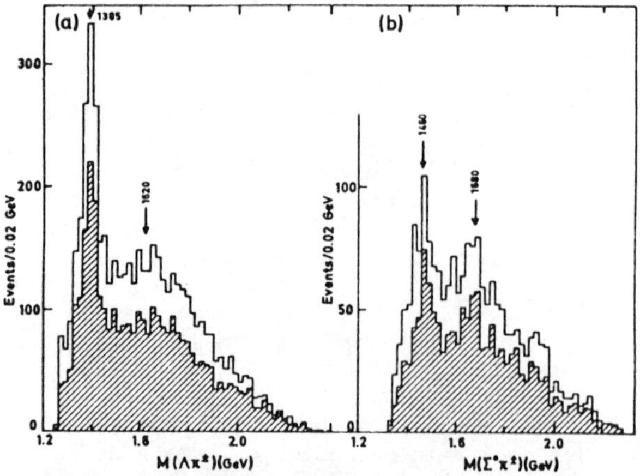

Fig. 2. Distribution of (a) the three $\Lambda\pi^{\pm}$ mass combinations in the $\Lambda\pi^{+}\pi^{-}\pi^{-}$ sample A′ (1644 events); and (b) the three $\Sigma^{0}\pi^{\pm}$ mass combinations in the $\Sigma^{0}\pi^{+}\pi^{-}\pi^{-}$ sample B (654 events). The two $\Lambda\pi^{-}$ or $\Sigma^{0}\pi^{-}$ combinations [in (a) and (b) respectively] are shaded.

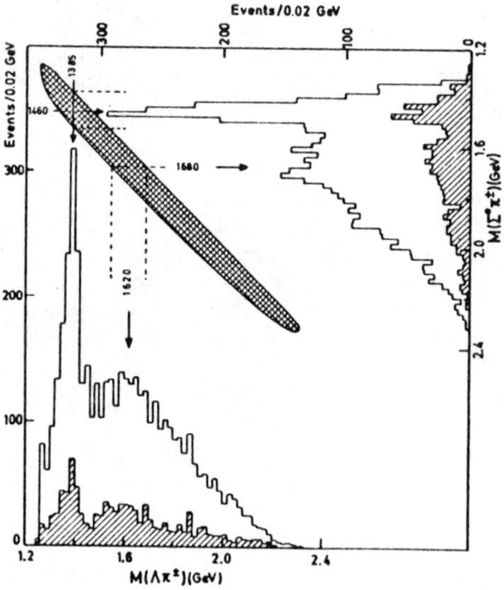

Fig. 3. Correlation between fitted $\Lambda\pi^{\pm}$ and $\Sigma^{0}\pi^{\pm}$ masses for 1517 events (samples A13 and B13) fitting both hypotheses (1) and (3) (three mass combinations per event). All but a few events lie in the narrow diagonal cross-hatched region. In the $\Lambda\pi^{\pm}$ and $\Sigma^{0}\pi^{\pm}$ mass projections, the 312 events of sample B13 [$MM^{2}(3\pi) > 1.4\ \mathrm{GeV}^{2}$] are shaded.

FIGURE 8

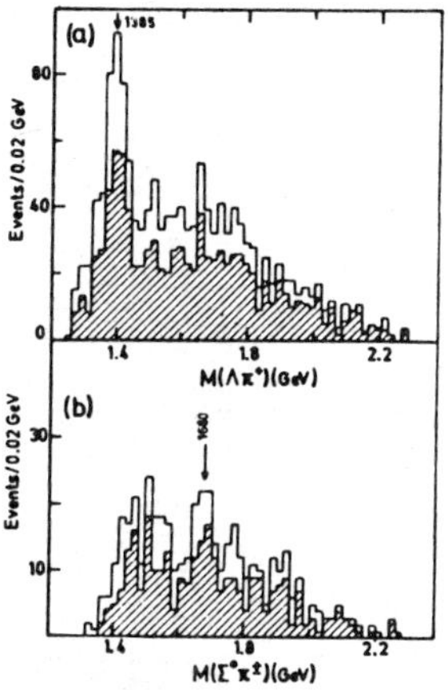

Fig. 4. (a) $\Lambda\pi^{\pm}$ mass distribution for 469 events (sample A1) fitting reaction (1) only; (b) $\Sigma^0\pi^{\pm}$ mass distribution for 152 events (sample B3) fitting reaction (3) only. Three mass combinations per event are plotted; the two $\Lambda\pi^-$ or $\Sigma^0\pi^-$ combinations [in (a) and (b) respectively] are shaded.

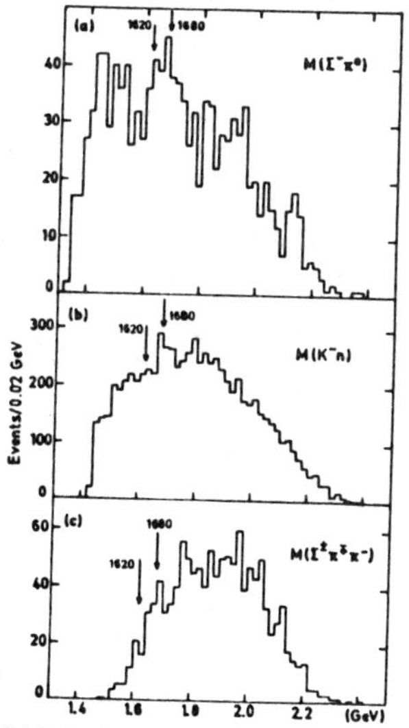

Fig. 5. (a) $\Sigma^-\pi^0$ mass distribution from 1171 $\Sigma^-\pi^0\pi^+\pi^-$ final states; (b) K^-n mass distribution from 7754 $K^-n\,\pi^+\pi^-$ final states; (c) $\Sigma^+\pi^{\pm}\pi^-$ mass distribution from 160 $\Sigma^-\pi^+\pi^-\pi^+\pi^-$ final states (4 mass combinations per event) and 184 $\Sigma^+\pi^-\pi^-\pi^+\pi^-$ final states (3 mass combinations per event). In (a) only events having visible spectator protons have been included.

FIGURE 9

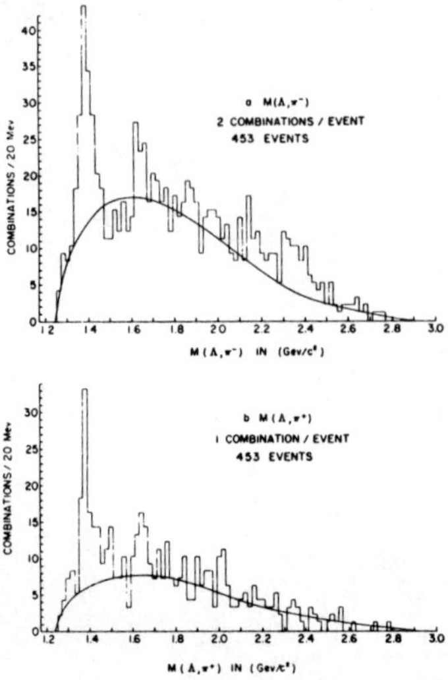

FIG. 1. (a) Effective-mass histogram of the two $\Lambda\pi^-$ combinations. (b) Effective-mass histogram of the $\Lambda\pi^+$. Both are for reaction (1).

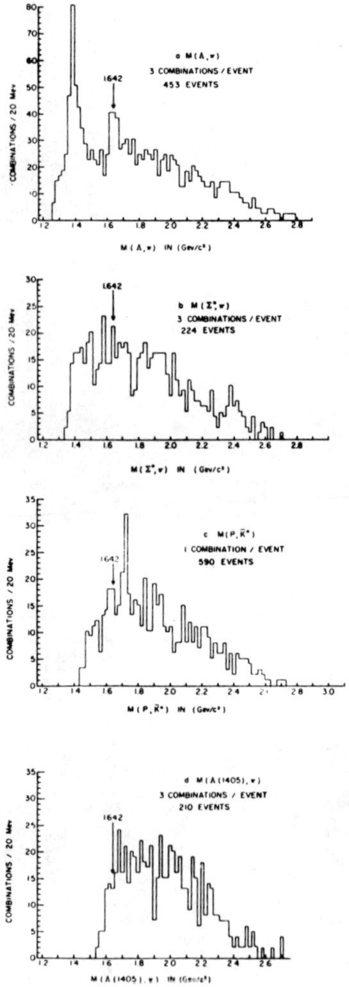

FIG. 2. (a) Effective-mass histogram of all three $\Lambda\pi$ combinations for Reaction (1); (b) of all three $\Sigma\pi$ combinations for Reaction (2); (c) of $\bar{K}^0p$ for Reaction (4); and (d) of all three "$\Lambda(1405)\pi$" combinations for Reaction (13). The $\Lambda(1405)$ region was taken from 1375 to 1435 MeV in hyperon missing mass (MM).

BRANCHING RATIOS

$$\frac{(\Sigma\,\pi)^+}{(\Lambda\,\pi)^+} < 1\cdot 1 \text{ AT } 90\% \text{ C.L.}$$

$$\frac{(\bar{K}\,N)^+}{(\Lambda\,\pi)^+} = \cdot 4 \pm \cdot 4$$

$$\frac{\Lambda(1405)\,\pi}{\Lambda\,\pi} = \cdot 7 \pm \cdot 4$$

MASS = 1642 ± 12 MeV

Γ = 55 ± 24 MeV

FIGURE 10

$\Lambda(1405) \rightarrow \Sigma^o\pi^o$ was looked for. The 1660 if it is produced in these data is also known to decay via this mode.

As can be seen in Fig. 10(f) no obvious structure exists. The conclusion is, therefore, that little $\Sigma(1660)$ is produced but a $(\Lambda\pi)$ state at 1640 is produced. Once again the production is not peripheral. This state does raise several problems however since it appears to decay with little $\Sigma\pi$ or $\bar{K}N$. A decuplet or $\overline{10}$ or 27 representation in SU3 requires a $\bar{K}N$ decay of magnitude a little less than $\Lambda\pi$, whilst an octet assignment with small $\bar{K}N$ would require a large $\Sigma\pi$ decay. The mass found by Ammann et al. although consistent with 1620 is also consistent with the 1660 region and conceivably may have some bearing on the varying branching ratios of the $\Sigma(1660)$ seen in other experiments.

The overall situation is still unclear experimentally, but analysis of other channels may well settle some of the problems.

To date this resonance has not been seen in $K^-p \rightarrow \Lambda\pi^-\pi^+\pi^o$ or $\Lambda\pi^-\pi^+$ although the expected magnitude of the signal in $\Lambda\pi^-\pi^+\pi^o$ is such that this is only minor evidence against the state.[13]

An important point which also bears on the discussion of the 1660 region is the KN and K^*N couplings of the resonances. In a normal peripheral one particle exchange process K exchange is forbidden in $K^-p \rightarrow Y_1^{*+}\pi^-$ and only K^* exchange is allowed. On the other hand in $K^-n \rightarrow Y_1^*\pi\pi$ both K and K^* exchange can take place. It is possible, therefore, that different states could be excited in differing magnitudes.

$Y_1^*(1660)$ Region

This is the mass region where the most confusion has arisen, centered mainly around the determination of branching ratios of the Y_1^* produced in K^-p interactions at various energies. As a standard reference we have the CERN Heidelberg Saclay (CHS) formation data[10] which provides branching ratios of the $Y_1^*(1660)$. This does not mean that their data necessarily represent the ultimate truth, but it has been a standard with which people have tried to reconcile the production data. This effort is shown on p. 187 of the January 1970 Reviews of Modern Physics[3] with a diagram implying that perhaps only experiments with small statistics disagree with the CHS numbers.

To summarize the CHS results: they find a resonant $3/2^-$ I = 1 partial wave with $\Gamma = 44 \pm 4$ and branching of 9 ± 2, 25 ± 5, $\simeq 28$ and < 10 percent into $\bar{K}N$; $\Sigma^o\pi^+$; $\Lambda\pi^+$; and $(\Sigma^\pm\pi^\mp\pi^+)$. In addition, very little $\Lambda(1405)\pi$ decay is seen.

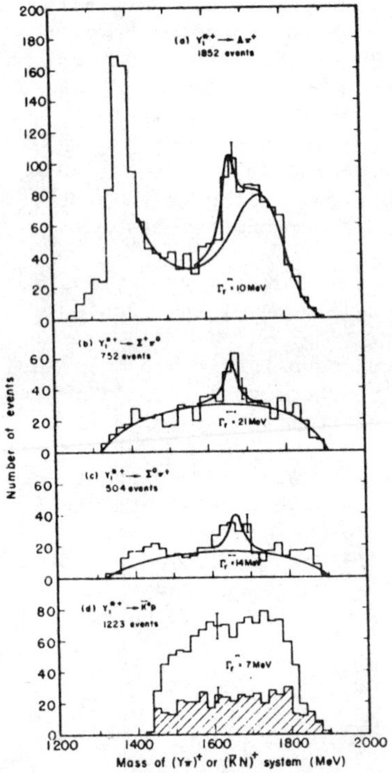

FIG. 2. Two-body decay modes of the 1660-MeV Y_1^{*+}. The dashed portion of Fig. 2(d) includes only non-K^* events [events satisfying the criterion 850 MeV < $M(K\pi)$ < 940 MeV have been removed]. The curves represent the best fits of the data to a smooth background curve plus a Breit-Wigner distribution. Here Γ_γ indicates the experimental resolution in the 1660-MeV region.

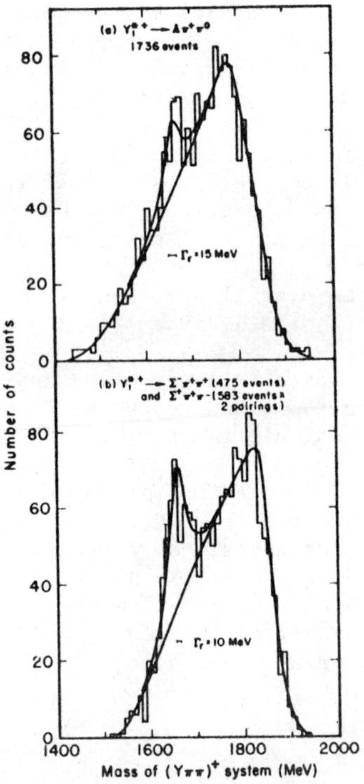

FIG. 3. Three-body decay modes of the 1660-MeV Y_1^{*+}. The curves represent the best fits of the data to a smooth background curve plus a Breit-Wigner distribution. Here Γ_γ indicates the experimental resolution in the 1660-MeV region.

FIGURE 11

The two experiments with largest statistics however, the 2.6 GeV data from Berkeley and the 4.6 GeV data from Brookhaven now agree that more than one object exists in this region. Before giving a review of the data on the 1660 I would like to make a few observations about the determination of branching ratios. One of the main sources of ambiguity arising is the Λ/Σ^o ambiguity which I feel may be quite serious as one goes to higher beam momentum. In general the Λ/Σ^o ambiguity occurs in a large fraction of the events and various authors use different means for cross checking their χ^2 selection criteria. For example, to look for missing energy or momentum carried off by the γ, and the distribution of γ ray in the Σ^o rest frame. These techniques in general cause events with small γ ray momenta to be selected as 'true' Λ events, and high momentum γ rays as 'true' Σ^o events, whilst the intermediate region being ambiguous is usually assigned to the Λ category. The angle test has some deficiencies also, in that outgoing particles have low transverse momentum and therefore measurement errors will give a missing momentum in the plane of the event. Now if a Σ^o is produced which gives rise to a low energy γ ray the fit may well move that γ ray toward the plane of production and destroy the Σ^o decay isotropy. The bearing this has on the 1660 is that for the decay $(\Sigma\pi)^+$ for example both $\Sigma^+\pi^o$ and $\Sigma^o\pi^+$ should be analyzed before a definitive $(\Sigma\pi)^+$ rate can be determined.

Now I would like to go through the data somewhat historically paying particular attention to the ratios $(\Sigma^\pm\pi^+\pi^\mp)$: $\Sigma^o\pi^+$: $\Lambda\pi^+$, so that we can see what consistency there is. In addition the cross section masses and widths as a function of energy, and also four momentum transfer could give an idea of the complexity of the analysis.

One of the original papers on the 1660 with reasonable statistics was the 1.5 GeV K^-p experiment of Alvarez <u>et al</u>.[14] They studied the final states

$\Lambda\pi^+\pi^-$	1
$\Lambda\pi^+\pi^-\pi^o$	2
$\Sigma^o\pi^+\pi^-$	3
$\Sigma^\pm\pi^+\pi^\mp$	4
$\Sigma^\pm\pi^\mp\pi^+\pi^-$	5
$\bar{K}^o\pi^-p$	6

and their data is shown in Fig. 11. They find

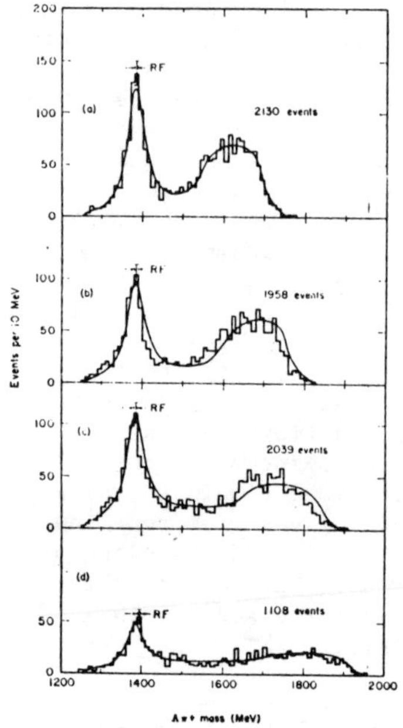

FIG. 11. Effective-mass distribution of the $\Lambda\pi^+$ pair from the reaction $K^- + p \rightarrow \Lambda + \pi^+ + \pi^-$ at (a) 1.15–1.30 BeV/c, (b) 1.30–1.45 BeV/c, (c) 1.45–1.60 BeV/c and (d) 1.60–1.75 BeV/c. The sample of events was selected by the criteria described in the text. The spacing marks labeled RF give the full width at half-maximum (FWHM) of the resolution function for the events in the Y_1^* peak. The solid curve on each graph is the expected shape, projected on this axis, of phase space and Y_1^{*+}, Y_1^{*-}, and ρ^0 resonances weighted according to the populations given in Table III.

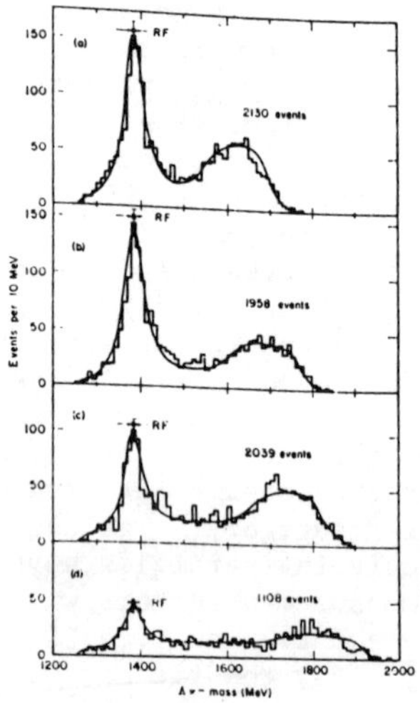

FIG. 12. Effective-mass distribution of the $\Lambda\pi^-$ pair. Caption is the same as that for Fig. 11.

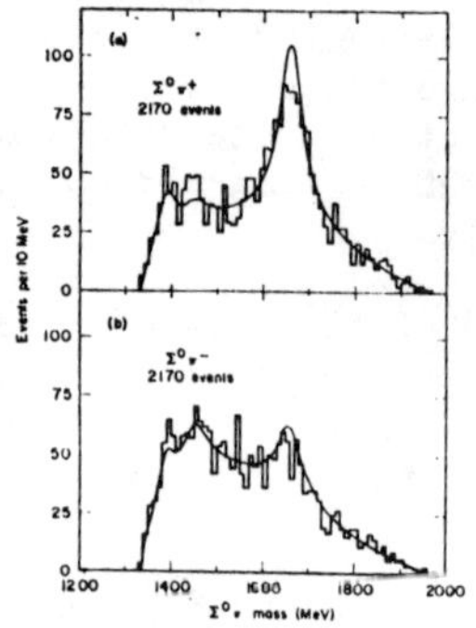

FIG. 31. Effective-mass distributions (a) of the $\Sigma^0\pi^+$ pair and (b) of the $\Sigma^0\pi^-$ pair from the reaction $K^- + p \rightarrow \Sigma^0 + \pi^+ + \pi^-$ at $P = 1.00$–1.75 BeV/c (includes special low-energy study). Criteria for selection of events are explained in the text. The solid curve is the sum of nonresonant-phase-space and Breit-Wigner shapes for the $Y_1^*(1385)$, the $Y_1^*(1660)$, and the $Y_1^*(1385)$ which is produced as a $\Lambda\pi$ state but fitted as a $\Sigma^0\pi$ state, in the proportions given by Table XII.

FIGURE 12

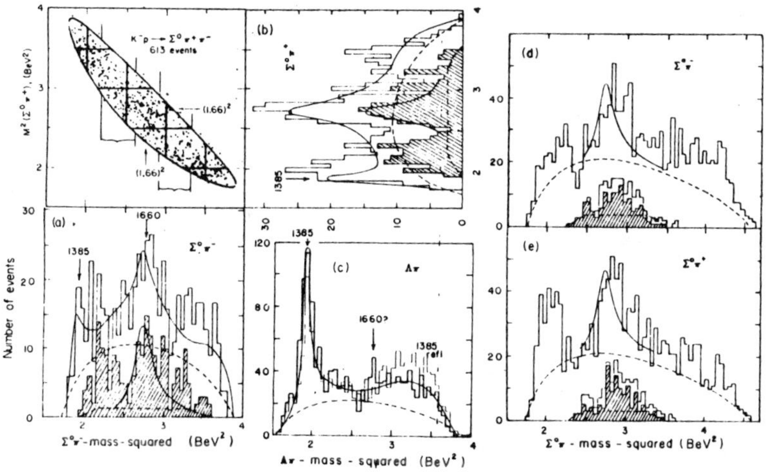

FIG. 1. Dalitz plot for Reaction (1) at 1.7 BeV/c, mass spectra for Reactions (1) and (2) at 1.7 BeV/c [(a), (b), and (c)], and mass spectra for (1) at 2.1 BeV/c [(d) and (e)]. Noninterference events are crosshatched. The heavy histogram in (c) represents the $\Lambda\pi^+$ mass spectrum; the light histogram, the $\Lambda\pi^-$ mass spectrum. With incoherent superposition of phase space plus resonances, fits were made simultaneously for the $\Sigma^0\pi^\pm$ data of (a) and (b), and for the $\Lambda\pi^\pm$ data of (c). In (d) and (e) the peaks represent the $Y^*(1660)$; the lower histograms present the noninterference events in a "nonresonant" $\Sigma\pi$ band from 2.92 to 3.32 BeV^2 and in a cleaner nonresonant band from 3.03 to 3.32 BeV^2 (crosshatched). Dashed lines represent phase space.

FIGURE 13

$$\Lambda\pi^+ : \Sigma^o\pi^+ : \Lambda\pi\pi : \Sigma\pi\pi : \bar{K}N$$
$$7 : 3 : 4 : 2 : \overline{<1}$$

with a total cross section to all channels of 400 μb. The errors on the individual partial cross sections are at least 25%.

This paper has been superceded by the paper of Huwe[15] who includes the above data plus other data in the range 1 → 1.7 GeV/c. He modifies the $\Sigma^o\pi^+/\Lambda\pi^+$ quoted above as .43 to 3.4 ± 1.5. As can be seen from Fig. 12 there is little evidence of a $\Lambda\pi^+$ decay of the 1660 but the $\Sigma^o\pi^+$ decay is very strong. At these low energies the analysis of the $\Lambda\pi$ state is complicated by $Y_1^{*-}(1385)$ production which overlaps the 1660^+ mass region. It appears on inspection of the data that the $\Lambda\pi$ decay could be much smaller than quoted. In addition the error quoted by Huwe is twice the statistical error. Therefore, although, 3.4 ± 1.5 could be consistent with 1 it appears to me that in fact, the $\Lambda\pi^+$ is much smaller than the $\Sigma^o\pi^+$.

The next paper at a slightly higher energy is by J. Button-Shafer[16] at 1.7 and 2.1 GeV looking only at reactions 1 and 3. The data is shown in Fig. 13 with quoted B. R. of $\Sigma^o\pi^+/\Lambda\pi^+ = 4.5 \pm 2.3$ and $\Sigma^o\pi^+/(\Sigma\pi\pi)^+ = .8 \pm .3$. The $\Sigma\pi/\Lambda\pi$ is quoted for the data as a whole whereas the $\Sigma\pi/\Sigma\pi\pi$ is for $\cos\theta^* < -.7$ (θ^* is the angle between K^- and Y^* in the CMS). This is a region in which only .25 ± .05 of the $\Sigma^o\pi^+$ signal is and, therefore, this branching ratio should be treated with some caution.

One of the best papers on this subject is the one written by Eberhard _et al._[17] from 2.6 GeV/c K^-p interactions. In their initial paper they have studied reactions 5, 3 and 4 in detail. Their results are presented in Fig. 14 where they show their data split up into various production angle ($\cos\theta^*$) regions. The results indicate a drastic change in the branching ratio as a function of $\cos\theta^*$ as can be seen from the table. The $\Sigma\pi\pi$ mode is very peripheral and is only really seen in their interval I($-.95 > \cos\theta^* > -1.0$) whereas the $(\Sigma\pi)^+$ decay mode appears stronger in the region $-.7 > \cos\theta^* > -.9$. The $\Sigma\pi\pi$ decay mode is consistent with being 100% $\Lambda(1405)\pi$ which is true of all production experiments. The data clearly show that more than one object exists in this region.[18] Most spin parity analyses of the 1660 state[19,20,21,22] have yielded $3/2^-$ for the J^P. This was true for the $\Sigma\pi\pi$ decay of Eberhard's data and also for the $\Sigma^o\pi^+$ decay of Ref. 16. This leads to the possibility proposed in Ref. 17 that the two states have the same spin parity and interfere strongly. The behavior of the peak as a whole would then be some linear combination of the two states and could account for the wild fluctuations seen in the behavior of this state.

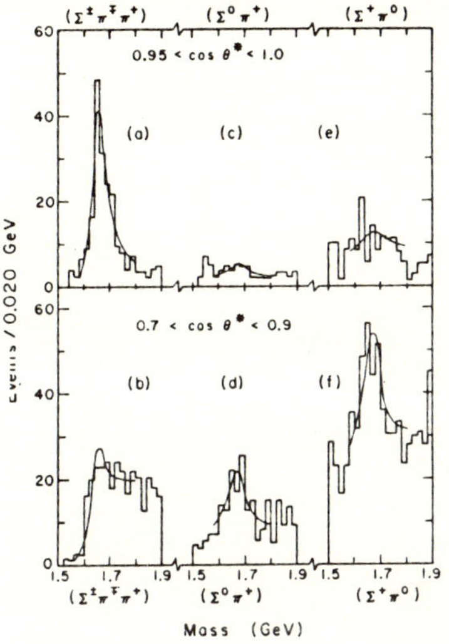

FIG. 1. Mass plots of the $\Sigma\pi\pi$ and $\Sigma\pi$ systems for various production $\cos\theta^*$ intervals. The events in the $(\Sigma^0\pi^+)$ spectra have not been weighted to correct for the unseen neutral decay of the Λ in the Σ^0 decay. The curves shown are the results from the fit described in the text.

Table I. Results of fit to $\Sigma(1660)$ production in $\Sigma\pi\pi$ and $\Sigma\pi$ modes.

Mode	Amount of $\Sigma(1660)$ in interval (μb/sr) Interval I $0.95 < \cos\theta^* < 1.0$	Interval II $0.9 < \cos\theta^* < 0.95$	Interval III $0.7 < \cos\theta^* < 0.9$	Resonance parameters used in fit: Mass (MeV)	Width (MeV)
$\Sigma^{\pm}\pi^{\mp}\pi^{+}$	57.0 ± 4.1	21.1 ± 3.8	8.1 ± 1.4	1651	70
$\Sigma^0\pi^+$	7.5 ± 4.6[a]	13.5 ± 5.5[a]	8.9 ± 2.5[a]	1667	80
$\Sigma^+\pi^0$	6.0 ± 5.5	6.4 ± 4.4	11.1 ± 2.7	1667[b]	80[b]

[a]Includes correction for unseen neutral decay of the Λ^0.
[b]Values taken from fit of $\Sigma^0\pi^+$ mass spectrum.

FIGURE 14

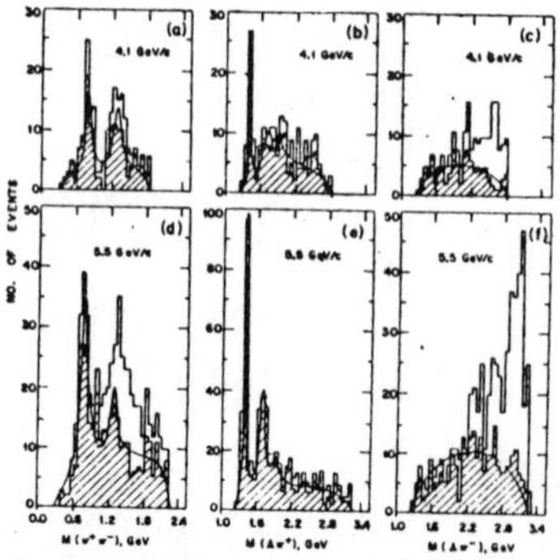

FIG. 4. Mass projections for the $\Lambda\pi^+\pi^-$ final state. In making fits to the Breit-Wigner resonance shape, the mass and width used for each resonance was that which gave a best fit to the combined data at 4.1 and 5.5 GeV/c. (a) The $\pi^+\pi^-$ mass at 4.1 GeV/c. The shaded area contains 149 events after subtraction of 64 $Y_1^{*+}(1385)\pi^-$ and $Y_1^{*+}(1695)\pi^-$ events. The solid curve is a best fit to the shaded histogram for phase space and two Breit-Wigner resonances with $M(\rho^0)=755$ MeV, $\Gamma(\rho^0)=140$ MeV, and $M(f)=1285$ MeV, $\Gamma(f)=140$ MeV. The relative contributions are 49% phase space, 29% ρ^0 and 22% f. (b) The $\Lambda\pi^+$ mass at 4.1 GeV/c. The shaded area contains 144 events after subtraction of 69 $\Lambda\rho^0$ and Λf events. The solid curve is a best fit to the shaded histogram for phase space and three Breit-Wigner resonances with $M(Y_1^{*+}(1385))=1383$ MeV, $\Gamma(Y_1^{*+}(1385))=31$ MeV, $M(Y_1^{*+}(1695))=1700$ MeV, $\Gamma(Y_1^{*+}(1695))=130$ MeV, and $M(Y_1^{*}(1910))=1910$ MeV, $\Gamma(Y_1^{*+}(1910))=60$ MeV. The relative contributions are 67% phase space, 14% $Y_1^{*+}(1385)$, 15% $Y_1^{*+}(1695)$, and 4% $Y_1^{*+}(1910)$. (c) The $\Lambda\pi^-$ mass at 4.1 GeV/c. The shaded area contains 101 events after subtraction of 112 $\Lambda\rho^0$, Λf, $Y_1^{*+}(1385)\pi^-$, and $Y_1^{*+}(1695)\pi^-$ events. The solid curve is phase-space normalized to the shaded histogram. (d) The $\pi^+\pi^-$ mass at 5.5 GeV/c. The shaded area contains 345 events after subtraction of 218 $Y_1^{*+}(1385)\pi^-$ events and $Y_1^{*+}(1695)\pi^-$ events. The solid curve is a best fit to the shaded histogram for phase space and the two Breit-Wigner resonances described in (a). The relative contributions are 60% phase space, 30% ρ^0, and 10% f. (e) The $\Lambda\pi^+$ mass at 5.5 GeV/c. The shaded area contains 432 events after subtraction of 131 $\Lambda\rho^0$ and Λf events. The solid curve is a best fit to the shaded histogram for phase space and the three Breit-Wigner resonances described in (b). The relative contributions are 45% phase space, 28% $Y_1^{*+}(1385)$, 25% $Y_1^{*+}(1695)$, and 2% $Y_1^{*+}(1910)$. (f) The $\Lambda\pi^-$ mass at 5.5 GeV/c. The shaded area contains 248 events after subtraction of 315 $\Lambda\rho^0$, Λf, $Y_1^{*+}(1385)\pi^-$, and $Y_1^{*+}(1695)\pi^-$ events. The solid curve is phase-space normalized to the shaded area.

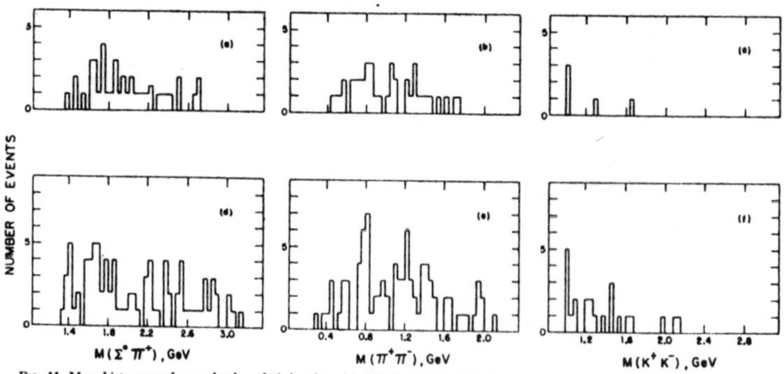

FIG. 11. Mass histograms for a selection of $\Sigma^0\pi^+\pi^-$ fits with $M^2<0$, as defined in $K^-+p\rightarrow\Lambda+\pi^++\pi^-+M$, after removal of events with four-constraint fits, as well as for a corresponding selection of $\Sigma^0K^+K^-$ fits. (a) The $\Sigma^0\pi^+$ mass for 41 $\Sigma^0\pi^+\pi^-$ events at 4.1 GeV/c. (b) The $\pi^+\pi^-$ mass for 41 $\Sigma^0\pi^+\pi^-$ events at 4.1 GeV/c. (c) The K^+K^- mass for 5 $\Sigma^0K^+K^-$ events at 4.1 GeV/c. (d) The $\Sigma^0\pi^+$ mass for 89 $\Sigma^0\pi^+\pi^-$ events at 5.5 GeV/c. (e) The $\pi^+\pi^-$ mass for 89 $\Sigma^0\pi^+\pi^-$ events at 5.5 GeV/c. (f) The K^+K^- mass for 23 $\Sigma^0K^+K^-$ events at 5.5 GeV/c.

FIGURE 15

Another experiment in this energy region by London et al.[23] at 2.24 GeV/c is poorer statistically than Eberhard's data, and the 2.6 GeV experiment provides the standard in this energy region. I now turn to the results of Derrick et al.[24,25] at much higher momenta 5.5 GeV/c published in a paper prior to Ref. 17, and claiming to see a 1660 decaying dominantly to $\Lambda\pi^+$ with no visible $\Sigma^o\pi^+$ decay. This they therefore said, was clear evidence for a new state near 1660. These data are shown in Fig. 15. The $\Lambda\pi^+$ signal (their figure 4e) is very strong and their $\Sigma^o\pi^+$ shows no obvious enhancement. One thing I should point out is that the number of $\Lambda\pi^+$ events displayed at 5.5 GeV/c is 563 whereas the number of $\Sigma^o\pi^+$ events is only 89. This gives a ratio of $\Sigma^o\pi^+/\Lambda\pi^+$ events of .16. This is abnormally low. For example the Brookhaven data at 4.6 Ref. 26 yield a $\Sigma^o\pi^+/\Lambda\pi^+$ of .3 which is twice as high. This probably represents a difference in the way Σ^o fits were selected rather than in a variation of the ratio as a function of energy. If this is true reaction 3 is very depleted in Ref. 25 and that leads to the question of which final state they were assigned to. Subsequent to the publication of this state the $\Sigma^{\pm}$ final states were published by the Illinois group.[27] They do not show the $\Sigma^+\pi^o$ mass spectrum from the final state $\Sigma^+\pi^-\pi^o$, though they say no significant structure was seen. The Σ^+ is subject to large scanning loss and this sample tends to be depleted, however, it would have been nice to see a combined $\Sigma^o\pi^+, \Sigma^+\pi^o$ spectrum. The final states $\Sigma^{\pm}\pi^{\mp}\pi^+\pi^-$ are shown in Fig. 16. Clear 1660 signals can be seen which come from the $(\Sigma\pi\pi)^+$ combinations. A cross section of 9 ± 4 μb is quoted for this decay mode giving

$$\frac{\Lambda\pi^+}{(\Sigma\pi\pi)^+} = \frac{25 \pm 5}{9 \pm 4} = 2.8 \pm 1.4$$

Subject to what has been said above no $(\Sigma\pi)^+$ decay mode was observed. A subsequent experiment by Colley et al.[28] which is much weaker statistically gave supporting evidence (Fig. 17). There is also the question of the mass and width which for Derrick et al. was 1695 ± 10 Γ = 120. This is clear evidence of inconsistency with M = 1660 Γ = 44. We now come to the BNL data at 3.9, 4.6 and 5.0 GeV/c. These data together with those of Eberhard should provide very good parameters and characteristics of the 1660 region at two quite different momenta. Several publications have come from the BNL data and a more detailed study of cos θ^* dependence has been submitted for publication.

The paper by Primer et al.[29] was on a partial analysis of this data and appeared to support a $\Lambda\pi(1695)$ effect. This comes mainly from observing a broad Γ = 105 $\Lambda\pi$ enhancement centered at 1694; $\Sigma\pi$ and $\Sigma\pi\pi$ signals were observed but the branching into $\Lambda\pi^+$ seemed too high and the mass and width were shifted. These

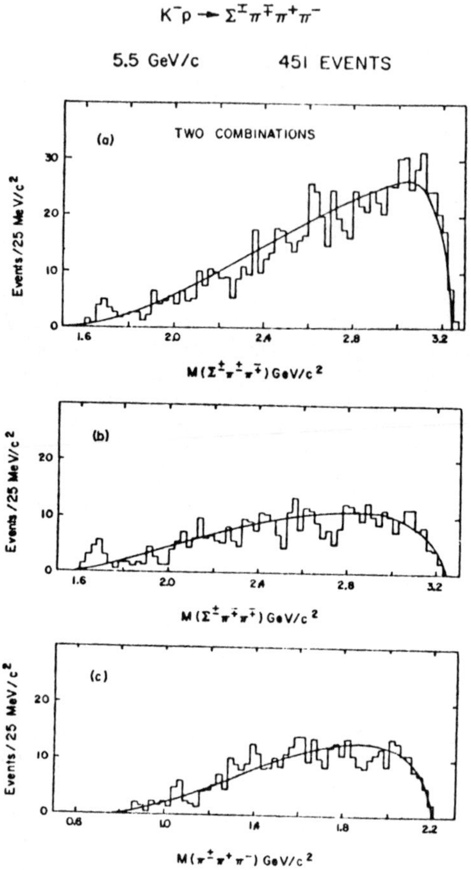

FIG. 9. Invariant-mass spectra for singly charged combinations from $K^-p \rightarrow \Sigma^{\pm}\pi^{\mp}\pi^+\pi^-$. (a) $\Sigma^{\pm}\pi^{+}\pi^{\mp}$, two combinations; (b) $\Sigma^{+}\pi^{\mp}\pi^{\mp}$; and (c) $\pi^{\pm}\pi^{+}\pi^{-}$. The solid curves are obtained from a maximum-likelihood fit (see text).

FIGURE 16

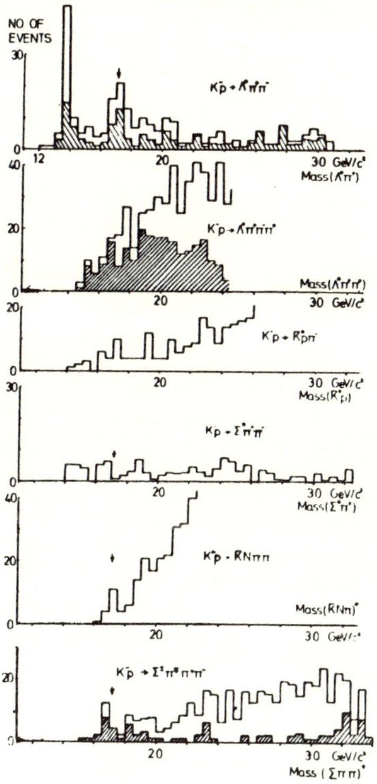

Fig. 1. Effective mass histograms for the reaction $K^- + p \rightarrow Y^{*+} + \pi^-$ where Y^{*+} are possible resonant states with decay modes (a) $\Lambda\pi^+$ (b) $\Lambda\pi^+\pi^0$ (c) $\bar{K}^0 p$ (d) $\Sigma^0\pi^+$ (e) $(\bar{K}N\pi)^+$ and (f) $(\Sigma\pi\pi)^+$. Shown shaded are: (a) Unique fits to $\Lambda^0\pi^+\pi^-$, i.e. no ambiguity with Σ^0. (b) Events with one $\Lambda\pi^{\pm}$ mass in the region of the Y_1^* (1385). (f) Events with one $\Sigma^{\pm}\pi^{\mp}$ mass in the region of the Y_0^* (1405).

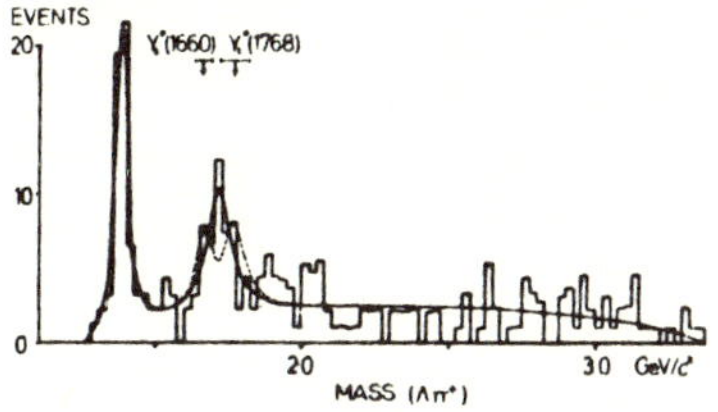

Fig. 2. Effective mass histogram for $\Lambda\pi^+$ from the reaction $K^- + p \rightarrow \Lambda\pi^+\pi^-$. Curves shown are maximum likelihood fits using (solid line) Y_1^* (1385) plus Y_1^* (1715) and (dotted line) Y_1^* (1385) plus Y_1^* (1660) and Y_1^* (1768).

FIGURE 17

data have been fully analyzed now with a paper submitted to the Boulder Conference last summer,[26] and a detailed analysis contributed to this conference and submitted to PRL.[30]

The Boulder paper which supplied the branching ratios shown in Ref. 3 has been superceded by the new contribution. I think there is no doubt that the table of branching ratios given in Ref. 3 is misleading judging by the complexity of the BNL and Berkeley data.

The BNL data at 3.9 and 4.6 GeV/c is an exposure of ~ 14 events/μb. The $\Sigma^o\pi^+$ mass spectrum from reaction 3 and the $\Lambda\pi^+$ mass spectrum from reaction 1 are shown in Fig. 18. The $\Lambda\pi^+$ shows a very broad enhancement extending over the 1600 → 1800 Mev region. In contrast the $\Sigma^o\pi^+$ shows a well defined peak at 1660 on a fairly low background. In Fig. 19 the $\Lambda(1405)\pi^+$ mass spectrum is shown for the final state $\Sigma^+\pi^-\pi^+\pi^-$. A very significant peak is observed at 1660. The state $\Sigma^-\pi^+\pi^+\pi^-$ has not been analyzed since the Σ^- can form a Y^* with either π^+. These $\Lambda(1405)$ bands overlap in the 1660 region and interference effects could be very strong. The main analysis has been, therefore, on the $\Sigma\pi\pi$ and $\Sigma\pi$ decay modes where the signal to noise is very good. The data strikingly confirm the results of Ref. 17 as to the differing characteristics of the two decay modes. This is shown in Fig. 20 where the $\Sigma^o\pi^+$ decay is shown for cos $\theta^* > -.7$ and $< -.7$. As can be seen the 1660 is produced about equally in each region. This is also shown in Fig. 21 where $M(\Sigma\pi)$ is plotted against cos θ^*. The 1660 band extends over the whole cos θ^* range with significant u channel production. On the other hand there is no $\Sigma\pi\pi(1660) > -.7$ (Fig. 22), the $\Sigma\pi\pi$ decay being very peripheral. To make a quantitative comparison, the BNL group have calculated $\alpha = B/F+B$ where B(F) is the number of $\Sigma(1670)$ events with cos $\theta^* \geqq -.7$ ($< -.7$). For the $\Sigma\pi\pi$ $\alpha = 0.0 \pm .04$ and for the $\Sigma^o\pi^+$ $\alpha = 0.42 \pm .08$. In addition, they observe a different energy dependence for the two modes (Table 1b). No $p\bar{K}^o$ mode is seen with a two standard deviation upper limit of $N\bar{K}/\Sigma\pi = 0.2$. No $\Sigma(1765)$ decay into $p\bar{K}^o$ is seen either in contrast to the Berkeley data. Table I shows the fitted parameter for the $\Sigma^o\pi^+\pi^-$ final state and a mass of 1668 ± 10 and $\Gamma = 135 {}^{+40}_{-30}$ was found for the $\Lambda(1405)\pi$ decay.

The $\Lambda\pi^+$ mode is still being analyzed, although the width of the signal seen seems to point toward at least more than one effect existing. Similar effects to these are seen at 2.6 GeV in the Berkeley data.[18]

I have not covered other decay modes such as $\Lambda\pi\pi$ or KN because the data is less definitive. The $\Lambda\pi\pi$ mode seen in Ref. 14 for example, is complicated because of the $Y_1^*(1385)$ bands

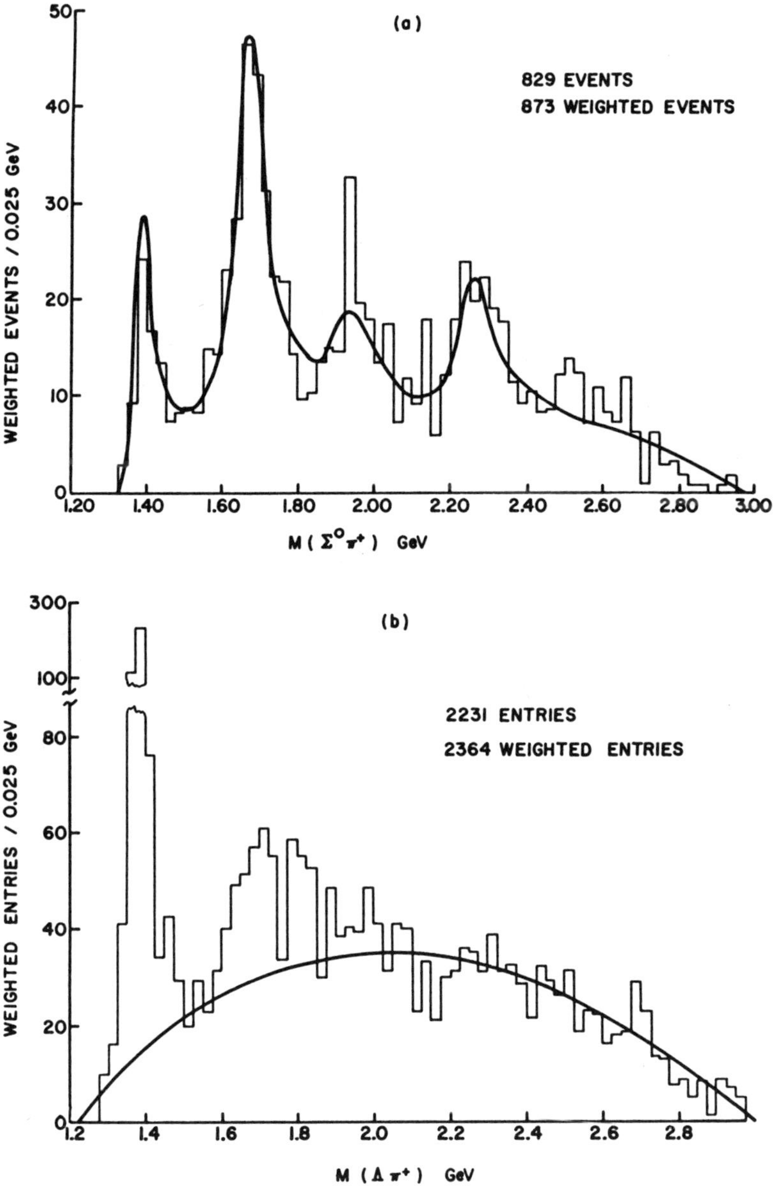

(a)
829 EVENTS
873 WEIGHTED EVENTS
WEIGHTED EVENTS / 0.025 GeV
M (Σ° π⁺) GeV
(b)
2231 ENTRIES
2364 WEIGHTED ENTRIES
WEIGHTED ENTRIES / 0.025 GeV
M (Λ π⁺) GeV

FIGURE 18

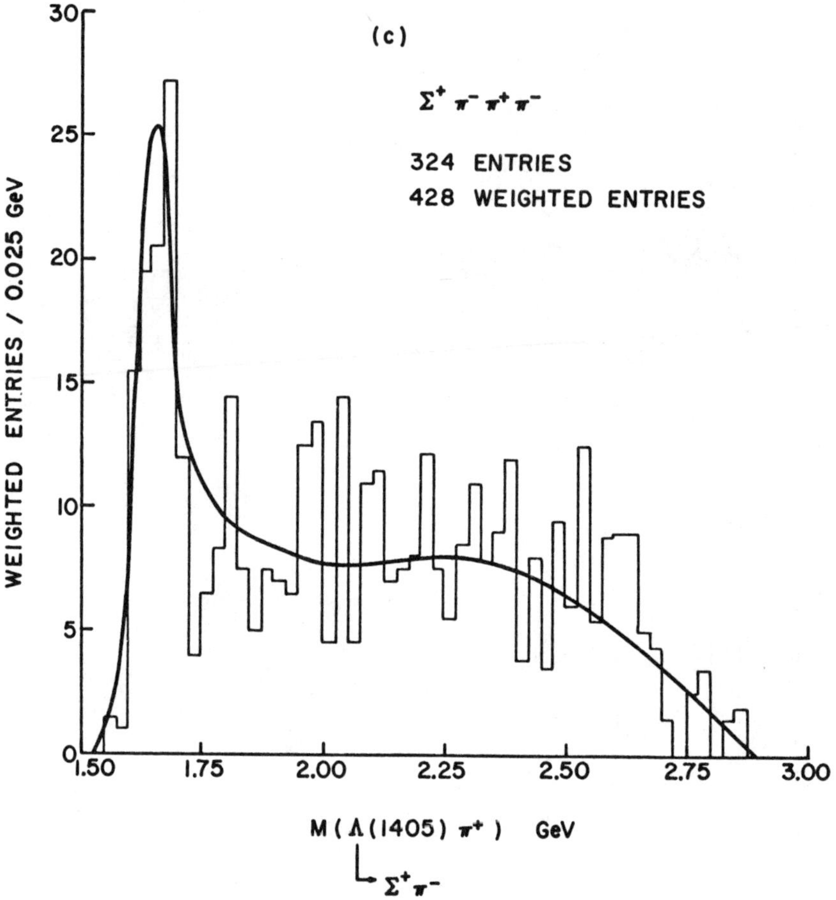

(c)
$\Sigma^+ \pi^- \pi^+ \pi^-$
324 ENTRIES
428 WEIGHTED ENTRIES
WEIGHTED ENTRIES / 0.025 GeV
0
5
10
15
20
25
30
1.50
1.75
2.00
2.25
2.50
2.75
3.00
M (Λ(1405) π^+) GeV
$\Sigma^+ \pi^-$

FIGURE 19

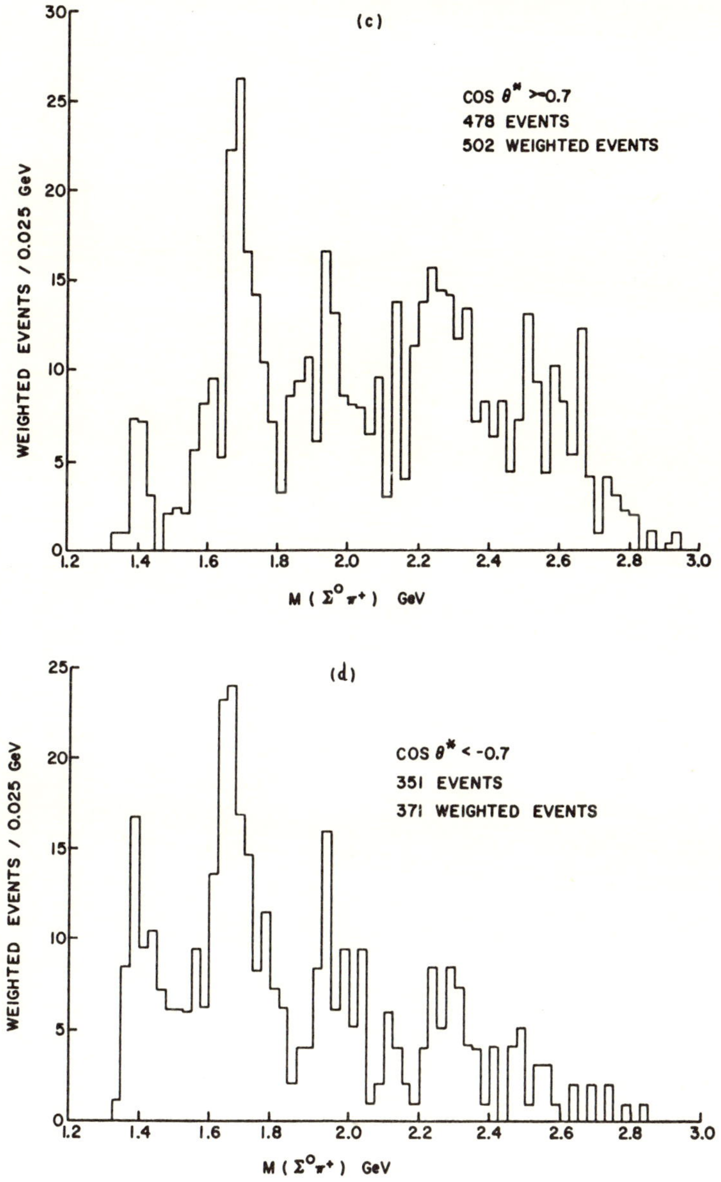
(c)
COS θ* > -0.7
478 EVENTS
502 WEIGHTED EVENTS
WEIGHTED EVENTS / 0.025 GeV
M (Σ⁰π⁺) GeV
(d)
COS θ* < -0.7
351 EVENTS
371 WEIGHTED EVENTS
WEIGHTED EVENTS / 0.025 GeV
M (Σ⁰π⁺) GeV

FIGURE 20

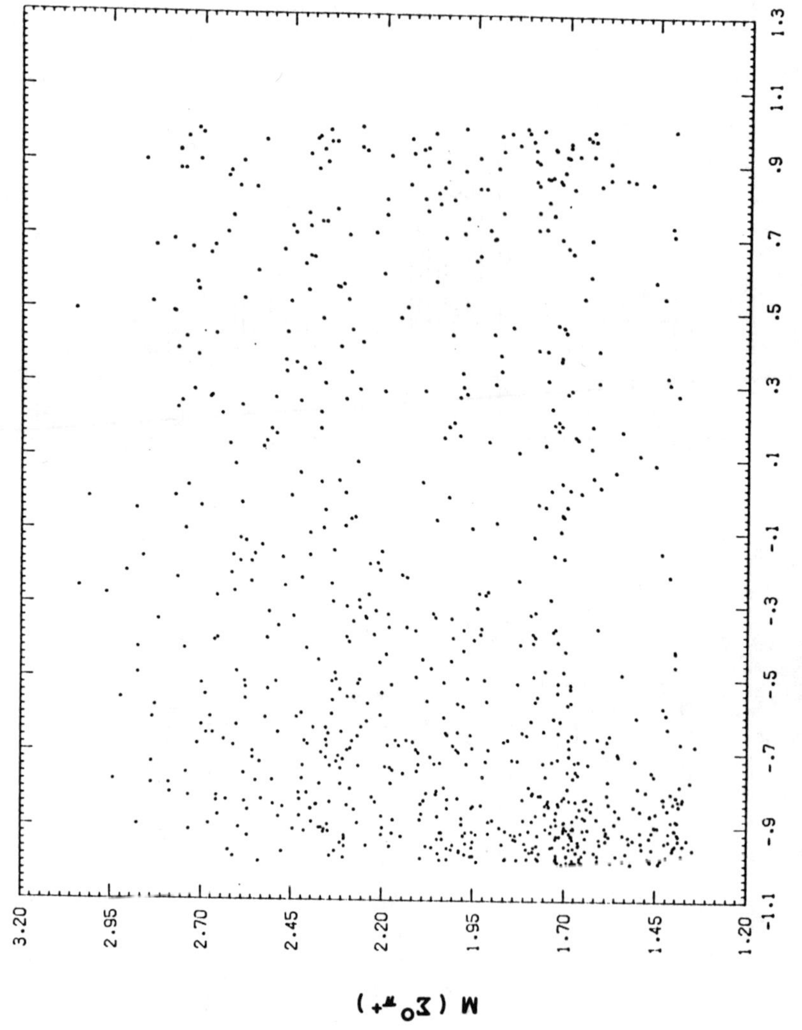
M (Σ₀π⁺)
3.20
2.95
2.70
2.45
2.20
1.95
1.70
1.45
1.20
-1.1
-.9
-.7
-.5
-.3
-.1
.1
.3
.5
.7
.9
1.1
1.3
cos θ*

FIGURE 21

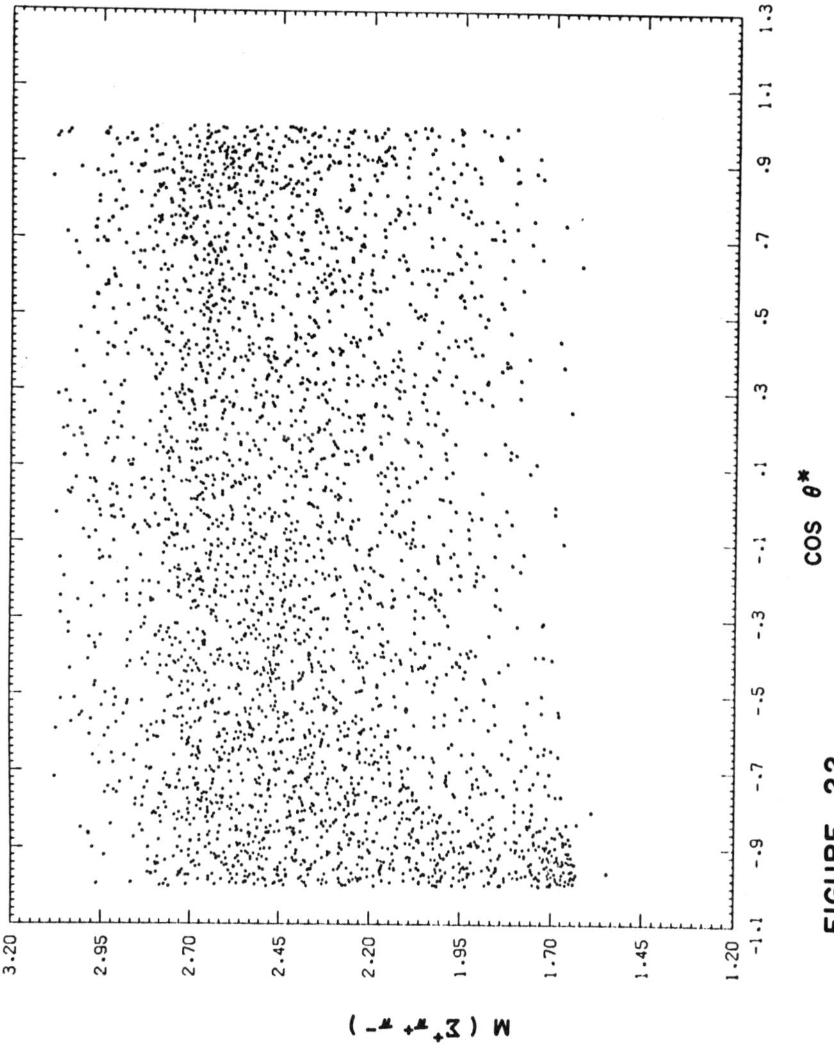
3.20
2.95
2.70
2.45
2.20
1.95
1.70
1.45
1.20
M (Σ⁺π⁺π⁻)
-1.1
-.9
-.7
-.5
-.3
-.1
.1
.3
.5
.7
.9
1.1
1.3
COS θ*

FIGURE 22

Table I

			Percentages		
Resonances	Mass	Width	Total Sample	$\cos\theta^* < -0.7$	$\cos\theta^* > -0.7$
$\Sigma(1385)$	1390 ± 6	33 ± 12	5.6 ± 1.1	11.7 ± 2.4	2.7 ± 1.0
$\Sigma(1670)$	1670 ± 6	110 ± 12	25.9 ± 2.3	37.2 ± 3.5	18.9 ± 2.9
$\Sigma(1940)$	1940 ± 11	90 ± 20	4.5 ± 1.6	3.7 ± 2.3	6.2 ± 2.3
$\Sigma(2280)$	2280 ± 14	100 ± 20	7.8 ± 1.8	6.4 ± 2.4	9.9 ± 2.8
ρ	756 ± 9	157 ± 22	29.6 ± 2.7	22.1 ± 4.0	33.9 ± 3.8
f^o	1265 ± 16	141 ± 35	10.2 ± 2.2	3.2 ± 2.7	13.7 ± 3.0

MODE	CROSS SECTION	MOMENTUM
$\Sigma^0\pi^+$	28 ± 5μb	3.9
$\Sigma^+\pi^+\pi^-$	22 ± 4μb	3.9
$\Sigma^0\pi^+$	25 ± 4μb	4.6
$\Sigma^+\pi^+\pi^-$	7 ± 3μb	4.6

FIGURE 23

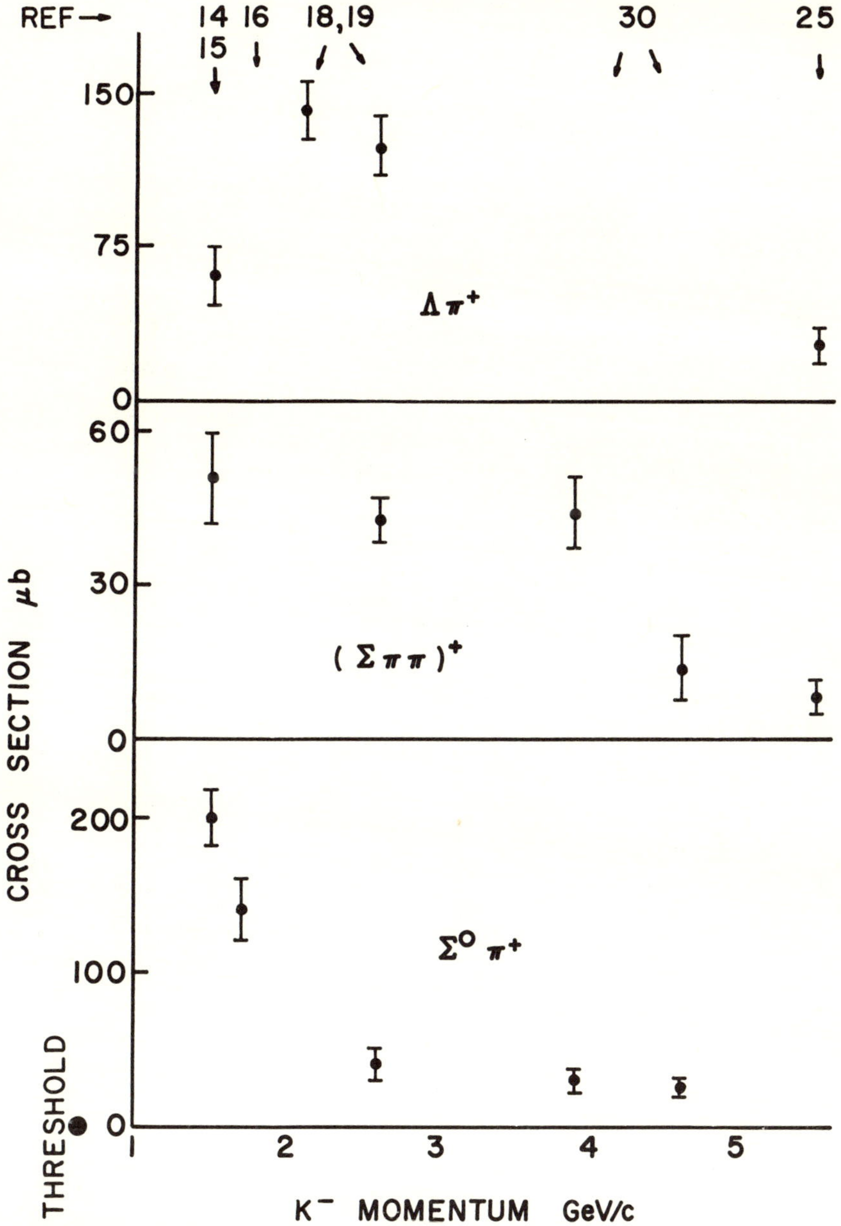
REF →
14
15
16
18,19
30
25
150
75
0
60
30
0
200
100
0
$\Lambda\pi^+$
$(\Sigma\pi\pi)^+$
$\Sigma^0\pi^+$
CROSS SECTION μb
THRESHOLD
1
2
3
4
5
K⁻ MOMENTUM GeV/c

FIGURE 24

crossing in the 1660 region. Certainly the KN coupling will be an interesting quantity to compare with the formation experiments.

The questions remain, therefore, as to how many objects exist in this mass region and what are their spin-parities. Both Berkeley and BNL mention the possibility of two $3/2^-$ resonances interfering. This could give varying properties of the 1660 region as a function of momentum. Because of the complexity shown above, I think it makes little sense to compare branching ratios for the published data before the number of states is unravelled and understood. Apart from the CHS formation data, the production experiments are consistent for all decay modes with a mass of $\sim$ 1670 and $\Gamma \sim 80 \rightarrow 100$ MeV. More data is needed to see if Γ is a function of cos θ^* for example. It is rather remarkable to me that the $\Sigma^o\pi^+$ is so non peripheral even up to 4.6 GeV/c and drops off so fast in cross section not to be seen at 5.5 GeV/c. Figure 24 shows the production cross sections of the various decay modes I have been able to extract as a function of energy. The variation as a function of beam momentum is not particularly abnormal for any of the modes. Detailed fitting will be required to separate the different states and interference effects and to reconcile production and formation data.

Resonances M > 1700

In the paper by Barnes <u>et al</u>.[31] in addition to the 1660 they observe $\Sigma\pi$ decay modes of the $\Sigma(1940)$ and $\Sigma(2280)$. This is shown in Fig. 25 together with the branching ratios. The 1940 is observed in the Barnes data to decay strongly into $\Sigma\pi$, a necessary condition for inclusion in the $J^P = 5/2^+$ octet. The most recent masses and widths from these data are shown in Fig. 23.

Another state seen clearly in production experiments is the 1815 with a strong $\bar{K}N$ signal. The best data, I think, come from the SABRE collaboration[33] (see Fig. 26) who see approximately 60 μb of 1815 into K^-p from the reaction $K^-n \rightarrow K^-p\pi^-$ at 3 GeV/c, at 2.24 GeV/c no obvious signal is seen with 552 events.[34] Similarly at 5.5 GeV/c with about 500 events no signal is seen.[35] Therefore, no conclusion can be drawn about the behavior as a function of energy.

SUMMARY

The number of states observed in production experiments is small compared to the phase shift data. As can be seen from the above review, the situation is also very unclear nearly a decade after, for example, the 1660 was first seen. It remains to be

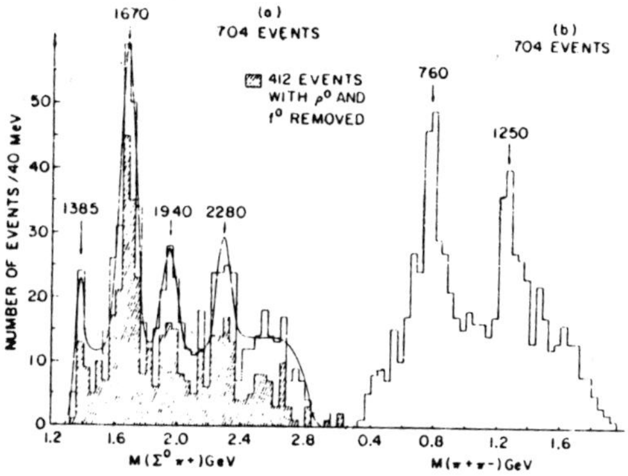

FIG. 2. (a) $\Sigma^0\pi^+$ mass spectrum for Reaction (A). The solid curve is a fit with four resonances. (b) $\pi^+\pi^-$ effective-mass distribution for Reaction (A).

Table I. Tabulation of the $\Lambda\pi/\Sigma\pi$ and $\bar{K}N/\Sigma\pi$ branching ratios for the $\Sigma(1660)$, $\Sigma(1940)$, and $\Sigma(2280)$ hyperons. The quoted upper limit corresponds to 1 standard deviation.

Hyperon	Branching ratios $\Lambda\pi/\Sigma\pi$	$\bar{K}N/\Sigma\pi$
$\Sigma(1660)$	0.52 ± 0.09	<0.06
$\Sigma(1940)$	<0.28	<0.37
$\Sigma(2280)$	<0.18	<0.18

FIGURE 25

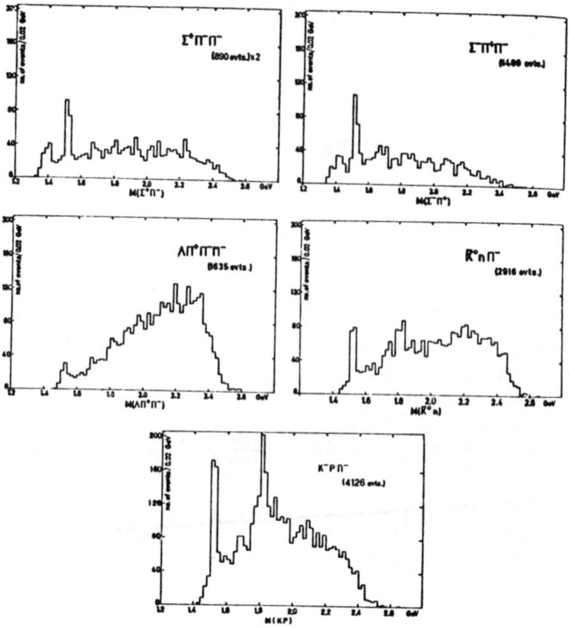

Fig. 1. Effective mass of:

$\Sigma^+\pi^-$ in the reaction $K^-d \to \Sigma^+\pi^-\pi^-(p_S)$
$\Sigma^-\pi^+$ in the reaction $K^-d \to \Sigma^-\pi^+\pi^-(p_S)$
$\Lambda\pi^+\pi^-$ in the reaction $K^-d \to \Lambda\pi^+\pi^-\pi^-(p_S)$
$\bar{K}^0n$ in the reaction $K^-d \to \bar{K}^0n\pi^-(p_S)$
K^-p in the reaction $K^-d \to K^-p\pi^-(p_S)$.

Table 1
Branching fractions of the $Y_0^*(1520)$.

Decay mode	K^-p	$\bar{K}^0n$	$\Sigma^-\pi^+$	$\Sigma^+\pi^-$	$\Sigma^0\pi^0$	$\Lambda\pi^+\pi^-$	$\Lambda\pi^0\pi^0$
Number of events	287	125	186	163		55	
Corrected cross section b) (μb)	65 ± 7	75 ± 11	47 ± 5	52 ± 7	(50 ± 10) a)	18 ± 4	(9 ± 2) a)
Branching fraction in %							
This experiment	44 ± 3		47 ± 3			9 ± 2	
ref. [3]	30 ± 4		55 ± 7			15 ± 2	
ref. [4]	47 ± 9		45 ± 4			8 ± 2	
ref. [5]	52 ± 4		37 ± 4			11 ± 2	
ref. [7]	47 ± 5		42 ± 5			11 ± 2	

a) The $\Sigma^0\pi^0$ and $\Lambda\pi^0\pi^0$ cross sections have been deduced from those for $\Sigma^\pm\pi^\mp$ and $\Lambda\pi^+\pi^-$ respectively.
b) A possible systematic overall error is not taken into account.

FIGURE 26

seen whether the disentangling of the production results will yield new physics e.g. the existence of two $3/2^-$ states overlapping. The need for more data over larger energy regions and perhaps in other reactions such as π nucleon is clear.

ACKNOWLEDGMENTS

I would like to thank the BNL group, particularly Drs. N. Samios, R. Eisner and V. Barnes for sending me their data prior to publication. I also appreciate the help they have given me in discussing their data. I would like to thank Dr. M. Pripstein for discussions concerning the $\Sigma(1660)$. In addition, I acknowledge the help of Dr. M. Derrick and my colleagues at Purdue for a critical review of this paper.

NEW RESULTS

The CHS group now quotes $\Lambda\pi^{+} \sim 11\%$, $\Sigma\pi \sim 44 \pm 10\%$ for the D13(1660) with $\Gamma = 44 \pm 4$ MeV. Possible resonances now seen in this region by various phase shift analyses are a $P_{11}(1660)$ coupling mainly to $\Sigma\pi$, a $S_{11}(1730)$ coupling to $\Lambda\pi^{+}$ in addition to the D_{13} and $D_{15}(1765)$. The production experiments may include contributions from all these states.

REFERENCES

1. R. Levi Setti in Proceedings of the Lund International Conference on Elementary Particles, 1969.
2. Robert D. Tripp, UCRL 19361 to be published in the Proceedings of the Third Topical Conference on Particle Physics, Honolulu, Hawaii, 1969.
3. Review of Particle Properties, Reviews of Modern Physics 42, 87 (1970).
4. Yu-Li Pan and Fred L. Forman, PRL 23, 806 (1969).
5. Yu-Li Pan and Fred L. Forman, PRL 23, 808 (1969).
6. Alberto Benvenuti, Erwin Marquit and Frank Oppenheimer, PR 187, 1852 (1969).
7. David J. Crennell, William C. Delaney, Enzo Flaminio, Uri Karshon, Kwan Wu Lai, Wesley J. Metzger, John S. O'Neall, J. Michael Scarr, Alan M. Thorndike, Phillip Baumel, Robert M. Lea, Alexander Montwill and Thomas G. Schumann, PRL 21, 648 (1968).
8. David J. Crennell, Uri Karshon, Kwan Wu Lai, John S. O'Neall, J. Michael Scarr, Phillip Baumel, Robert M. Lea, Thomas G. Schumann and Ernest M. Urvater, paper presented at the Lund International Conference on Elementary Particles 1969 (unpublished).
9. R. Barloutaud, D. W. Merrill, J. C. Scheuer, W. Hoogland, J. C. Kluyver, G. Giacomelli, A. Minguzzi-Ranzi, P. Serra, Y. Eisenberg, G. Yekutieli, J. Goldberg, A. Rouge and G. Lamidey, NP B16 201 (1970).
10. R. Armenteros, P. Baillon, C. Bricman, M. Ferro-Luzzi, D. E. Plane, N. Schmitz, E. Burkhardt, H. Filthuth, E. Kluge, H. Oberlack, R. Barloutaud, P. Granet, J. Meyer, J. P. Porte and J. Prévost, PL 28B, 521 (1969).
11. See References 24 and 25.
12. A. C. Ammann, A. F. Garfinkel, D. D. Carmony, L. J. Gutay, D. H. Miller and W. L. Yen, PRL 24, 327 (1970).
13. In addition the analysis of the $\overline{\Lambda}\pi^-\pi^+\pi^0$ final state is complicated by copious resonance production.
14. Luis W. Alvarez, Margaret H. Alston, Massimiliano Ferro-Luzzi, Darrell O. Huwe, George R. Kalbfleisch, Donald H. Miller, Joseph J. Murray, Arthur H. Rosenfeld, Janice B. Shafer, Frank T. Solmitz and Stanley G. Wojcicki, PRL 10, 184 (1963).
15. Darrell O. Huwe, PR 181, 1824 (1969).
16. J. Button-Shafer, PRL 21, 1123 (1968).
17. Phillipe Eberhard, Jerome H. Freidman, Morris Pripstein and Ronald R. Ross, PRL 22, 200 (1969).
18. In addition to the published results on the 2.6 GeV/c Berkeley data a paper will be presented at this meeting discussing the $\Sigma\pi\pi$, $\Sigma\pi$, $\Lambda\pi$ and $\overline{K}N$ decays of the 1660. The results indicate a very peripheral $\Sigma\pi\pi$ mode, and a $\Sigma\pi$ mode spread over all cos θ^*. The overall $\Lambda\pi^+/\Sigma^0\pi^+$ branching is

about 1.6 ± .6 but this varies depending on the cos θ^* interval. There is observable 1660^- production and also Σ(1765) production in both $\bar{K}N$ and $\Lambda\pi^+$. The overall analysis is, therefore, extremely complicated and I refer you to their paper for details. The above is a private communication from Dr. M. Pripstein.

19. Phillipe Eberhard, F. T. Shively, Ronald R. Ross and Daniel M. Siegel, J. R. Ficenec, Robert I. Hulsizer, D. W. Mortara, Morris Pripstein and William P. Swanson, PRL 14, 466 (1965).
20. P. Eberhard, M. Pripstein, F. T. Shively, U. E. Kruse and W. P. Swanson, PR 163, 1446 (1967).
21. A. Leveque, M. Ville, P. J. Negus, W. M. Blair, A. L. Grant, I. S. Hughes, R. M. Turnbull, A. A. Z. Ahmad, S. Baker, L. Celnikier, S. Misbahuddin, I. O. Skillicorn, J. G. Loken, R. L. Sekulin, J. H. Mulvey, A. R. Atherton, G. B. Chadwick, W. T. Davies, J. H. Field, P. M. D. Gray, D. E. Lawrence, L. Lyons, A. Oxley, C. A. Wilkinson, C. M. Fisher, E. Pickup, L. K. Rangan, J. M. Scarr and A. M. Segar, PL 18, 69 (1965).
22. See Reference 10.
23. G. W. London, R. R. Rau, N. P. Samios, S. S. Yamamoto, M. Goldberg, S. Lichtman, M. Primer and J. Leitner, PR 143, 1034 (1966).
24. M. Derrick, T. Fields, J. Loken, R. Ammar, R. E. P. Davis, W. Kropac, J. Mott and F. Schweingruber, PRL 18, 266 (1967).
25. J. Mott, R. Ammar, R. Davis, W. Kropac, D. Slate and B. Werner, S. Dagan, M. Derrick, T. Fields, J. Loken and F. Schweingruber, PR 177, 1966 (1969).
26. V. E. Barnes, S. U. Chung, R. L. Eisner, E. Flaminio, J. B. Kinson, N. P. Samios, M. Goldberg, K. Jaeger, BNL 13823, presented at the Bounder Conference on Elementary Particles, Boulder, Colorado, 1969.
27. U. Kruse, J. S. Loos, E. L. Goldwasser, PR 177, 1951 (1969).
28. D. C. Colley, F. Macdonald, B. Musgrave, W. M. R. Blair, I. S. Hughes, R. M. Turnbull, S. J. Goldsack, K. Paler, L. K. Sisterson, W. Blum, W. W. M. Allison, D. H. Locke, L. Lyons, P. J. Finney, C. M. Fisher, and A. M. Segar, PL 24B, 489 (1967).
29. M. Primer, M. Goldberg, K. Jaeger, V. Barnes, P. Dornan, I. Skillicorn and Jack Leitner, PRL 20, 610 (1968).
30. M. Aguilar-Benitez, V. E. Barnes, D. Bassano, S. U. Chung, R. L. Eisner, E. Flaminio, J. B. Kinson, N. P. Samios and K. Jaeger, submitted to Phys. Rev. Letters.
31. V. E. Barnes, E. Flaminio, L. Montanet, N. P. Samios and I. O. Skillicorn, PRL 22, 479 (1969).
32. Birmingham-Glasgow-London (I.C.)-Oxford-Rutherford Collaboration, PR 152, 1148 (1966).
33. J. C. Scheuer, D. Merrill, A. Verglas, S. A. De Wit, A. G. Tenner, A. Minguzzi-Ranzi, A. M. Rossi, U. Karshon, A. Shapira, E. Huffer and G. Lamidey, NP B8, 503 (1968).

34. K. F. Galloway, E. D. Alyea Jr., J. P. Chandler, R. R. Crittenden and H. J. Martin Jr., NP B8, 545 (1968).
35. W. Kropac, R. Ammar, R. E. P. Davis, B. Werner, H. Yarger, Y. Cho, M. Derrick, B. Musgrave, J. J. Phelan, T. P. Wangler, Argonne Report (unpublished).

DISCUSSION

Derrick: Let's have no discussion about the Σ (1660) until after the next three papers.

Rosenfeld: I would like to make one remark about our tables if I may. You may not like exactly what was said in the note in the listings but as far as the tables themselves are concerned, I quote to you: "Σ (1660) D_{13}. The branching ratios as reported here are from formation experiments. Production experiments still confused. See note in listings." Isn't that consistent with your point of view.

Miller: I think that's very clear but personally I think it was a mistake to include that branching ratio table when the Eberhard data was so clear that the $\Sigma\pi\pi$ was so different from the $\Sigma\pi$. Once you are given that fact, then you have to worry about the $\Lambda\pi$ and associating some of it with the $\Sigma\pi$. And, of course, in the meantime the Brookhaven paper was never published.

Barbaro-Galtieri: Let me comment on what happens in the back of the book. In the back of the book there is this plot which includes the BNL data that might have been retracted but I didn't know at the time. I took it from a widely circulated preprint.

Goldberg: That curve is wrong by the way. That curve originally came from Barnes but it obviously came to you through another route and is mistaken. The thing is that the branching ratio is really $\Lambda\pi$ to $\Sigma^o\pi^o$, not to $\Sigma\pi$ total.

Barbaro-Galtieri: That's all right. You just have to put the $\Sigma^+\pi^o$ in.

Goldberg: But many people don't realize that.

Barbaro-Galtieri: But that doesn't change the relative disagreement of the data. What it says in the back of the book is that "the evidence for a large ratio is now based on experiments with small statistics. Still unexplained is a small value of R at low incident K^- energy and the variation of R' with momentum transfer." So the statement is not that only those experiments with small statistics do not agree with the formation experiments, which is what you said. I said that the large ratio is based on experiments with small statistics.

Miller: I certainly agree. Unfortunately you imply that those $\Lambda\pi$ to $\Sigma\pi$ branching ratios are correct and that is not true. The situation is much more complicated than that. People have handled their data in very different ways using different widths and so on. Brookhaven has different numbers -- particularly for the $\Lambda\pi$.

Barbaro-Galtieri: There is no implication in this plot or the note which goes with it. Just the fact that the situation is confused. People have quoted the numbers and here they are.

Derrick: The thing that really upset some people, including me, was that you showed a graph which is incorrect and never published. And you took $\frac{1}{4}$ of a page even if it is in the back of the book to show this implying that it is a very important thing.

Barbaro-Galtieri: Well I was trying to put together all the information there is and this information is the $\Sigma\pi$ to $\Lambda\pi$ ratio.

Goldberg: There's one other thing. Alvarez's data is included there in addition to Huwe's data as two separate pieces of data which I hardly think is fair since in Huwe's thesis (D. Huwe, P.R. 181, 1824(1969)) he claimed that Alvarez's data is wrong and that his was an updating of Alvarez's data.

Derrick: I think that's enough beating on the wallet card people.

Kalmus: We have data pertaining to the $Y_1^*(1480)$ found by Pan and Forman. We have about 60% of the data of Pan and Forman in $\pi^+p \to \Lambda K^+ \pi^+$ and $\Sigma^{+}_{0} K^{0\,+}_{+\,0}$ above $Y_1^*(1480)$ threshold. We do not see any enhancement at 1480 in the $\Lambda \bar{K}\pi$ channel. Their data already was not very statistically significant and ours is even less. If we add our data together, it reduces the significance of the bump in the total data. We also did the calculation which you did for the shape of the background if the $N^*(1688)$ decays into ΛK, and indeed it does peak in that region. The significant thing in the data is really the dip and not the peak. You could get an enhancement there easily. In the $\Sigma K\pi$ the statistics are very poor. With some imagination there is a split between a 1385 peak and a higher peak.

London: What about polarization?

Kalmus: Their data on Σ polarization I think was wrong by a factor of 3. I have a feeling the Σ^0 polarization went to 1.0 and I think it should only go to 0.3. In the Λ we have fewer events, therefore the errors are larger. The other thing is that the curve that is shown through the polarization points is a convenient way for the eye to follow the points. You can draw it differently and it doesn't look as significant. (See Figure 2 in Miller's Review).

Miller: I was surprised that Pan's error bars were so small. Particularly in the Σ^0 case, it is very difficult to get good polarization measurements and the number of events is not great.

Kalmus: I think that their graph is wrong.

Tripp: That's correct. According to the dashed line the Σ^o polarization is three times bigger than it could possible be.

Pan: (Comment added in proof.) As one can see from both Fig.2A and the figure caption in our Physical Review Letter article, we plotted the Λ^o polarization from the Σ^o decay. We stated in the text that the Σ^o polarization is related to the Λ^o polarization by the relation $\overline{P}(\Sigma^o) = -3\overline{P}(\Lambda^o)$. We also emphasized in the figure caption that the dashed line on the figure under discussion is just a free-hand curve drawn through the experimental data. The errors were calculated from maximum likelihood technique.

It is true that the point near 2 GeV^2 in the $\Sigma^o \pi^+$ mass-square gave a Λ^o polarization of approximately -0.9 with an error of approximately $\pm$0.3. (This value is only two-standard-deviations from a value of -0.3.) This would result in a Σ^o polarization of approximately +2.7 in this mass-square region. Such a large value can be due to two reasons. It can be due to the lack of statistics or due to the fact that the data do not fit the linear relationship used in the polarization calculation. I feel it is the lack of statistics in our case.

How one presents one's data is to some extent a personal thing. Thus, it would be ludicrous to say someone's result is wrong because one does not like the way it is presented. I like data to be presented with as little processing as possible. One can always take the data and make more calculations but not vice versa. I feel the way we presented the data gives one more feel as to the real significance of the results than if we had processed the data more and gave the Σ^o polarization.

There is nothing wrong with our Λ^o polarization data.

Barloutaud: I would like to comment on your explanation why the Σ (1620) is not produced in association with one pion. You said it could be strongly coupled to $\overline{K}N$. In this case we should see its decay to K^-p as well as its presence in formation experiments.

Miller: Yes, that's true. Of course the 1620 seen in the CHS is now in some disrepute -- at least since yesterday. I'm not sure whether it's the same particle Kim wants to be a P_{11} near 1660 which is not what the CHS people want.

Kraemer: What's the situation on the 1620 for just the events with measured spectators? Can you combine your data with Kwan Lai's data?

Miller: We have not done so. We don't have enough statistics on our own. If we just take our events when we see spectators they look the same but, of course, the statistical significance is down by a factor of three. I have not seen his $\Sigma^o \pi$ spectrum or his $\overline{K}N$ spectrum. He's collecting more data so presumably there will be an answer to this problem.

Ma: In the 1620 case you say that your branching ratio for $\Sigma\pi$ to $\Lambda\pi$ is something like 1.1. Is it consistent with the Brookhaven data that's 0.2?

Miller: That's an upper limit. It's very difficult with our data. The Σ (1620) sits on very high backgrounds and it's very difficult to get good branching ratios out. Really the only place that any signal has been seen is in the $\Lambda\pi$ spectrum and so you look at all other spectra and try to estimate how much you see. They're all upper limits and it's very difficult because you have large background to deal with.

OBSERVATION OF Σ RESONANCES IN THE 1670 MEV MASS REGION*

M. Aguilar-Benitez, V. E. Barnes,† D. Bassano, S. Chung,‡
R. L. Eisner, E. Flaminio, J. B. Kinson, and N. P. Samios

Brookhaven National Laboratory, Upton, New York 11973

K. Jaeger

Syracuse University, Syracuse, New York 13210

We have studied the reaction $K^-p \to \Sigma(1670)^+\pi^-$ at incident beam momenta of 3.9 and 4.6 GeV/c in the BNL 80-inch hydrogen bubble chamber for the following final states:

$$\Sigma^+\pi^+\pi^-\pi^- \quad (1)$$

$$\Sigma^-\pi^-\pi^+\pi^+ \quad (2)$$

$$\Sigma^0\pi^+\pi^- \quad (3)$$

$$p\bar{K}^0\pi^- \quad (4)$$

$$\Lambda\pi^+\pi^- \quad (5)$$

We observe enhancements in the hyperon mass combination produced opposite the π^- at about 1670 MeV in all reactions except (4).[1] Recently an LRL group[2] has reported that the production angular distribution for the $\Sigma^+(1660)$ produced in reaction (3) is less peripheral than that for $\Sigma^+(1660)$ in reactions (1) and (2). We present confirmation of that result at higher incident beam momenta.

In Fig. 1a we plot the mass of the $\Sigma\pi\pi$ from reactions (1) and (2) against $\cos\theta^*$ (where θ^* is the angle that the hyperon system makes with the beam direction in the overall center of mass). It is clear that there is resonance production near a mass of 1670 MeV which is very peripheral. The absence of a signal in Fig. 1b, in which only events having $\cos\theta^* > -0.7$ are plotted, confirms this observation. Since this $\Sigma(1670)$ peak is consistent with being all $\Lambda(1405)\pi$, we display the $\Lambda(1405)\pi^+$ mass spectrum for reaction (1)[3]

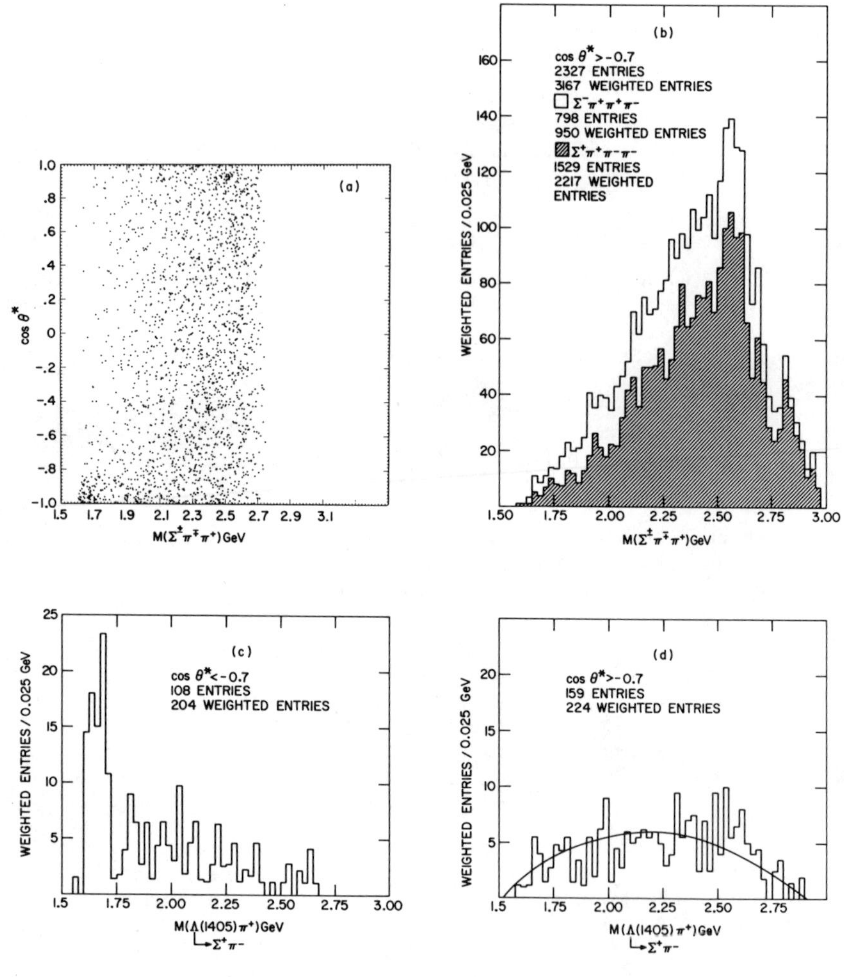

Figure 1

a) Two-dimensional distribution of $\cos\theta^*$ vs. $M(\Sigma^{\pm}\pi^{\mp}\pi^{+})$ for events of reactions (1) and (2). b) $\Sigma^{\pm}\pi^{\mp}\pi^{+}$ effective mass spectrum from reactions (1) (shaded) and (2) with the restriction $\cos\theta^* > -0.7$. c) $\Sigma^{+}\pi^{-}\pi^{+}$ effective mass spectrum from reaction (1) with the $\Sigma^{+}\pi^{-}$ required to form a $\Lambda(1405)$ and $\cos\theta^* < -0.7$. d) Same as c) but with $\cos\theta^* > -0.7$.

for the $\cos\theta^* < -0.7$ ($\cos\theta^* > -0.7$) region in Fig. 1c (1d). It is clear that the entire enhancement is produced peripherally. If we define α as the ratio of the number of $\Sigma(1670)$ events with $\cos\theta^* > -0.7$ to total $\Sigma(1670)$, we obtain $\alpha(\Sigma\pi\pi) = 0.0 \pm 0.04$.

We now turn to the $\Sigma^o\pi^+$ final state from reaction (3). The most prominent feature in the $\Sigma^o\pi^+$ mass spectra seen in Fig. 2a is the large peak at 1670 MeV with a width of about 110 MeV. From Fig. 2b, which displays a 2-dimensional plot of the $\Sigma^o\pi^+$ mass versus its production angle, we see that the 1670 peak has contributions from all $\cos\theta^*$ regions. This is shown more clearly in Figs. 2c and 2d which contain those events with $\cos\theta^* > -0.7$ (non-peripheral), and $\cos\theta^* < -0.7$ (peripheral) respectively. In contrast with the $(\Sigma\pi\pi)^+$ case, we see that $\Sigma^o\pi^+$ peak at 1670 is produced non-peripherally as well as peripherally. We calculate $\alpha(\Sigma^o\pi^+) = 0.42 \pm 0.08$ which is greater than 4-standard deviations away from $\alpha(\Sigma\pi\pi) = 0.0 \pm 0.04$.

We now discuss reactions (4) and (5). Fig. 3a shows the $p\bar{K}^o$ invariant mass plotted for reaction (4). We notice some possible indication of $\Sigma(1765)$ production. There is no signal in the 1660 region in agreement with other experiments which find very small $\Sigma(1660)\bar{K}N$ coupling. Fig. 3b shows the $\Lambda\pi^+$ effective mass distribution from the $\Lambda\pi^+\pi^-$ final state in which we observe a broad amorphous structure in the mass region 1600 to 1900 MeV. It is therefor difficult to know how to draw the background, especially in light of the fact that there may be many resonances in this region as we have seen from phase shift analyses presented at this Conference. If we estimate the background by normalizing phase space to the number of events in the region $M(\Lambda^o\pi^+) > 2.0$ GeV (solid curve in Fig. 3b) we get a ratio of $\Lambda\pi^+/(\Lambda\pi^+ + \Sigma^o\pi^+) = 0.55$. Since one might very easily conceive of a background which is higher than the phase space curve, the value of 0.55 for this ratio should be considered close to an upper limit.

In conclusion we find that there are at least two resonances in the 1670 MeV mass region as indicated by the different production angular distributions for the $\Sigma^o\pi^+$ and the $\Sigma\pi\pi$ final states. There still exists the possibility of a third resonance[4] decaying mainly to $\Lambda\pi^+$ but we are not able to decide whether this is the case. Complicated interferences between two resonances can conceivably produce branching ratios which vary as a function of incident beam momentum.

---

* Work performed under the auspices of the U.S. Atomic Energy Commission.

† Present address: Purdue University, Lafayette, Indiana.

‡ Present address: CERN, Geneva, Switzerland.

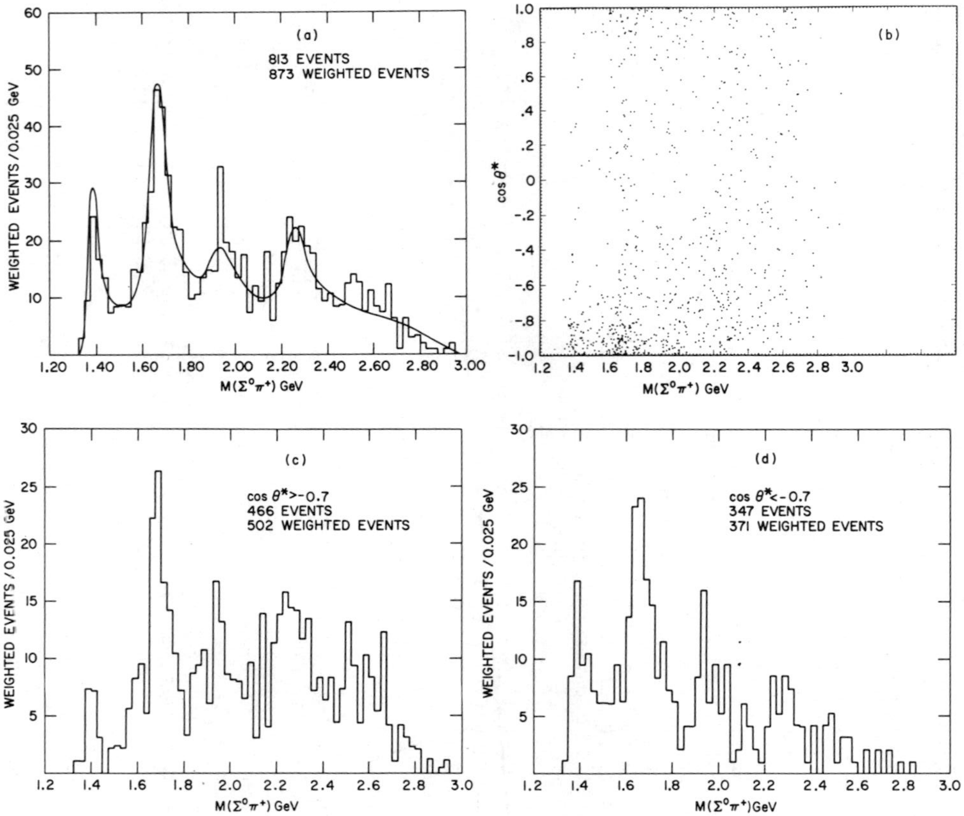

Figure 2

a) $\Sigma^o\pi^+$ effective mass spectrum from reaction (3).

b) Two-dimensional distribution of $\cos\theta^*$ vs. $M(\Sigma^o\pi^+)$ for events of reaction (3).

c) $\Sigma^o\pi^+$ effective mass spectrum from reaction (3) with $\cos\theta^* > -0.7$.

d) Same as c) but with $\cos\theta^* < -0.7$.

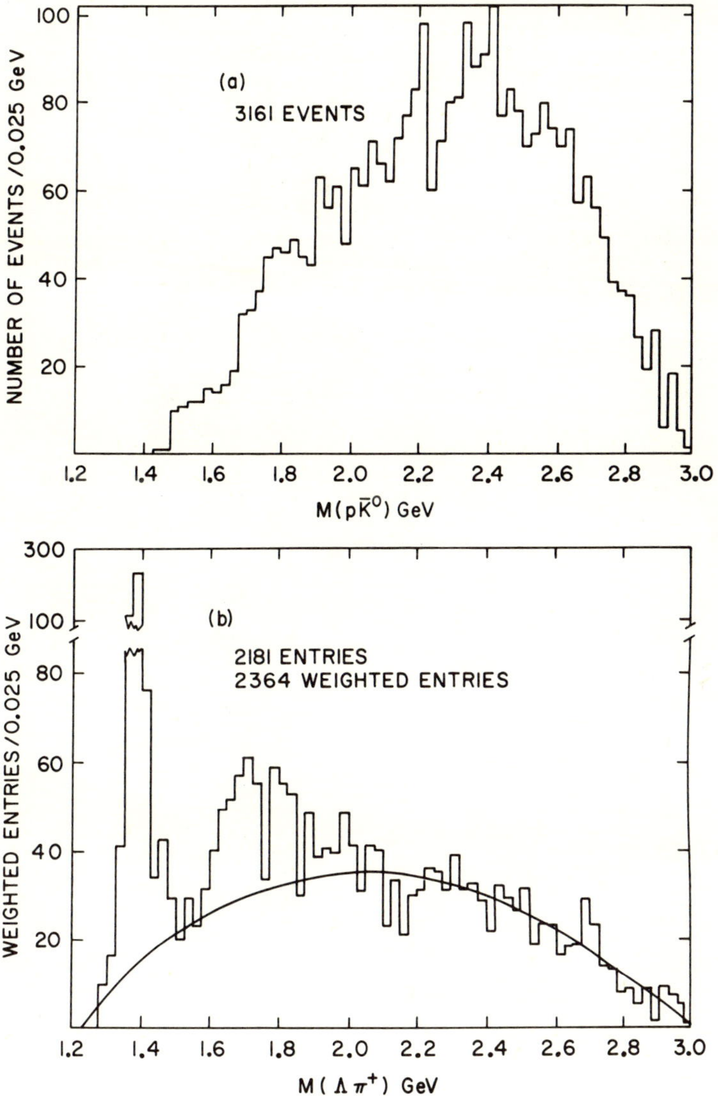

Figure 3

a) p$\bar{K}^0$ effective mass spectrum from reaction (4). b) $\Lambda\pi^+$ effective mass spectrum from reaction (5).

Footnotes

1. A more complete discussion of this analysis will be published (Aguilar-Benitez et al., submitted to Phys. Rev. Letters). Also see Barnes et al., Phys. Rev. Letters 22, 479 (1969).

2. P. Eberhard et al., Phys.Rev. Letters 22, 200 (1969).

3. Since the two $\Lambda(1405)$ bands of reaction (2) overlap on the $\Sigma^-\pi^+\pi^+$ Dalitz plot, the $\Sigma(1670)$ peak may be caused by constructive interference between the two resonance amplitudes. This problem is not present in reaction (1) which is therefore more appropriate for investigating the $\Sigma(1670)$ properties.

4. M. Derrick et al., Phys. Rev. Letters 18, 266 (1967).

REMARKS ON THE EXISTENCE OF TWO Σ(1660) RESONANCES

Robert D. Estes, Jean Duboc[†], Philippe H. Eberhard,
Jerome H. Friedman, Morris Pripstein, and Ronald R. Ross

Lawrence Radiation Laboratory, University of California,
Berkeley, California

We are in the process of analyzing the following reactions:

$$K^-p \to \Sigma^+\pi^+\pi^-\pi^- \quad (1)$$

$$\to \Sigma^-\pi^+\pi^+\pi^- \quad (2)$$

$$\to \Sigma^0\pi^+\pi^- \quad (3)$$

$$\to \Sigma^+\pi^0\pi^- \quad (4)$$

$$\to \Lambda\pi^+\pi^- \quad (5)$$

for the purpose of studying the quasi-two-body reaction

$$K^-p \to X^+\pi^- \quad (6)$$

with special emphasis on the X^+ mass region near 1660 MeV.

The data were obtained from an exposure of the Berkeley 72-inch hydrogen bubble chamber to a K^- beam from the Bevatron at momenta of 2.1, 2.58, 2.61, and 2.70 GeV/c. The K^- pathlength equivalent at 2.1 GeV/c is 5.8 events/μb and for the combined upper three moments it is 12.8 events/μb.

In a previous publication[1] we discussed the data of reactions (1) through (4) at the upper three momenta for various center-of-mass production cosine (cos Θ*) intervals, where Θ* is the angle between the target proton and X^+. These intervals were: $.95 < \cos\Theta^* < 1.0$ (region I), $.9 < \cos\Theta^* < .95$ (region II), and $.7 < \cos\Theta^* < .9$ (region III).

[†]Permanent address: Institut de Physique Nucleaire, 9 Quai St. Bernard, Paris V, France.

We measured the production differential cross section in each of these intervals for a $\Sigma(1660)$ resonance decaying via the modes $\Sigma^0\pi^+$, $\Sigma^+\pi^0$ $\Sigma^\pm\pi^\mp\pi^+$ under the assumption of no interference between resonance and background amplitudes. The measurements showed that the ratio $\Sigma^0\pi^+/\Sigma^+\pi^0$ is consistent with unity in all three intervals. However, the ratio of $(\Sigma\pi\pi)$ to $\Sigma\pi$ is different in interval I by more than 3 standard deviations from its value in interval III.

For this conference, we have performed a similar analysis at 2.1 GeV/c. These data exhibit the same general behavior, thus corroborating the results obtained at the upper momenta. As in reference 1, we interpret the above phenomenon to imply the existence of two $\Sigma(1660)$ resonances. The existence of two such resonances allows a variety of possibilities of which the following are three extreme cases:

1. There are two $\Sigma(1660)$ resonances, essentially degenerate in mass and width, but whose amounts are never affected by interference effects between them after integration over all decay angles.

2. There are two $\Sigma(1660)$ resonances, degenerate in mass and width, interfering after integration over decay angles. This of course can happen only if they have the same spin-parity.

3. There are two $\Sigma(1660)$'s of the same spin-parity, one of mass and width similar to that observed on the mass plot, and another one at a different mass and very broad, but overlapping the 1660 region. Both would have fixed branching ratios into $\Sigma\pi\pi$, $\Sigma\pi$, and $\Lambda\pi$ states and would be interfering.

Case 2 and 3 are considered in addition to case 1 because there exists in the literature separate spin-parity determinations of $3/2^-$ for a $\Sigma(1660)$ resonance in the $\Sigma\pi\pi$ (Reference 2) and $\Sigma\pi$ (Reference 3) modes of decay. However, for the production experiments, the analyses assumed the presence of one resonance only.

We plan to add information from reaction (5) and employ a new analysis [4] based on three distinct channels of decay in order to further help distinguish between these three cases.

We intend to measure for various incident energy and production cosine intervals the following partial branching ratios of the 1660-MeV enhancement: $R_1=(\Sigma\pi)^+/T$, $R_2=\Lambda(1405)\pi^+/T$, and $R_3=\Lambda\pi^+/T$, where $T=(\Sigma\pi)^+ + \Lambda(1405)\pi^+ + \Lambda\pi^+$. Only two of the three ratios are independent since $R_1 + R_2 + R_3=1$ for any experiment.

The results of any experiment measuring the three rates can be recorded as a point in a plane with Cartesian axes for the variables R_1 and R_2. The physical domain for the case R_1, R_2, $R_3 > 0$ is a triangle as shown in Fig. 1.

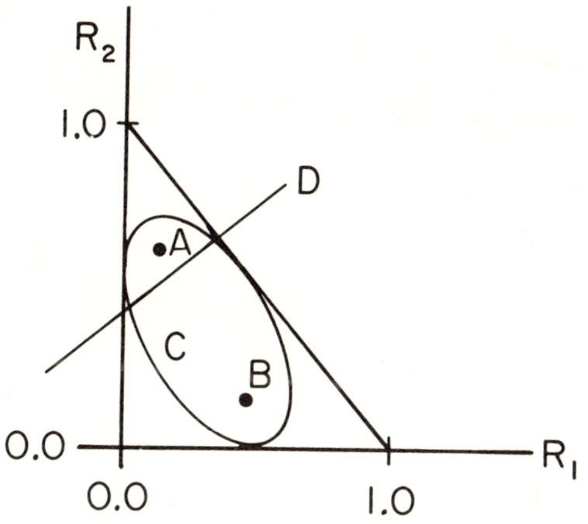

Fig. 1. Illustration of a partial branching ratio plot.

For a single resonance, all experiments must give answers consistent with a single point in the plane, such as point A in Fig. 1.

Two noninterfering resonances (case 1) having different branching ratios corresponding to points A and B on Fig. 1 would give rise to experimental results that could lie anywhere on the line joining points A and B.

If there exist two interfering resonances, it is possible to have different results, depending on the relation between the masses and widths. Cases 2 and 3 are the extreme possibilities. Case 2 (interfering resonances, degenerate in mass and width) would give rise to experimental results anywhere within an ellipse that is tangent to all three sides of the triangle, such as curve C. The other extreme case (case 3) could yield the same results as case 2. However, if the different decay amplitudes are relatively real, then the results would be constrained to lie on a line passing through the triangle. It is even possible in this case for experiments to show negative amounts in some of the channels, and therefore the corresponding points would lie outside the triangle, such as D of Fig. 1.

At this conference, a plot based on preliminary results was presented for purposes of illustrations. The points were such that no conclusion could be drawn about the relative likelihood of cases 1, 2, and 3. However, the results were only preliminary and are subject to future modifications. When definitive results are obtained, we hope to be able to differentiate between some of those cases.

We thank Werner Koelner for his help in fitting of experimental distributions and R. Bock, A. Sterling, and D. Siegel for help in the analysis of these data at a preliminary stage of the experiment.

References

1. Philippe H. Eberhard, Jerome H. Friedman, Morris Pripstein, and Ronald R. Ross, Phys. Rev. Letters 22, 200 (1969).

2. P.H. Eberhard, M. Pripstein, F.T. Shively, U.E. Kruse, and W.P. Swanson, Phys. Rev. 163, 1446 (1967).

3. R. Armenteros, P. Baillon, C. Brieman, M. Ferro-Luzzi, D.E. Plane, N. Schmitz, E. Burkhard, H. Filthuth, E. Kluge, H. Overlack, R.R. Ross, R. Barloutand, P. Granet, J. Meyer, J.P. Porte, and J. Prevost, Nucl. Phys. B8. 223. (1968); and Janice Button Shafer, Phys, Rev. Letters 21, 1123 (1968).

4. Philippe Eberhard, Lawrence Radiation Laboratory Physics Notes 707, April 16, 1970 (unpublished).

PROPERTIES OF $\Sigma(1660)$ OBSERVED IN FORMATION EXPERIMENTS

D. E. Plane

CERN

MASS AND WIDTH FROM PARTIAL-WAVE ANALYSES OF $K^-p \rightarrow \Sigma\pi$

(a)	C.H.S.*	M = 1661	Γ = 44	Nucl. Phys. B8, 223 (1968)
(b)	B.N.L.*	M = 1672	Γ = 34	Phys. Lett. 30B, 430 (1969)

BRANCHING FRACTIONS

(a)	$\overline{K}N^*$ $x = 0.09 \pm .02$	C.H.S.	Nucl. Phys. B8, 195 (1968)
(b)	$\Sigma\pi^*$ $x' = 0.44 \pm .10$	C.H.S.	Nucl. Phys. B8, 223 (1968)
		B.N.L.	Phys. Lett. 30B, 430 (1969)
(c)	$\Lambda\pi^*$ $x' \simeq 0.11$	C.H.S.	Unpublished, submitted to Duke Conference

(d) $\Sigma\pi\pi$

Take $(\Sigma\pi\pi)$ in Ispin = I and $(\Sigma\pi)$ in Ispin T: $[(\Sigma\pi)_T\pi]_T$

$I = 1$ $(\Sigma\pi)_0\pi$ $x' < .06$	C.H.S.	Nucl. Phys. B8, 216 (1968)
$T = 1$ $(\Sigma\pi)_0\pi$ $x' = .08$		Florida contribution to Duke Conference

The above numbers should contain any $\Sigma(1660) \rightarrow \Lambda(1405)\pi$

$I = 1$ $(\Sigma\pi)_1\pi$ $x' < 0.14$	C.H.S.	
$\Lambda(1405)\pi$ $x' < 0.08$	B.N.L.	Phys. Lett. 30B, 430 (1969)

(e) $\Lambda\pi\pi$

$x' < 0.11$ Nearly all $\Sigma(1385)\pi$	C.H.S. 28B, 521 (1969)
$\Sigma(1385)^*\pi$ $x' = 0.34 \pm .08$	Florida, Tufts, Brandeis Phys. Rev. Lett. 21, 1413 (1968)

*Effects observed in $3/2^-$ wave.

DISCUSSION

Button-Shafer: I believe I can speak for Darrell Huwe as well as myself in suggesting that the $\Lambda\pi$ /$\Sigma\pi$ ratio rather than its inverse be quoted for $Y^*(1660)$ decay. In Huwe's data from 1.0 to 1.7 BeV/c and the later run at 1.7 BeV/c for which I was responsible, the 1660 $\rightarrow\Lambda\pi$ decay mode was consistent with zero. I think Huwe's $\Lambda\pi$ / $\Sigma\pi$ value from our earlier data was about 0.25 ± 0.10, whereas the number I reported on the later 1.7 BeV/c data was 0.21 ± 0.10 approximately. (These numbers have nearly Gaussian errors.) My data sample was not only three times Huwe's data at 1.7 GeV/c (where the largest 1660 signal had been seen); it was 2 to 3 times the Primer-Goldberg et al. data and still larger compared with Derrick et al. or Colley et al. Our 0.2 ratio of $\Lambda\pi$ / $\Sigma\pi$ is nicely consistent with the 0.1 value the CHS group is now finding for $\Lambda\pi$ / total in their formation study of the $Y_1^*(1660)$.

Derrick: From Kim's analysis it appears that $\Lambda\pi$ is also small although he tells me it will be another four weeks before he's willing to give a number.

Tuan: I gather from Dr. Ross's presentation that two Σ (1660)'s with $J^P = 3/2$ are possible in this mass region. I would like to ask him or any other member of this audience whether in this mass region there exists another Λ(besides Λ (1690)) with $J^P = 3/2^-$. Such an additional Λ(~1660 MeV) can then complete two nonchiral SU(2) x SU(2) doublets (belonging to the (1/2, 1/2) representation) in the "clashing" internal symmetries scheme (S.F. Tuan PRL 23, 1198(1969)).

Hemingway: Can I ask Ron Ross how you got your rates?

Ross: Yes. Do you want to know now publically?

Hemingway: Yes.

Ross: I have some slides I'll show you. These are slides of the fits to the data and you can judge for yourself on whether we understand it at this time. Let me show you slide 4 which is the overall fit to the data at 2.6 GeV. This is the $\Lambda\pi^+$ data. These are 20 MeV bins. Here you see the 1385 which is cut off. It goes up to 461 events in the central bin. This is the 1660 signal. This is the 1765 in here. It was measured and it is consistent with the $\bar{K}N$ branching ratio as seen in Terry Friedman's thesis. Here is $\Sigma^0\pi^0$ data. Here we have a 1385 plus a reflection of the 1385 due to contamination from the $\Lambda\pi$ data which we have put in as a fake resonance at 1405 which we consider as something relatively well understood. Here is the 1660 signal which shows very well and here is the 1900 or whatever you want to call it.

Hemingway: Can I ask you if you assumed a fixed mass and width. Did you assume it was a single resonance in the region of the 1660 in $\Lambda\pi^+$, for example?

Ross: For the $\Lambda\pi^+$ I can say the following. There are non-interferring resonances in the fit of the following forms. There are ρ, $Y^{*\pm}$, $\Sigma^{\pm}(1660)$, $\Sigma^{\pm}(1765)$, and the possibility of a linear mass dependent background.

Derrick: He made the usual assumptions.

Goldberg: In this data there is still one thing that worries me. I looked in Siegel's thesis for the separation of the $\Lambda\pi\pi$, $\Sigma\pi\pi$, and $\Lambda\pi\pi\pi$. Whereas I find that the $\Lambda\pi\pi$ and $\Sigma\pi\pi$ are quite well separated, the $\Sigma\pi\pi$ and $\Lambda\pi\pi\pi$ are quite difficult to separate and as far as I can see in Siegel's thesis all that was achieved was a very crude statistical separation. You actually pick the best fit after multiplying by some confidence level factor to symmetrize the mass spectrum which is only a statistical separation. It's certainly not clear from that that you can't be having contaminating effects from $\Lambda\pi\pi\pi$ production which, by the way, has a much larger cross-section at these energies than $\Sigma\pi\pi$. I just wondered if you could comment on that.

Ross: I accept part of what you say and part I don't. But I don't think it can be discussed here. It is written down in several pages and takes a logic on which we cannot possibly come to agreement with respect to this problem in this meeting. It is not a well established procedure. I can give a personal preference and that is all I can do. I happen to have confidence in the work that was done to the stated accuracy in the paper.

Goldberg: In other words you ask us to accept evidence for some new effect on something you haven't written down for anybody to see. Is that correct?

Ross: No. I'm saying with respect to the separation of the $\Sigma\pi\pi$ from the $\Lambda\pi\pi\pi$ that I will stand behind the separation of Dan Siegel in this thesis. I believe that it is published as need be with this kind of work and I'm willing to discuss the details of this kind of separation and what the errors are that are stated and what they mean. And I think that they mean that this data has statistical significance and does measure things in the $\Sigma^0\pi^+\pi^-$ events. (We also analyze the $\Sigma^+\pi^0\pi^-$ events as a check on the $(\Sigma\pi)^+$ state.)

Hemingway: There is a question that you can answer. In the $\Sigma^0\pi^+\pi^-$ events shown, what fraction come from $\Lambda\pi\pi\pi$?

Ross: I don't remember. (Note added in proof: Less than 20% of the $\Sigma^0\pi^+\pi^-$ events in our sample have an ambiguous fit with $\Lambda\pi^+\pi^0\pi^-$.)

Derrick: I think this is very important but I think maybe you could discuss it privately. It sure is nice to have several groups working the same field.

Garfinkel: I have two questions for Dave Plane about the Σ (1660) $\rightarrow \Sigma\pi\pi$. To what extent can you rule out there being two D_{13} states in this region? In other words can you put some limits on a $\overline{K}N$ branching ratio for a second state -- I'm sure that's quite difficult. The other question is suppose the second state is not a D_{13}, can the people doing the phase shift analysis in the $\Sigma\pi\pi$ system suggest another way it might be resonating?

Plane: Well, the first question about whether there could be two D_{13}'s -- we haven't tried it. We get a reasonable fit to the data with one D_{13} in $\overline{K}N$ and $\Sigma\pi$ channels. We tried to make the simplest hypothesis and got a reasonable fit and find no evidence from that point of view for a second D_{13}. Are there any other I=1 waves resonating there? From what I presented yesterday, the nearest one was about 1620 -- a possible P_{11} mainly in the $\Sigma\pi$. As regards the $\Sigma\pi\pi$ work, there's been no measurement of the spin-parity information experiments of the state which decays into Λ (1405)$\pi\rightarrow\Sigma\pi\pi$. So it could have any spin-parity. We assume that it was Σ (1660) with $3/2^-$.

Ross: I have a comment with respect to this same question. I think a formation experiment would not see two Σ (1660)'s with the same spin-parity if either of two things obtained. If two resonances are degenerate in mass and width roughly, a formation experiment would not see them and they would still obey the unitarity conditions. Or if there is one 1660 with the accepted mass and width and another 1660 that is very broad, they would not be seen. The broad resonance could be seen as background.

Meshkov: When you do the phase-shift analysis, is there a way of putting in the possibility of seeing two resonances; if you don't put in that possibility ahead of time can you get it out if the resonances are of the same spin-parity?

Ferro-Luzzi: But there are much simpler explanations like, for example, taking any of the states you have seen yesterday on all those Argand plots. It's much simpler to use one of those amplitudes than to introduce two amplitudes resonating at the same mass with the same quantum number.

Derrick: The point is that it's very difficult to measure the spin and parity in production experiments.

Kim: The K-matrix analysis can take care of two resonances in the same spin-parity. The K-matrix is really similar to the R-matrix in nuclear physics. For instance the 4 x 4 K-matrix can take care of 4 resonances. It can take care of two resonances very easily if there is a strong effect. I didn't constrain that there is only one D_{13}

resonance. I didn't constrain anything. All you see is what comes out of it. By looking at that it is seen that D_{13} wave is first of all small and only one small thing is trying to come out.

Derrick: After coffee we'll have these two short papers on Y^*'s.

FORWARD PRODUCTION OF I = 0 HYPERONS IN THE REACTION $K^- n \rightarrow Y \pi^-$ AT 3 GeV/c

S.A.B.R.E. Collaboration

At 3 GeV/c K^- incident momentum, the reaction $K^- n \rightarrow Y \pi^-$ where Y is either a I = 0 hyperon resonance, such as $\Lambda(2100)$, $\Lambda(1815)$, $\Lambda(1690)$, $\Lambda(1520)$ or the $\Lambda(1115)$, presents a backward peak in the CM angular distribution of the outgoing π^-. If one interprets this property in terms of a pure proton exchange mechanism, the relative value of the coupling constants YK^-p can be extracted from the differential backward cross sections.

In a 7 ev/ub K^-d bubble chamber experiment [1], the following reactions have been analysed :

$$K^- n \rightarrow \pi^- K^- p \qquad (1)$$

$$K^- n \rightarrow \pi^- \bar{K}^0 n \qquad (2)$$

$$K^- n \rightarrow \pi^- \pi^\mp \Sigma^\pm \qquad (3)$$

$$K^- n \rightarrow \pi^- \Lambda \qquad (4)$$

In reaction 1 (and 2) the production of $\Lambda(1520)$, $\Lambda(1690)$, $\Lambda(1815)$ and $\Lambda(2100)$ is observed whereas in reactions 3 only $\Lambda(1405)$, $\Lambda(1520)$ and $\Lambda(1690)$ can be detected (Fig. 1a and 2a). Except in case of $\Lambda(1405)$ these peaks emerge more strongly above background when one selects the events associated with positive values of the squared transfer momentum u between the incident K^- and the outgoing Y system as seen in Fig. 1b, 1c, 2b. The mass and width of these resonances shown in Table 1 are in good agreement with the accepted values [2].

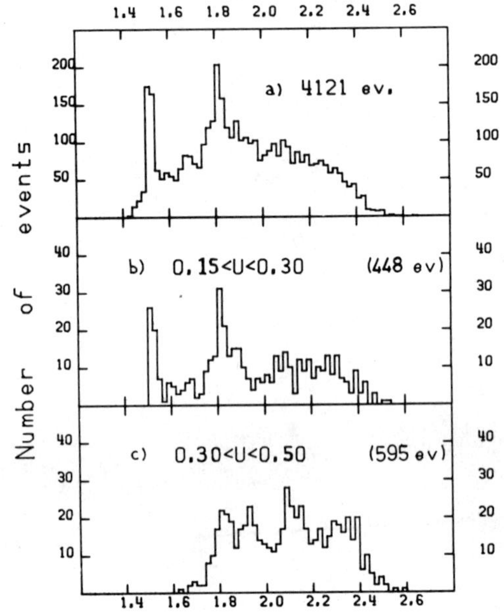

Fig. 1

K^-p mass distributions (GeV/c^2)
in reaction $K^-p\,\pi^-$

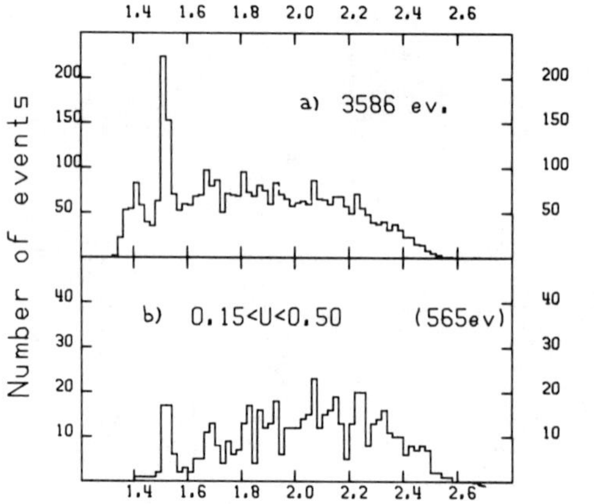

Fig. 2

$\Sigma^{\pm}\pi^{\mp}$ mass distributions (GeV/c^2)
in reactions $\Sigma^{\pm}\pi^{\mp}\pi^-$

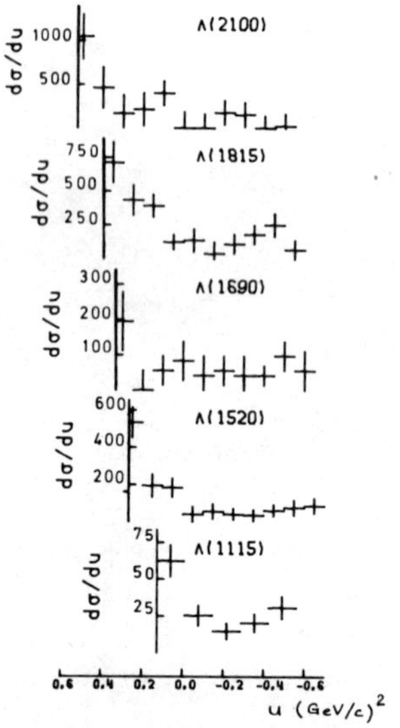

Fig. 3

$d\sigma/du$ distributions for reactions (5) as obtained from final states (1) and (4). All differential cross-sections have been corrected for the unseen decay modes

An interesting feature of the differential cross sections $\frac{d\sigma}{du}$ of the reaction

$$K^- n \rightarrow Y \pi^- \qquad (5)$$

is the presence of a dip near $u \simeq -0.15$ (Fig. 3). This feature can be related to the dip observed in backward $\pi^+ p$ scattering between 5 and 16 GeV/c [3] ; in this reaction this dip has been interpreted by many authors [4] as due to the exchange of a reggeised nucleon. As in the u channel reaction (4) is simply related to $\pi^+ p$ elastic scattering by SU3, it seems natural to make the hypothesis than the backward scattering ($u \gtrsim -0.15$) in reaction (5) is dominated by proton exchange since I=3/2 exchange in the u channel is forbidden. In this case the decay angular distribution of the produced Y^* should be respectively

$$1 + P_2 (\cos \theta^*), \text{ and } 1 + \frac{8}{7} P_2 (\cos \theta^*) + \frac{6}{7} P_4 (\cos \theta^*)$$

for $\Lambda(1520)_{J=3/2}$ and $\Lambda(1815)_{J=5/2}$ respectively.

Fig. 4 shows that the agreement with this prediction is quite good for both. With this same hypothesis of a pure proton exchange the differential cross section $d^2\sigma/dM_Y du$ for production of a Y system having a mass M_Y can be related to the total $K^- p$ cross section off mass shell σ_{off} by the usual relation [5]

$$\frac{d^2\sigma}{dM_Y du} \approx f(u)\, M_Y^2 \quad q_{off} \quad \sigma_{off}(M_Y)$$

where f(u) is the propagator and q_{off} the incident K^- momentum in the Y rest system. In similarity with the DÜRR PILKHUN off mass shell corrections which where applied successfully in the one pion exchange model [6], the off and on mass shell cross sections have been related by

$$\frac{(q\sigma)_{off}}{(q\sigma)_{on}} = \frac{B_{\ell\ off}}{B_{\ell\ on}}$$

where $B_{\ell\ off}$ and $B_{\ell\ on}$ are the usual centrifugal barrier respectively calculated with q_{off} and q_{on} . If one inserts the relation between the elastic width Γ_{YK^-p} and the squared coupling constant $G^2_{YK^-p}$

$$\Gamma_{YK^-p} = G^2_{YK^-p} \, B_{\ell \, on} \, \frac{M_p}{M_Y} \, q_{on}$$

used by TRIPP [7], the differential cross section of a resonance Y having a spin J can be written

$$\frac{d\sigma(Y)}{du} \sim (J+1/2) \; f(u) \; M_Y \; B_{\ell \; off} \; G^2_{YK^-p}$$

Thus in a given u range the coupling constant $G^2_{YK^-p}$ is proportional to $\frac{d\sigma(Y)}{du}/(J+1/2) \; M_Y \; B_{\ell \; off}$. These quantities have been evaluated for all the Y(I=0) states by calculating $B_{\ell \; off}$ with R = 1F or R = ∞ ($B_{\ell \; off}$=1). After normalisation to $G^2_{\Lambda(1815)K^-p}$ they are presented Table 1. For $\Lambda(2100)$, $\Lambda(1690)$ and $\Lambda(1520)$ a reasonable agreement is found between $G^2_{YK^-p}/G^2_{\Lambda(1815)K^-p}$ thus determined and the corresponding quantities obtained in formation experiments [8] or expected from SU3 invariance by using the g_1, g_d, g_f coupling constants recently published [8]. However there is a large discrepancy for $\Lambda(1405)$, which is not produced in the forward direction [9] : this phe-

Y J^P	Mass (MeV)	Width (MeV)	Cross Sections(μb)		$G^2_{YK^-p}/G^2_{\Lambda(1815)K^-p}$			$G^2_{YK^-p}/G^2_{\Lambda(1815)K^-p}$	
			Total	u > 0	u range	with $B_{\ell \, off}$ = 1	with $B_{\ell \, off}$(1F)	formation exp	SU3
$\Lambda(2100)$ $7/2^-$	2120 ± 20	90 ± 30	210 ± 70	150 ± 50	0 - 0.375	0.43	0.6	0.7	0.44
$\Lambda(1815)$ $5/2^+$	1815 ± 10	50 ± 15	370 ± 50	156 ± 30		1	1	1	1
$\Lambda(1690)$ $3/2^-$	1680 ± 10	45 ± 15	100 ± 60	13 ± 13	0 - 0.325	0.40	0.27	0.11	0.18
$\Lambda(1520)$ $3/2^-$	1520 ± 10	< 20	290 ± 25	73 ± 15	0 - 0.275	1.25	1.5	0.68	0.68
$\Lambda(1405)$ $1/2^-$	1400 ± 10	60 ± 15	100 ± 25	< 3	0 - 0.2	< 0.10	< 0.15	~ 4	0.5
$\Lambda(1115)$ $1/2^+$			300 ± 30	10 ± 2	0 - 0.125	2.0	1.6		

TABLE 1

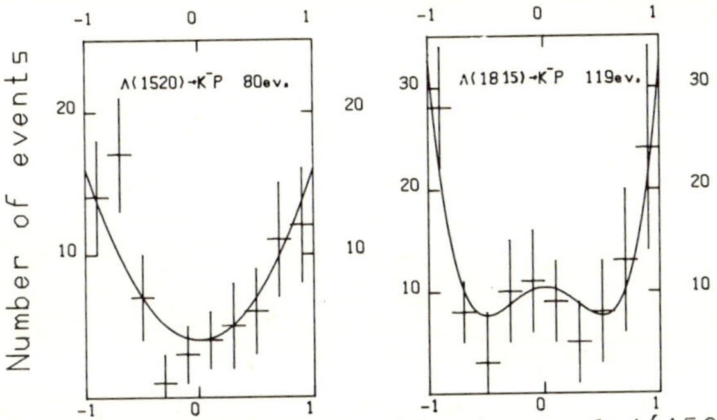

Fig. 4 Decay angular distributions of Λ(1520) and Λ(1815), as obtained in reaction 1, with $u > -0.15$. A background evaluated on each side of the Λ(1520) and Λ(1815) peaks has been substracted. The curve represents the distribution predicted by a pure proton exchange mechanism

nomenon is not presently understood. Finally the coupling constant $G^2_{\Lambda K^- p}$ is by this way directly related to the others $G^2_{YK^- p}$.

In the exchange degeneracy model of SCHMID [10] the resonances Λ(2100), Λ(1815), Λ(1520), Λ(1115) which lie on a degenerate Regge trajectory should have the same coupling constant to K^-p. The result here obtained is in qualitative agreement with this prediction.

References

[1] D.MERRILL et al, to be published in Nuclear Physics
[2] Particle Data Group, Rev.Mod.Phys. 41, 109, (1969)
[3] See Ref. 1 and 2 of V.BARGER, D.CLINE, Phys. Rev. Letters 21, 392, (1968)
[4] C.B.CHEW, J.D.STACK, Phys. Rev. 153, 1575, (1967) V.BARGER, D.CLINE, Phys. Rev. 155, 1792, (1967)
[5] E.FERRARI, F.SELLERI, Supp. Nuovo Cimento 24, 453, (1962)
[6] H.P.DÜRR, H.PILKUHN, Nuovo Cimento 40, 889, (1965) T.G.TRIPPE et al, Report UCLA-1024
[7] R.D.TRIPP et al, Nuclear Physics B3, 10, (1967)
[8] R.LEVI-SETTI, Proc. of the Lund Conference (1969)
[9] J.BADIER et al, Report CEA R-3037, (1966)
[10] C.SCHMID, Lettere al Nuovo Cimento 1, 165, (1969)

THE REACTION $K^{-}n \to Y^{*o}\pi^{-}$ AT 5.0 TO 5.5 GeV/c

Argonne National Laboratory, Carnegie Mellon University,

Case Western Reserve University*

This paper is a compilation of data on low mass (< 2.2 GeV) neutral hyperons which decay to $\Sigma^{-}\pi^{+}$, $\Sigma^{+}\pi^{-}$, pK^{-}, or $n\bar{K}^{o}$, with emphasis on production and decay of the $\Lambda(1520)$. The data are taken from A.N.L. 30" and B.N.L. 80" deuterium bubble chamber experiments. No weights have been applied to the data shown. Figure 1 (center) shows the $\Sigma\pi$ invariant mass distribution in which the $\Lambda(1405)$ and $\Lambda(1520)$ are clearly evident. Smaller accumulations are seen at $\sim$ 1680 and 1825 MeV. The corresponding $N\bar{K}$ mass plot, Figure 1 (top), also shows a clear $\Lambda(1520)$ as well as a broader peak at $\sim$ 1825 MeV. The relative decay branching ratio $\Sigma\pi/N\bar{K}$ deduced from these plots indicates that the 1825 peaks should be associated with the $\Lambda(1815)$ rather than the $\Lambda(1835)$, but the difference in shape between the two peaks, and the considerable background precludes further analysis at this stage. The accumulation in the $\Sigma\pi$ plot at $\sim$ 1680 may be associated with $\Lambda(1670)$, $\Lambda(1690)$ and/or $\Sigma(1670)$.

Figure 1 (bottom), a scatter plot of the c.m.s. production cosine ($\hat{K}_{inc} \cdot \hat{Y}^{o*}$) versus the hyperon mass, and Figure 2, six projections of the scatter plot, show the following features of the production mechanism: a.) the $\Lambda(1405)$ is produced essentially entirely peripherally; b.) the $\Lambda(1520)$ is produced predominantly peripherally, but has a clear antiperipheral contribu-

---

*Participants in this work are: R. Ammar, W. Kropac, H. Yarger (Kansas); Y. Cho, M. Derrick, D. Johnson, B. Musgrave, T. Wangler (A.N.L.); R. Davis, B. Werner (Northwestern); F. DiBianca, M. Einschlag, R. Endorf, A. Engler, H. Fisk, R. Kraemer, G. Yekutieli (C.M.U.); W. Fickinger, S. Gourevitch, J. Korpi, D. Robinson (C.W.R.U.)

tion; c.) the 1.80 to 1.85 GeV mass cut also shows both backward and forward production.

The projection for the 1.50 to 1.54 GeV mass cut has been shaded to show that the $\bar{K}^o n$ combinations (from two labs) do not agree with the rest of the data, in that peripheral events are missing. This may be associated with scanning or reconstruction difficulties for one prong events with fast forward π^-.

Figure 3 is concerned with the production and decay of the $\Lambda(1520)$, where a mass cut of 1480-1550 MeV has been applied. In 3a is shown the Stodolski-Sakurai polar angle $(\hat{M}\cdot[\hat{n} \text{ X } (\hat{K} \text{ X } \hat{\pi})])$ where the unit vectors are as follows: $\hat{M}$ = meson from hyperon, $\hat{n}$ = target neutron, $\hat{K}$ = incident K^-, $\hat{\pi}$ = nonresonating π^-. The histogram is for the peripheral hemisphere, $(\hat{K}\cdot\hat{Y}^*)<0.$, and the curves are for vector dominance M2 and E1 transitions. The data are consistent with an M2 transition. Figure 3b. shows the corresponding azimuthal angle.

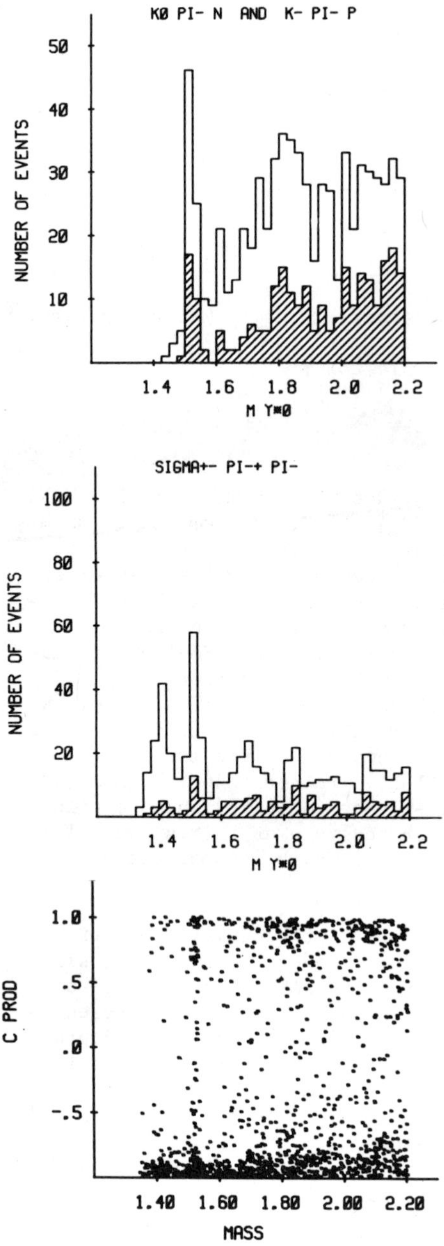

← Figure 1. Upper: KN effective mass distribution; Center: $\Sigma^{\mp}\pi^{\pm}$ effective mass distribution (shaded events antiperipheral); Bottom: Scatter plot of c.m.s. production cosine versus hyperon mass for all events (peripheral at CPROD = - 1.0).

Figures 3c and 3d show the scattering and Treiman-Yang angular distribution, $(\hat{M}\cdot\hat{K})$ and $\cos^{-1}[(\hat{K} \times \hat{M})\cdot(\hat{n} \times \hat{\pi})]$, for the events in the antiperipheral hemisphere: $(\hat{K}\cdot\hat{Y}^*)>0$. The curve shown is $1 + 3\cos^2\theta$, corresponding to proton K scattering through a spin 3/2 hyperon state. The agreement with the data is not as good as that in a similar comparison made with the SABRE data (3 GeV/c).

Figure 3e shows, for the events in the range indicated, the $t - t_{min}$ distribution (still for the 1520 region) indicating a smooth peripheral production behavior. In Figure 3f is shown the $u - u_{min}$ distribution (antiperipheral events), which indicates possible structure in dN/du, with a dip at $u - u_{min} \simeq -0.5$ $(GeV/c)^2$.

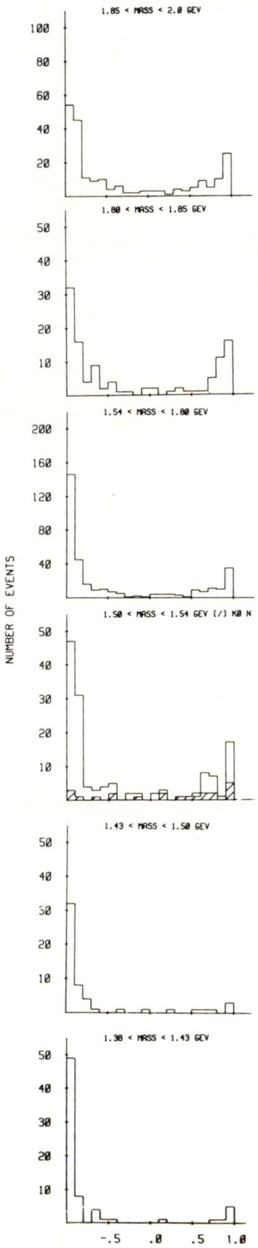

← Figure 2. Center of Mass production cosine distributions for indicated cuts on hyperon mass (peripheral at - 1.0). Shaded events in $Y^{*o}(1520)$ region are from the $\bar{K}^on$ sample.

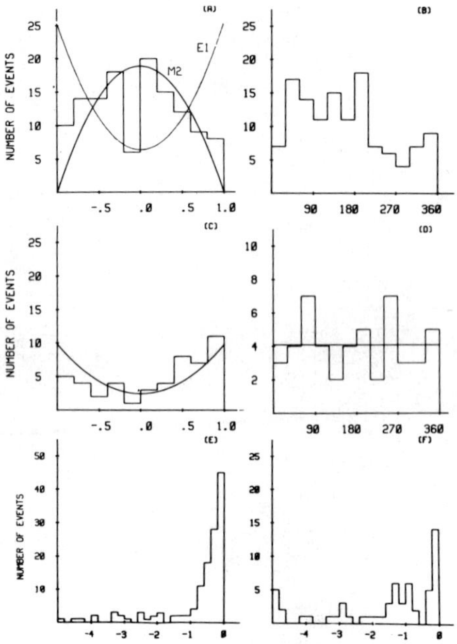

Figure 3. Production and decay distributions for $Y^{*o}(1520)$. Mass cut: 1480 to 1550 MeV. (a. and b.) Stodolski-Sakurai polar and azimuthal angles for events in peripheral production hemisphere. (c. and d.) Polar and azimuthal decay angles for events in the antiperipheral hemisphere. (e.) $t - t_{min}$ distribution for all events in the range $|(t - t_{min})| < 5.0\ (GeV/c)^2$. (f.) $u - u_{min}$ distribution for all events in the range $|(u - u_{min})| < 5.0\ (GeV/c)^2$.

DISCUSSION

Yodh: I would like to ask Dr. Barloutaud that if you cannot get coupling constants which are meaningful for the 1520 and 1820 from your analysis, what significance can you put on coupling constants for the 2100 where the background is moved higher?

Barloutaud: We have meaningful coupling constants for the 1520.

Yodh: They were not in agreement with the formation experiments.

Barloutaud: I think they were in agreement. We do not put any error on the coupling constants since the error is about 50 percent. The formation coupling constants are not known to better than 20-30 percent.

Yodh: Let me put my question another way. If you were to put an error or estimate the uncertainty in the theory, is the answer good to a factor of 2?

Barloutaud: Yes, I think it's good to a factor of 2.

Rosenfeld: Unless I'm confused, the last two talks analyze mainly a $K^- p$ vertex in a production experiment -- something which it seems to me is completely accessible easier in a formation experiment.

Barloutaud: Oh, yes. The object is not to compete with formation experiments. It is just to show that in a production experiment one can have a good idea of the coupling constant. This is useful because some coupling constants like the Λ Kp cannot be obtained by formation experiments.

Rosenfeld: Yes, below threshold is a place where I can understand.

Derrick: You can turn this all around and say you take the data from formation experiments and this is then very interesting with respect to the characteristics of production of these Y^*'s.

Butterworth: There's an interesting effect in the failure of the 1405.

COMMENTS ON THE Σ (1620)

Arthur F. Garfinkel

Purdue University, Lafayette, Indiana 47907

Recently Ammann et al.[1] confirmed the existence of this $\Lambda\pi$ enhancement at a mass of 1642 GeV/c^2. The enhancement was first observed by Crennell et al.[2] (BNL), who presented additional data at the Lund Conference where they show a mass of 1619 GeV/c^2. Both signals were observed in the reaction

$$K^-n \rightarrow p_s\ \Lambda^o\pi^+\pi^-\pi^- \qquad (1)$$

at the neighboring energies of 4.6 (Purdue) and 3.9 (BNL). More recently, Barloutaud et al.[3] studied reaction (1) at 3.0 GeV/c and observed no signal other than a possible reflection of a Σ(1660) from misidentified events from the reaction

$$K^-n \rightarrow p_s\Sigma^o\pi^+\pi^+\pi^- \qquad (2)$$

The analysis of reactions (1)and (2) for the Σ(1620) are complicated by the following facts:

1. For a significant fraction of the events, the reactions (1) and (2) are not easily distinguishable.
2. The production mechanism is not simple.

The Purdue experiment assigned 453 events to reaction (1), 224 events to reaction (2), and considered 219 events ambiguous. Figure(a) shows the $\Lambda\pi$ spectrum and Figure (b) the $\Sigma^o\pi$ spectrum for that data.

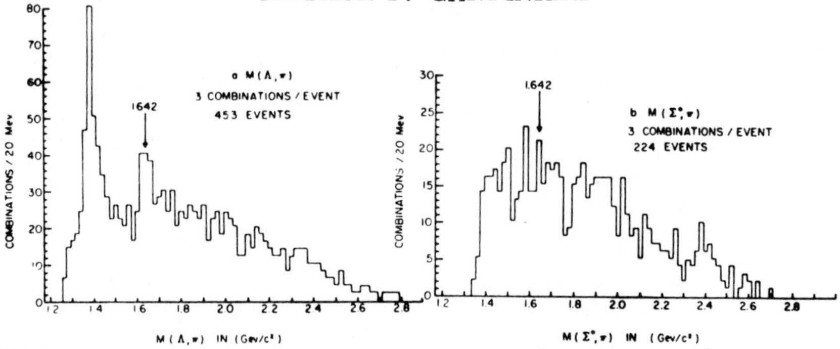

With no structure on the $\Sigma\pi$ spectrum for the definite Σ^o events, it appears that one need not worry about reflections from misidentified $\Sigma(1660)$s at that beam momentum.

The fact that the $\Sigma(1620)$ is not produced in a simple quasi two body final state makes the analysis unusually difficult. One cannot make the usual t cuts to enhance the signal to noise, and the problem of reflections is not at all trivial. Nor can one make a reliable a priori guess as to the excitation function.

If one accepts the validity of SU(3), a test of the resonant nature of the object can be made by looking for alternate decay modes. The branching ratios related to $\Lambda\pi$ are, of course, $\Sigma\pi$ and $\overline{K}N$. The BNL group set a limit of

$$\frac{\text{fraction } \Sigma(1620) \rightarrow \overline{K}^o p}{\text{fraction } \Sigma(1620) \rightarrow \Lambda\pi^+} = 0.0 \pm 0.2$$

but did not have an estimate of the $\Sigma\pi$ mode. The Purdue group set an upper limit of

$$\frac{\text{fraction } \Sigma(1620) \rightarrow \Sigma\pi}{\text{fraction } \Sigma(1620) \rightarrow \Lambda\pi} < 1.1$$

at 95% confidence level. Thus a signal has not been detected in either of these modes. These branching ratios rule out a decuplet assignment. They are just on the edge of being compatible with the octet assignment. In fact, taking the $\Sigma\pi/\Lambda\pi$ decay fraction to be equal to the assigned upper limit, implies a $\overline{K}^o p$ decay fraction one and one half standard deviations greater than the BNL estimate of zero. A smaller branching fraction to $\Sigma\pi$ would imply a higher ratio to $\overline{K}^o p$ and worse agreement. This coupling to $\overline{K}N$ implies that the state is succeptible to study in formation experiments. The phase shift analysis of Armenteros et al.[4] do indicate the existence of $\Lambda\pi$ states in this mass region. Further production data to reduce the errors in the branching fractions, and higher statistics formation experiments may both be required to settle the question of the resonant nature of the $\Sigma(1620)$. Evidence showing that the $\Sigma(1620)$ is not one of the $\Sigma(1660)$ states is discussed in reference (1).

References

(1) A. C. Ammann, A. F. Garfinkel, D. D. Carmony, L. J. Gutay, D. H. Miller and W. L. Yen, PRL 24, 327 (1970).

(2) D. J. Crennell, U. Karshon, K. W. Lai, J. S. O'Neall, J. M. Scarr, P. Baumel, R. M. Lea, T. G. Schumann and E.M. Urvater, Lund Conference (1969) and D. J. Crennell, W. C. Delaney, E. Flaminio, U. Karshon, K. W. Lai, W. J. Metzger, J. S. O'Neall, J. M. Scarr, A. M. Thorndike, P. Baumel, R. M. Lea, A. Montwill and T. G. Schumann, PRL 21, 648 (1968).

(3) R. Barloutaud, D. W. Merrill, J. C. Scheuer, W. Hoogland, J. C. Kluyver, G. Giacomelli, A. Minguzzi-Ranzi, P. Serra, T. Eisenberg, G. Yekutieli, J. Goldberg, A. Rouge and G. Lamidey, NP B16, 201 (1970).

(4) R. Armenteros, P. Baillon, C. Bricman, M. Ferro-Luzzi, H. K. Nguyen, V. Pelosi, D. E. Plane, N. Schmitz, E. Burkhardt, H. Filthuth, E. Kluge, H. Oberlack, R. R. Ross, R. Barloutaud, P. Granet, J. Meyer, J. L. Narjoux, F. Pierre, J. P. Porte, and J. Prevost, NP B8, 186 (1968).

SESSION VI

Ξ^* PRODUCTION

Discussion Leader: Malcolm Derrick

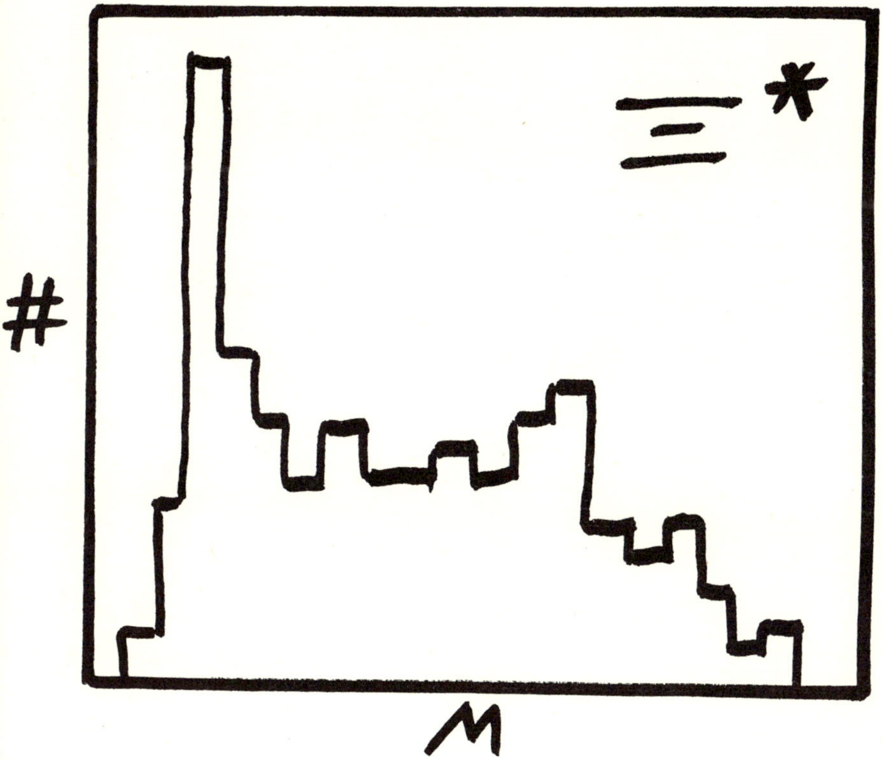

A REGGE POLE ANALYSIS OF THE REACTION $K^- p \rightarrow \Xi^o$ (1530)K^o

Brandeis-Maryland-Syracuse-Tufts Collaboration[††]

Recently a $(\Sigma_\alpha, \Sigma_\gamma)$ reggeized baryon exchange model has been used to account for backward scattering in the reaction $\pi^- p \rightarrow \Lambda^o K^o$[1]. This reaction has allowed resonances (or trajectories) in all three channels. We have applied a similar model analysis to the reaction $K^- p \rightarrow \Xi^o(1530) K^o$ which is forbidden to proceed via single t-channel exchanges. We are able to describe the full angular distributions in terms of $(\Sigma_\alpha, \Sigma_\gamma)$ exchanges alone.

A dominant feature of the reaction $K^- p \rightarrow \Xi^- \pi^+ K^o$ from threshold to 3.0[2-4] is copious $\Xi^o(1530)$ production via the two body reaction

$$K^- p \rightarrow \Xi^o(1530) K^o \quad (1)$$

with a minimal background contamination (mainly via the reaction $K^- p \rightarrow \Xi^- K^{*+}(890)$)[5]. The production characteristics of the two body reaction (1) are: (as shown in Figures 1 and 2)

(1) It is dominated by a large backward peak $(\cos\theta_{K^- \cdot K^o} = -1)$[1]

(2) The cross section shows a momentum dependence characteristic of strange baryon exchanges[6] above 1.7 GeV/c.

(3) No s-channel resonant effects are discernible.

(4) All angular distributions show a dip in $d\sigma/du$ at $u = u_{max}$.

(5) A minimum in $d\sigma/du$ around $u = -.9$.

(6) No appreciable polarization of $\Xi^o(1530)$ has been observed.

In this paper we present a phenomenological Regge pole analysis of reaction (1) from 1.7 to 2.87 GeV/c in terms of Σ baryon exchanges. As the reaction is forbidden to proceed via single t-channel exchanges[7] we have attempted to fit a reggeized Σ model to the entire angular distribution, motivated by the concept of duality. First we describe the model and then make a fit to the experimental data. We follow the general procedure of Barger et al.[9]

The Regge Trajectories which can contribute are Σ_α^8, Σ_γ^8, Σ_δ^{10} and Σ_β^8. From the Chew-Frautschi plot[8] the trajectory parameters are obtained to be Σ_α: $\alpha = -.75 + 0.9u$, Σ_γ: $\alpha = -.98 + 0.9u$ Σ_δ: $\alpha = -.25 \pm .9u$ and Σ_β: $\alpha = -.34 + 0.9u$. Thus the trajectories cluster into two approximately degenerate groups $(\Sigma_\alpha, \Sigma_\gamma)$ and $(\Sigma_\delta, \Sigma_\beta)$. The observed dip of differential cross sections at u_{max} and energy dependence p^{-4} is characteristic of the $(\Sigma_\alpha, \Sigma_\gamma)$ trajectories. The dip can be due to wrong signature nonsense zero of Σ_α trajectory at $\alpha = -1/2$ and the intercept $\alpha(0) = -.78$ can give the correct energy dependence. Therefore we use a reggeized baryon exchange model with $(\Sigma_\alpha, \Sigma_\gamma)$ trajectories.[10]

In order to fit the experimental data we parameterized the residues for Σ_α and Σ_γ trajectories consistent with the absence of lowest MacDowell reflected particles[11] as follows:

$$\gamma(\Sigma_\alpha) = A(\sqrt{u} + M_\alpha)e^{Bu}$$
$$\gamma(\Sigma_\gamma) = C(\sqrt{u} + M_\alpha)e^{Du}$$

The four parameters were allowed to vary to fit the data.

A simultaneous fit to all data points was obtained with a χ^2 of 32.9 for 36 degrees of freedom. The best values of A, B, C and D were A = 22.6 ± 1.0, B = 1.63 ± .12, C = 3.68 ± .37 and D = - 0.28 ± .05. The resulting fits to the angular distributions are shown in Fig. 1 and the energy dependence in Fig. 2.

The regge pole model used herein accounts for the data very well. The dip at $u = u_{max}$ is produced by the Σ_α wsns point, the minimum at $u = -.9$ is due to the vanishing of Σ_α contribution and the rise of the Σ_γ contribution and the energy dependence is correctly predicted. The model predicts negligible polarization of $\Xi^o(1530)$.

We wish to acknowledge informative discussions with Drs. V. Barger, O. W. Greenberg and S. Meshkov.

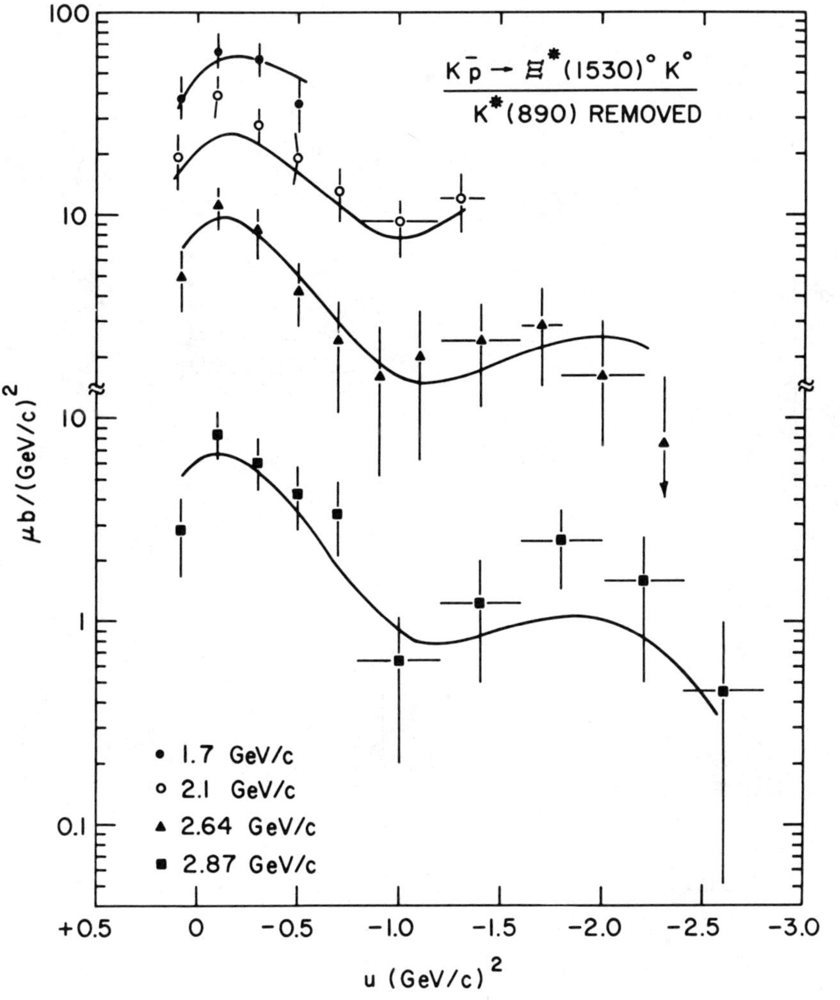

Figure 1 Differential cross sections $d\sigma/du$ for $K^-p \to \Xi(1530)K^0$ from 1.7 to 2.87 GeV/c. The curves are the best fit to all the data in terms of reggeized Σ-baryon exchange model described in the text. The u-range spans the complete angular range $-1 \leq \cos\theta \leq 1$.

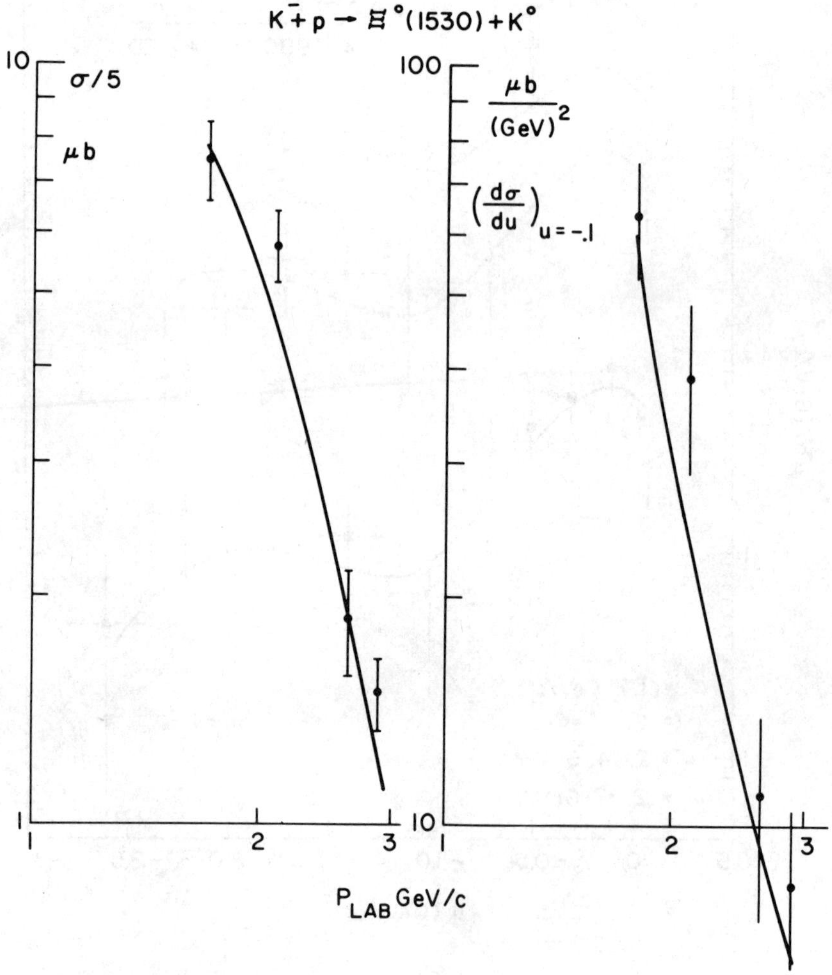

Figure 2 Energy dependence of σ and $(d\sigma/du)_{u = -.1}$. The curves are the model fits.

References

[†]Supported in part by U.S. A.E.C.

[††]S. Apsell, N. Barash-Schmidt, L. Kirsch, P. Schmidt - Brandeis; C. Y. Chang, R. Hemingway, B. Khoury, A. Stottlemyer, G. B. Yodh - Maryland; S. Glickman, M. Goldberg, S. Jacobs, K. Jaeger, C. McCarthy, B. Meadows, G. C. Moneti - Syracuse; J. Bartley, R. M. Dowd, J. Schneps, G. Wolsky - Tufts.

[1]V. Barger, D. Cline and J. Matos, Phys. Letters 29B, 121 (1969).

[2]P. M. Dauber et al., Phys. Rev. 179, 1262 (1969).

[3]S. Apsell et al., (Brandeis-Maryland-Syracuse-Tufts Collaboration) Phys. Rev. Letters 23, 884 (1969).

[4]J. Badier et al., Phys. Letters 16, 171 (1965).

[5]The only background under $\Xi^o(1530)$ mass band is due to the reaction $K^-p \to \Xi^- K^{*+}(890)$. The percentage of pure $\Xi^o(1530)$ has been estimated to be 95%, 70%, 80% and 85% at 1.7, 2.1, 2.64 and 2.87 GeV/c respectively (see reference 1 and 2 for details). A uniform subtraction from $\Xi^o(1530)$ angular distribution was made based on these percentages.

[6]D.R.O. Morrisson, Proceedings of Stony Brook Conference on High Energy Two Body Reactions (1966) and CERN report No. 66-20 (1966) (unpublished).

[7]The quantum numbers of the t-channel would require a S = -2 baron for single exchanges. No evidence for such states exists.

[8]V. Barger and C. Michael, "Experimental Aspects of Dual Theories for Baryons" Univ. of Wisconsin Preprint #C00-238, May 1969 (unpublished); R. Dolen, D. Horn and C. Schmid, Phys. Rev. 166, 1768 (1968); H. Harari, Phys. Rev. Letters 22, 562 (1969).

[9]V. Barger, Phys. Rev. 179, 1371 (1968).

[10]The Σ_α trajectory passes thru $\Sigma(1189, 1/2^+)$ and $\Sigma(1910, 5/2^+)$ and Σ_γ trajectory passes through $\Sigma(1660, 3/2^-)$. They have same τp but opposite τ (and p).

[11]V. Barger and D. Cline, "Phenomenological Theories of High Energy Scattering", W. Benjamin (1969). We have treated $\Xi^o(1530)$ as a spin 1/2 Fermion in this formula.

[12]R. A. Muller, UCRL-19372, August 1969 (unpublished); G. Burgun et al., Nuclear Physics B8, 447 (1968).

DISCUSSION

Ward: Is it possible in fitting the charged cascade production to determine any of the Λ exchange parameters from the K^+ p backward scattering analyses?

Yodh: I don't know. Among the mysteries of Regge pole fitting that I learned was that in the K^+ p backward scattering fitting led by Barger et al. the residues -- you're allowed Λ_α and Λ_γ -- are assumed to have the form, apart from McDowell killing factors, of

$$\gamma = A\ e^{Bu} + Ce^{-Du}$$, B and C being positive,

and then you get some fit, with some value of the residue function. I'm not sure that I'm quite prepared to use SU(3) or something and go on to the other reactions as of the moment until I understand better this kind of residue function -- what do they mean? Maybe David can tell us a little bit about these residue functions. Is it more than fitting the data, a phenomenological representation of the data?

Cline: It's sort of low energy for Regge pole fits, and there's a large t-channel overlap, so I don't take too much stock in these complications up to 3 BeV or 4 BeV. It's a little bit low energy with the large t-channel.

Rosenfeld: I have a simple phenomenological question which involves pointing at your slide. You see there are four trajectories.

Yodh: Four parameters.

Rosenfeld: Right. But you picked two out of the four trajectories, and in fact their mass only shows up in two places. Now supposing I don't take all that trajectory business very seriously, and I just tried some other two trajectories, so I changed only the masses to give the different intercepts. Would you get just as good a fit? Is it really sensitive to that?

Yodh: I will answer by saying that I don't know the answer to your question; in my fit it makes no difference.

Rosenfeld: Did you leave those as variables or did you pick them?

Yodh: The M's I picked to be 1660 in one case and 1189 in the other.

OBSERVATION OF Ξ^* RESONANCES IN 5.5 GeV/c K^-p INTERACTIONS*

P. F. Schultz

University of Illinois

Urbana, Illinois 61801

We have studied a 4events/μb exposure of 5.5 GeV/c K^-p interactions for evidence of S = -2 resonances decaying into a Ξ^- hyperon. The total sample of 594 identified Ξ^- events consists of 418 events in the 'kink and vee' topologies and 176 events in 'kink-no vee' topologies. The selection and identification of the events is described elsewhere.[1] Final states of all multiplicities, including missing mass events, have been examined. However, only the 4- and 5- body final states show evidence for Ξ^* production.

Figure 1 shows the $\Xi^-\pi^{+,o}$ invariant mass distribution for the 4- and 5- body final states. The $\Xi^*(1530)$ is strong; in fact, it appears in almost every possible final state. The $\Xi^*(1930)$ is also evident in this distribution. In an attempt to reduce the background in the 1930 MeV region, we plotted only those $\Xi\pi$ combinations for which Ξ^- is not in the $\Xi^{*o}(1530)$ when combined with a different π or the π is not in the $K^*(890)$ when combined with the K^+ or K^o. The cross hatched distribution in Fig. 1 is the result of this selection. This procedure results in a sharper signal. In the $\Xi^*(1930)$ region, we fitted these selected combinations to a hand drawn background plus a simple Breit-Wigner resonance function. The mass width obtained from the fit are 1955 ± 14MeV and 56 ± 26MeV, respectively, which agree well with those found by Apsell et al.[2] There is no evidence for other structure in Fig. 1.

The $\Xi^-\pi^+\pi^-$ mass distribution in the 4- and 5- body final states were investigated in a similar manner. Figure 2 shows this distribution; the singly cross hatched distribution is for those $\Xi^-\pi^+\pi^-$ combinations which do not have a conflicting $\Xi^{*o}(1530)$ or $K^*(890)$ mass combination and the doubly cross hatched distribution is for those selected $\Xi^-\pi^+\pi^-$ which are $\Xi^{*o}(1530)\pi^-$ combinations. The

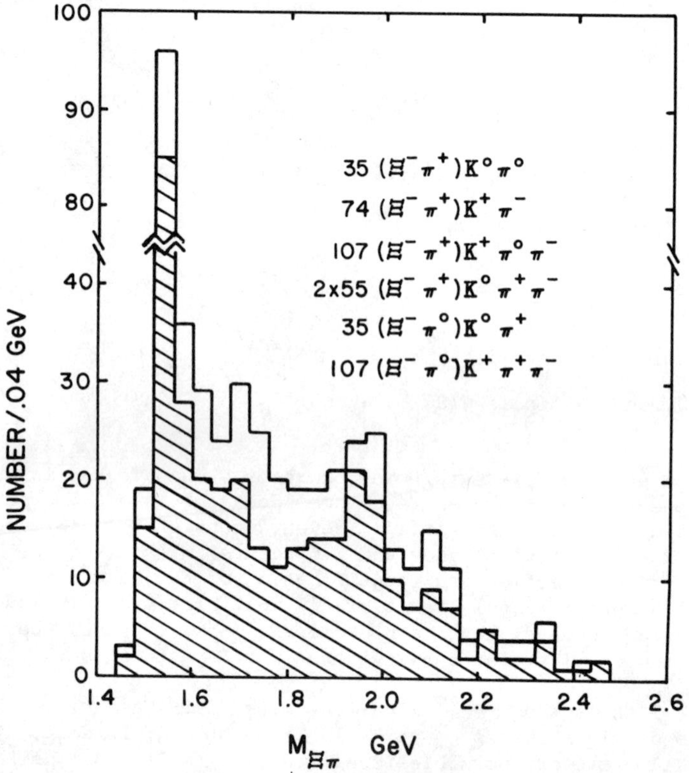

Fig. 1. Mass of the $\Xi^-\pi^+$ and $\Xi^-\pi^0$ combinations in the 4- and 5-body final states. The hatched events are those $\Xi\pi$ combinations which do not have a conflicting $\Xi^{o}(1530)$ or K (890) combination.

most significant structure is the sharp spike at ~2300 MeV in the singly cross hatched events corresponding to ~3 standard deviations above background. It is very narrow; our mass resolution is approximately 10 MeV in this region and the width of this structure is comparable. If this spike is a resonance its mass and width are 2295 ± 15MeV and < 30 MeV. As shown in Fig. 2, the Ξ^* decay mode appears to constitute only a small fraction, perhaps less than a third, of the structure. The $\Xi^-\pi^+\pi^0$ system does not show any significant structure in this mass region.

The doubly-cross hatched distribution, the $\Xi^{*o}(1530)\pi^-$ mass combinations, shows that the $\Xi^*(1820)$ is also produced in this experiment. The number of events in the $\Xi^{*o}(1530)\pi^-$ decay mode, estimated to be ~6 above a background of ~2, is small and we have not attempted a mass and width determination.

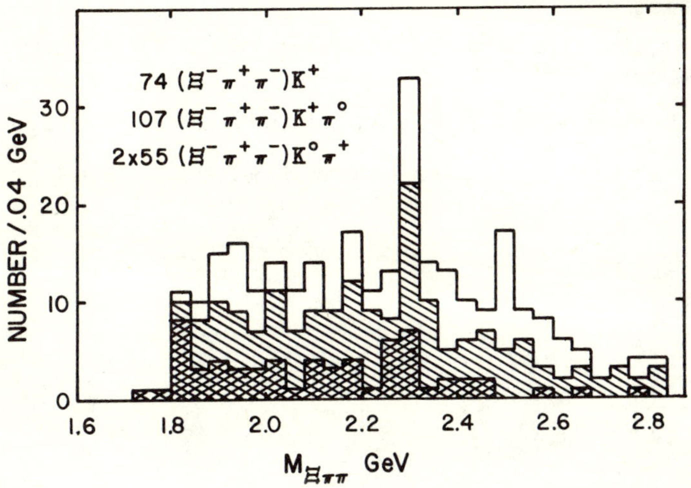

Fig. 2. Mass of the $\Xi^-\pi^+\pi^-$ combinations in the 4- and 5- body final states. The singly hatched events are those $\Xi\pi\pi$ combinations which do not have a conflicting $\Xi^o(1530)$ or $K^o(890)$ combination. The doubly hatched events are the $\Xi^o(1530)\pi$ subset of the singly hatched events.

To summarize, we have observed production of the following previously observed Ξ* resonances: Ξ*(1530), Ξ*(1930) and Ξ*(1820). In addition we observe a very narrow structure, $\Gamma < 30$ MeV, at a mass of 2295 ± 15 MeV in the $\Xi^-\pi^+\pi^-$ mass combinations; the statistical significance is 3 standard deviations above background. In our data, we do not find evidence for the Ξ^- being a decay product of any Y* resonance (Y*→Ξ or ΞKπ) or any I = 3/2 Ξ* resonance.

1. P. F. Schultz, Ph.D. Thesis, University of Illinois, 1969 (unpublished).

2. S. Apsell, N. Barash-Schmidt, L. Kirsh, P. Schmidt, C.Y. Chang, R. J. Hemingway, B. V. Khoury, A. R. Stottlemeyer, H. Whiteside, G. B. Yodh, S. Glickman, M. Goldberg, S. Jacobs, K. Jaeger, C. McCarthy, B. Meadows, G. C. Moneti, J. Bartley, R. M. Dowd, J. Schneps, and G. Wolsky, Phys. Rev. Letters 24, 777 (1970).

* Work supported in part by Atomic Energy Commission.

DISCUSSION

Yodh: I would like to ask you the percentage of the 1933 and errors on the percentage, if you have it.

Schultz: I don't have that right now, but I can get it for you.

Derrick: He'll calculate it during the next half-hour.

Ξ RESONANCES BELOW 2 GeV*

Brandeis-Maryland-Syracuse-Tufts†

In part I we present further results of an extensive search for Ξ resonances below 2 GeV. Of the channels presented, $\Lambda\bar{K}$ and $\Sigma\bar{K}$ have not been previously reported. Part II compares this data with published work from other experiments and presents a critique of the situation below 2 GeV.

PART I

The data comes from a 30 event/μbarn exposure of 2.9 GeV/c K^- in the BNL 31" HBC. Preliminary results for the reactions $K^-p \to \Xi^-\pi K$ and $K^-p \to \Xi\pi\pi K$ have been published recently(1,2). Herein the total data sample is presented except for the reactions $K^-p \to K^+K^-\Lambda^o$ and $K^-p \to K^+K^-\Sigma^o$ where approximately 1/3 of the data has been accumulated to date.

Fig. 1 shows our weighted $\Xi^-\pi^+$ mass distribution for the reaction $K^-p \to K^o\pi^+\Xi^-$ (831 events) and the $\Xi(1530)\pi$ mass distribution for the reaction $K^-p \to K\pi\Xi(1530)$ (271 events). Resonance parameters are given in Table I. The smooth curves indicate the results of maximum likelihood fits to the data.

Fig. 2 shows the Dalitz plots for the reactions $K^-p \to K^o\bar{K}^o\Lambda^o$ (581 events) and $K^-p \to K^-K^+\Lambda^o$ (303 events). In both cases, ϕ production is dominant. Only the $K^o\bar{K}^o\Lambda^o$ final state, where we have plotted that ΛK combination most forward in the overall CMS, shows evidence for further structure in the 1860 MeV region of the $(K\Lambda)_f$ mass projection. A fit to the complete Dalitz plot, including both ΛK combinations, yields 12 ± 4%, 1871 ± 11 MeV and 58 ± 39 MeV for the rate, mass and width of the ΛK enhancement.

FIG. I $\Xi^-\pi^+$ AND $\Xi(1530)\pi$ MASS SPECTRA AT 2.9 GeV/c

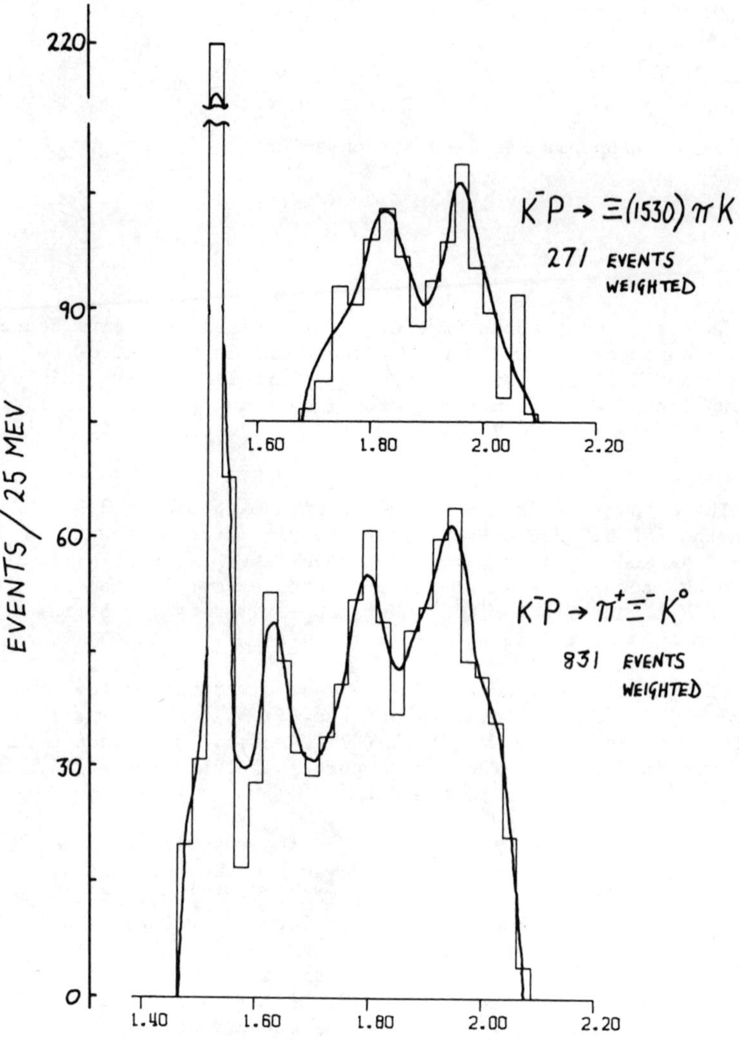

FIG. 2 FINAL STATES $K^-K^+\Lambda^o$ AND $K^o\bar{K}^o\Lambda^o$ AT 2.9 GeV/c

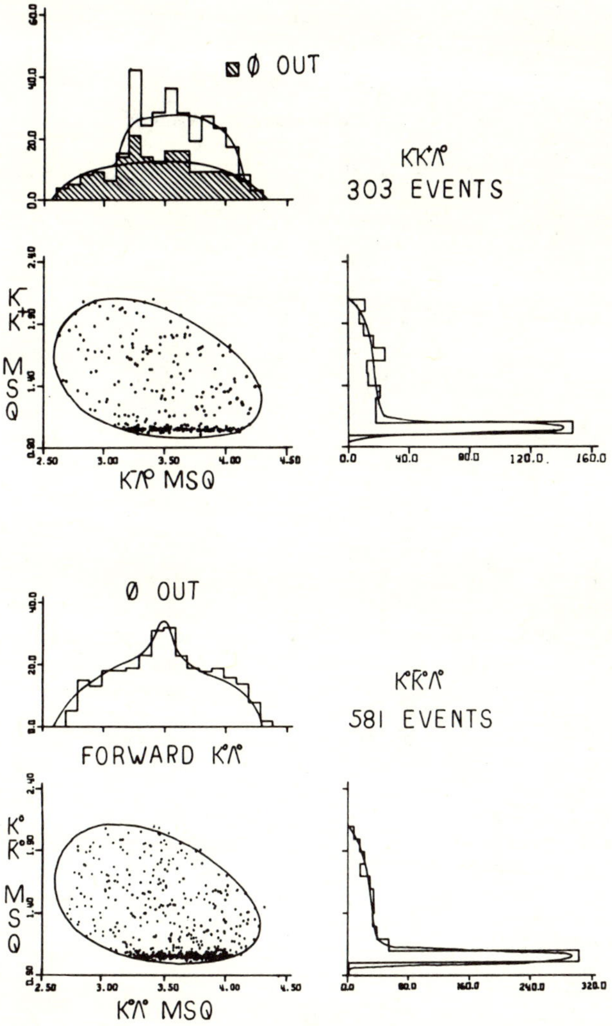

Fig. 3 shows the Dalitz plots for the reactions $K^-p \rightarrow K^-K^+\Sigma^o$ (105 events), $K^-p \rightarrow \bar{K}^oK^+\Sigma^-$ (92 events) and $K^-p \rightarrow K^-K^o\Sigma^+$ (55 events). Of the S = −2 projections, only $\Sigma^-\bar{K}^o$ shows significant structure. Resonance parameters are 39 ± 11%, 1854 ± 12 MeV and 56 ± 14 MeV.

Table I summarizes the signals observed. All are >3σ. The region 1800-1900 is certainly confused. Although it seems unlikely that 1795 ± 10, 1820 ± 12, 1854 ± 12 and 1871 ± 11 represent different mass determinations of the same resonance, we do not know which signals to associate. On the other hand there is no reason why the 1954 and 1961 signals cannot be associated at this time.

TABLE I

Summary of Resonances Observed

Mass(MeV)	Observed Width(MeV)	Decay Mode	No. Events
1531 ± 1	13 ± 2	$\Xi^-\pi^+$	200 ± 14
1635 ± 10	57 ± 18	$\Xi^-\pi^+$	40 ± 12
1795 ± 10	99 ± 31	$\Xi^-\pi^+$	65 ± 20
1820 ± 12	82 ± 42	$\Xi(1530)\pi$	55 ± 13
1854 ± 12	56 ± 14	$\Sigma^-\bar{K}^o$	35 ± 10
1871 ± 11	58 ± 39	$\Lambda^oK^o + \Lambda^o\bar{K}^o$	65 ± 21
1954 ± 9	129 ± 30	$\Xi^-\pi^+$	110 ± 25
1961 ± 8	42 ± 35	$\Xi(1530)\pi$	40 ± 12

What about branching ratios? Can we quote meaningful numbers? Fig. 4 indicates the situation for this experiment. We have plotted all out available S = −2 data in such a way that equal areas represent equal numbers of events, neutral channels on the left and negative channels on the right. The fully-corrected cross sections for these data are $\Xi^-\pi^+$ (1189 ωt. events, 52μb), $\Lambda\bar{K}^o + \Lambda\bar{K}^o$ with ϕ removed (215 events, 17μb), $K^-\Sigma^+$ (55 events, 16μb), $\Xi^o(1530)\pi^o$ (55 events, 6μb), $\Xi^-(1530)\pi^+$ (119 events, 28μb), $\Xi^-\pi^o$ (339 events, 25μb), ΛK^- with ϕ removed (169 events, 24μb), Σ^oK^- with ϕ removed (53 events, 8μb), $\bar{K}^o\Sigma^-$ (92 events, 11μb) and $\Xi^o(1530)\pi^-$ (98 events, 11μb). Because of the confused 1800-1900 region and the absence of signals in many channels we feel that it is impossible to quote reliable branching ratio information.

FIG. 3 FINAL STATES $K^-K^+\Sigma^o$, $\bar{K}^oK^+\Sigma^-$ AND $K^-K^o\Sigma^+$ AT 2.9 GeV/c

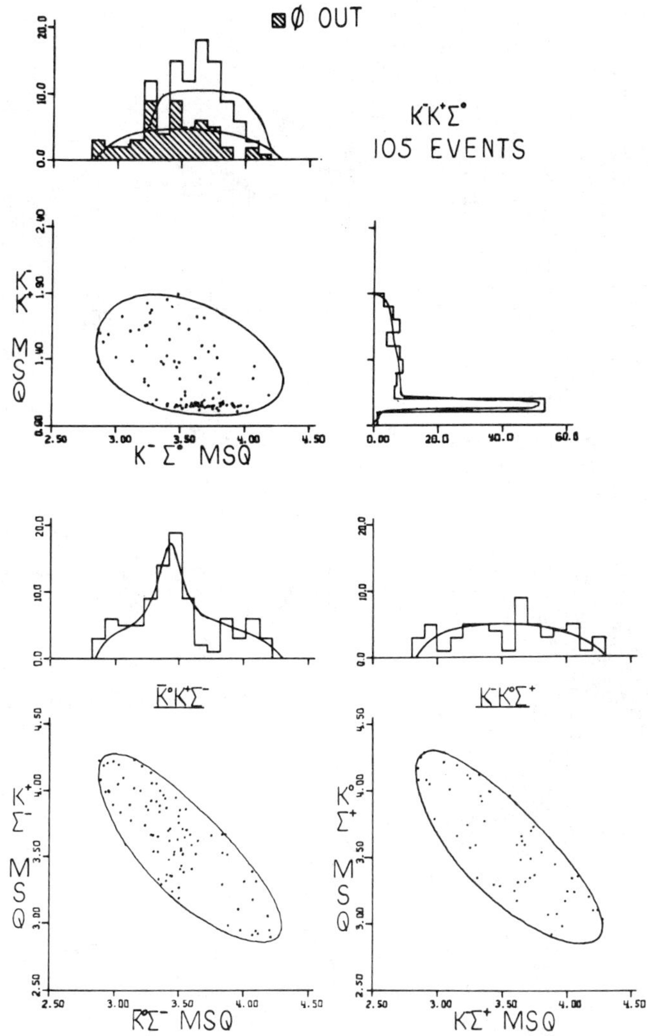

FIG. 4 BRANCHING RATIOS AT 2.9 GeV/c?

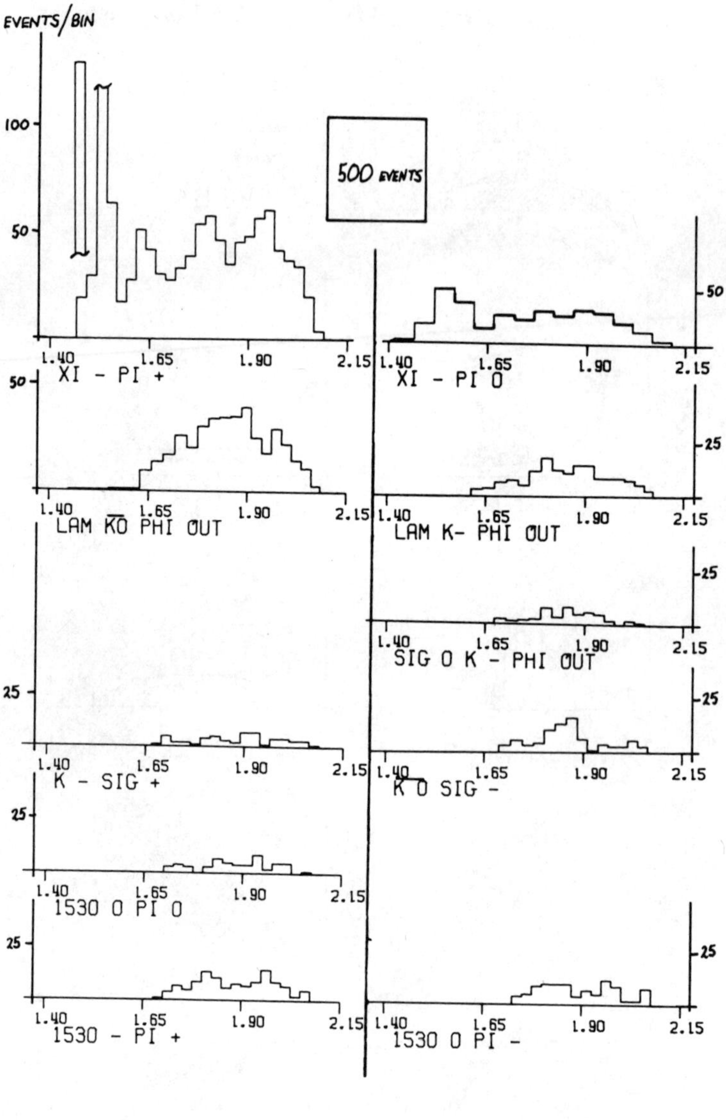

PART II

In comparing and contrasting our results with those of previously published experiments we wish to point out the various reported signals. We have taken the liberty of replotting old data so that a direct comparison can be made. The data have been drawn in such a way that

(i) Horizontal mass scales are the same for each experiment.
(ii) Equal areas represent equal numbers of events.

The histograms presented are those where resonant effects/excesses of events/bumps have been reported. Many of the effects are $>3\sigma$ and we warn the reader to be cautious in the interpretation of the histograms. Small bumps do not necessarily imply small significance. Vertical scales are approximate in cases where $(\text{mass})^2$ histograms have been used. Numbered arrows refer to the comments below and quotations are from original papers.

$\Xi\pi$ System

Figure 5 shows a compilation of data for the $\Xi\pi$ system.

A. Reported Signals

(1) 'A small bump is present at $M^2(\Xi\pi) \approx (1700\ \text{MeV}/c^2)^2$. Smith and Lindsey combined the bump in these data with a similar enhancement observed in the $\Lambda\bar{K}$ spectrum to suggest a $\Xi^*(1705)$'. P. Dauber et al. PR 179, 1262 (1969).

(2) 'Assuming a single resonance, ... $M = 1894 \pm 18$ MeV, $\Gamma = 98 \pm 23$ MeV', P. Dauber et al. 179, 1262 (1969).

(3), (4) and (5) From Table I the resonance parameters are 1635 ± 10, 57 ± 18; 1795 ± 10, 99 ± 31 and 1954 ± 9, 129 ± 30 respectively.

(6) 'An accumulation appears in the $\Xi^-\pi^+$ squared mass spectrum in the region of 3.8 GeV^2. $M = 1933 \pm 16$ MeV, $\Gamma = 140 \pm 35$ MeV.' J. Badier et al. PL 16, 171 (1965).

(7) 'Results suggest the existence of a possible new resonant state of the Ξ. If this state does exist its mass is $\sim$1.77 GeV, its width probably less than 80 MeV, and its isospin is 1/2.' A. Halsteinslid et al. SIENNA (1963).

(8) 'A three standard deviation effect at a $(\Xi\pi)$ mass of 1930 ± 20 MeV and width 80 ± 40 MeV is seen.' J. Alitti et al. PRL 21, 1119 (1968).

(9) 'No evidence for other structures apart from a possible broad enhancement near 1650 MeV.' J. Bartsch et al. PL 28B, 439 (1969).

B. Discussion

The 1530 is always seen. Only the 2.9 GeV/c experiment resolves the high mass excess/bump. The 1705 has gone. The 10.0 GeV/c experiment has a possible enhancement near the 1635 resonance

FIG. 5 DATA COMPILATION FOR $\Xi\pi$ SYSTEM

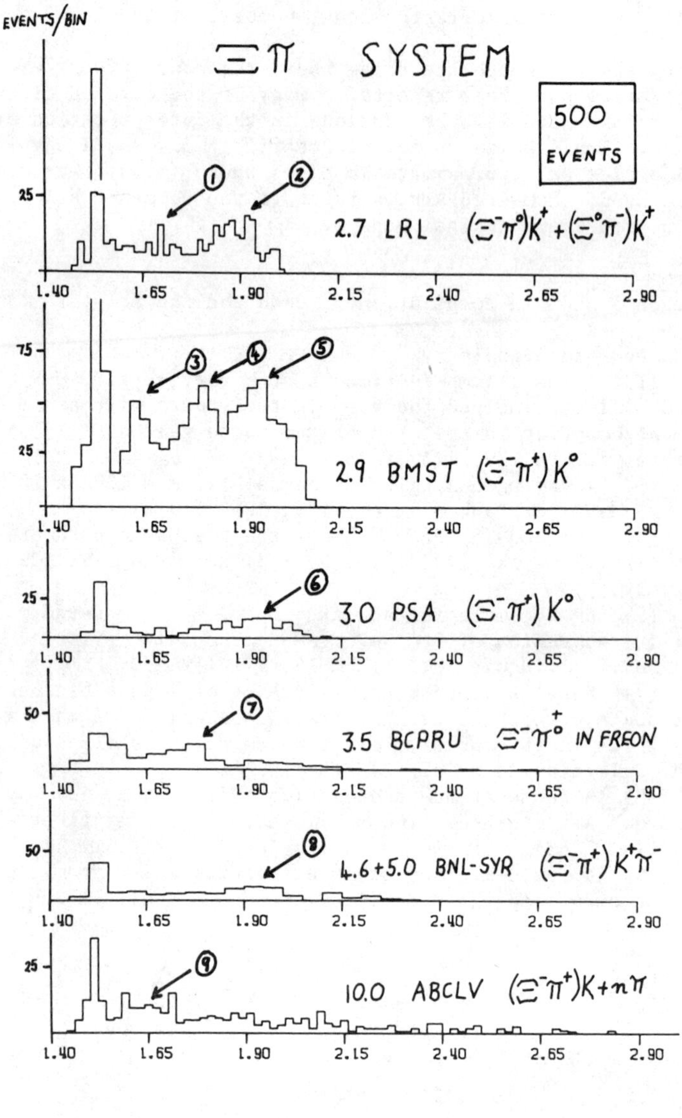

reported by the BMST group. However, using the multiplicity distribution of final particles, (J. Bartsch et al., NP B4, 326 (1968)) it can be shown that this bump is exactly at the place where phase space peaks.

$\Lambda\bar{K}$ System

Figure 6 shows a compliation of data for the $\Lambda\bar{K}$ system.

A. Reported Signals

(10) The 1705 'no longer thought to exist'. Review of Particle Properties, Jan. 1970.

(11) 'A new resonance M = 1817 ± 7 MeV' 'The previously quoted value of the width of the state (≈60 MeV) was overestimated by the Berkeley group; we now arrive at a value of $\Gamma \approx 30 \pm 17$ MeV, not in violent disagreement with the PSA value of 12 ± 4 MeV' (see (13)) G. A. Smith et al. ATHENS (65).

(12) From Table I M = 1871 ± 11, Γ = 58 ± 39 MeV.

(13) 'The probability that our bump...is due to a statistical fluctuation is 1/200' 'Our best estimates for the parameters of the Ξ^* are: M = 1814 ± 4 MeV, observed width 12 ± 4 MeV' J. Badier et al. PL 16, 171 (1965).

(14), (15) and (16) 'Observation of the decay of all three Ξ resonances (1830, 2030, 2430) into the ΛK^- final state restricts their isospin to 1/2' J. Alitti et al. PRL 22, 79 (1969).

(17) 'a peak is observed at the position of the Ξ(1820)' and (18) 'We conclude that the existence of Ξ(2050) in the ($\Lambda\bar{K}$) decay mode has been confirmed' J. Bartsch et al. PL 28B, 439 (1969).

B. Discussion

LRL apart, there is certainly no confirmatory evidence in the $\Lambda\bar{K}$ system for a resonance around 1705.

The 2.63 - 2.70 GeV/c LRL experiment has a significant peak around 1815 in both ΛK^- and $\Lambda K^o + \Lambda\bar{K}^o$. The 2.9 GeV/c BMST experiment has nothing definite in ΛK^-, but the $\Lambda K^o + \Lambda\bar{K}^o$ bump is broader and shifted to higher mass. (Warning: N^* decay into $\Lambda^o K^o$ will contribute to the $\Lambda^o K^o + \Lambda^o\bar{K}^o$ plot).

The PSA single bump in $\Lambda^o K^o + \Lambda^o\bar{K}^o$ can hardly be considered as evidence for a resonance. The inclusion of their mass and width with such small errors in compilations erroneously weights averages toward their value.

The BNL-SYR and ABCLV data for this channel alone is not convincing, and indeed, in the former experiment, the existence of three resonances at 1830, 2030 and 2430 is derived by combining $\Lambda\bar{K}$ and $\Sigma\bar{K}$ channels, and examining forward production.

FIG. 6 DATA COMPILATION FOR $\Lambda\bar{K}$ SYSTEM

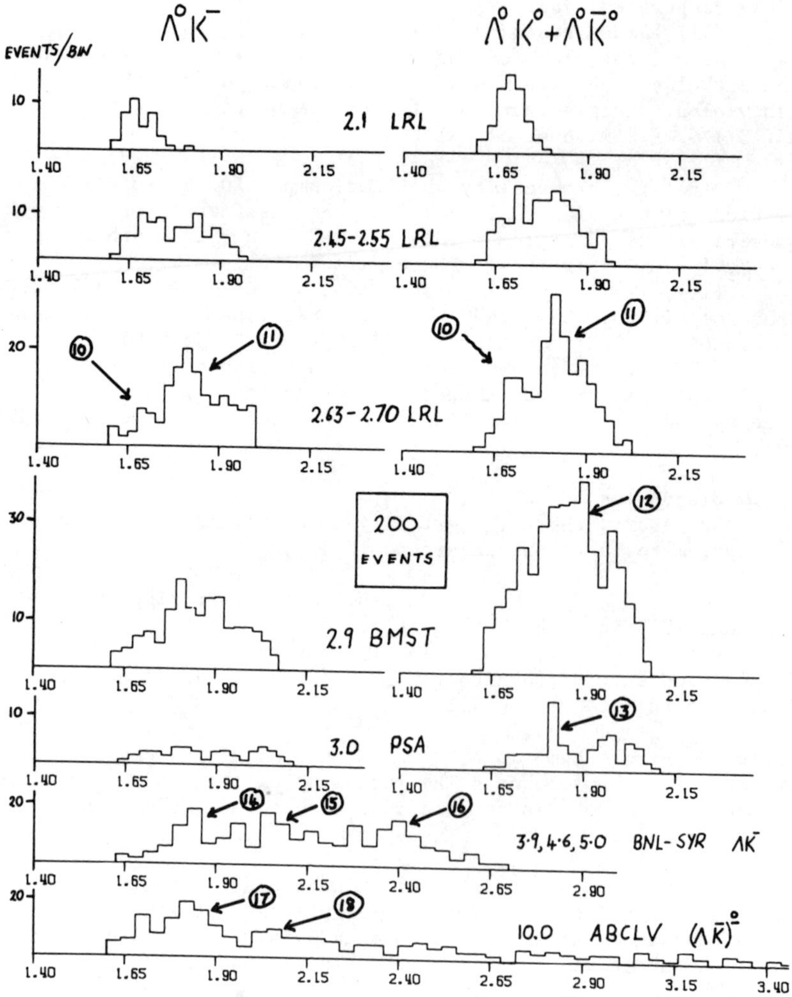

$\Sigma^- \bar{K}^o$ and $\Xi(1530)\pi$ Systems

A. Reported Signals

Fig. 7 shows data for the $\Sigma^- \bar{K}^o$ and $\Xi(1530)\pi$ channels.

(19) 'statistically our bump is not convincing' 'these data provide only very weak evidence for a possible $\Xi\pi\pi$ decay mode of the claimed resonance' P. Dauber et al. PR 179, 1262 (1969).

(20), (21) From Table I,

$M = 1820 \pm 12$, $\Gamma = 82 \pm 42$ and $M = 1961 \pm 8$, $\Gamma = 42 \pm 35$.

(22) 'only the $\Xi(1830)$ is clearly observed' J. Alitti PRL 22, 79 (1969).

(23) From Table I,

$M = 1854 \pm 12$, $\Gamma = 56 \pm 14$.

B. Discussion

Only the 2.9 GeV/c experiment has demonstrated significant signals in $\Xi(1530)\pi$ and $\Sigma^- \bar{K}^o$.

Conclusions

Taking the $\Xi\pi$ and $\Xi(1530)\pi$ systems together, there are four resonances below 2 GeV viz. 1530, 1635, $\approx$1815, $\approx$1960. The $\Sigma\bar{K}$ and $\Lambda\bar{K}$ systems provide complications - some data indicate the 1815 $\to \Lambda\bar{K}$, while the BMST group claim an object $\approx$1860 $\to \Sigma\bar{K}$ and $\Lambda\bar{K}$. This could indicate a new resonance in the 1800-1900 region.

Footnotes & References

*Supported in part by the National Science Foundation and the U.S. Atomic Energy Commission.

†S. Apsell, N. Barash-Schmidt, L. Kirsch, P. Schmidt - Brandeis Univ.; C. Y. Chang, R. J. Hemingway, B. V. Khoury, A. R. Stottlemyer, G. B. Yodh - Maryland Univ.; S. Glickman, M. Goldberg, K. Jaeger, S. Jacobs, C. McCarthy, B. Meadows, G. C. Moneti, J. Sahouria - Syracuse Univ.; J. Bartley, J. Brenner, R. M. Dowd, J. Schneps, G. Wolsky - Tufts Univ.

1. S. Apsell et al., PRL 23, 884 (1969).
2. S. Apsell et al., PRL 24, 777 (1970).
3. J. Bartsch et al., NP B4, 326 (1968).
4. J. Bartsch et al., PL 28B, 439 (1969).
5. P. M. Dauber et al., PR 179, 1262 (1969).
6. J. Badier et al., PL 16, 171 (1965).
7. J. Alitti et al., PRL 21, 1119 (1968).
8. J. Alitti et al., PRL 22, 79 (1969).
9. A. Halsteinslid et al., SIENNA Conference 1963, pg. 173.
10. G. A. Smith & J. S. Lindsey, ATHENS Conference 1965, pg. 251.

FIG. 7 DATA COMPILATION FOR $\Sigma^-\bar{K}^0$ AND $\Xi(1530)\pi$ SYSTEMS

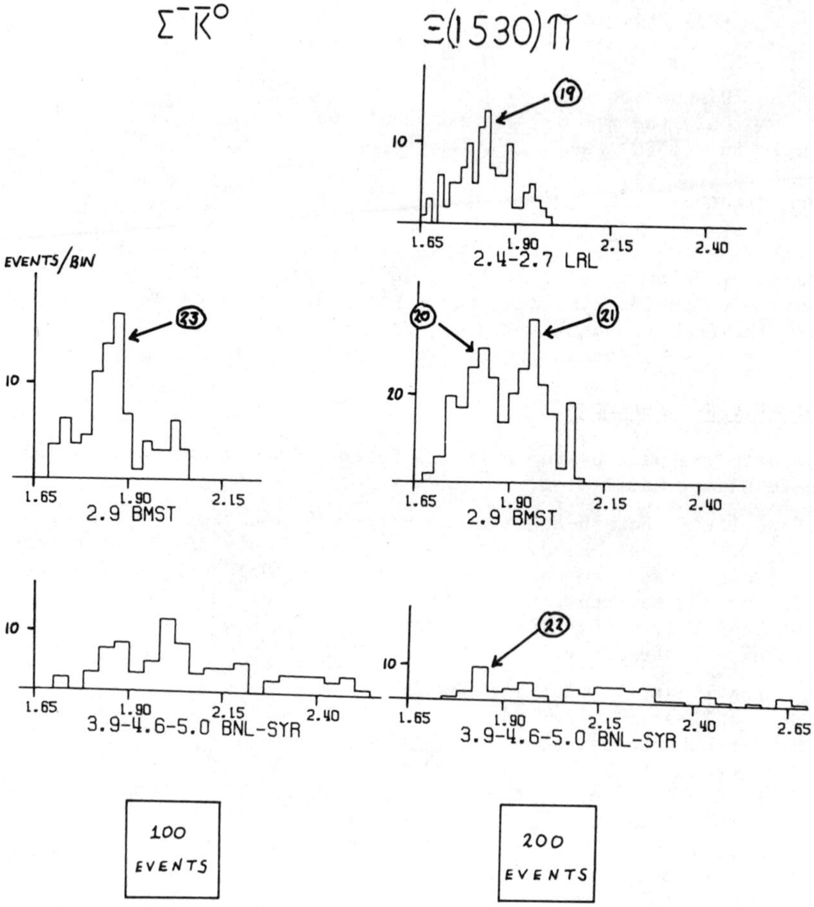

DISCUSSION

Meshkov: With your monstrous display of data, would you care to tell me what the spins and parities of any of these states are?

Hemingway: I'll take a few minutes to explain what we've done. There is only one channel for which we have any spin and parity discrimination whatsoever. This is the channel

$$K^- p \rightarrow \Xi^*(1535)\, K$$

and we have the Dalitz plot of $K\pi$ vs. $\Xi^*(1535)\,\pi$. There you can see a dominant K^* bump and in fact this is the 1815 going across, and the 1960 going up. The analysis has been pointed out by Janice Button-Shafer and alternatively by some other authors, but we're following the work of Button-Shafer because she actually did it for the Berkeley data where they didn't really have any effect. The reaction is:

$$\begin{array}{llll} K^- p \rightarrow & \Xi^{**} & K & \\ & J^P & 0^- & \\ & \hookrightarrow \Xi^* & \pi & \\ & \quad 3/2^+ & 0^- & \\ & \qquad \hookrightarrow \Xi & \pi \\ & \qquad\quad 1/2^+ & 0^- \end{array}$$

where we cannot determine the spin and parity of the parent Ξ^{**}, and there are two strong decays, and what happens is that you look at the angular distribution of this Cascade in the rest frame of this $\Xi^*(1530)$ and the next slide shows what predictions Button-Shafer has made for particular values of the spin-parity of the parent resonance. The angular distribution of the $1/2^+$ object in the rest frame of the Ξ^* with respect to its line of flight is given by the very simple distribution:

$$I(\psi) = 1/3\,(1 + \mathcal{B}\, P_2(\psi))$$

where all you have to determine is the coefficient $\mathcal{B}$, and $\mathcal{B}$ is predicted to have a set of values for particular values of spin and parity. I think this is a very powerful technique in that even if you have large errors it may be easy to discriminate between a particular series with all parities positive and the other series, which are all negative. The next slide shows our angular distributions for the region of 1815, where we plot $\cos\psi$, and the dashed curve represents our fit to $1 + \mathcal{B}\, P_2(\psi)$. Here is the series we tend to favor:

$$3/2^-,\ 5/2^+,\ 7/2^-$$

and here is the series which is not favored:

$$3/2^+,\ 5/2^-,\ 7/2^+.$$

This dotted curve represents the best fit you could get, assuming that the value of $\mathcal{B}$ was -0.5 for this series. In fact, we are certainly less than about 2.5 standard deviations away from the second series. For the 1960 object the data are somewhat more convincing. Our fit is over here; -0.5 is three standard deviations away from the second series, so both these resonances tend to favor $3/2^-$, $5/2^+$, $7/2^-$ but we're not basing our reputations on it. One could argue: "What about the K^* that was dominant in that Dalitz plot?" and in an attempt to see if that has any effect, the next slide will show what happens. This is a diagram representing the Dalitz plot. Here's the K^* and if you use the method of Eberhard and Pripstein, you can take either, say, the events in the lower half of that Dalitz Plot and analyze those, which we have done(we took all the events in the lower half of that Dalitz plot), and calculate a β from this. This is the value that comes out. If you take all events, which include any possible effects of the K^*, then in fact you'll get the same value. So we don't think the K^* is killing us.

Tripp: For a number of years, there has been an outstanding discrepancy with SU(3)in the cascade width compared to the Δ width of a factor of 2. Now, I noticed that your observed width is 13 MeV. Could you say what your true width is for the 1530?

Hemingway: No. There is really only one determination of the 1530 width which was by a UCLA group where they got 7 ± 2. Since then all experiments have in fact not determined the 1530 width; especially Berkeley has not determined the 1530 width; what they have done was to fold in their experimental resolution with a true Breit-Wigner of 7 and said that they can get an acceptable fit, but in those cases they have to multiply their errors by a factor that in some cases went up to 1.5, so I would suggest that new determinations of the 1530 width, the real width, are in fact needed, and we intend to do them for this experiment.

Ross: What is your mass resolution in the $\Xi^*(1530)$ region?

Hemingway: $\pm$ 3.5 MeV in the region of the 1530.

Barnes: You observed a $\Sigma^- K^o$ decay mode of an object which you give a mass of 1854. However you should, correcting for the visibility of neutral decays, see equal amounts of $\Sigma^o K^-$ decay mode and you see nothing.

Hemingway: That is correct. I forgot to mention that for the $\Sigma^o K^- K^+$ channel we are reporting only 1/3 of the data available in the exposure, as compared with the cascade channels. So the data could be consistent. At the moment we don't seem to get a good $\Sigma^o K^-$ signal, but that's not inconsistent with 1/3 of the rate, and we'll adjust it.

Ξ STATES

George R. Kalbfleisch

Brookhaven National Laboratory,

Upton, New York 11973

The study of Ξ states is fraught with the difficulties of priorities (who discovered it first), statistical significance (few events, many channels, combining of data, etc.), and the interpretation of marginally significant data, since the data are so very hard to obtain. I have been asked to review these states, since it was believed that I'm a relatively disinterested party regarding these difficulties. I will attempt to do this. I would like to review these states with a somewhat different viewpoint than previously used and then present a list of Ξ states as certain, reasonable, or indicated candidates. This list of Ξ states could then presumably "update" the "Particle Data Group" listing of such states. I will try to avoid the political questions of priorities. I must certainly acknowledge the help of Yodh's review paper[1] of last year and of the BMST (Brandeis, Maryland, Syracuse, Tufts) paper presented earlier by Dr. Hemingway.[2] I will not discuss cascade production mechanisms nor comment on Professor Yodh's previous paper.

Before presenting the list of potential candidates, let's remark on the question of the importance of cascade states. The two well-known and certain states are the Ξ(1320) and Ξπ(1530). The Ξ(1320) was reasonably important in completing the Ξ I=1/2 doublet to confirm the Gell-Mann Nishijima strangeness scheme. The Ξπ(1530) was the first excited cascade state which along with the Δ(1238) and Σ(1385) led to the prediction of the Ω^-(1675). However, looking back, this prediction could possibly have rested almost as well on the Δ and Σ states alone. However the Ξπ(1530) was discovered and studied rapidly at about the same time as the Σ(1385) was being well understood. The Σ(1385) had been known for quite a few years but its spin parity was established just about the time that the $\Xi\pi(1530)$ state was found. The $\Xi\pi(1530)$ spin parity was shown to be consistent with $3/2^+$ soon thereafter.[3] The question of the importance of cascade states now is less obvious. A large number are needed to fill out symmetry multiplets. SU3 predictions of decay rates are interesting but not necessarily crucial to our understanding of the strong interactions. The question of the presence or absence of certain SU3 representations such as $\overline{10}$, 27, etc. is of course very important; it may or may not be answered by the finding of some I=3/2 cascade states. We have none yet. However I believe we all feel that from the experimental point of view at least, we should continue attacking this problem. For the future this attack will require very large exposures. Exposures of less than 30-50 events per micro-barn will not be particularly useful for the study of cascade states. Some experiments at ten times this level may be done in the not too far distant future.

Lets now discuss the Ξ states. The small statistics relating to Ξ states has led to much combining of different charge states, of different decay modes, and even of different final states (three-body plus four-body, etc.). In view of the difficulties with the vast multiplicity of Y^* states in the 1660 region with each final state being possibly a different resonance, we will adopt a conservative viewpoint. This viewpoint is that each mode should be considered on its own. Branching ratios and comparisons with SU3 predictions should only be made after some semblance of order on the masses, widths, and isotopic spins is made. Ideally the spin-parity should also be determined before associating various decay modes but this is probably precluded for cascade states. This is difficult enough for N, Δ, Λ, and Σ states at this time. By following such a procedure, I think it will be clear what the status of these states really is. I will now present a list of Ξ states and their properties and finally conclude with a new list from this review. We will discuss (a) some representative evidences for these states, (b) their isotopic spin, and (c) a little regarding SU3 predictions for these states. None of this will be done in much detail; just enough to give the overall flavor of this field. Table I gives a list of Ξ states as previously given by Yodh[1] and as listed in the January 1970 Particle Data Group tables.[4]

We now present some representative data. Not all data are shown. We will comment on these as we go. The $\Xi\pi(1930)$ first claimed by Badier et al,[5] is shown in Fig. 1. A general lumping at large masses is seen. It is clear from the Dalitz plot that interference with the $K^*(890)$ may be very important; in particular, events with the K^* removed shows possibly a narrower bump at 1930 MeV. There may be other structures complicating these data. Establishment of this state was made in my opinion by Alitti, et al.. Figure 2 shows their $\Xi^-\pi^+$ mass combinations. About a three standard effect is seen in this $\Xi\pi K\pi$ channel; the events that are $\Xi\pi K^*$ are

Table I - Previous Tables of Ξ States

Yodh		Particle Data Group
- - -		Ξ(1320)
Ξ(1530)	OK	Ξ(1530) → Ξπ (Ξ*)
Ξ(1630)	OK	- - -
Ξ(1705)	NO GOOD	- - -
Ξ(1820)	OK	Ξ(1820) → $\Lambda\bar{K}$, $\Sigma\bar{K}$, Ξπ, Ξ*π
Ξ(1940)	OK	Ξ(1930) → Ξπ
Ξ(2030)	OK	Ξ(2030) → $\Sigma\bar{K}$, $\Lambda\bar{K}$
Ξ(2250)	?	Ξ(2250) → various
Ξ(2450)	?	Ξ(2500) → various
I=3/2 Ξ*	none seen	- - -

Ξπ (1930)

BADIER et al PL 16, 171 (65).

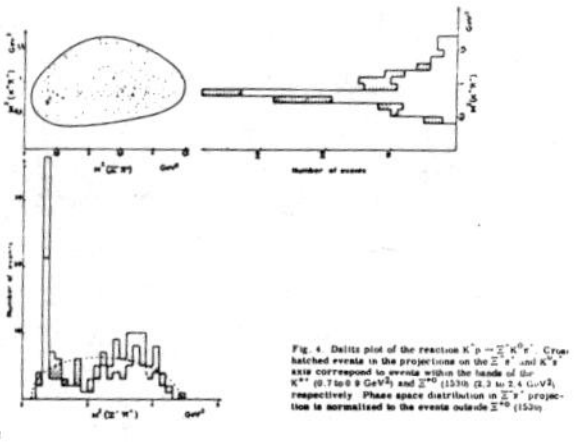

Fig. 1. Original evidence for Ξπ(1930). 3 GeV/c $K^-p \rightarrow \Xi^- K^o \pi^+$

Ξπ (1930)

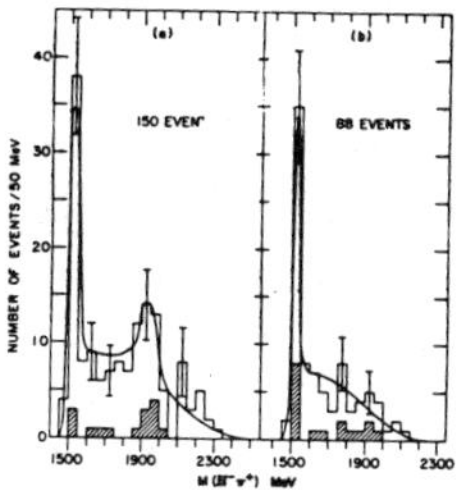

Fig. 2. BNL-S data establishing Ξπ(1930). 3.9,4.6,5.0 GeV/c $K^-p \rightarrow \Xi^-\pi^+K^+\pi^-$

shown shaded and make some contribution to the 1930 region. Both channels together, $\Xi\pi K\pi$ and $\Xi\pi K^*$, give evidence of a $\Xi\pi(1930)$ at about three standard deviations significance. In general we will see that there are very few effects in the Ξ studies that are better than about three standard deviations. Figure 3 shows the $\Xi\pi$ mass combinations from the $\Xi\pi K$ final states at low momentum; 2.1 and 2.7 GeV/c K^-p from LRL[7] and 2.9 GeV/c K^-p from Brandeis, Maryland, Syracuse, Tufts (BMST) collaboration.[8] On the left are shown the $\Xi^-\pi^+$ combinations and on the right the $\Xi^-\pi^o$ combinations. The strongest $\Xi\pi$ information generally comes from the $\Xi^-\pi^+$ state. This is the favored combination if the Ξ^*'s isospin is 1/2. The $\Xi(1530)$ is clearly seen in all plots. In addition the BMST collaboration claims an effect at 1630 MeV which in their estimation is of the order of three standard deviations. It is noted in this plot of course that there is some evidence for the 1630 spike also in the LRL data. The 2.9 $\Xi^-\pi^o$ data may also indicate it. This small effect should be remembered as we will refer to it later in the isotopic spin dicussion. The 2.7 and the 2.9 data also have some indications, certainly not convincing, that structure is possible at the 1800 and 1950 regions also. The problem here is how to draw the background. You can assume resonances are present but there is certainly no compelling evidence here. In Fig. 4 we show the combined $\Xi\pi$ plot (borrowed from Yodh) of 2.1, 2.4, 2.7, and 2.9 GeV/c LRL and BMST data. This seems to show a nice 1640 peak. At least here you see very clearly how to estimate the background. The statistical significance is three standard deviations, so that the $\Xi\pi(1640)$ is a reasonable candidate to be entered into the tables.

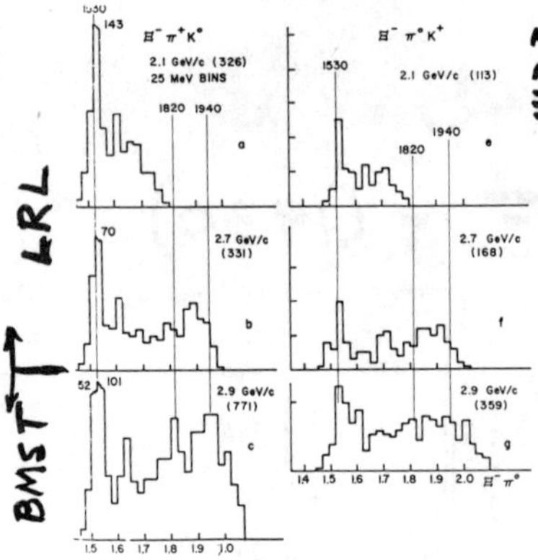

$\Xi\pi(\vec{p})$

Fig. 3. $\Xi\pi$ spectra at various momenta (2.1, 2.7, 2.9 GeV/c) for $K^-p \rightarrow \Xi^-K^o\pi^+$ and $\Xi^-K^+\pi^o$. $\Xi\pi(1630)$ shows much better in later plots (see ref. 2 and 8).

$\Xi\pi$ (1640)

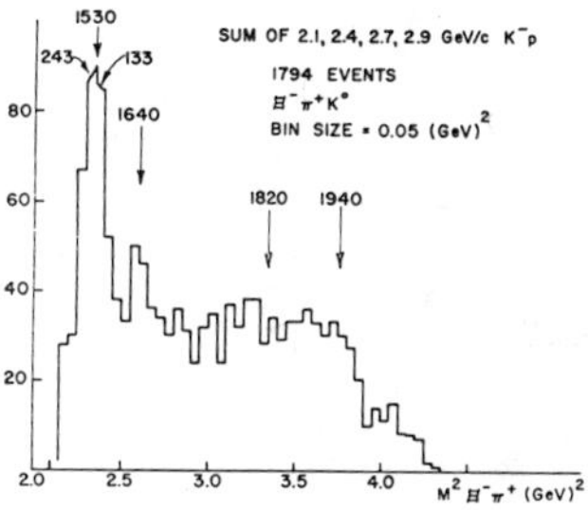

Fig. 4. "Combined" $\Xi\pi$ data from $K^-p \rightarrow \Xi^- K^o \pi^+$ to look at $\Xi\pi(1630)$.

In Fig. 5 we see the $\Lambda\bar{K}$ states. The original evidence for the 1810 is shown on top in the LRL[7] data where they have at 2.6 GeV/c both $\Lambda\bar{K}^o$(or ΛK^o) and ΛK^-. The BMST collaboration[8] has some evidence of the effect in $\Lambda\bar{K}^o$ (or ΛK^o) but it is somewhat shifted and broader, and the signal to moise ratio is not as good as the LRL data at 2.6. They have at this time no evidence for ΛK^- as only 1/3 of that data has been processed; they will have more to say about this at some future date I'm sure. The Brookhaven-Syracuse (BNL-S) collaboration at the 3.9, 4.6, and 5 GeV has some evidence for ΛK^- also;[9] it is certainly not overwhelming, being about two standard deviations; but it agrees with the earlier LRL observations. It would seem that one has a $\Lambda\bar{K}$ cascade state at 1810 MeV. In Fig. 6 there is $\Sigma^-\bar{K}^o$ data which seems to show a striking peak at 1850 MeV. Earlier BNL-Syracuse and earlier LRL analyses were looking for other modes of the 1810 $\Lambda\bar{K}$ state. It appears however that both BNL/Syracuse[10] and BMST[8] indicate a higher mass for the $\Sigma\bar{K}$ state. I would say that we have a $\Sigma\bar{K}$ state at 1850; whether or not this is partly due to 1810 and partly due to the 1930 with some kind of complicated production mechanism making a single bump appear at 1850 MeV, cannot be determined here. I would at this time classify it as a separate state distinct from the $\Lambda\bar{K}$. The evidence for $\Xi\pi\pi$ states at 1830 and 1930 are shown in Fig. 7 with the 2.7 LRL[11] data on the left and 2.9 BMST data on the right. There are indications of structure; but it's not clear that various reflections in the complicated four body state at not responsible. It will require a great deal more statistics, I believe, before one can really understand what's happening here. A search for I=3/2 Ξ states has been made by K.W. Lai at BNL in K^-n interactions at 3.9 and 3.6 GeV/c.[12] I will only show his $\Xi^-\pi^-$ data. Figure 8 shows

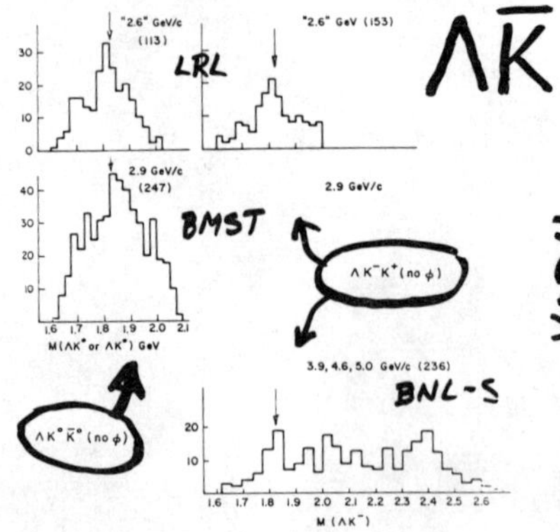

Fig. 5. $\Lambda\bar{K}$ data from $K^-p \to \Lambda K^o\bar{K}^o$ and ΛK^+K^-.

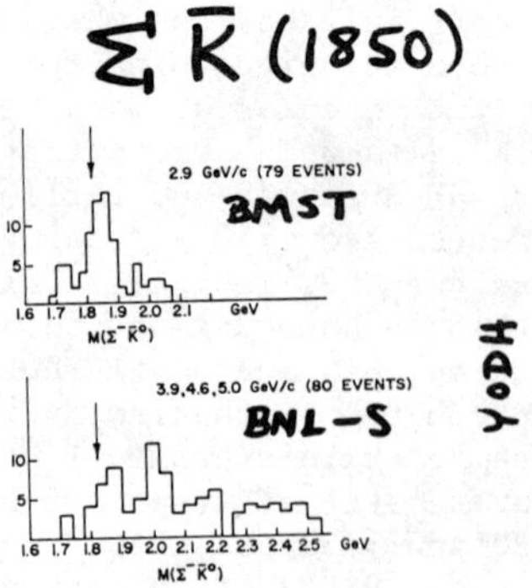

Fig. 6. $\Sigma\bar{K}$ data from the reaction $K^-p \to \Sigma^-\bar{K}^oK^+$.

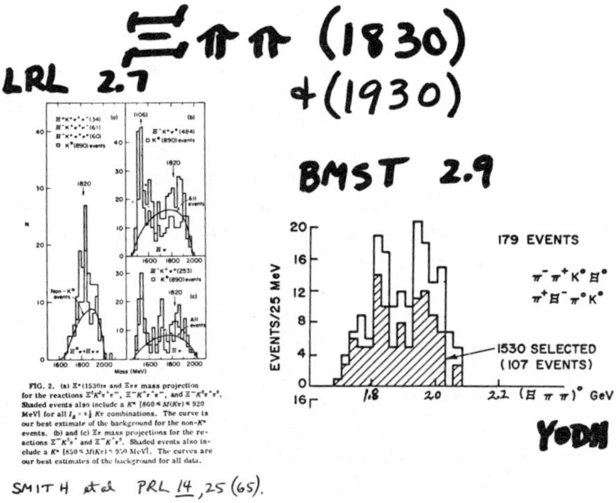

Fig. 7. LRL and BMST Ξππ data.

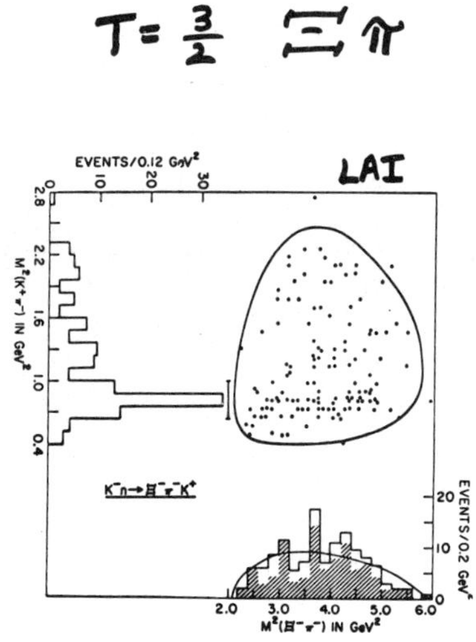

Fig. 8. BNL (K.W. Lai) data on $\Xi^-\pi^-$ from 3.6 and 3.9 $K^-n \rightarrow \Xi^-\pi^-K^+$.

the Dalitz plot and its projections for the reaction $K^-n \rightarrow \Xi^-\pi^-K^+$. One can see no significant deviations from the phase space curve on the $\Xi^-\pi^-$ projection; the shaded events are in the K*(890) band. Apart from Ξ^-K^* there is no structure. Although this data looks statistically meager, it represents a large piece of information against existence of 3/2 $\Xi\pi$ states, since it represents some 25 events/μ b. This data implies that no I=3/2 $\Xi\pi$ state with a width of 40 or 80 MeV exists at the 1 micro-barn level. Lets look at the 1630 region in particular. There are no excess events at 2.6 mass squared; especially no excess outside the K* band; I would say that in this particular region the upper limit is 0.3μ b; remember this, as we will use this information later.

Looking at all these data collectively we see how weak the statistics are if one does not combine data. Also, if we had labeled the abscissa of these graphs as $N\pi$ or $\Lambda\pi$ or $\Sigma\pi$, you would probably have laughed at most of these data saying, "What do they show?" And that just shows how difficult it is to study Ξ states and why one is pushed to reach conclusions about these data. In the 3 body final states, the K^*/Ξ^* interference problem may be significant in confusing the results. The 4-body Ξ^*K^* states may be a better place to look in general.

Next, lets discuss the isotopic spin of these states (the $\Lambda\bar{K}$ states are obviously I=1/2). For example, lets take the $\Xi\pi$(1630). If its isotopic spin is 3/2, we would have the reaction $\bar{K}N \rightarrow \Xi^*K$ proceeding through the I=1 channel; call this cross-section σ_1. Then the cross section for $K^-n \rightarrow \Xi^-\pi^-K^+$ would be 3/4 σ_1, $K^-p \rightarrow \Xi^{*-}K^+$ would be 1/4 σ_1 and $K^-p \rightarrow \Xi^{*o}K^o$ (where K^o is both seen and not seen) would also be 1/4 σ_1. The subsequent decay $\Xi^{*-} \rightarrow \Xi^-\pi^o$ would be 2/3 and $\Xi^{*o} \rightarrow \Xi^-\pi^+$ would be 1/3, opposite to the usual circumstance. Thus the ratio $(\Xi^-\pi^o/\Xi^-\pi^+)$ would equal 2 and that clearly was not the case (remember Fig. 3, the $\Xi^-\pi^o$ had a small effect; it was clearly not double the $\Xi^-\pi^+$; if anything it was at most one-half). Alternatively, one can use the $\Xi^-\pi^-$ data. You see that the $(\Xi^-\pi^-/\Xi^-\pi^+)$ ratio would be 3/4 over (1/3 times 1/4) or 9. However we have experiments done at different energies. According to Morrison,[13] the cross section (well above threshold) as a function of energy would probably go down like the 4th power of the lab momentum, like p^{-4} say. For momenta of 2.1, 2.7, 2.9, and 3.9, $(p/2.9)^{-4}$ gives the values of 4, 1.3, 1, and .35 for the relative cross-sections. The $\Xi^-\pi^+$(1630) cross section is of the order of 2, 1, and 1μb at the 3 lower energies and the cross section for $\Xi^-\pi^-$ at 3.9 GeV/c in the 1630 mass region is less than 1μb; I would say (as mentioned above) 0 ± 0.3 μb would be a reasonable value. The 1 μb cross-sections at 2.7, 2.9 GeV/c would predict 9 micro-barns for $\Xi^-\pi^-$ at 2.8 GeV/c. Scaling this by the p^{-4} would predict 3 micro-barns at 3.9 GeV/c. The absence of $\Xi^-\pi^-$ thus implies again that I ≠ 3/2 for $\Xi\pi$(1630) (if it really exists). Similarly one might analyze the $\Xi\pi(1930)$ state; I believe it would be inconclusive at this time, however.

Finally, a few words about SU3. The $\Lambda\bar{K}(1815)$ state has been tentatively assigned to a $J^P=3/2^-$ octet. This requires a large $\Sigma\bar{K}$ mode;[14] but the $\Sigma\bar{K}$ mode appears to be at a mass of 1850 and probably a different state as I remarked earlier. Likewise, if the $\Xi\pi(1930)$ belongs to a $J^P= 5/2^-$ octet[14] a reasonably strong $\Sigma\bar{K}$ mode is needed.

In conclusion my list of Ξ states is given in Table II. Several statements about Table II are in order. The certain states Ξ(1320) and (1530) should clearly be included in any table. The next four states $\Xi\pi$ (1630), $\Lambda\bar{K}(1810)$, $\Sigma\bar{K}$ (1850), and $\Xi\pi$ (1920) seem like reasonable candidates; these cascade states should probably be included in the Particle Data Group tables.[4] In my opinion, however, the remaining states are only indicated; they're certainly not at the three standard deviation level without some interpretive assumptions. These states should remain in the footnotes or listings of the Particle Data Group at this time. There will always be arguments as to where one should "draw the line." However I hope that we have shown that different decay modes should not be grouped at this time, and that this list has the better candidates at the top and the poorer ones at the bottom. A couple of comments might be made about states classified as <u>certain</u>. The spin and parity of the Ξ(1320) hasn't been firmly established. The spin is less than 3/2 by like 2-1/2 standard deviations, and the parity has not really been done at all. This doesn't bother us, of course, because theoretically we know how to classify this particle. Similarly, the spin parity of the Ξ*(1530) has not been too well determined.[3] The spin has reasonably been shown not to be 1/2, the odds are maybe 10 to 1 that the parity is positive for spin 3/2 and that the indicated sequence is favored. Theoretically we believe that it's $3/2^+$ of course. Thus the status of these two states seems certain combining our experimental and theoretical knowledge. I believe the width of the $\Xi\pi(1530)$ should be listed as 10 ± 3 MeV. The 7 ± 2 MeV value was obtained by UCLA originally.[3] This value depends on some assumptions put into a likelihood function. The actual data have a width at the half maximum point three bins wide (of 4 MeV); I would think 10 ± 3 MeV agrees better with other observations on the $\Xi\pi(1530)$. I think this also fits in better with SU3 predictions and branching ratio calculations.

What about the future? Maybe we shouldn't do Ξ physics in the future, as we remarked earlier. However, I am sure that it will be attempted; experimentally it seems desireable to forge ahead some more; exposures at the 100 to 1000 event/μb level will undoubtedly be attempted. Studies via Ξ^- beams will also possibly be made. If further experiments should not be persued, someone should present the arguments now since it will take large expenditures of time, money, and effort to do this. We know that nucleon, delta, sigma, and lambda searches have shown extreme diversity, and much time and effort over the past ten years has been expended acquiring that data. It will certainly be impossible to reach similar levels in the search for cascade states, and in the study of their properties -- but undoubtedly we will press on anyway.

Table II - Ξ States

State	Width	I	J^P Exp.	J^P Theory	Status
Ξ(1320)	---	$\frac{1}{2}$	< 3/2	$\frac{1}{2}$ +	Certain
Ξ(1530)	10±3	$\frac{1}{2}$	$\frac{3+}{2}$, $\frac{5-}{2}$ ...	$3/2^+$	Certain
Ξπ(1630)	~50	$\frac{1}{2}$	?	?	Reasonable*
$\Lambda\bar{K}$(1810)	~50	$\frac{1}{2}$	Indications	$3/2^-$?	Reasonable
$\Sigma\bar{K}$(1850)	~50	?	?	?	Reasonable
Ξπ(1920)	~80	?	?	$5/2^-$?	Reasonable
Ξπ(1.8 GeV)					Indicated
Ξππ(1.8) or $\Xi^*\pi$(1.8)					Indicated
Ξππ(1.9) or $\Xi^*\pi$(1.9)					Indicated
$\Sigma\bar{K}$(2.0)					Indicated
$\Lambda\bar{K}$(2.4)					Indicated
$\Sigma\bar{K}$(2.4)					Indicated

* At least 3σ in one experiment and seen elsewhere. The future status may improve or it may decline.

References

1. G.B. Yodh, "Ξ Spectroscopy," Univ. of Maryland Technical Report 991 (ORO-2504-140), June 1969.

2. Brandeis-Maryland-Syracuse-Tufts, "Ξ Resonances below 2 GeV," paper presented to this conference.

3. P.E. Schlein, et al, Phys. Rev. Letters 11, 167 (1963).

4. A. Barbaro-Galtieri, et al., Rev. Mod. Physics 42, 87 (1970).

5. J. Badier, et al., Phys. Letters 16, 171 (1965).

6. J. Alitti, et al., Phys. Rev. Letters 21, 1119 (1968).

7. G.A. Smith and J.S. Lindsey, in Proceedings of the Athens Second Topical Conference on Resonant Particles, edited by B.A. Munir (Ohio Univ. Press, Athens, Ohio, 1965), p. 251.

8. S. Apsell, et al., Phys. Rev. Letters 23, 884 (1969); Phys. Rev. Letters 24, 777 (1970). The $\Xi\pi$(1630) data is much prettier in the first paper of this reference and that of Reference 2, than in Fig. 3 of this review.

9. J. Alitti, et al., Phys. Rev. Letters 22, 79 (1969).

10. J. Alitti, et al., "Confirmation of the Ξ(1815) and Evidence for a New Ξ Resonance with Mass 2030 MeV," BNL report 12808, (Submitted to XIVth International Conference on High Energy Physics, Vienna, Austria, 1968).

11. G.A. Smith, et al., Phys. Rev. Letters 14, 25 (1965).

12. D.J. Crennell, et al., "Study of Ξ* Resonances in K^-n Interactions at 3.6 and 3.9 GeV/c," BNL report 14050 (submitted to Phys. Rev.).

13. D.R.O. Morrison, Phys. Letters 22, 528 (1966).

14. M. Goldberg, et al., Il Nuovo Cimento 45A, 169 (1966); R.D. Tripp, et al., Nuclear Phys. B3, 10 (1967); E. Flaminio, et al., "Boson and Baryon Spectra and SU(3)," BNL report 14572, March 1970.

DISCUSSION

Rosenfeld: First, before I answer George, let me make a disclaimer about the "Rosenfeld tables." As you all know, we now have ten authors and ask in every edition to be called the Particle Data Group.

Next, I don't want to kill anybody. These last two talks have been among the best I've heard in a long time. In fact I would like strongly to associate myself with this sceptical point of view. If you look back at the April 1969 issue of Physics Today you will see a letter to the Editor by Phil Dauber and me, complaining about all the fuss raised by the Samios group over the Ξ (1930). So I think I'm on your side.

To show further that we are on the side of thoughtful scepticism, let me remind you that members of the Particle Data Group have worked harder than most to question or explain doubtful bumps. In our Jan. 1967 Rev. Mod. Phys. article we devoted an appendix to questioning the kappa and H mesons, and I think really killed them at the 1968 Philadelphia Conference.

Let me make two remarks about the problems of compiling data. The first is to admit that in the case of Ξ^*'s we included too many doubtful resonances in the table; we'll change that, and we thank you gentlemen for pointing it out. The second is to point out that most of the remarks that have been made today are about entries in the listings; not in the tables, and I would just like to point out our problem. In the tables, you find a lot of arrows, and then if you look at the bottom of the tables, you find the following comment: "An arrow at the left of the table indicates a candidate that has been omitted because it is open to considerable question. See listings for information . . ." In other words, the listings are supposed to be complete, including unsubstantial claims; the tables are supposed to be critically evaluated. We think there is a need for both. Would you want to re-write physics by omitting those claims which we feel to be unconvincing? Is it not better to catalog them, and then to ignore them? If some of you feel otherwise, we'd be happy to get your suggestions.

Hemingway: Can I ask a quick question. What defines a claim? If somebody says "apart from a possible broad enhancement" is that a claim, or what is it? That goes into the tables at the moment.

Rosenfeld: I think you mean the listings. We don't have any firm criteria. If a claim is too trivial we tend to ignore it. My personal criterion is this: If a bump is convincing enough that somebody is likely to remember it, and want to go back and find the reference when he himself finds an interesting nearby effect, then I list it.

Galtieri: I have to say one thing. As far as the Ξ^*'s go, I have always been tempted to eliminate all except the Ξ (1530) which is the only one on which there is any information. However, as George said, a lot of work went into this, a lot of papers have been based on it , many million pictures have been looked at, so what can we do? We have to list something. So we have to decide between very conflicting information as to what we want to put in the tables. Now people keep talking about averaging in the data. We have never done any averaging in the data except for the Ξ (1530). If you look, there are numbers with no errors in the tables except for Ξ (1530). The listings are a useful thing for people who want to look at what has been reported. But we give meaning to errors in the listings only when we report them in the tables. And the errors are in the tables with scale factors only as a guide to the reader to show that the numbers are in disagreement. But if the authors do give a number with an error, it goes into the listings. The meaning we give to it is reflected in what appears in the tables. Since the time we decided to include some states in addition to Ξ (1530), there are new experiments. The one reported by Hemingway is larger than the others. It is unfair now that you have more information, to just say "the tables are all wrong". You can say it if you want, but we did our best with what was available at the time.

Kalbfleisch: I would like to see each mode listed separately in the tables.

Rosenfeld: We already do that for the mesons.

Button-Shafer: If I understand George correctly, he suggested that history was a little different from what we commonly believe. I may have misunderstood. But the $Y^*(1385)$ spin-and-parity was definitely established by January of 1963 (as reported at the Physical Society Meeting, and appeared in Phys. Rev. Letters just a few months later) with the use of normal and magic direction components of polarization. The Byers-Fenster analysis formalism was still being developed; it was not available for Schlein to use on the $\Xi^*(1530)$ until some months later.

Kalbfleisch: I tried to say they were nearly simultaneous. Here is the PRL paper. "Spin- Parity Determination of the $\Xi\pi$ Resonance at 1530," Schlein et al., received 8 July 1963.

Button-Shafer: George, our sample had 0 $\pm$ 4 percent background and was reported in Phys. Rev. Letters some months before that Ξ^* study. One other comment. The $\Xi^*(1530)$ spin-and-parity was not completely established by Schlein in as much as I would say the spin was established to slightly better than three standard derivations, but the parity had a confidence level for the rejected hypothesis of D 3/2 which was .035. One further comment. The $\Xi^*(1530)$ has been, in fact very elusive, and there was spin parity analysis which we tried on the data that I worked on and Deane Merrill worked on; it led, unfortunately, once again to rejection of spin 1/2, but to no discrimination that was very convincing for the parity. It was roughly again a few percent confidence level. I understand that Maryland also has a 3 percent confidence level for the opposite parity from that we would like to believe.

Hemingway: We have no spin or parity discrimination in the final state you are referring to.

Button-Shafer: So we still don't really have the $\Xi^*(1530)$ parity established. Also, on the 1820, since you made reference to that in your talk as not having been determined, we always said that the slight suggestion that we had for $3/2^-$, $5/2^+$, etc., was subject to considerable reservations in as much as the data were 40 events or so, 15 events of which were background.

Tuan: There has always been a problem of interest to theoreticians, that is, we have had resonance in the ηN, $\eta\Lambda$, and $\eta\Sigma$ channels which fortunately experimentally you can look at in formation experiments and apparently these resonances are strongly coupled to the η-baryon system. An interesting question is: should there or should there not be a similar η-cascade resonance near threshold, that is, at about 1870 MeV. Now, I'm going to ask the experimentalists what is the feasibility of looking at it in a production experiment, like $K^-p \to (\eta\ \Xi^-)K^+$.

Hemingway: The only experiment which has got enough energy available to look for that decay mode is the BNL-Syracuse 3.9, 4.6, and 5 GeV/c, and as far as I know, when I asked them, they said they had no evidence for a cascade-eta decay mode of anything.

Tuan: But could this be looked at, because it's quite interesting?

Hemingway: It's certainly accessible in a bubble chamber experiment.

Bassano: I'd like to make one comment about Hemingway's slide, where he showed all the data. I think the important thing is not the absolute number of events. The important thing is how many standard deviations the events are above background. I'd like to point out that at the lower energies there are large numbers of events, since the cross-sections are higher, but also the phase space is restricted and the backgrounds are much higher, so that again the important thing is the number of standard deviations above the background. Now at higher energies, like at BNL, and even higher, although the cross-sections are lower and the absolute number of events are lower, the phase spaces are spread out, and still the standard deviations are the important thing, and I think a three standard deviation effect at a high energy is just as good as a three standard deviation effect at a low energy.

Moneti: I wonder how George could make up his mind about the reasonableness of the various bumps. He has accepted as reasonably evident the $\Xi\pi$ (1630) which is seen as about a three standard deviation effect in essentially only one experiment, and also the $\Sigma^-\bar{K}^0$ (1850) for which there are about 30 events in the mass interval out of a total of 79 $\Sigma^-\bar{K}^0 K^+$ events and which still lacks confirmation in the $\Sigma^0 K^-$ system. On the other hand, he has rejected the $\Xi\pi$ (1810) and $\Xi\pi$ (1810) while keeping the $\Xi\pi$ (1950). I agree that a few tens of events in between the 1810 and 1950 can change the double bump structure into a single large bump, but I cannot see how one can prefer the 1950 bump to the 1810 one: either we keep both or we lump everything together.

Kalbfleisch: See Fig.4 for $\Xi\pi$ (1630) where the background level is evident. See Fig. 6 for $\Sigma^-\bar{K}^0$ (BMST and BNL-S). $\Xi\pi$ (1930) is indicated in the $\Xi K\pi$ channels but was established in $\Xi\pi$ $K\pi$ (see Fig. 2). Evidences for $\Xi\pi$ (1830), $\Xi\pi\pi$ (1830), and $\Xi\pi\pi$ (1930) are much weaker.

SESSION VII

Z^* PRODUCTION

Discussion Leader: Gerson Goldhaber

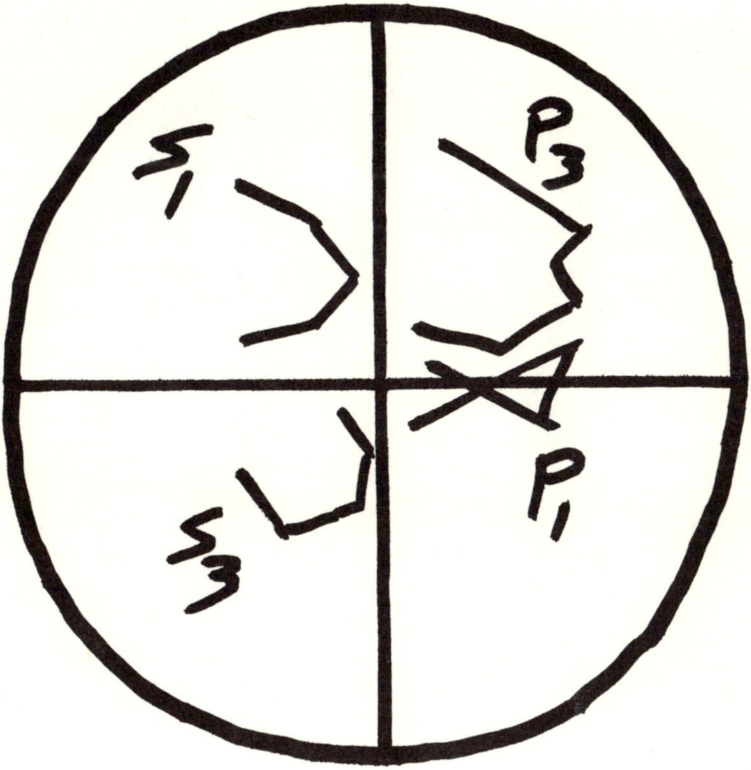

SEARCH FOR STRANGENESS S=+1 BARYON STATES*

Vernon W. Hughes
(with R.D. Ehrlich, A. Etkin, P. Glodis, K. Kondo, D.C. Lu, S. Mori, R. Patton, G.A. Rebka Jr., J.E. Rothberg, P.A. Thompson, and M. Zeller.)
Gibbs Laboratory, Yale University

Over the past three years a group of us at Yale have been interested in the question of the existence of strangeness S=+1 baryon states, which were first suggested experimentally by resonance-like structures in the total cross sections for K^+-p and K^+-d scattering.[1,2] With this interest we have done two experiments:

1) The first was a photoproduction experiment at the 6 BeV Cambridge Electron Accelerator in which the excitation curve for the photoproduction of K^- mesons in the reaction:

$$\gamma + p \rightarrow K^- + (\text{other things}) \qquad (1)$$

was measured.

2) The second is an experiment at the Brookhaven National Laboratory in which the asymmetry and differential cross section for the elastic scattering of K^+ mesons by polarized protons was measured.

I should like to discuss these two experiments briefly.

PHOTOPRODUCTION EXPERIMENT[3]

If the photoproduction reaction has a two-body final state, the particle accompanying the K^- meson is a doubly charged strangeness S=+1 baryon denoted by Z^{++}:

$$\gamma + p \rightarrow K^- + Z^{++} \qquad (2)$$

This reaction has a threshold photon energy, E, for the production of a K^- meson, which depends on the kaon laboratory momentum p_K, the kaon production angle θ_K, and the mass, m_Z, of the Z-particle:

$$E(P_K, \theta_K) = \frac{m_Z^2 - m_p^2 - m_K^2 + 2m_pE_K}{2P_K \cos\theta_K - 2E_K + 2m_p} \qquad (3)$$

in which m_p and m_K are the proton and K^--meson masses and E_K is the K^- total energy. If the K^- meson alone is observed, the K^- yield per equivalent quantum plotted versus bremsstrahlung end point energy will be a step function with its step at the energy E.

Figure 1 shows the experimental arrangement. A bremsstrahlung beam is produced from an internal target in the electron synchrotron and strikes a liquid hydrogen target. A single arm spectrometer, which consists of quadrupole magnets, dipole magnets, gas Cerenkov counters and scintillation-counter hodoscopes, detects the K^- meson and measures its momentum and production angle. The spectrometer had an acceptance solid angle of 1.5 msr and a momentum acceptance of 16%; the mean momentum was determined to 0.5% and the momentum resolution was 1%; the mean production angle was determined to 0.1°.

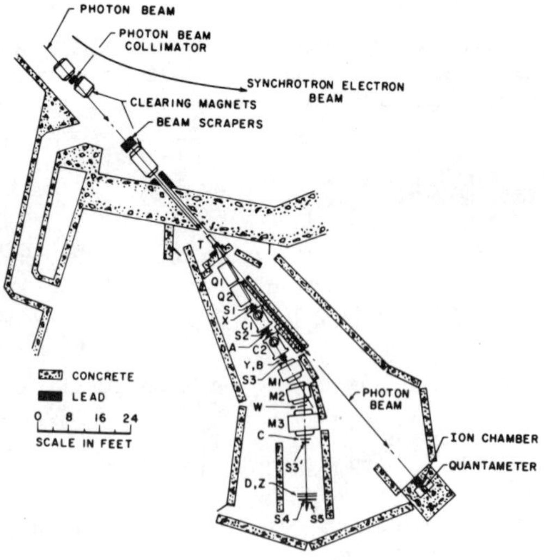

Fig. 1. Experimental arrangement for photoproduction experiment.

Figure 2 illustrates the operating characteristics of the spectrometer. The K^+ excitation function for the reactions: $\gamma + p \rightarrow K^+ + \Lambda$ and $\gamma + p \rightarrow K^+ + \Sigma^o$, obtained as a function of bremsstrahlung end point energy and plotted as a function of missing mass, is shown. Production of Λ and Σ^o are clearly resolved.

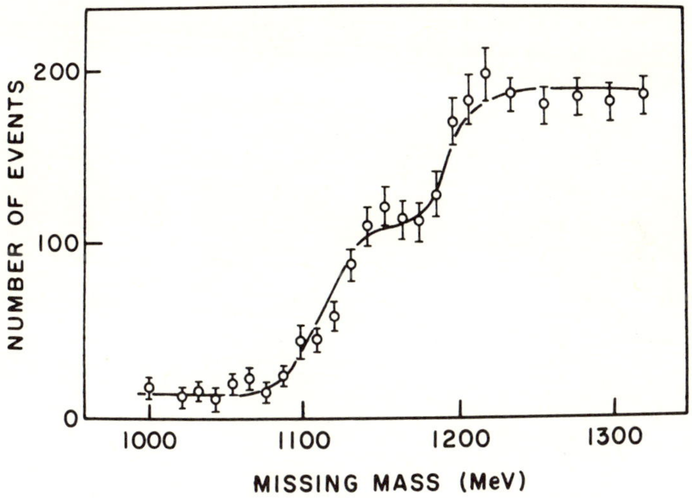

Fig. 2. K^+ excitation function for P_K=2.5 BeV/c and $\theta_K = 10^o$, showing the reactions $\gamma + p \rightarrow K^+ + \Lambda$ and $\gamma + p \rightarrow K^+ + \Sigma^o$.

For K^- photoproduction we measured the excitation function for P_K = 2.5 BeV/c and for two laboratory production angles, $\theta_K = 10^o$ and $\theta_K = 15^o$.

Photoproduction reactions with three- and many- body final states, including one K^- meson, constitute the background in our experiment. The principal background reactions are:

$$\begin{aligned} &\gamma + p \rightarrow K^+ + Y^{*o} \rightarrow K^+ + p + K^- \\ &\gamma + p \rightarrow K^{*+} + Y^{*o} \rightarrow K^{*+} + p + K^- \\ &\gamma + p \rightarrow p + \phi \rightarrow p + K^+ + K^- \\ &\gamma + p \rightarrow p + K^+ + K^- \end{aligned} \qquad (4)$$

Figure 3 shows experimental points and fitted yields plotted as a function of missing mass for P_K = 2.5 BeV/c and $\theta_K = 10^o$.

Figure 3(a) shows the experimental points and the dashed curve is a third-order polynomial background function fit to the data. Figure 3(b) illustrates the fit to the data with a third-order polynomial background function plus three Breit-Wigner resonances. The experimental points shown are the differences between the observed yield values and the background function, and correspond to about 20% of the total yield. The dashed curve is a third-order polynomial background function which provides a very poor fit to the data, corresponding to 3 ½ standard deviations. The solid curve is a fitted function for three Breit-Wigner resonances, and represents an excellent fit with masses of 1870 ± 20 MeV, 2120 ± 20 MeV, and 2270 ± 20 MeV.

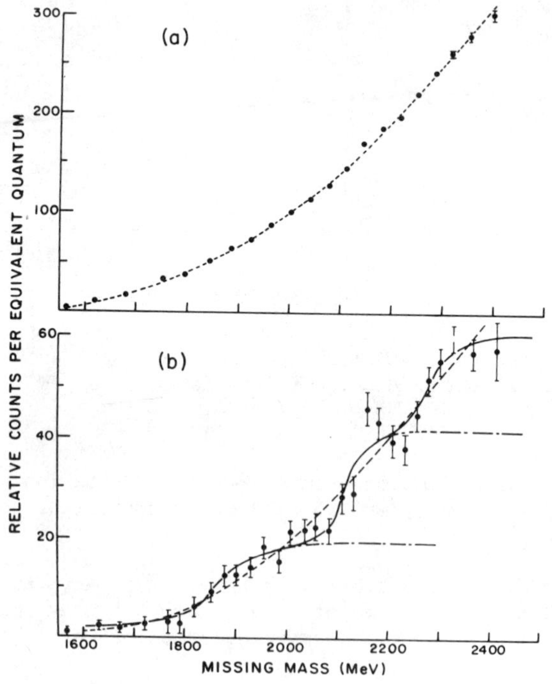

Fig. 3. K^- yield and fitted functions for $P_K = 2.5$ BeV/c and $\theta_K = 10^o$.

Figure 4 shows experimental points and fitted yields plotted as a function of missing mass for p_K = 2.5 BeV/c and θ_K = 15°, similar to Figure 3 for the θ_K = 10° data. An adequate fit to the experimental points is provided by the third-order polynomial background function alone without the addition of Breit-Wigner resonances. A principal aim in studying the K^- yield at θ_K = 15° was to observe the kinematic behaviour of the structures suggested by the θ_K = 10° data. However, no resonant structures were observed in the θ_K = 15° data.

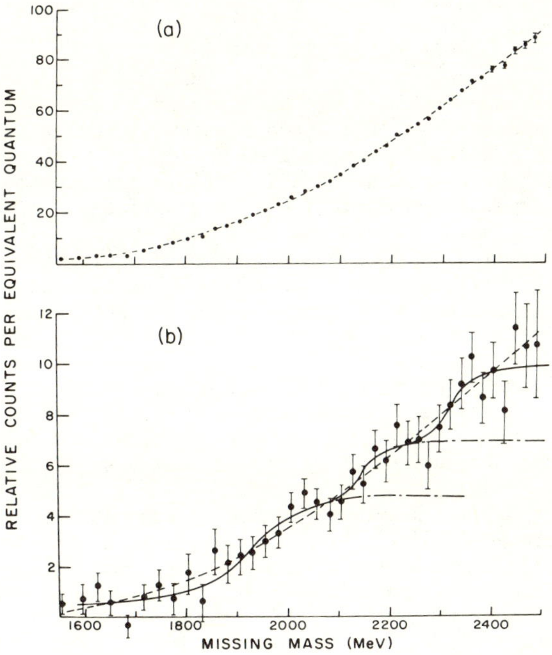

Fig. 4. K^- yield and fitted functions for P_K = 2.5 BeV/c and θ_K = 15°.

Hence the results of this experiment are inconclusive with regard to establishing the existence of Z particles. One possibility is that the structures observed in the θ_K = 10° data are resonances, but that the sensitivity of our experiment was not sufficient to observe these resonances at $\theta_K = 15^o$, where the production cross sections would be expected to be smaller. Indeed a decrease of the differential cross section as e^{at} with $a>3(BeV/c)^2$ (t = square of the four-momentum transfer) is rapid enough so that structures would not have been observed at $\theta_K = 15^o$, and such a decrease in differential cross section is not unexpected. A second possibility is that the structures in the $\theta_K = 10^o$ data are due to complicated, rapid variations in the large background. Considerably greater sensitivity in the experiment is required to distinguish these two possibilities.

ELASTIC SCATTERING OF K^+ MESONS BY POLARIZED PROTONS

The second experiment on the measurement of the differential cross section and asymmetry (or polarization P) in the elastic scattering of K^+ mesons from polarized protons contributes to the question of the existence of a Z particle through the second traditional approach of a formation-type experiment. If a Z particle exists and is coupled to the elastic K^+-p channel, the approach of measuring $d\sigma/d\Omega$ and P for elastic scattering, of performing a phase shift analysis, and of studying the resulting Argand diagrams is a powerful approach for the determination of resonances.[4]

A brief report on some early, preliminary results of our Brookhaven experiment has been published.[5] Since then we have had a major 4-month period of data-taking, which just finished a week before this Conference on Hyperon Resonances.

Figure 5 shows a schematic diagram of the beam channel of partially separated beam No. 5 in the G-10 area at the AGS, which has as its last element the quadrupole magnet Q8. The beam separators are electrostatic separators. A K^+ beam with a momentum spread $\Delta p/p$ of $\pm$ 2% was used.

Figure 6 shows the experimental arrangement. K^+ mesons were identified either by the ethylene gas differential Cerenkov counter, CB1, or by the liquid Cerenkov counter, CB2, and were focussed onto a target, T, 3cm x 2cm x 2.5cm (thick). The K^+ intensity at the target varied from 3,000 K^+ per pulse at the lowest momentum of 1.3 BeV/c to 100,000 K^+ per pulse at the highest momentum of 2.5 BeV/c, and the π/K ratio was between 2 and 3. A 10 x 10 hodoscope of scintillation counters, HB, located beam particles at its plane to $\pm$ 5mm. The polarized proton target, T, was a lanthanum magnesium nitrate (LMN) target containing protons with 0.6 polarization. The scattered K^+ mesons and protons were

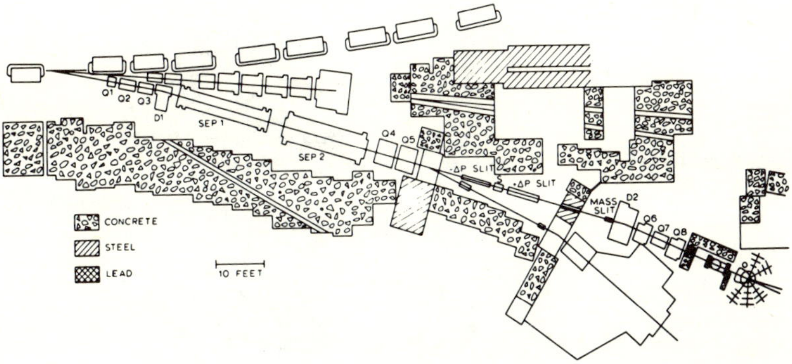

Fig. 5. Secondary beam channel which produces the K^+ meson beam.

detected by six banks of wire spark chambers with magnetostrictive readout- four banks having 3 chambers each and two banks having 2 chambers each-which had a spatial resolution of about 1mm FWHM. The angular range covered in the laboratory by the spark chambers was from 7° to 120°. One pair of water threshold Cerenkov counters, CS1 and CS2, were used to distinguish scattered K^+ mesons from protons in the approximately 20° range of angles in the laboratory for which kinematic distinction alone was ambiguous for incident K^+ momenta from 1.3 to 1.8 BeV/c. For incident momenta above 1.9 BeV/c the pair of total reflection lucite counters, CS3 and CS4, were used for distinction of K^+ mesons from protons.

A signal S1S2S3S4$\overline{5}$ from the telescope of scintillation counters opened the gates of a dc logic system to an array of 62 scintillation counters LI, LII, LIII, LIV, RI, RII, RIII, and RIV, as well as hodoscope HB. Signals from the array were formed into two logic tables, a sector table set to require one and only one particle on either side of the incident beam, and a coplanarity table set to require the event to be coplanar within 10°. When these conditions were met as well as coincident signals CB and $\overline{S6}$, which were derived from scintillation counters covering the poles of the polarizing magnet, then the spark chambers were triggered. The spark chamber outputs, the state of the dc logic observing HB, and the pulse heights in all Cerenkov counters

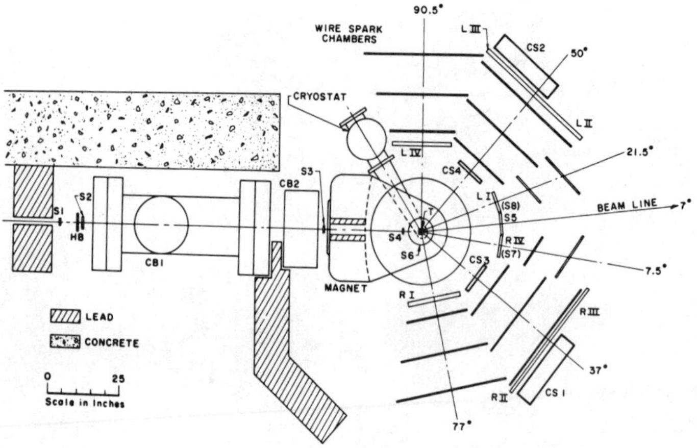

Fig. 6. Experimental arrangement for measurement of asymmetry in elastic scattering of K^+ mesons by polarized protons.

were entered into a PDP-8 computer, which edited and transferred the data to magnetic tape and the BNL on-line PDP-6 computer for analysis. The spark chamber recovery time, which was about 4msec, together with the PDP-8 computer processing time, determined the maximum data acquisition rate. Typical rates for an incident K^+ momentum of 2.0 BeV/c were the following:

K^+ intensity = 50,000 K^+ per pulse
Spark chamber trigger rate = 25 per pulse
Elastic scattering rate = 1 per pulse

Two kinematic reconstructions were computed for each event. The first reconstruction assumed that a K^+ meson had been elastically scattered to the left. From the coordinates established by the spark chambers, the trajectory of each scattered particle was traced through the field of the polarizing magnet to determine the point of interaction in the target to about 6mm. Using these trajectories and the point of interaction, the horizontal and vertical scattering angles of each scattered particle were established, the scattering angles expected for a recoil proton on the right were computed, and the differences between the actual angles for the particle on the right and these predicted angles, $\Delta\theta_H$ and $\Delta\theta_V$, were determined. The reconstruction was then repeated assuming that a K^+ meson had been elastically scattered to the right, and both solutions were given equal weight in the analysis to follow.

For identification of the particles and separation of elastic events from background, reconstructed events were combined in each range of $\Delta(\cos\theta_K^{cm}) = 0.1$ to form a density of events as a function of $\Delta\theta_V$ and $\Delta\theta_H$. This density contained a sharp peak at the origin corresponding to correctly identified elastic events and a second broad peak displaced in $\Delta\theta_H$ corresponding to incorrectly identified elastic events. The average density was computed ignoring the region of the misidentification peak, except where the two peaks merged, and kinematic ambiguity was resolved by one of the pair of Cerenkov counters CS1, CS2 or CS3, CS4.

Figure 7 shows a plot of density of events versus deviation from elastic prediction for $P_K = 1.96$ BeV/c for a forward and backward range of cos $_K^{cm}$. The horizontal scale is the value of $[(\Delta\theta_H)^2 + (\Delta\theta_V)^2]^{\frac{1}{2}}$, and the vertical scale is a normalized number of events per unit area, $\Delta\theta_H \times \Delta\theta_V$. The elastic peak is clearly evident and the angular width at half maximum is about 1.0^o in the forward range and 0.6^o in the backward range. The narrow width and the large ratio of elastic signal to background are achieved due to the use of spark chambers. The larger width in the forward direction is due to the scattering of the low momentum proton. The larger background in the backward direction is due primarily to the misidentified forward quasielastic scattering.

Figures 8 through 14 show the results for asymmetry versus $\cos\theta^{cm}$ for incident K^+ momenta of $P_K = 1.30$, 1.40, 1.50, 1.60, 1.78 1.87, 1.96, 2.05, 2.14, 2.24, 2.34, 2.44 and 2.53 BeV/c. About 20,000 elastic events were obtained for each K^+ momentum. The error bars are statistical counting errors only. The relatively larger errors in the backward scattering direction are due, of course, to the smaller differential cross section. The rather large errors in the range $-0.55 < \cos\theta^{cm} < -0.35$ are due to lack of complete angular coverage by the spark chambers. The polarization of the target protons averaged to about 0.57 and was measured to an accuracy of ± 0.03. The target polarization was reversed about four times for each kaon momentum and hence the asymmetry was determined on each side of the apparatus independently, so acceptance and efficiency factors are not involved in determining the asymmetry provided they are constant. Systematic errors in the subtraction of the background are not shown; these are now being studied and are expected to amount to about $\pm$ 0.02.

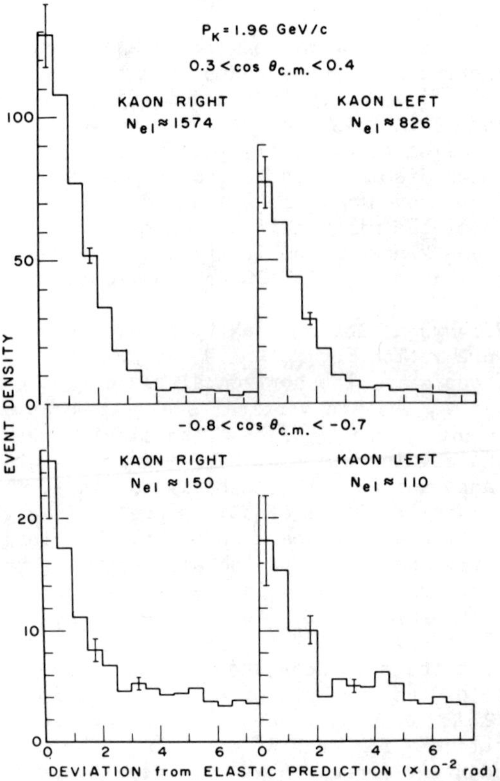

Fig. 7. Density of events versus deviation from elastic prediction.

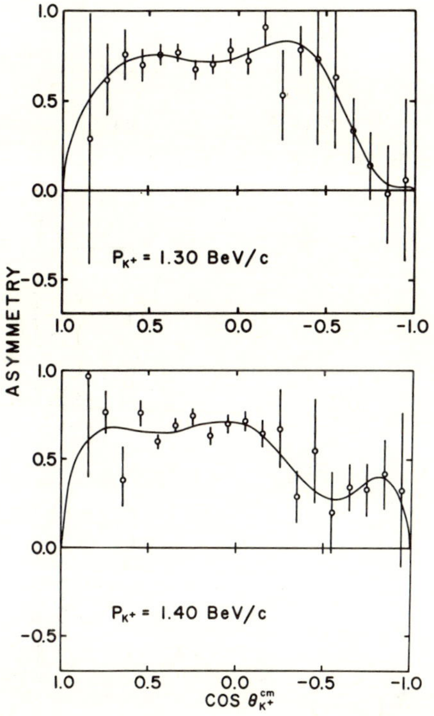

Fig. 8. Asymmetry versus $\cos\theta^{cm}$ for incident kaon momenta P_K = 1.30 and 1.40 BeV/c. The experimental points are indicated, together with one standard deviation error bars. The solid curve is an associated Legendre polynomial fit to the experimental points using P_ℓ^1 $(\cos\theta^{cm})$ for $\ell \leq 6$.

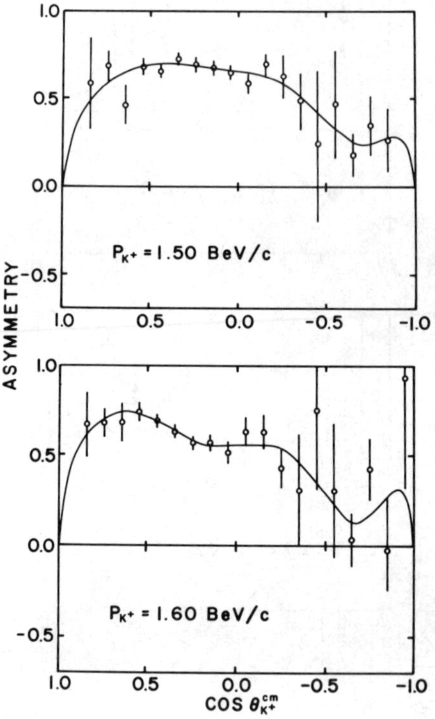

Fig. 9. Asymmetry versus $\cos\theta^{cm}$ for P_K = 1.50 and 1.60 BeV/c.

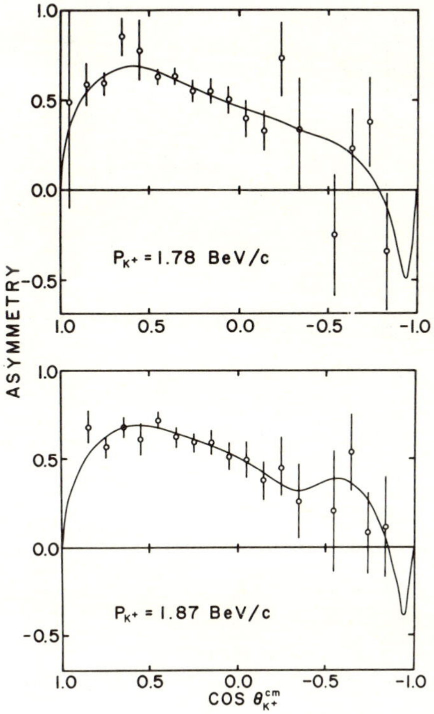

Fig. 10. Asymmetry versus $\cos\theta^{cm}$ for P_K = 1.78 and 1.87 BeV/c.

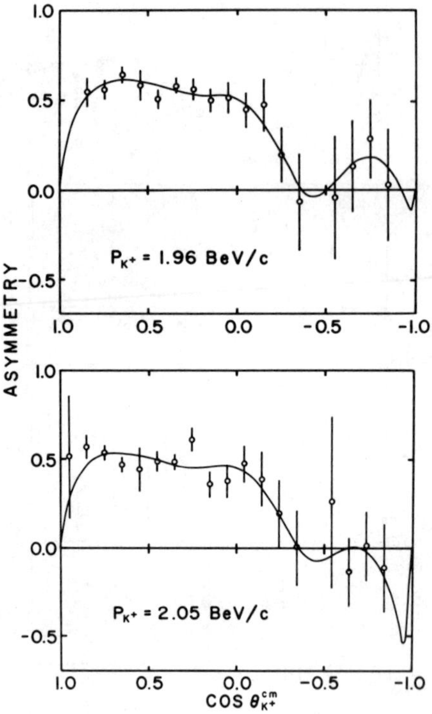

Fig. 11. Asymmetry versus $\cos\theta^{cm}$ for P_K = 1.96 and 2.05 BeV/c.

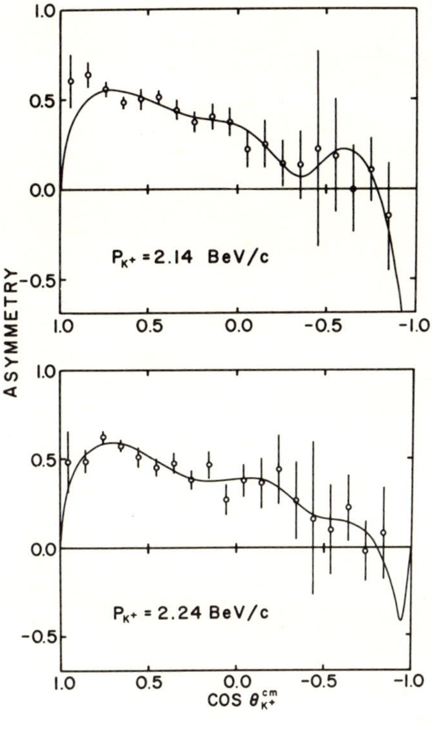

Fig. 12. Asymmetry versus $\cos\theta^{cm}$ for P_K = 2.14 and 2.24 BeV/c.

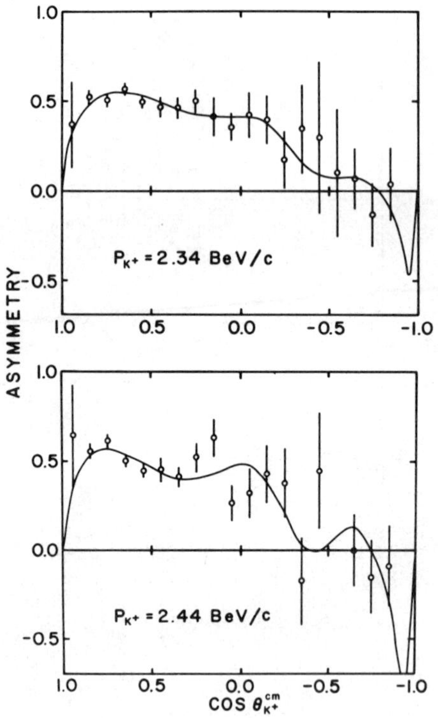

Fig. 13. Asymmetry versus $\cos\theta^{cm}$ for P_K = 2.34 and 2.44 BeV/c.

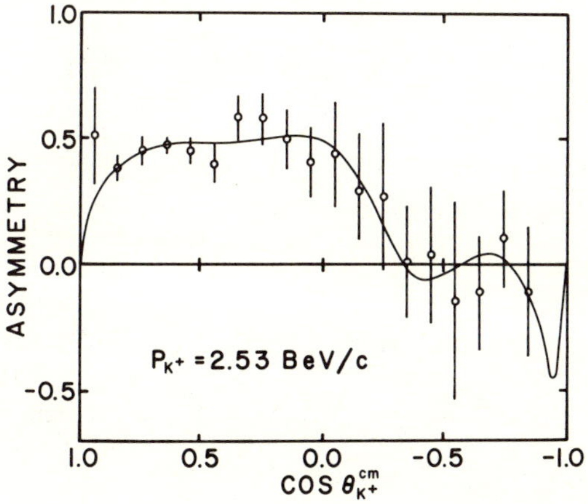

Fig. 14. Asymmetry versus $\cos\theta^{cm}$ for P_K = 2.53 BeV/c.

Values of $d\sigma/d\Omega$ will be obtained from our measurements after further analysis of the data.[6]

Comparison of our data on asymmetry with other published data from CERN[7,8] and Argonne[9], which is available for P_K in the range 0.89 to 1.89 BeV/c, indicates reasonable agreement. Our present data add considerable new information on asymmetry, particularly in the high momentum range from P_K = 1.9 to 2.5 BeV/c. Our data have significantly higher ratios of elastic signal to background than those of the other experiments due to our use of wire spark chambers, and our data cover a wider range of P_K values and scattering angles in a single experiment.

We are now undertaking a phase shift analysis of K^+-p scattering, which will include our present data on asymmetry and differential scattering cross section as well as all other relevant published data. Our earlier analysis[5] had the exciting feature of indicating a possible resonance behaviour for the $P_{3/2}$ partial wave amplitude. From that analysis, asymmetry data above 2.0 BeV/c seemed to be needed. We hope that our extensive present data will help to clarify the question of the existence of S = +1 baryon states.

References

*This research was supported in part by the U.S. Atomic Energy Commission.

1) R.L. Cool, G. Giacomelli, T.F. Kycia, B.A. Leontic, K.K. Li A. Lundby, and J. Teiger, Phys. Rev. Letters 17, 102 (1966).

2) R.J. Abrams, R.L. Cool, G. Giacomelli, T.F. Kycia, B.A. Leontic, K.K. Li, and D.N. Michael, Phys. Rev. Letters 19, 259 (1967).

3) S. Mori, J.S. Greenberg, V.W. Hughes, D.C. Lu, R.C. Minehart, J.E. Rothberg, P.A. Thompson, and J. Tyson, Phys. Rev. 185, 1687 (1969).

4) R.D. Tripp, "Baryon Resonances" in Proceedings of the International School of Physics (Enrico Fermi)-Strong Interactions, edited by L.W. Alvarez, Academic Press, New York (1966) p. 70.

5) G.A. Rebka, Jr., J. Rothberg, A. Etkin, P. Glodis, J. Greenberg, V.W. Hughes, K. Kondo, D.C. Lu, S. Mori, and P.A. Thompson, Phys. Rev. Letters 24, 160 (1970).

6) A Compilation of K^+N Reactions, L.R. Price, N. Barash-Schmidt, O. Benary, R.W. Bland, A.H. Rosenfeld, and C.G. Wohl, UCRL-20000 K^+N, September 1969.

7) S. Andersson, C. Daum, F.C. Erné, J.P. Lagnaux, J.C. Sens, and F. Udo, Phys. Letters 28B, 611 (1969).

8) S. Andersson, C. Daum, F.C. Erné, J.P. Lagnaux, J.C. Sens, F. Udo, and F. Wagner, Phys. Letters 30B, 56 (1969).

9) J.G. Asbury, J.D. Dowell, S. Kato, D. Lundquist, T.B. Novey, A. Yokosawa, B. Barnett, D.F.M. Koehler, and P. Steinberg, Phys. Rev. Letters 23, 194 (1969).

DISCUSSION

Barnett: Did you have time to do a pp normalization of your target polarization? I mean, how well do you know your target polarization

Hughes: It is, I believe, known to about 5 percent, but not by pp normalization.

K^+p PHASE-SHIFT ANALYSIS AND A Z^*

A. Yokosawa

Argonne National Laboratory

I would like to summarize the work carried out up to present in K^+p polarization measurements and phase-shift analysis aimed at clarifying the status of a Z^*. The names of those who participated in this investigation, either in full or in part, are J.G. Asbury, B. Barnett, G. Burleson, J.D. Dowell, D. Eartly, D. Hill, A. Laasanen, D. Lundquist, S. Kato, P.F.M. Koehler, T.B. Novey, K. Pretzl, P. Steinberg, and A. Yokosawa (ANL-Maryland-Northwestern-NAL Collaboration).

In our recent publication,[1] we described results of a phase-shift analysis in K^+p elastic scattering. In our energy-independent analysis, we started with about 80 sets of initial-guess values which consist of typically η_ℓ = 1.0, 0.95, 0.90, ..., 0.50 and $\delta_\ell = 0, \pm 5^\circ, \pm 10^\circ, \ldots, \pm 60^\circ$. Such an analysis was performed at 0.86, 0.96, 1.21, 1.37, 1.45, 1.70, and 1.95 GeV/c.[2] After reaching the minimum χ^2 value in each set, we performed random searches in the vicinity of that minimum. In this random search, we ended up with the same minimum most of the time, and thus we assumed that we chose sufficient sets of initial-guess values. We found about 50 solutions ($\chi^2/N_{DF} \leq$ 2.5) at each momentum. By examining all the partial waves simultaneously, we found it possible to form several groups on the Argand diagram. Such a grouping made the performance of the next steps possible. We next applied continuity conditions, with respect to the incident momenta, of each partial wave, which were given by the sum of the changes in the gradients of the vectors of the Argand diagram between adjacent momentum points. Because of the limited number of data points, we did not adopt a short-path method. To avoid using numerous combinations of the above solutions at seven different momenta, we first investigated the continuity in intervals of four momenta at a time. We also checked for continuity of the calculated differential cross sections and polarization at given value of $|t|$ in a similar way. It may also be important to apply other constraints, such as limits on the minimum step size between adjacent momentum points on the curves, but we could not find a suitable method of applying this. We finally came up with three solutions[3] by the above procedure.

We have also investigated the effect of partial waves with angular momenta higher than the $\ell_{max} = 3$ by using higher partial waves obtained from the Regge-pole model of Blackmon and Goldstein,[4] which are tabulated in Table I. This model is supposed to be good only in the forward region, but it is also reasonable in the backward region; in any event, we found that higher partial waves are not affected by the backward-scattering amplitudes. The analysis with the Regge-pole model also leads to four solutions, which do not differ significantly from those obtained in the above analysis.

Our preliminary polarization data[5] from 1.60 to 2.10 GeV/c at backward angles, $-0.8 > \cos > -1.0$, are roughly consistent with zero. If this were also true at 1.89 GeV/c, solutions I and III give more satisfactory values than II, as shown in Fig. 1, in which plots of polarization calculated from the above three solutions are given. If we adopt phase shifts of solution I at 1.95 GeV/c for solutions II, solution II then practically becomes solution III. Figures 2 and 3 show all the partial wave amplitudes for solution I and III, respectively, plotted in the complex plane. Those points at 2.10 GeV/c[5] are tentative. The errors are estimated both from the error matrix obtained in the minimization procedure and from the size of the group to which the solution belongs.

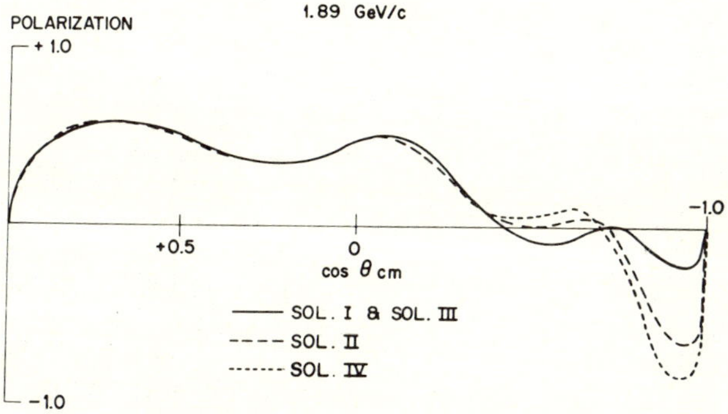

Fig. 1 Plots of the polarization parameter calculated from solutions I, II, and III at an incident K^+ momentum of 1.89 GeV/c.

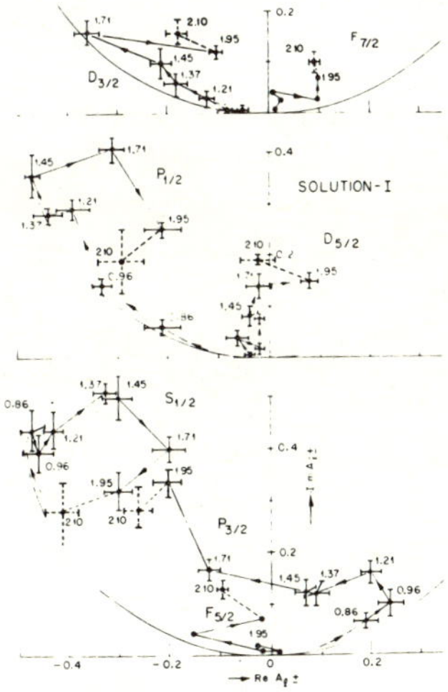

Fig. 2 Argand diagram of the complex scattering amplitude for all the partial-waves of solution I.

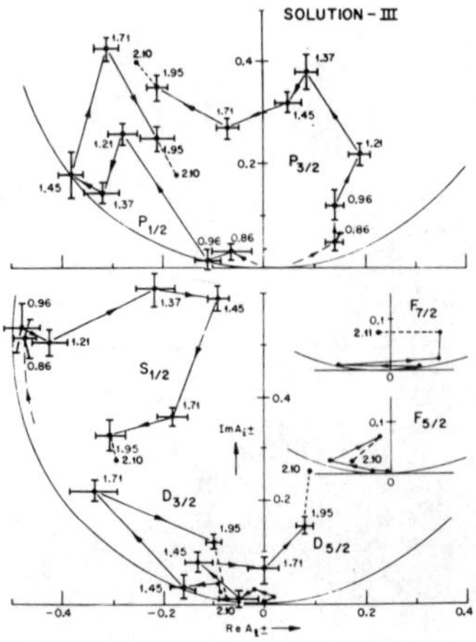

Fig. 3 Argand diagram of the complex scattering amplitude for all the partial-waves of solution III.

The partial total cross section of the $P_{3/2}$ partial wave, $\sigma_T(P_{3/2})$, of solution I has a peak at around 1.21 GeV/c, as shown in Fig. 4. This momentum coincides with the occurrence of the first bump in the total cross section[6] and the peak in the KΔ production cross section.[7] If we assume a background in the $\sigma_T(P_{3/2})$ curve as shown by the dotted line in Fig. 4, we can fit the remainder after background subtraction[8] to a Breit-Wigner shape. The properties of such a Z^* resonance are shown in Table II. This interpretation would imply that the first bump in the total cross section is due, at least partly, to a resonance effect.

Solution III gives a Z^* with a somewhat higher resonance mass and a much higher elasticity. The partial total cross section, $\sigma_T(P_{3/2})$, is shown in Fig. 5. The elasticity of a resonance is determined by assuming a background[8] as shown in Fig. 5. The parameters of possible resonance for solution III is also shown in Table II.

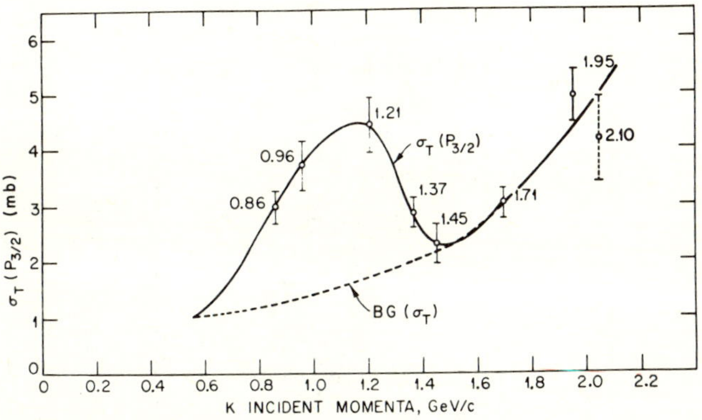

Fig. 4 Partial total cross-section, $\sigma_T(P_{3/2})$, of the $P_{3/2}$ partial wave from solution I. The dotted line indicates the assumed background.

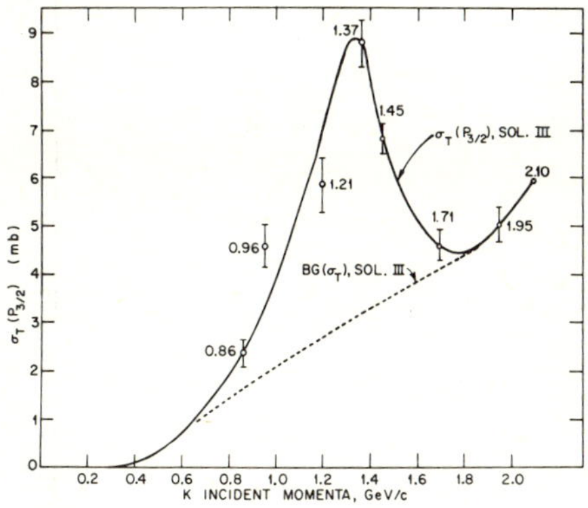

Fig. 5 Partial total cross-section, $\sigma_T(P_{3/2})$, of the $P_{3/2}$ partial wave from solution III. The dotted line indicates the assumed background.

TABLE I

Higher Partial Waves[a] From the Regge-pole Model

Momentum (GeV/c)	η_{3+}	η_{3-}	η_{4+}	η_{4-}	η_{5+}	η_{5-}	δ_{3+}	δ_{3-}	δ_{4+}	δ_{4-}	δ_{5+}	δ_{5-}
									(In degrees)			
1.21	.988	.946	.995	.988	1.000	.996	-.24	-1.69	.02	-.45	-.04	-.02
1.37	.980	.932	.993	.979	.999	.993	-.34	-2.30	-.01	-.68	.05	-.08
1.45			.992	.975	.998	.991			-.03	-.81	-.05	-.13
1.70			.986	.961	.995	.986			-.12	-1.26	-.006	-.32
1.95			.979	.947	.992	.979			-.24	-1.73	-.09	-.56
2.10			.974	.938	.989	.974			-.31	-2.00	-.12	-.73

[a] Partial scattering amplitudes are defined as $A_{\ell\pm} = [\eta_{\ell\pm} \cdot \exp(2i\delta_{\ell\pm}) - 1]/2i$

TABLE II

Solution No.	Mass (MeV)	Gamma Total Width, Γ_{tot} (MeV/c)	Elasticity
I	1880 ± 90	~ 130	~ 0.10
III	1950 ± 100	~ 180	~ 0.28

References

(1) S. Kato, P. Koehler, T. Novey, A. Yokosawa G. Burleson, Phys. Rev. Letters 24, 615 (1970).

(2) We note that we combine 1.89 GeV/c polarization data and 1.95 GeV/c differential cross section data at 1.95 GeV/c.

(3) Phase-shifts for these solutions are given in the Appendix, ANL/HEP 6926 (1969), S. Kato *et al.*

(4) M. L. Blackmon and G. R. Goldstein, Phys. Rev. 179, 1480 (1969).

(5) A. Laasanen, B. Barnett, G. Burleson, D. Eartly, D. Hill, T. B. Novey, K. Pretzl, P. Steinberg, and A. Yokosawa, to be published.

(6) R. L. Cool et al., Phys. Rev. Letters 17, 102 (1966).

(7) R. W. Bland *et al.*, Phys. Rev. Letters 18, 1077 (1967); R. W. Bland *et al.*, UCRL Report No. UCRL-18758, 1969 (unpublished).

(8) In this subtraction we neglected an interference term which is shown in an expression of the resultant partial scattering amplitude, $\vec{A} = \vec{A}_{BG} + \vec{A}_{RES} \times$ (Phase Factor). As we can see in Fig. 2 and 3, we need more energy points to determine the background term, $\vec{A}_{BG}$.

DISCUSSION

Rosenfeld: Is the situation this simple, that if you take your solutions and compare them with the total cross section measurements, you can just rule out second solution?

Yokosawa: No, no . . . I did not say that. We rule out the second solution because of new backward polarization data.

Rosenfeld: But you drew quite different bumps in the total partial cross sections.

Yokosawa: Yes, they are quite different but we did not rule out any solutions based on that.

Rosenfeld: Did you say that the one solution is inconsistent?

Yokosawa: No, no . . . all solutions are completely consistent.

Kim: You have analysed the K^*N reaction?

Goldhaber: No, actually we did $K\Lambda$ reaction. Wagner will talk on the K^*N reaction.

Steinberg: The thing that concerns me most is the question of error assignment, because that seems to be very critical here. I am very much concerned with the error assignment because they seem to be uniform in magnitude from one momentum to another, whereas the quality of data is very different from one momentum to another.

Yokosawa: We have not yet completed the full analysis, but the tentative error based upon the error matrix and the group size is given in our published work. Our final error assignment will be based on the method number 3.

Hemingway: What was wrong with method No. 1?

Yokosawa: Nothing wrong, actually. I may be biased here; the error matrix calculation will give you very small error.

STRUCTURE OF THE K^+p AMPLITUDES IN THE REGION OF THE "COOL" BUMP

F.C. Erné+), J.C. Sens+) and F. Wagner++)

CERN, Geneva, Switzerland

+) Visitor from the Foundation for Fundamental Research of Matter (FOM), The Netherlands.

++) Max-Planck-Institut für Physik und Astrophysik, München Germany

ABSTRACT

This paper presents an updated and extended energy-independent phase-shift analysis at the elastic K^+p channel from threshold to 1.9 GeV/c. Several solutions show a resonance-like behaviour in the P_3-wave in the region of the bump at 1.25 GeV/c in the total cross-section.

The compatibility of these solutions with other analysis of the elastic and inelastic channels, and with production experiments, is discussed.

1. INTRODUCTION

This paper is concerned with the question of whether or not the bump in the K^+p total cross-section at 1.25 GeV/c can be considered as the manifestation of a (Z^*) resonance. Firstly, an energy-independent partial-wave analysis with "shortest path" selection is presented on K^+p elastic differential cross-sections, polarization, and other available data from threshold to 1.9 GeV/c; this is mostly an extension to higher momenta of an analysis published earlier [1,2]. Secondly, we briefly discuss the consequences of the results found in the elastic channel for the question of Z^* formation in the inelastic channels. Finally, some comments will be made on partial wave analyses performed recently by other groups.

2. K^+p PARTIAL WAVE ANALYSIS FROM THRESHOLD TO 1.9 GeV/c

As input to the partial-wave analysis we have used the data on polarizations, differential cross-sections, inelastic cross-sections, and total cross-sections at 19 momenta between 140 and 1455 MeV/c (listed in reference 1), in addition to new data on differential cross-sections and polarizations, recently obtained at CERN[3] and at other laboratories[4]. At the new momenta, total[5], inelastic[4], and differential backward[6] cross-sections have been included wherever available[*].

The only theoretical input consisted of the real parts of the forward scattering amplitude, taken from dispersion relation calculations[7],[8], with a 10% error.

Below 650 MeV/c, only S- and P-waves have been assumed; between 650 MeV/c and 1400 MeV/c D-waves, and above 1400 MeV/c also F-waves have been admitted.

The minimization technique of the CERN II phase-shift programme[9] was used to find 200 different solutions at each momentum. A "solution" is a set of partial waves with a $P(\chi^2) > 10^{-3}$; this corresponds in nearly all cases to a true minimum in χ^2. One solution is different from another if the distance D, defined by

$$D = \sum_{\ell,j} \left| f_{\ell,j}^{(1)} - f_{\ell,j}^{(2)} \right|^2 \cdot (j+\tfrac{1}{2})$$

[*] The only new data not included in this analysis are the additional polarization measurements of Rebka et al.[4] and the differential cross sections of the University College London reported at this conference.

is larger than a specified value D_{min}, which is chosen such that after some 100 random searches, a recurrence of solutions begins to set in; the searches are continued until 200 solutions have been found. Between 0.86 and ~1.5 GeV/c, the solutions fall into distinct groups; an example is given in Fig. 1. At the higher momenta no obvious grouping is apparent; as an example, Fig. 2 shows the 200 solutions at 1.9 GeV/c.

The "shortest path" method[9] has been used to link the solutions at the different momenta. The shortest path is taken upwards from a zero solution at threshold. The calculation can be done either by minimizing the path between neighbouring momenta or by minimizing both the path and its derivative ("Smoothest shortest path"). A further option is that subsets of the 200 solutions at each momentum may be selected, e.g. sets with high values of $P(\chi^2)$. For simplicity we have removed 22 solutions at 0.86 GeV/c thereby eliminating shortest paths of type III[1,2] which would lead to dominant P-waves instead of S-waves near threshold.

Using these various options, a total of 11 different shortest paths, out of the 200 possible, appear to be continued through the entire range of momenta. They can be divided into three types: A, B, and C. Of the eleven paths, five belong to type A, five to type B, and one to type C. Three of the solutions of type A and one of each type B and C are shown in Figs. 3 to 8.

Solutions B can be regarded as unlikely because of the physically improbable hairpin structure of the P-waves and the irregular energy dependence of the D-waves. Of the other types, we regard solution A as "preferred", since most of the remaining shortest paths found are of this type; on the other hand, solution C is not ruled out by the analysis.

The types A, B, and C are nearly identical in all partial waves up to ~ 0.86 GeV/c and agree with the repulsive S-wave solution A_I- of Bland et al.[10]. Between 0.86 and ~1.45 GeV/c, A and C are still quite similar and show the same features as the "preferred" solution (solution II) of our earlier analysis[1,2]. Above ~1.45 GeV/c, A and C differ significantly.

The S-wave (Fig. 3) is nearly elastic and repulsive in all solutions up to ~1.45 GeV/c. Above this momentum, solutions A and C become strongly inelastic.

The P_1-wave (Fig. 4) is mostly elastic up to ~ 1.5 GeV/c; above this momentum P_1 changes rapidly in phase and becomes inelastic in A, and continues elastically with a slow phase change in C.

The P_3-wave (Fig. 5) is attractive at low momenta, turns inwards at ~ 0.86 GeV/c and becomes rapidly inelastic, but with only a slow

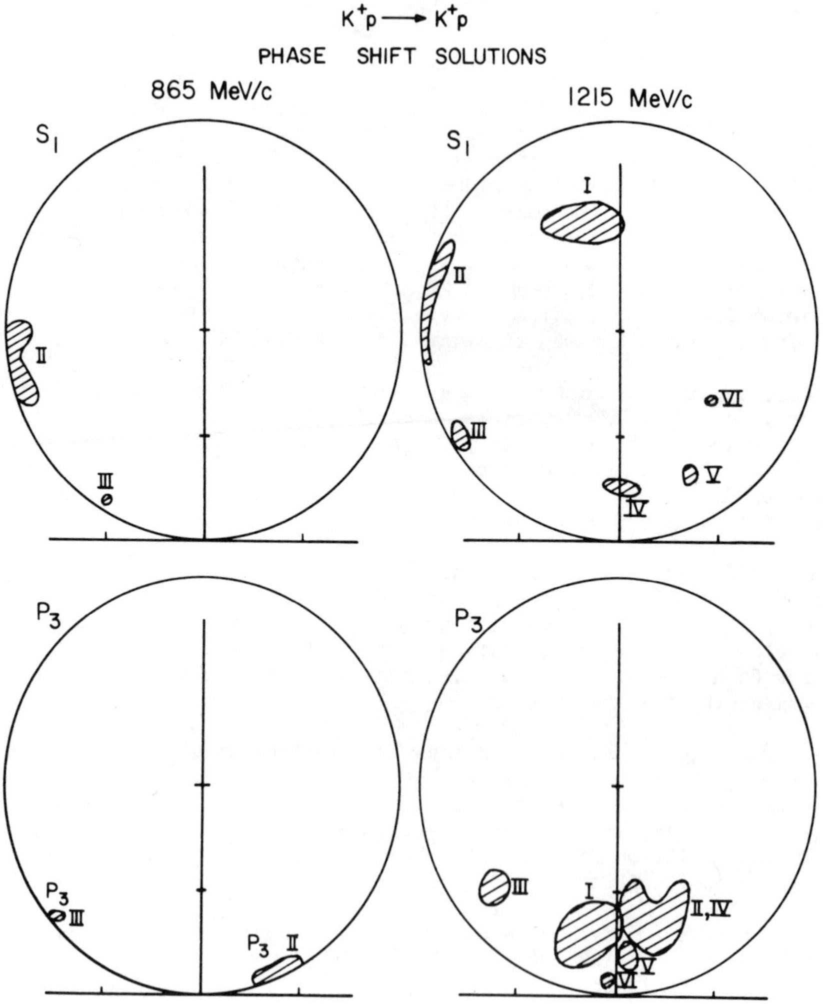

Fig. 1: Argand diagrams for S_1- and P_3-waves at 0.865 GeV/c and 1.215 GeV/c. Domains are shown in which the 200 solutions at these moments are contained. Only the domains II and III are connected by shortest paths.

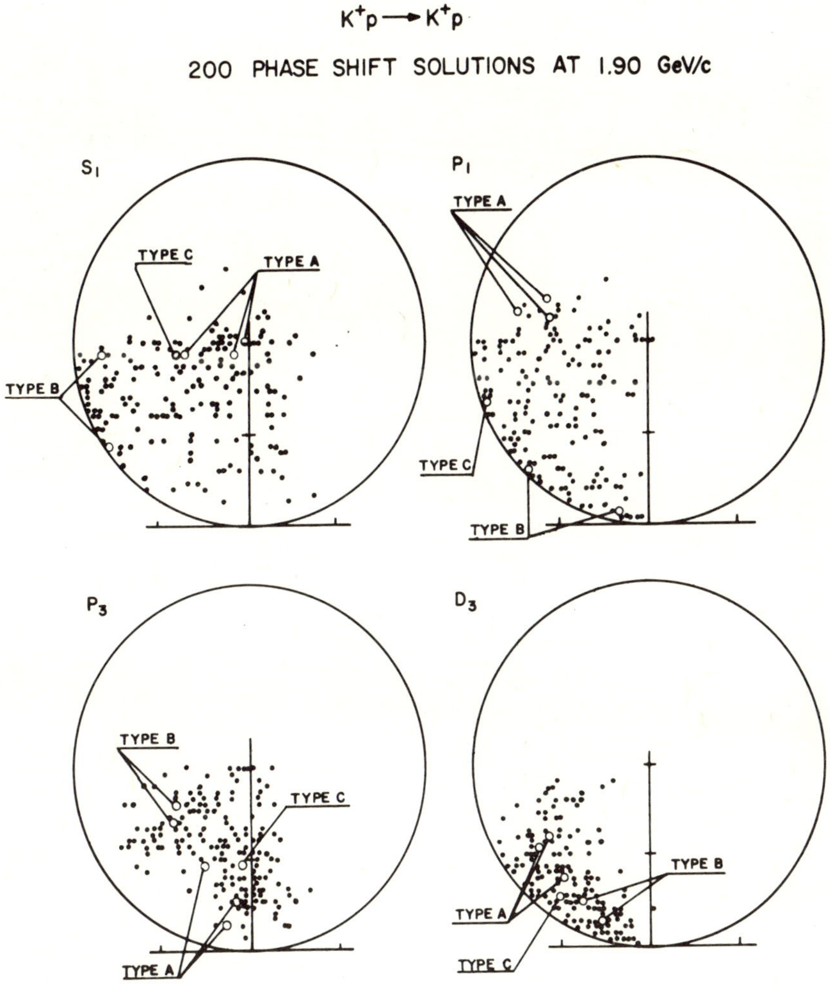

Fig. 2: Argand diagram for S_1-, P_1-, P_3-, and D waves at 1.9 GeV/c. 200 solutions are shown, obtained in 350 random searches.

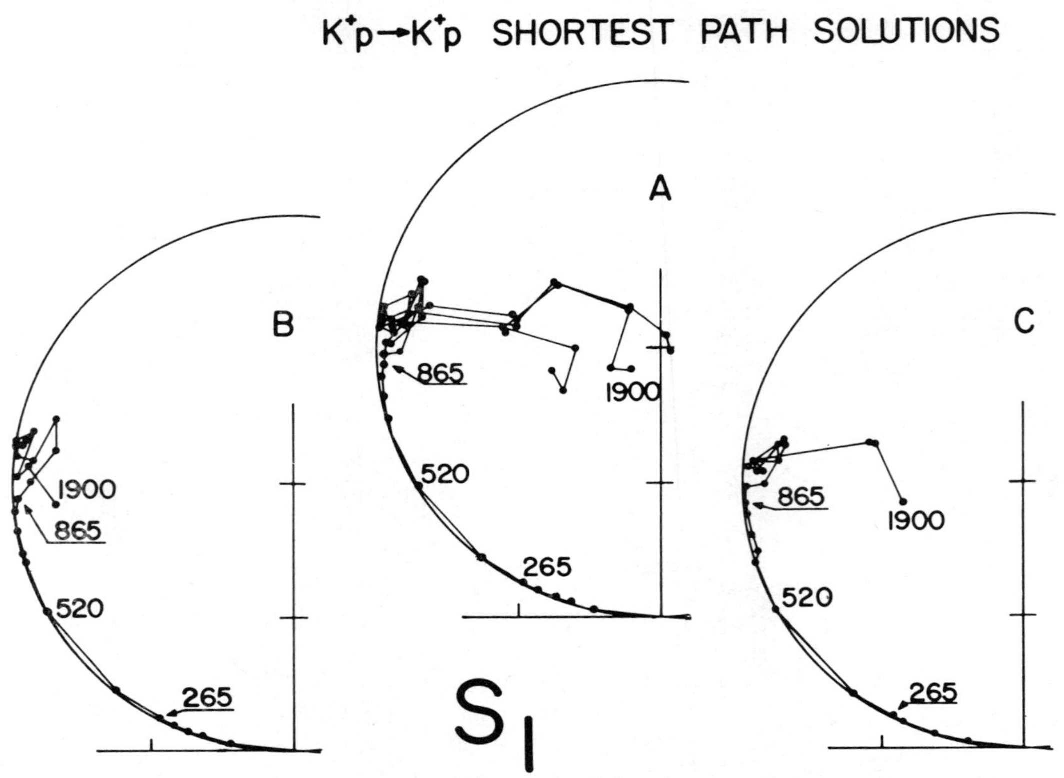

Fig. 3: Argand diagram for the S_1-wave, for three shortest paths, one of type A, one of type B, and one of type C, from threshold to 1.9 GeV/c.

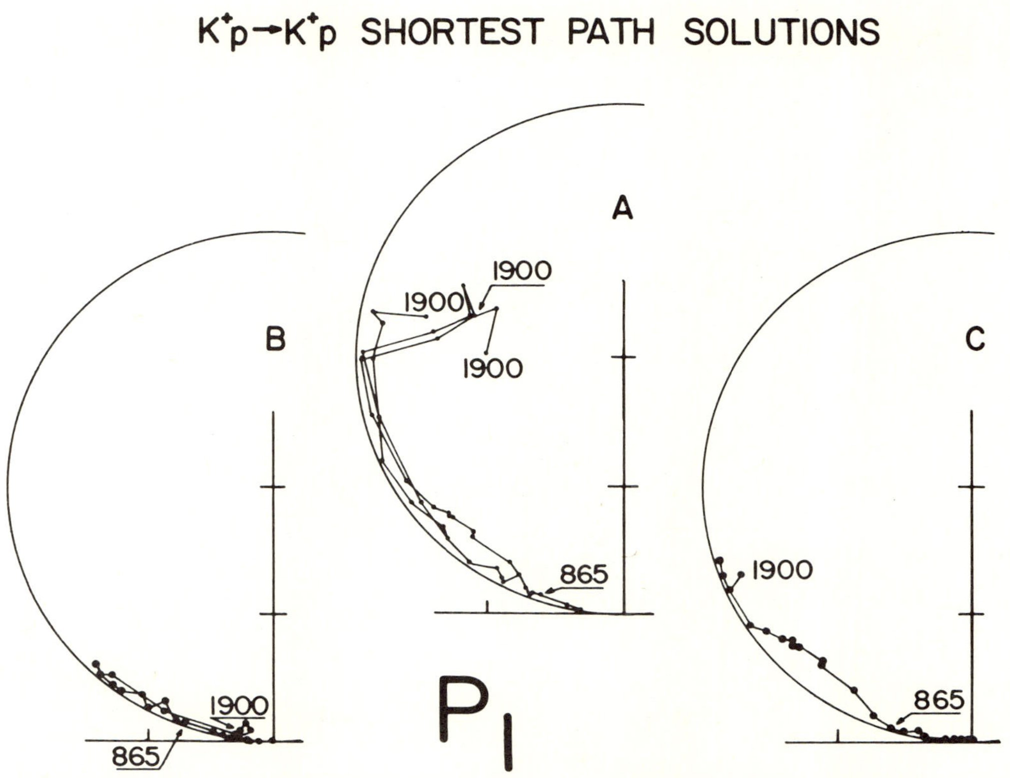

Fig. 4: Argand diagram for the P_1-wave, for three shortest paths, one of type A, one of type B, and one of type C, from threshold to 1.9 GeV/c.

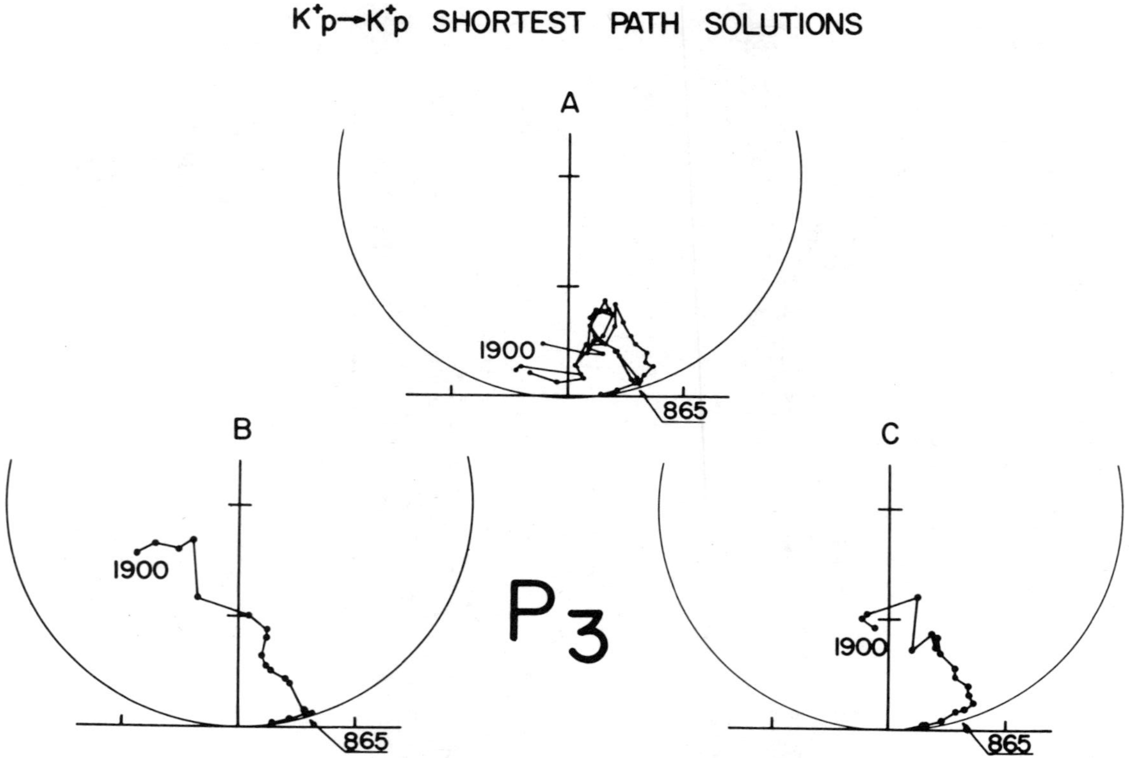

Fig. 5: Argand diagram for the P_3-wave, for three shortest paths, one of type A, one of type B, and one of type C, from threshold to 1.9 GeV/c.

change in phase, between 0.86 GeV/c and ~1.4 GeV/c in all solutions. Above ~1.4 GeV/c, solution A turns downwards; the new data above 1.45 GeV/c thus appear to complete the semi-circle, suggested by our earlier analysis[1,2], below this momentum. Contrary to this, solution C continues to move upwards and does not complete a resonance-like circle.

The D_3-wave (Fig. 6) remains small and mostly inelastic between 0.86 GeV/c and 1.45 GeV/c. Above ~1.45 GeV/c, solutions A and C continue to rise.

The D_5-wave (Fig. 7) is nearly zero up to 0.86 GeV/c and then rises along the imaginary axis. This occurs in all solutions. Above ~1.45 GeV/c, solution C becomes ill-determined.

The F-waves (Fig. 8) are admitted only above 1.4 GeV/c and are small. F_5 is repulsive and mostly elastic, F_7 attractive and absorptive in all solutions.

Figure 9 shows the total and elastic cross-sections for the partial waves of the solutions of type A as a function of momentum. The "error bars" indicate the range of values covered by the different solutions of this type. There is a decrease of $\sigma(S_1)$ with increasing momentum, running approximately parallel to the unitarity limit; $\sigma(P_1)$, $\sigma(D_3)$, and $\sigma(F_5)$ (F_7 is not shown, but is similar to F_5) show an energy dependence consistent with that of background waves. The most striking features are the energy dependence of the P_3- and the D_5-waves. The peak in $\sigma(P_3)$ is at 1.30-1.35 GeV/c, i.e. close to the bump in the total K^+p cross-section at 1.25 GeV/c; from the difference in energy dependence of $\sigma_{tot}(P_3)$ and $\sigma_{el}(P_3)$ it is clear that the resonance-like structure of this cross-section is almost entirely accounted for by the energy dependence of the absorptive part of the wave. A very striking behavior is also seen in the almost purely inelastic D_5-wave, which increases rapidly above the threshold for production of KN .

Figure 10 shows $\sigma(P_3)$ and $\sigma(D_5)$ for types A, B, and C. It is seen that the maximum in $\sigma(P_3)$ is only clearly present in type A; the momentum dependence of $\sigma(D_5)$, on the other hand, occurs in all solutions found.

Figure 11 shows the ratios α of the real to imaginary parts of the forward amplitude of phase-shift solution A, in comparison with the calculated values of Refs. 7 and 8, along with measured values of α[11]. Apart from a small systematic shift, there appears to be good agreement. Already at ~1 GeV/c, α approaches the value ~-0.5, also found at high energies.

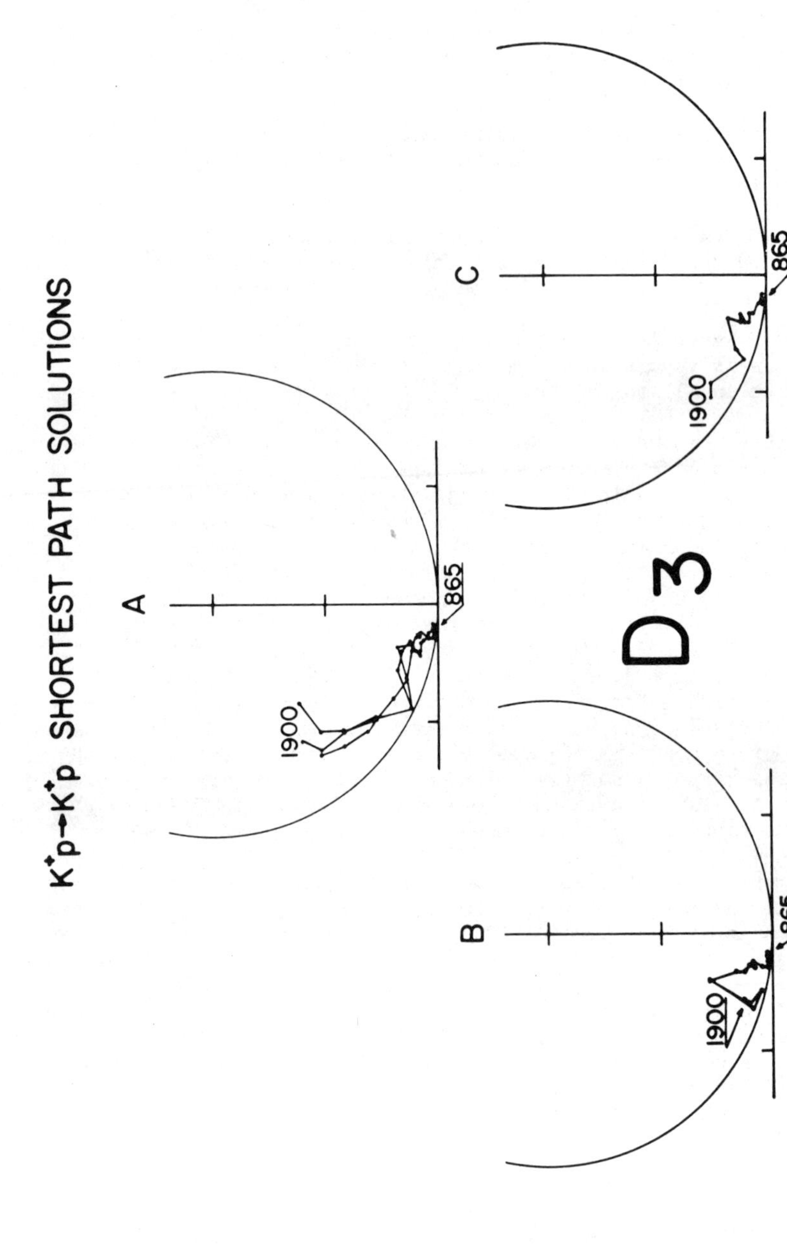

Fig. 6: Argand diagram for the D_3-wave, for three shortest paths, one of type A, one of type B, and one of type C, from threshold to 1.9 GeV/c.

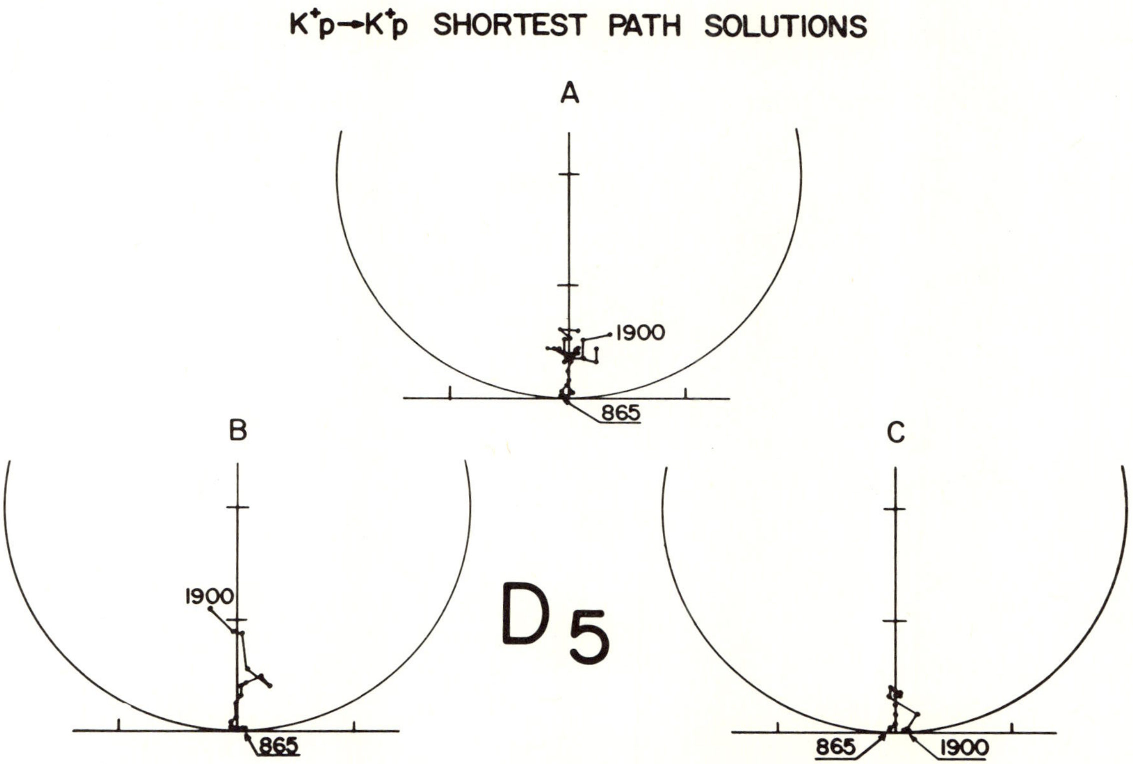

Fig. 7: Argand diagram for the D_5-wave, for three shortest paths, one of type A, one of type B, and one of type C, from threshold to 1.9 GeV/c.

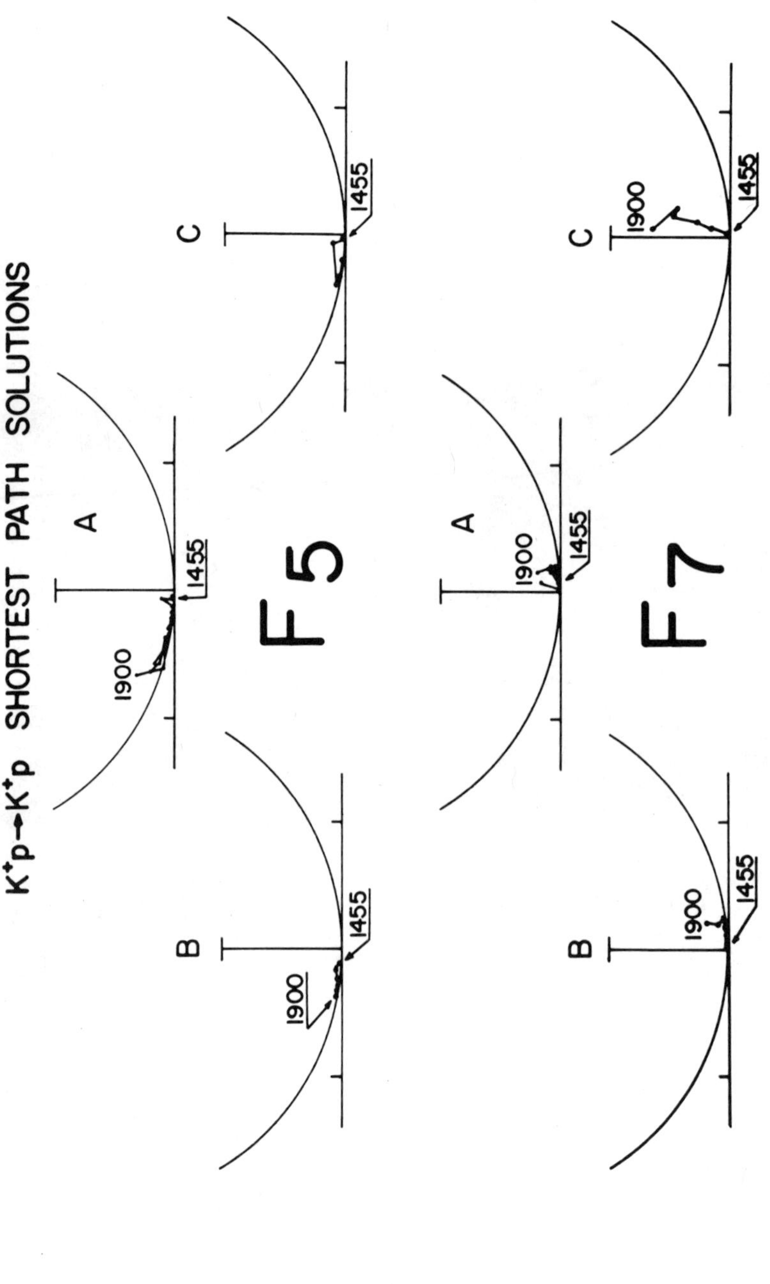

Fig. 8: Argand diagram for the F-waves, for three shortest paths, one of type A, one of type B, and one of type C, from threshold to 1.9 GeV/c.

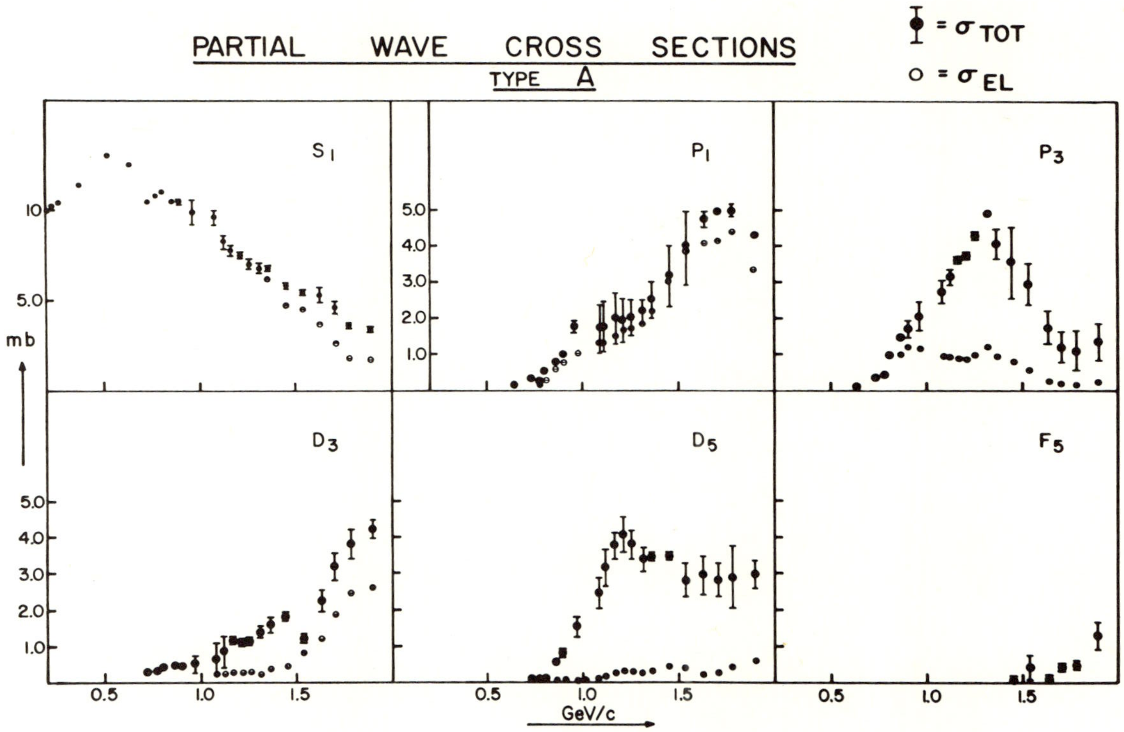

Fig. 9: Total and elastic cross-sections for the partial waves of the solutions of type A. The "error bars" indicate the range of values covered by the different solutions of this type.

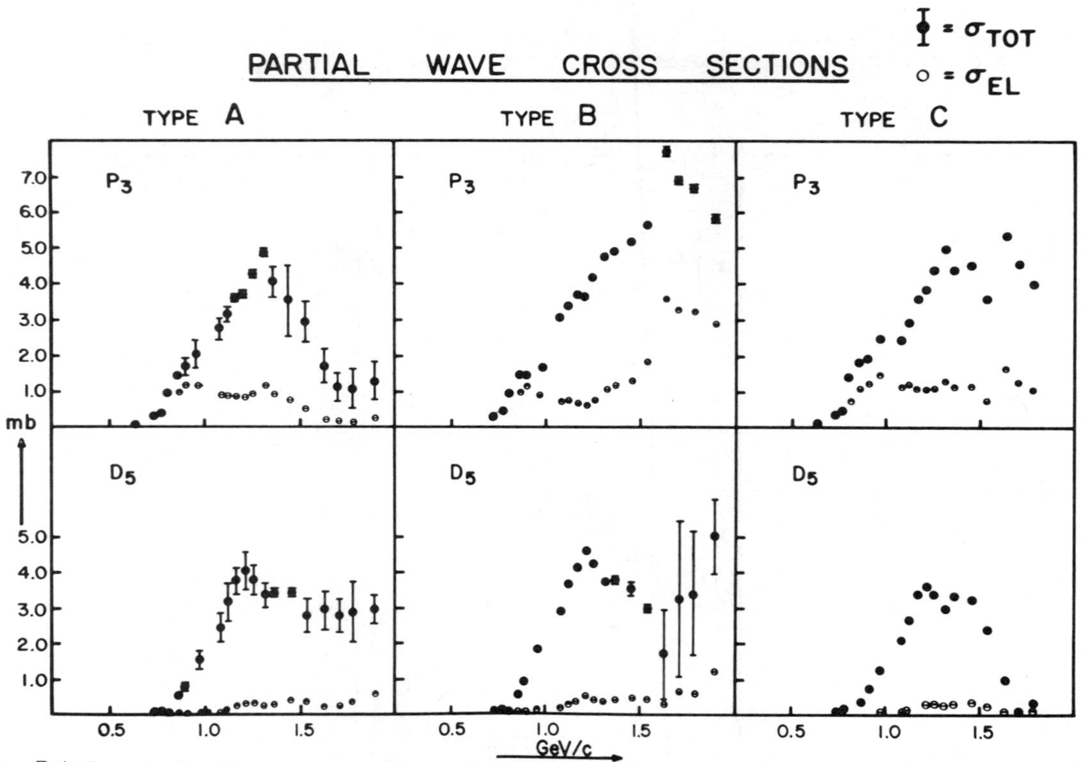

Fig. 10: Total and elastic cross-sections for the P_3 and D_5 partial waves for the solutions of type A, B, and C. The "error bars" indicate the range of values covered by the different solutions of this type.

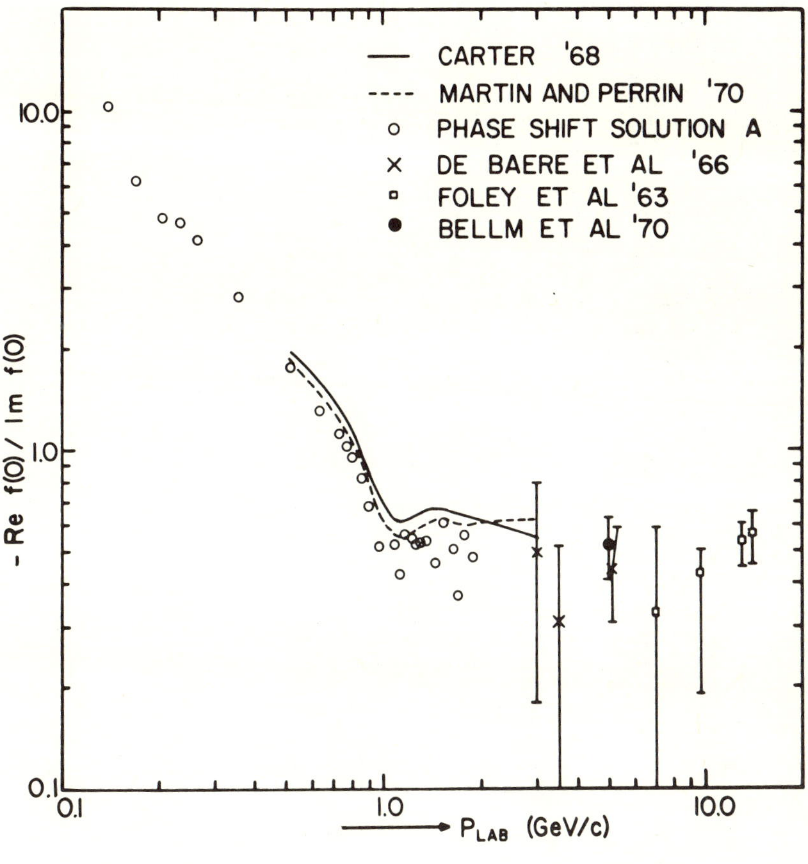

Fig. 11: Ratios of fitted real and imaginary forward amplitudes in a shortest path of type A, compared with forward dispersion relation calculations of Refs. 7 and 8, and measured values at momenta above 2 GeV/c[11].

3. COMMENTS ON OTHER EXPERIMENTS CONCERNING Z^*

As is well known, several inelastic channels in K^+p pass their threshold in the region of the bump at ~ 1.25 GeV/c in the total cross-section. Thus it is of interest to examine the consistency of the results discussed above with data on the inelastic channels, obtained in other experiments. This is of particular relevance if we are to decide between a "resonance" or a "threshold" interpretation of the data.

An extensive search for Z^* produced in K^+p and decaying into inelastic channels has been made by Bassompierre et al.[12]. The reactions studied were

$$K^+ + p \rightarrow K + N + m\pi \qquad (m = 1,2,3,4) \qquad (1)$$

at 3.0, 3.5, and 5 GeV/c incident momentum.

The authors conclude that the upper limits for seeing a bump in the mass region around 1.9 GeV is less than $\sim 30\,\mu b$.

The same data have been used to do a missing-mass-type analysis on reactions of the type:

$$K^+p \rightarrow Z^{*+} + \pi^+. \qquad (2)$$

Again no bumps are seen around 1.9 GeV and the data are well accounted for by phase space plus K^* and N^* production alone. The upper limits for the production cross-section in reaction (2) are 80 μb (15 μb) at 3 GeV/c (5 GeV/c). It may be noted, however, that production reactions are in general quite insensitive to resonances. The cross-sections for the production of the well-established $Y^*(1385)$ and $Y^*(1660)$ in the reactions $K^-p \rightarrow \pi^-Y^+$ or $\pi^+p \rightarrow K^+Y^{*+}$ are

	$Y^+(1385)$ [13]		$Y^+(1660)$ [14]
	3 GeV/c	5 GeV/c	4.6 GeV/c
	μb	μb	μb
K^-p	90	40	~ 33
π^+p	40	20	

In the 2 GeV mass region the cross-sections are an order of magnitude lower still, as is apparent from mass plots. Hence, the upper limits for the K^+ reaction (2) quoted above cannot be interpreted as evidence against the production of Z^*'s.

Other evidence, relevant to the question of the existence of Z^*'s, can be deduced from a partial-wave analysis of the formation reaction

$$K^+ + p \rightarrow K^0 + N^* \ (\rightarrow p\pi^+) \tag{3}$$

at five momenta between 0.86 GeV/c and 1.58 GeV/c, carried out by Bland[15].

There is considerable evidence that reaction (3) is dominated by ρ-exchange, and that the t-channel ρ-exchange amplitude results in a dominance of P-waves in the s-channel. The partial-wave analysis of Bland (see Fig. 12a) clearly confirms the P-wave character of the KN^* production channel. Both the P_1- and the P_3-waves, fed by the incident P_1- and P_3-waves, are large in magnitude at $\sim$1.2 GeV/c. The incident P_1-wave contributes 1.5 mb and the P_3-wave 1.4 mb to the KN^* cross-section. This is to be compared with the inelastic P_3-wave cross-section in $K^+p \rightarrow K^+p$, which is $\sim$3 mb at this momentum; for P_1 it is $\sim$0.6 mb. In view of the many assumptions, limited data, etc., this disagreement may not be too serious.

Having established that the KN^* channel is dominated by P-waves, one is then faced with the question whether the P_3-wave shows any resonance features in the region of energies, where the P_3-wave of the elastic channel appears to turn by 90° in the Argand diagram. Figure 12a shows that through the region of the bump in the total cross-section, the KN P_3-wave turns by 20° unless the major part of P_3 is assumed to be background; in the latter case a small resonant P_3-wave cannot be excluded.

K^+p elastic partial-wave analysis, similar to the one above, have been performed by Kato et al.[4] (ANL), by Rebka et al.[4] (BNL) and by Giacomelli et al. (BGRT Collaboration)[4]. All these analyses have been performed on subsets of the data input to the analysis above and their conclusions are thus not wholly independent.

The BGRT analysis extends from 0.14 GeV/c to 1.5 GeV/c. The phases δ_ℓ and absorption parameters η_ℓ are parametrized as functions of c.m. momentum, following older work by Lea et al.[16]. Such a parametrization can only be valid in a restricted energy range, since the energy dependence of δ_ℓ :

$$\delta_\ell \sim q^{2\ell+1} \cdot (\text{terms in } q^2 \text{ and } q^4)$$

is incompatible with any peripheral model which describes the data well down to $\sim$1 GeV/c, implying that δ_ℓ tends to zero at high energies. A further restriction of this analysis is the fact that only three partial waves are inelastic at the time. Since our solutions A and C are inelastic in more than three partial waves, it is perhaps not surprising that the four best BGRT solutions all

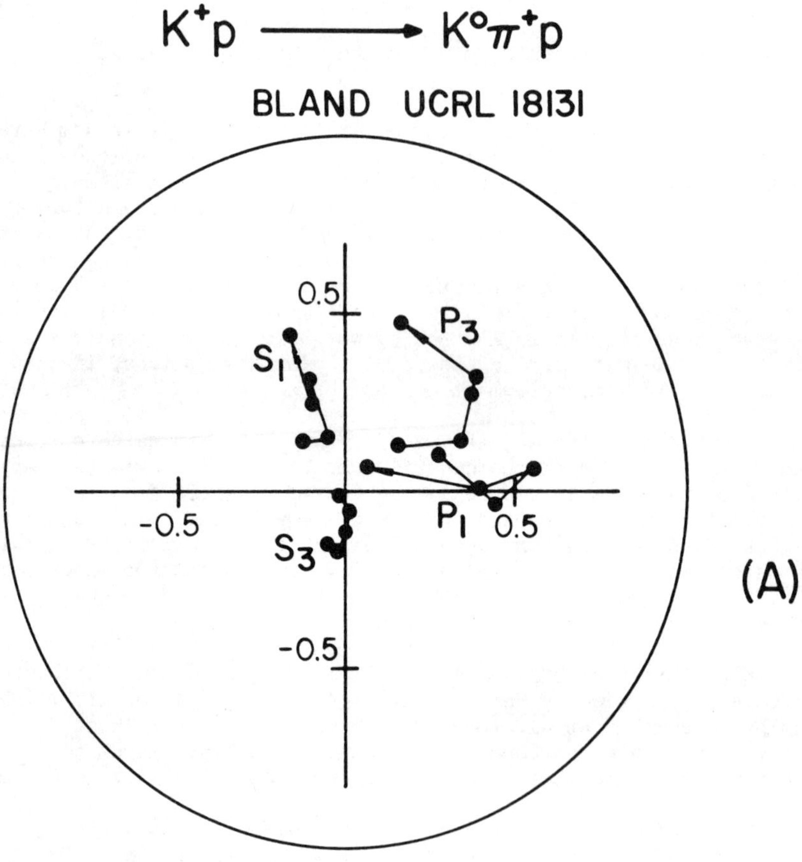

Fig. 12a: Argand diagram of the partial waves in the KN^* channel, as analysed by Bland (Ref. 15). The diagram has been turned upside down, for the sake of comparison with Fig. 12b.

The symbols S_3, P_1, P_3 refer to KN^* final states, fed by incident D_3-, P_1-, and P_3-waves, respectively. The symbol S_1 refers to S-waves between all particle pairs and is fed by the incident P_1-wave.

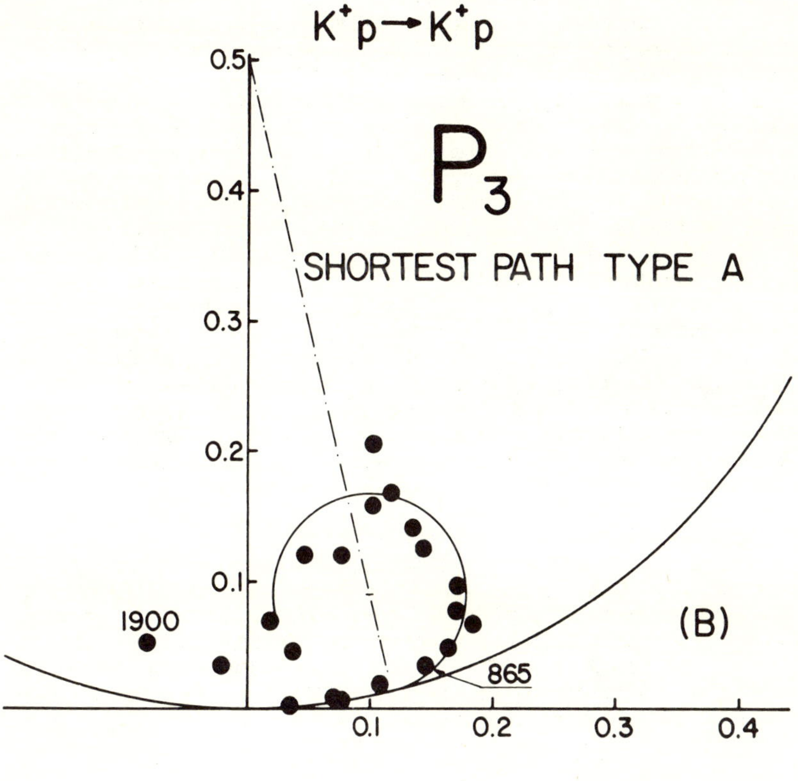

Fig. 12b: The P_3-wave of one solution of type A. Errors are roughly equal to the size of the points. The line is a Breit-Wigner fit to the points with parameters: Mass 1959 MeV, width 251 MeV, elasticity 0.15.

resemble our (rejected) solutions of type B with a nearly purely elastic P_1 forming a hairpin above the BGRT energy range and inelastic S_1-, P_3- and D_5-waves.

A curious feature in the ANL analysis is the apparent clustering of solutions at high momenta. At 1.95 GeV/c the solutions cluster in four quite small areas in the Argand diagram, while our analysis of the same data shows no grouping whatsoever. In our analysis, random searches result in 200 different solutions with $\chi^2/N_{DF} < 2$; the ANL analysis stops after 50 solutions with $\chi^2/N_{DF} < 2.5$. There is thus an apparent difference in randomizing the starting values in the minimization procedure. The ANL, which extends over nine momenta between 0.86 GeV/c and 1.95 GeV/c, results in four solutions, three of which show a counter-clockwise-turning P_3-wave, while the fourth turns clockwise in the P_3-wave, and counter-clockwise in the D_5-wave. In all their solutions the D_5-wave is rather strongly absorptive, although the data are too far apart in energy to indicate the energy dependence, characteristic of the D_5-waves in our analysis. The four solutions fit the polarization data well, but are in poor agreement with differential cross-section data, not included in the input data list.

The BNL analysis comprises data at ten momenta between 0.78 GeV/c and 2.0 GeV/c and, by the choice of starting values for the minimization, it is effectively energy dependent. One hundred solutions with $\chi^2/N_{DF} < 2.5$ were obtained. A minimum-path technique selects three solutions, which all turn counter-clockwise in the P_3-wave. Here again the agreement with data not included in the fit is poor.

Various analyses are at present being performed, in which theoretical models are used as input. An interesting example is an analysis by Cutkosky and Deo[17)] in which the analytic structure of the amplitude is used to get a series expansion which is more strongly convergent than an expansion in Legendre polynomials, albeit at the price of difficulties with unitarity and the inaccurately known $Kp\Lambda$ coupling constant. This model is in qualitative agreement with polarization data at 1.2 GeV/c but has thus far not been compared with polarization data at other momenta.

4. CONCLUSIONS

The partial-wave analysis of the elastic K^+p channel indicates that there is good, although not unambiguous, evidence that the P_3-wave goes through a counter-clockwise change of phase of $\sim 90°$ in the region of the Cool bump in the total cross-section. This conclusion is reached by four groups, each employing a different procedure and working on different subsets of the data. Although objections can be formulated against each of these analyses, it is felt that this combined conclusion is now on a rather firm footing.

The evidence is not in contradiction with the low upper limits for Z^* production found in the inelastic channels. In the conventional sense, one should thus call this a resonance; a Breit-Wigner fit to P_3 results in a mass 1959 MeV, width ~ 25 MeV and elasticity $x = 0.15$.

However, the fact that just in this energy region a new channel (KN^*), strongly excited by the same wave, sets in, suggests that more than the conventional criteria should be applied to test its existence; the evidence for Z^*'s is not complete until its main decay mode has been established. The most likely candidate for this decay mode, the KN^* channel, does not at present appear to be produced by Z^* formation.

The issue is further troubled by the presence of an unexpectedly strong imaginary D_5-wave. This wave has the same energy dependence as the KN production cross-section, while the combined inelasticities of P_3 and D_5 more than fully account for its magnitude. An initial D_5-wave leads to a final D_5-wave in both the KN^* and K^*N channels. It is thus surprising that the D_5-wave shows this behaviour, rather than, for example, the D_3-wave, which would produce KN^* and K^*N in a relative S-state (without angular momentum barrier) and would therefore be a more natural candidate for a "background" wave (with respect to the P-waves, demanded by exchange) than the D_5-wave.

There are, in our opinion, two ways in which to proceed, apart from filling the gaps between 1.5 GeV/c and 2 GeV/c and at low momenta:

i) a new partial-wave analysis of the KN^* (and K^*N) channel between 0.9 GeV/c and 1.5 GeV/c on better data, with the P- and D-waves freely varying, and with the ρ -exchange model replaced by a Regge model with ρ and Y exchange.

ii) a measurement of the Wolfenstein A and R parameters in elastic K^+p scattering. The predictions for our three types of solutions are indicated in Fig. 13, at 1.64 GeV/c.

Acknowledgements

We would like to thank Dr. C. Lovelace for discussions and the BGRT group for providing us their data prior to publication.

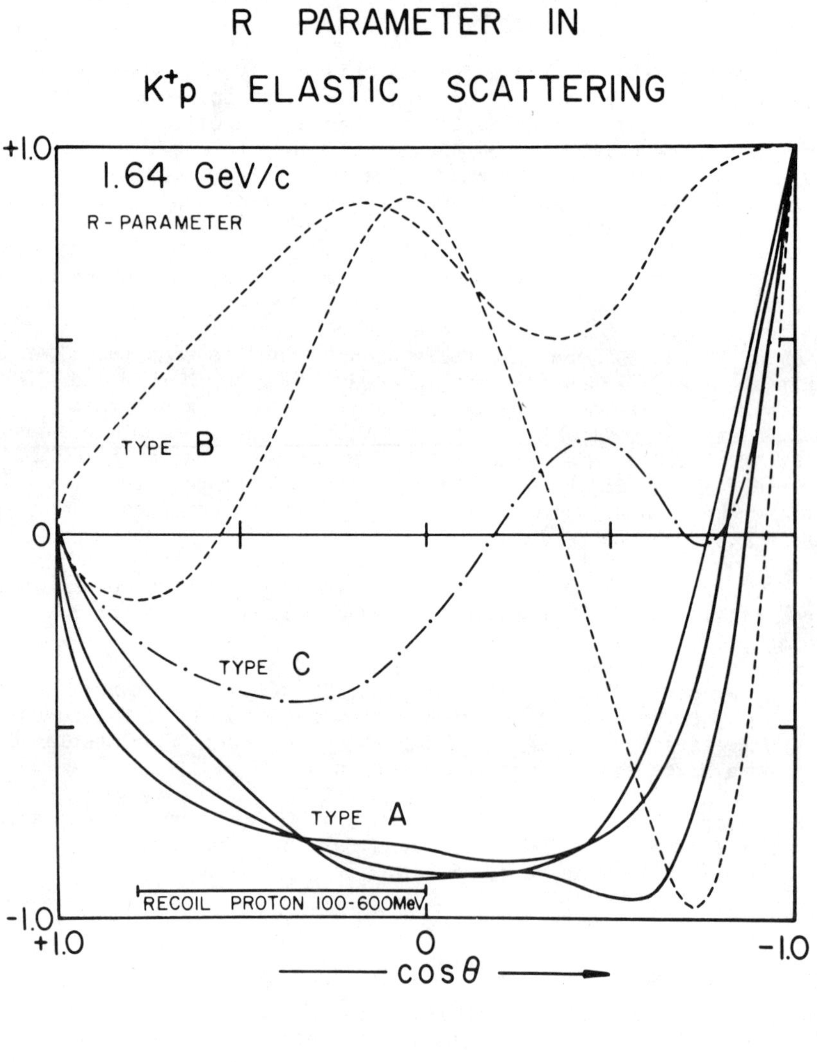

Fig. 13: Predictions for the Wolfenstein R parameter corresponding to solutions A, B, and C at 1.64 GeV/c.

REFERENCES

1) S. Andersson, C. Daum, F.C. Erné, J.P. Lagnaux, J.C. Sens, F. Udo and F. Wagner, contributed paper to the Int. Conf. on Elementary Particles, Lund (1969).

2) S. Andersson, C. Daum, F.C. Erné, J.P. Lagnaux, J.C. Sens, F. Udo and F. Wagner, Phys. Letters 30 B, 56 (1969).

3) M. Albrow, S. Andersson, B. Bosnjaković, C. Daum, F.C. Erné, Y. Kimura, J.P. Lagnaux, J.C. Sens, F. Udo and F. Wagner, Differential cross-sections and polarizations at 1.32, 1.54, 1.64, 1.71 and 1.79 GeV/c, CERN preprint, to be published.

4) S. Kato, P. Koehler, T. Novey, A. Yokosawa and G. Burleson, Phys. Rev. Letters 24, 615 (1970).
G.A. Rebka, Jr., J. Rothberg, A. Etkin, P. Glodis, J. Greenberg, V.W. Hughes, K. Kondo, D.C. Lu, S. Mori and P.A. Thompson, Phys. Rev. Letters 24, 160 (1970).
G. Giacomelli, P. Lugaresi-Serra, G. Mandrioli, A.M. Rossi, F. Griffiths, I.S. Hughes, D.A. Jacobs, R. Jennings, B.C. Wilson, G. Ciapetti, V. Costantini, G. Martellotti, D. Zanello, E. Castelli and M. Sessa, Differential elastic and inelastic cross-sections at 0.90, 0.97, 1.06, 1.13, 1.21, 1.25, 1.32, 1.38 and 1.48 GeV/c, preprint, University of Glasgow, 1970, submitted to Nuclear Physics B.
W. Chinowski, G. Goldhaber, S. Goldhaber, T. O'Halloran and B. Schwarzschild, Phys. Rev. 139 B, 1411 (1965).

5) R.L. Cool, G. Giacomelli, T.F. Kycia, B.A. Leontić, K.K. Li, A. Lundby and J. Teiger, Phys. Rev. Letters 17, 102 (1966).
D.V. Bugg, R.S. Gilmore, K.M. Knight, D.C. Salter, G.H. Stafford, E.J.N. Wilson, J.O. Davies, J.D. Dowell, P.M. Hattersley, R.J. Homer, A.W. O'Dell, A.A. Carter, R.J. Tapper, and K.F. Riley, Phys. Rev. 168, 1466 (1968).

6) A.S. Carrol, J. Fischer, A. Lundby, R.H. Phillips, C.L. Wang, F. Lobkowicz, A.C. Melissinos, Y. Nagashima, C.A. Smith and S. Tewksbury, Phys. Rev. Letters 21, 1282 (1968).

7) A.D. Martin and R. Perrin, preprint, University of Durham, England (1970).

8) A. Carter, Rutherford report, REHL-68/10 (1968).

9) C.H. Johnson and H.M. Steiner, UCRL-18001 (1967).
C. Lovelace and F. Wagner, Proc. 14th Int. Conf. on High-Energy Physics, Vienna (1968) (CERN, Geneva, 1968), abstract 254.

10) R.W. Bland, G. Goldhaber, B.H. Hall, J.A. Kadyk, V.H. Seeger, G.H. Trilling and C.G. Wohl, UCRL-18323.

11) W. de Baere, J. Debaisieux, P. Dufour, F. Grard, J. Heughebaert, L. Pape, P. Peeters, F. Verbeure, R. Windmolders, R. George, Y. Goldschmidt-Clermont, V.P. Henri, B. Jongejans, D.W.G. Leith, A. Moiseev, F. Muller, J.M. Perreau and V. Yarba, Nuovo Cimento 45 A, 885 (1966).
K.J. Foley, S.J. Lindenbaum, W.A. Love, S. Ozaki, J.J. Russell and L.C.L. Yuan, Phys. Rev. Letters 11, 503 (1963).
Th.H.J. Bellin, H. de Jonge, R.W. Meyer, A.G. Tenner, P. Vons and W. Winzeler, Nuovo Cimento Letters 3, 389 (1970).

12) G. Bassompierre, Y. Goldschmidt-Clermont, A. Grant, V.P. Henri, R. Jennings, B. Jongejans, D. Linglin, F. Muller, J.M. Perreau, R. Sekulin, W. de Baere, J. Debaisieux, P. Dufour, F. Grard, J. Heughebaert, J. Pape, F. Peeters, F. Verbeure and R. Windmolders, Phys. Letters 27 B, 468 (1968).

13) J. Kirz, Proc. 3rd Conf. on High-Energy Collisions (Stony Brook, Sept. 1969).

14) M. Primer, M. Goldberg, K. Jaeger, V. Barnes, P. Dornan, I. Skillicorn and J. Leither, Phys. Rev. Letters 20, 610 (1968).

15) R.W. Bland, thesis, UCRL-18131 (1968).

16) A.T. Lea, B.R. Martin and G.C. Oades, Phys. Rev. 165, 1770 (1968).

17) R.E. Cutkosky and B.B. Deo, preprint, Carnegie-Mellon University, Pittsburgh, Pa.

DISCUSSION

Steinberg: What about the D5 solution? It looks like there is a structure in the D5 solution.

Wagner: Yes, if the D5 is so large there may be another structure in it.

Steinberg: I would like to clarify our position with respect to our analysis on the question of error assignment. We feel that the effect we observe is barely above noise and, as such, one has to take the result with a grain of salt. But this returns us to the basic question of error assignment. For example, in your solution which appears to resonate, you have error bars on your partial cross sections. You also show the clusters that are fairly large. If you have assigned the errors on the basis of cluster sizes what kind of error would that have propogated to the partial cross sections?

Wagner: I don't know.

Lynch: What experiments can be done in order to remove some of the ambiguities in the analysis that have been discussed so far?

Goldhaber: I will talk about that later, at the end of this session.

Tripp: My question is directed to Dr. Yokosawa. The method you use to determine the mass and width of the P_{13} resonance neglects the behavior of the real part of the amplitude.

Yokosawa: Well, I don't know any other way of estimating the background, but this is a very important point and we'll look into this question when we have more points.

Cool: With regard to the real amplitude which are not very well known, how sensitive are your results to that? Also do you have any other ways which are not so subjective to put errors on the resonant amplitude?

Wagner: I have no idea really how to deduce errors. I can't answer that.

Feld: I have the same objection which Dr. Tripp raised, that is, the method of just subtracting the background from the partial cross section does not take into account any possibility of interference between the background and the Breit-Wigner.

Wagner: Yes.

Feld: The other point that I would like to raise is something that people forget all the time, this is the possibility of having a bump coming from the threshold effect, the cusp effect. It seems to be that you are very close to the inelastic threshold at the point where resonances are occurring.

Wagner: I know roughly what it means to have a cusp effect in S-wave. I really don't know what you mean by a threshold effect in a P-wave. What does it mean?

Biedenharn: I can answer this very quickly. In the S-wave you have a discontinuity which gives you the Wigner cusp phenomena. In the P-wave the discontinuity appears in the derivative. More precisely, for uncharged particles the discontinuity is in the Lth (energy) derivative for an L-wave so that the discontinuity will be smoothed out.

Tripp: In the extreme situation, if you take the CERN-Heidelberg-Saclay result for the upside down resonance, i.e. the S_0 resonance at 1670 MeV, there you get a dip in the cross section, not a bump. So if you try to determine the resonances solely based on bumps, it may be misleading.

K^+p ELASTIC SCATTERING AT 2.53, 2.76 AND 3.20 GeV/c[‡]

G. S. Abrams[+], L. Eisenstein, J. Kim, T. A. O'Halloran,
W. Shufeldt*, and J. Whitmore
University of Illinois

Urbana, Illinois 61801

The MURA 30" HBC was exposed to an electrostatically separated K^+ beam at momenta of 2.53, 2.76 and 3.20 GeV/c. The beam momenta extend over an invariant K^+p mass from 2.44 to 2.68 GeV and were selected to study the enhancement reported by Abrams et al. in the K^+p total cross section at 2.505 GeV[1]. In this note we present the results of a measurement of K^+p elastic scattering in the backward direction ($\cos\theta \leq -0.84$) at all 3 momenta and the preliminary results of a phase shift analysis at 2.53 GeV/c.

To study K^+p elastic scattering in the backward direction, a special template scan[2] yielded 78, 96, and 52 elastic events at 2.53, 2.76 and 3.20 GeV/c respectively. The differential cross sections shown in Figure 1 were determined by normalizing the elastic events to the observed tau-like decays of the K^+. The most apparent feature of Figure 1 is the difference, both in shape and magnitude, between the 3.20 GeV/c data and the data at the two lower energies. Parameterizing the distributions in the form $\frac{d\sigma}{d\Omega} = \frac{d\sigma}{d\Omega}(180°) \exp[A(1+\cos\theta)]$ or $\frac{d\sigma}{du} = \frac{d\sigma}{du}(u=0) \exp(Bu)$ permits a more quantitative analysis of the momentum and u dependence. Figures 2 and 3 are the results of such an analysis of the present experiment and previously published data[3]. It is apparent that in the momentum interval 2 - 3 GeV/c both the differential cross section at 180° and the slope of the backward peak deviate from a

---

‡Work supported in part by the U. S. Atomic Energy Commission
+Present Address: Lawrence Radiation Laboratory, University of California, Berkeley, California
*Present Address: University of Wisconsin at Green Bay, Green Bay, Wisconsin

monotonic energy dependence. We have fit the data excluding the momentum interval 2 - 3 GeV/c with an exchange degenerate Λ_α, Λ_γ u channel amplitude[4/]. It is clear that such a model gives an adequate parameterization of the data from $\sim$1 to 7 GeV/c except for the momentum interval 2 - 3 GeV/c. We note that this interval coincides with the enhancement reported by Abrams et al. at 2.505 GeV.

As a further study of the possible existence of KN resonances, a partial wave analysis has been performed on the complete differential cross section at 2.53 GeV/c. In addition to the measured cross section the polarization data of Anderson et al. at 2.48 GeV/c was also used[5/]. Five solutions were obtained and are shown in Figure 4 as the shaded areas on an Argand diagram. While no quantitative continuation from lower energy analyses[6/] is feasible at the present time, the results obtained here are not inconsistent with possible extrapolations from 2 GeV/c. In particular, several of the solutions would appear to be consistent with continuations of a counter-clockwise semicircle observed in the $p_{3/2}$ wave by phase shift analyses below 2 GeV/c.

---

1. R. J. Abrams et al. Phys. Rev. Letters 19, 259 (1967).

2. Preliminary results at 2.76 GeV/c have already been published in G. S. Abrams et al., Phys. Rev. Letters 21, 1407 (1968). The bias observed at that energy due to the template scan was corrected and was not found in the 2.53 and 3.20 GeV/c data.

3. J. Banaigs et al., Phys. Letters 24B, 317 (1967) and Nucl. Phys. B9, 640 (1969); D. Cline et al., Phys. Rev. Letters 19, 675 (1967); A. S. Carroll et al., Phys. Rev. Letters 21, 1282 (1968); W. F. Baker et al., Phys. Letters 28B, 291 (1968).

4. The amplitude used is similar to that of E. L. Berger and G. C. Fox, Phys. Rev. 188, 2120 (1969) with $f(s,u) \sim \gamma\ B(1-\alpha_t,\ 1/2-\alpha_u)$ where $B(x,y) = \frac{\Gamma(x)\Gamma(y)}{\Gamma(x+y)}$.

5. S. Andersson et al., Phys. Letters 28B, 611 (1969).

6. J. G. Asbury et al., Phys. Rev. Letters 23, 194 (1969); G. A. Rebka et al., Phys. Rev. Letters 24, 160 (1970); S. Kato et al., Phys. Rev. Letters 24, 615 (1970).

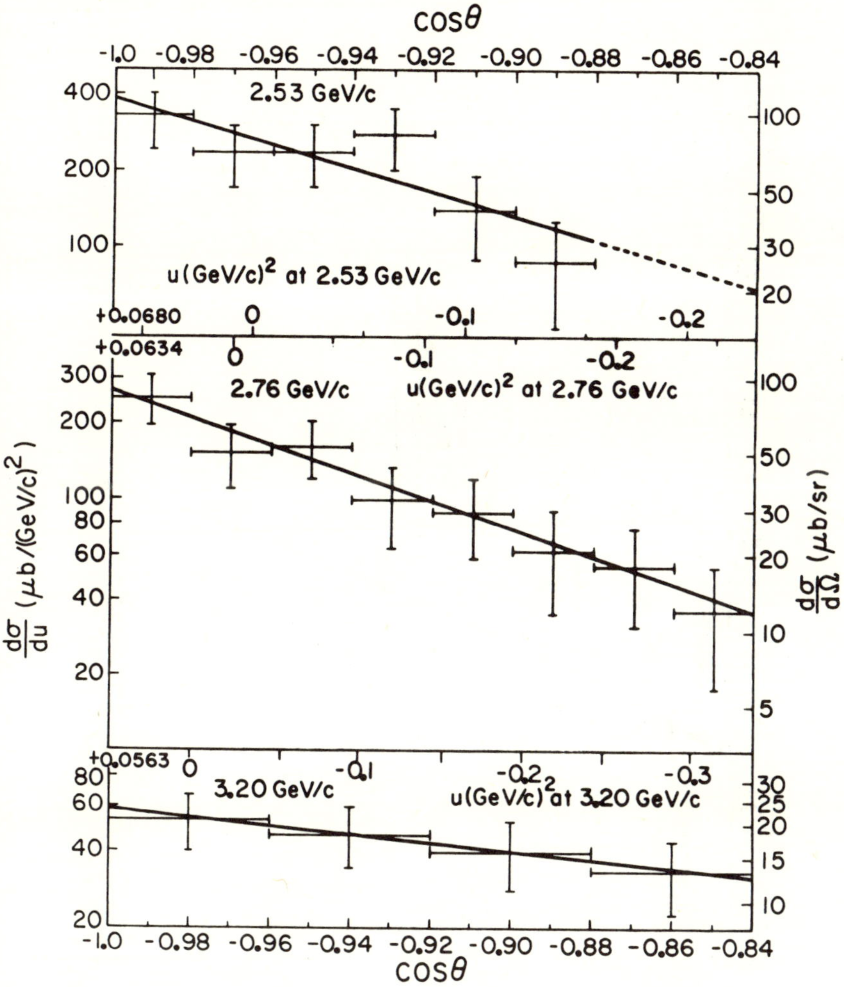

Fig. 1. Backward differential cross sections as a function of u. The solid line curves are the results of the best fit to the form $d\sigma/du = A \exp(Bu)$, as described in the text.

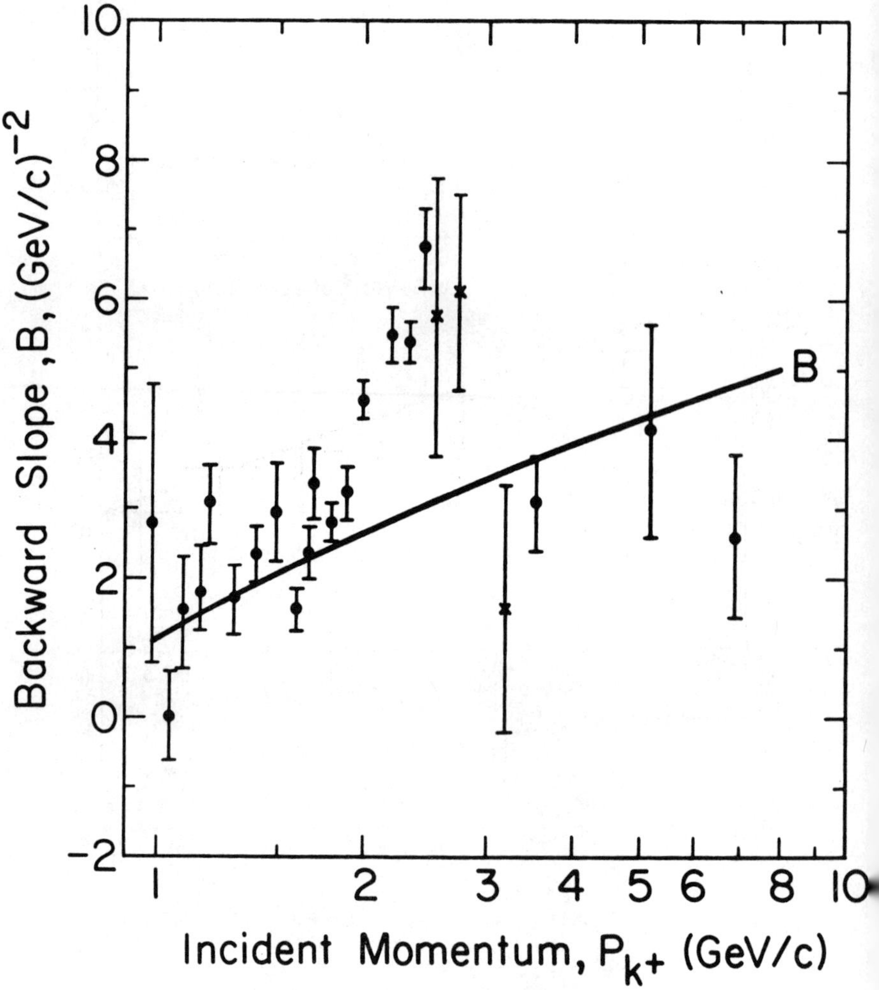

Fig. 2. Slope of the backward peak as a function of incident kaon momentum. The curve is the result of the fit described in the text.

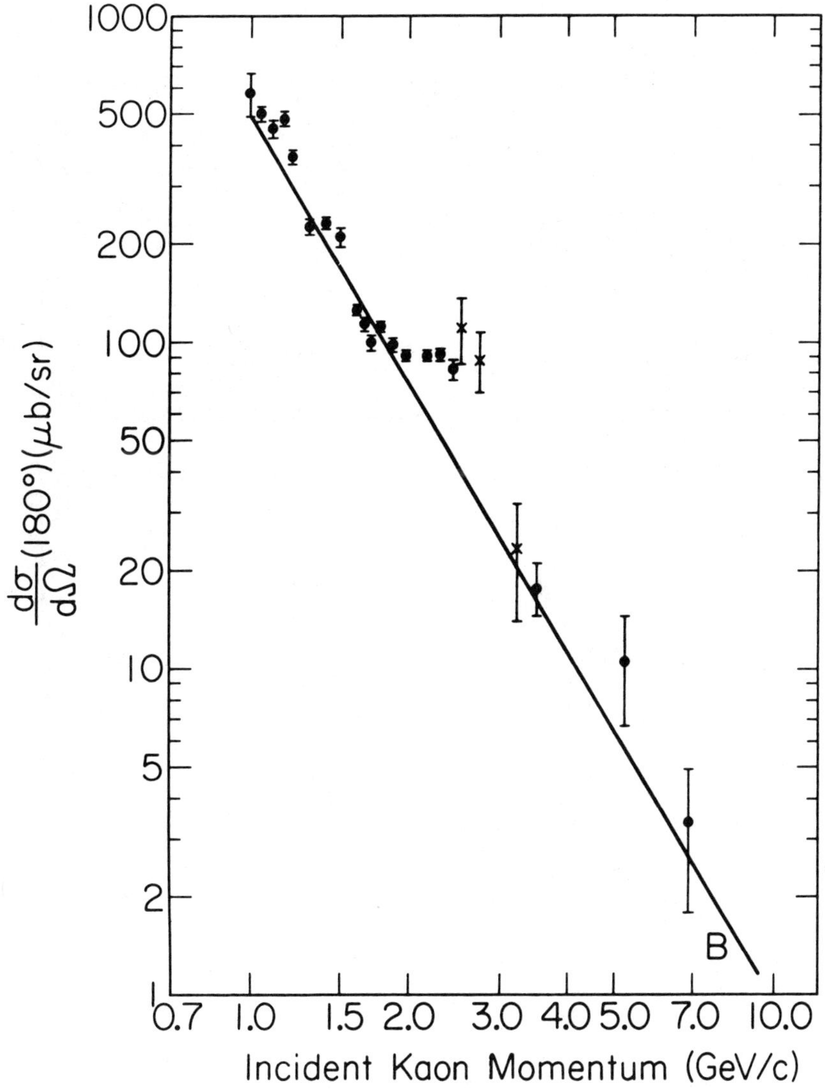

Fig. 3. The variation of the backward differential cross section $d\sigma/d\Omega$ (180°) as a function of incident kaon momentum. The curve is the result of the fit described in the text.

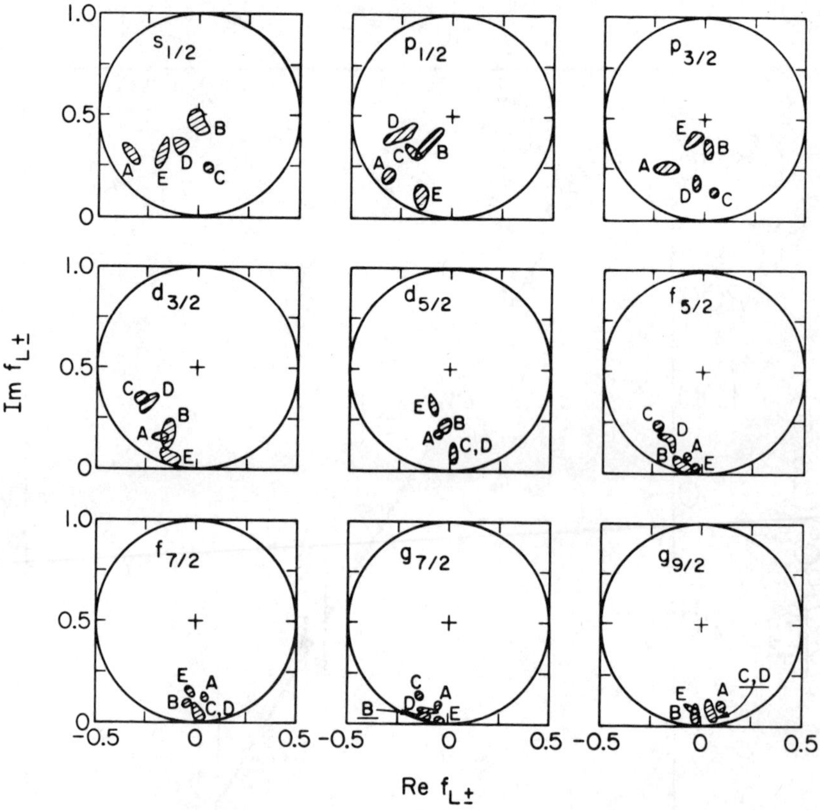

Fig. 4. The Argand diagrams for the partial wave amplitudes at 2.5 GeV/c. The shaded areas indicate the regions in which the solutions tended to occur.

DISCUSSION

Yokosawa: I would like to make just one comment. It is that we have a very interesting feature around 2 GeV. This is the same thing as in the π p reaction. So this Kp mystery might be something other than resonances.

O'Halloran: This is not necessarily true. That π^-p is well associated with the N^* production.

THE SEARCH FOR Z^*'s

Gerson Goldhaber

Department of Physics and Lawrence Radiation Laboratory

University of California, Berkeley, California 94720

At the 1962 CERN Conference Gell-Mann and Ne'eman were each considering the idea that the $\Delta(1238)$, the $Y^*(1385)$ and the $\Xi^*(1535)$ could belong to either a <u>10</u> or a <u>27</u> SU(3) group. The absence of any KN resonance phenomena of the magnitude of the then known resonances swayed them in favor of the <u>10</u> assignment.[1] This led to the prediction of the Ω^- whose exciting experimental confirmation is well known to all.

The possibility of the existence of KN resonances was raised again in 1966 by Cool et al. in a series of very beautiful precision K^+p and K^+d cross-section measurements at Brookhaven.[2] These measurements indicated a distinct bump in the K^+p cross section as well as in the $I = 0$ part of the K^+n cross section obtained after unfolding of the experimental K^+d cross section. If the resonance interpretation is adopted these would correspond to rather high inelasticity resonances: Z_1^* at 1910 MeV and Z_0^* at 1863 MeV. Confirmation of these measurements come shortly thereafter from work at the Rutherford Laboratory by Bugg et al.[3]

Since that time there has been a great effort on: inelastic cross-section measurements,[4] elastic K^+p scattering,[5] observation of further structure in total cross sections,[6] observation of steps in K^- production yields for γp reactions,[7] and more recently polarization measurements in K^+p scattering with polarized targets.[8] It is these later measurements which place more restrictions on phase shift analyses and limit the number of possible ambiguities, which eventually can lead to a unique solution and answers to the question of the existence of exotic Z^* resonances.

I will now report on three new papers submitted to this Conference and then summarize the present status on Z^*'s.

I. K^+p ELASTIC SCATTERING MEASUREMENTS FROM 1.4 TO 2.3 GeV/c

I'd like to report on the following results based on brand new data from University College, London, and Rutherford Laboratory (Barber et al.).[9] They have measured the K^+p elastic cross section between 1.4 and 2.3 GeV/c at 26 momenta at Nimrod with about 4,000 events at each momentum. The method is a Čerenkov counter and time of flight for particle identification and wire chambers for kinematics. Since this is new data there are only preliminary results available at present. They find that the Legendre polynomial expansions give a smooth variation of the coefficients with momentum. Figure 1 gives typical examples of the differential cross section.

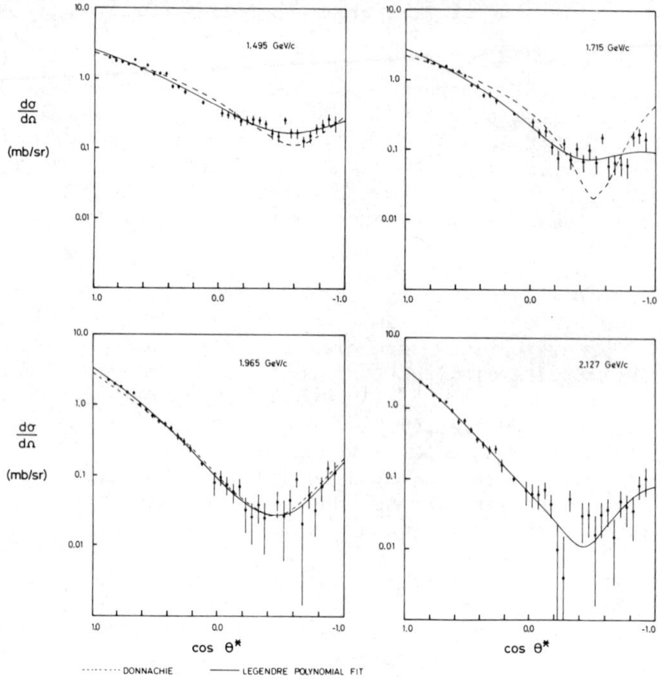

Fig. 1. Differential cross section sample of 4 out of 26 momenta studied by Barber et al.

Four out of the 26 momenta they have measured, 1.5, 1.7, 1.9, and 2.1 GeV/c roughly are shown. These distributions give you some idea of the quality of the data. The solid curves are Legendre polynomial fits. The dotted curves are fit according to a Regge pole model by Carreras and Donnachie[10] which was not calculated from these data, but from fitting all other existing data. This is the next communication I am going to mention. The Regge pole model was actually only fitted at momenta which are roughly equal to the ones given in Fig. 1. In between it does not fit so well, but it does fit where there had been earlier data at 1.5 and 1.96 GeV/c. Figure 2 shows the normalized Legendre coefficients A_1 to A_4 between 1.4 and 2.3 GeV/c. These are fairly smooth all the way; this is essentially true over the entire interval studied so far. At the higher momenta even higher coefficients than those shown as an example are needed. The authors have performed a phase shift analysis but they want to emphasize that it is very preliminary, because their data is so recent.

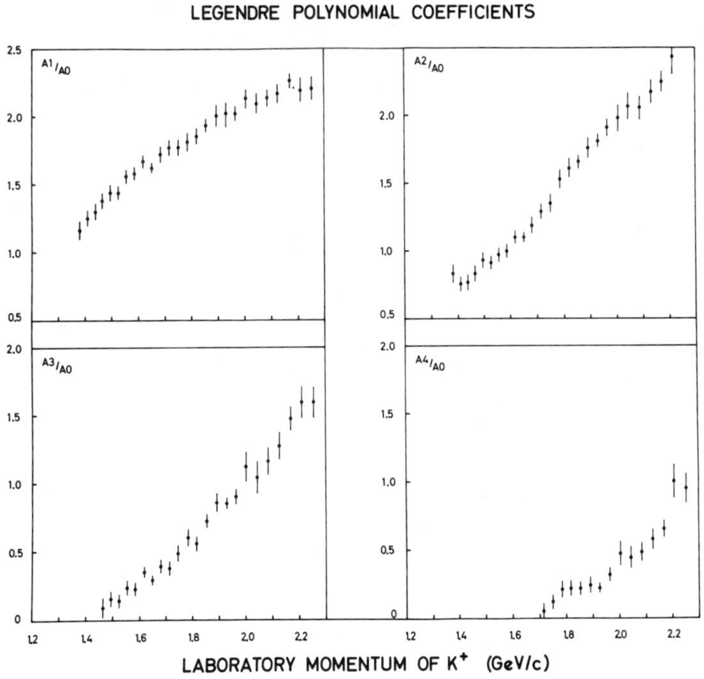

Fig. 2. Fit to data of Barber et al.

Figure 3 shows the results. This phase shift analysis was tied onto the Lee, Martin, Oades[11] solution up to 0.8 GeV/c, but there isn't much ambiguity up to 0.8 GeV. Beyond that it is an independent search at each momentum using starting values close to existing solutions. It is however not a complete random search. They find essentially that the main waves (S_{11}, P_{11} and P_{13}) have a tendency to go

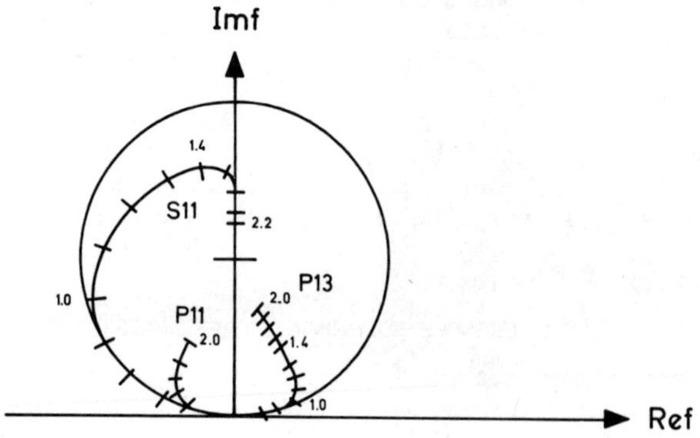

Fig. 3. Phase shift analysis by Barber et al.

towards the center of the Argand plot; namely the negative waves have the tendency that when they reach a certain phase they stay constant in phase and the inelasticity increases. This is true for the S_{11}, for the P_{11} and also for the positive phase shift the P_{13} which is so relevant to all of our discussion today.

II. A REGGE POLE ANALYSIS

I now want to go on to another contribution to this Conference. This is a Regge pole model by Carreras and Donnachie from Daresbury and also Kirsopp[12] who has done a phase shift analysis from the Regge pole model. In this Regge pole model they have attempted something different. Rather than trying the phase shift fits which other people have attempted, they try to fit all data above 0.7 GeV. That even includes data up to 12 GeV (whatever was available in the high energy region). They have also included all of the polarization data (including that at high energy). They carried out a Regge expansion; they mentioned a specific form, I won't go into the details, but its a parameterization which Das has used before. Furthermore, they allowed specifically for $K\Delta$ and K^*N thresholds.

They get χ^2 on this large amount of data of 1750 for 1000 or so degrees of freedom. By and large it looks like a reasonable fit. There are three basic results. The first is surprising and needs to be looked at carefully. This is that they get a positive S wave instead of a negative S wave. I will mention later the previous basis for the negative S wave. They get a negative P_{11} wave, as is the case for all the other solutions we have seen. Finally, they get a positive P_{13} wave. Figure 4 shows the positive S wave solution, in the Argand plot, you note that they have actually two solutions. They show both of them because though they prefer one, they cannot rule out the other. In both the S wave has a positive circle which

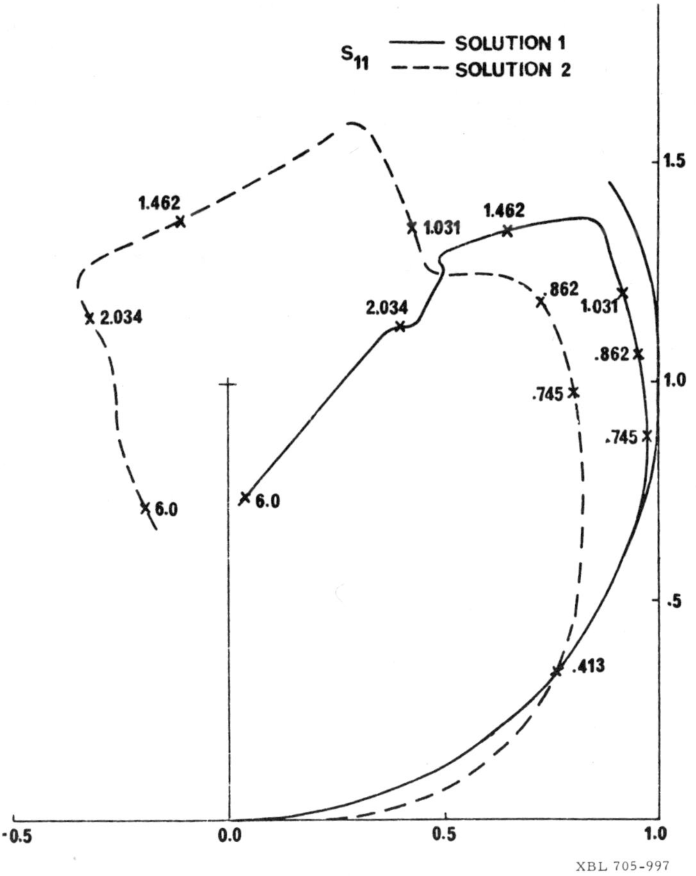

Fig. 4. The positive S wave solutions obtained from the Regge pole analysis of Carreras, Donnachie, and Kirsopp.

varies very slowly with momentum; it should be noted that the curves go up to 6 GeV. Figure 5 gives the P_{11} solutions. As I said it looks as though it has the same sign as all other solutions. Figure 6 gives the P_{13} solution which again is positive; however, the authors say that it has no resonant features. Now consider the

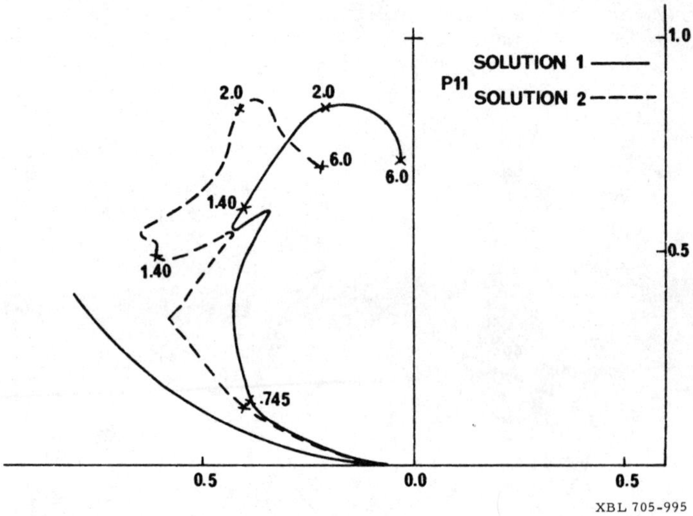

Fig. 5. The corresponding P_{11} solutions.

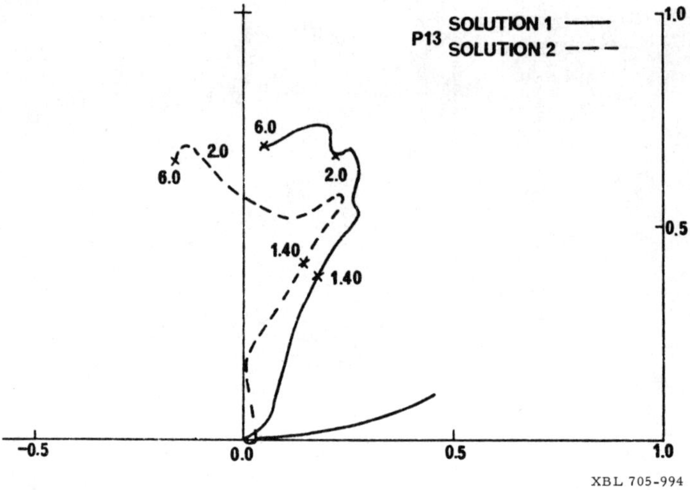

Fig. 6. The corresponding P_{13} solutions.

question of the negative versus the positive S wave. When we got this paper last week, we went back to our phase shift program and tried it for two momenta at 0.86 and 0.96 GeV/c with a positive instead of a negative S-wave phase shift. The answer is that either S wave fits well. Taking any given momentum you can fit with the positive S wave just as well as with a negative S wave.

A. The Old Evidence for the Sign of the S Wave Phase Shift

The reason for which most people use the negative S wave comes from an experiment which we have done nearly 10 years ago which showed interference between the elastic scattering and Coulomb scattering and which displayed constructive interference; namely, that the S wave is repulsive, just as the K^+p Coulomb interaction is repulsive. Now, in order better to answer the question about the sign of the S wave, I will show you some of our very old data. Figure 7 is from the experiment by Sula Goldhaber and co-workers, a collaborative experiment between LRL and UCLA. Figure 7 shows the elastic scattering from 140 MeV/c which was our lowest energy point up to 640 MeV/c.[13] Here the solid curve is a fit with a negative S wave while the dashed curve is a fit with a positive S wave. You see very clearly there are several standard deviations difference at 140 MeV/c between these fits. This data was divided into eight separate momenta and essentially for each one of them these features are observed. The positive S-wave solution is not shown at each momentum however it behaves similar to the 140 MeV/c data. By 640 MeV/c it becomes difficult to distinguish with the available data, but there are six momenta in between which all show the feature of constructive interference. Now the point which Carreras and Donnachie make is to say that if they take these data and add all other results and then compare with their solution the few points showing the interference make little difference in the overall chi square and this effect is lost. This is perfectly true but it means ignoring a very clear systematic effect! While I feel that the negative S wave is actually pretty firmly established, I would like to urge other experimenters to think about checking the result. At the moment it is based entirely on one single experiment carried out ten years ago!

Figure 8 is again from this old experiment just to show you the argument as to why we are dealing with an S wave at low momenta. If you have a flat distribution it can be $S_{1/2}$ or it can be $P_{1/2}$ or it can be a mixture of $P_{1/2}$ and $P_{3/2}$. However, only the S wave is going to give you a linear dependence between the phase shift and the momentum. The P wave would give you a p^3 dependence which is very different from the observed result in Fig. 8. Again the point which is used as the basis for all the dispersion relation calculations (which Dr. Wagner mentioned) coming out with a negative sign for the real part of the scattering amplitude also uses as an input the negative scattering length which is obtained from these data. If indeed there were a positive S wave at low energy, quite a number of relations would have to change very drastically.

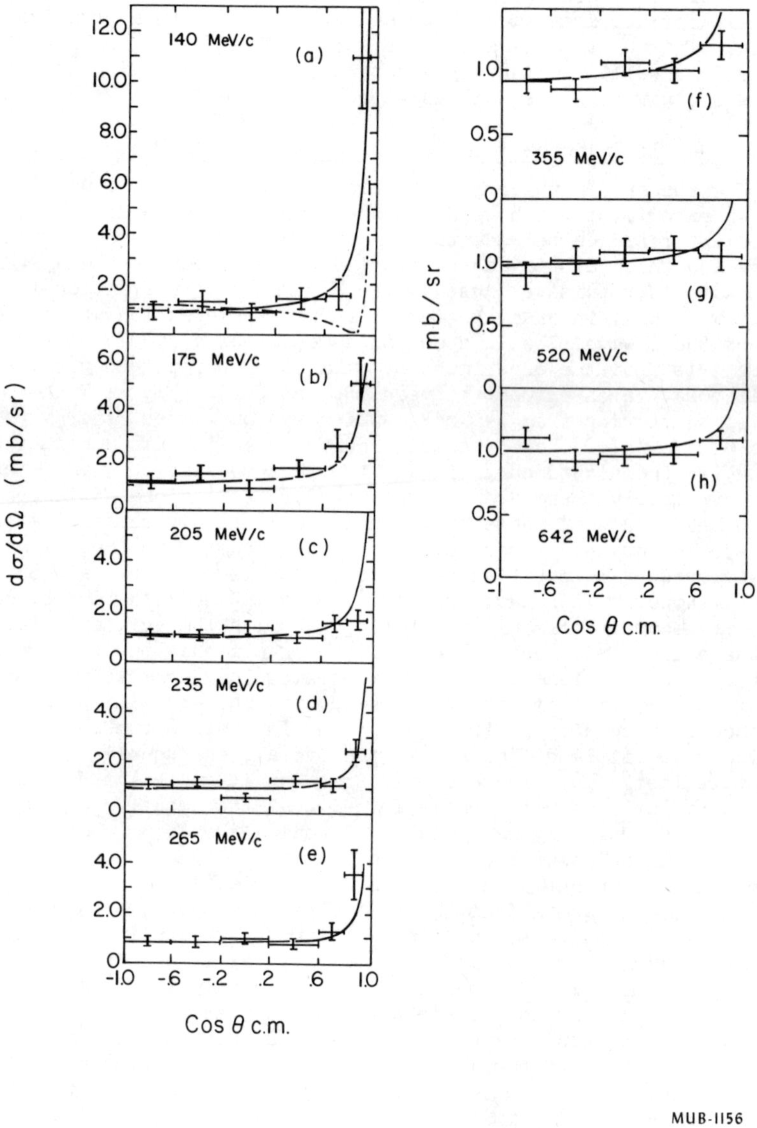

Fig. 7. The low energy differential cross section data of S. Goldhaber et al. giving the evidence for a negative S wave phase shift.

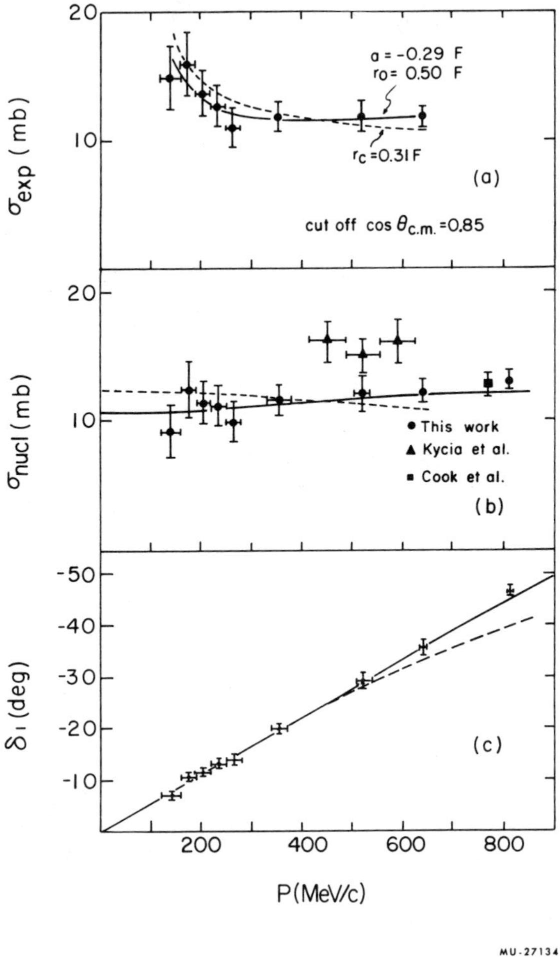

Fig. 8. Cross section and phase shift for the low energy data showing the characteristic S wave-linear momentum dependence and value for the scattering length.

III. K^* PRODUCTION FROM THE I = 0 K^*N STATE

Now I want to come to a third contribution. This is from our group, Hirata et al.,[14] and it is new data on the deuterium experiment. These experiments are fairly complicated so although the exposures have been taken some time ago we are only now completing the analysis of the deuterium data. This is in the momentum interval

860-1580 MeV/c which covers the region of the Cool bump. Now in this particular result I am going to show you the data on K^* production specifically. There we can isolate the K^* production in the $I = 0$ state. From isospin configurations you get the following results:

$$\sigma_0(KN \rightarrow K\pi N) = 3[\sigma(K^+n \rightarrow K^0\pi^+n) + \sigma(K^+n \rightarrow K^+\pi^-p) - \sigma(K^+p \rightarrow K^+\pi^0p)] \quad . \tag{1}$$

This is true for whatever happens between the K-nucleon and π. In most of our data all of these three cross sections are available so one can just take this algebraic sum and obtain the $I = 0$ cross section. For the data at the highest momentum 1580 MeV/c the last cross section is not available to us, and one can use an alternate formula where you specifically say that you are considering K^* production:

$$\sigma_0(KN \rightarrow K^*N) = 3[\sigma(K^+n \rightarrow K^{*+}n \rightarrow K^0\pi^+n) + \sigma(K^+n \rightarrow K^{*0}p \rightarrow K^+\pi^-p) - \tfrac{1}{2}\sigma(K^+p \rightarrow K^{*+}p \rightarrow K^0\pi^+p)] \ . \tag{2}$$

The main difference is that for the last term one can use the cross section $K^+p \rightarrow K^{*+}p$ with $K^* \rightarrow K^0\pi^+$. In Fig. 9 we show the algebraic sum of these three cross sections at 1210 MeV/c as an example. Here the first cross section, $K^+\pi^-$, as you see, has large K^* production; in the $K^0\pi^+$ cross section you can see a K^* as well. The $K^+\pi^0$ cross section also has a K^*. Now we add the first two and subtract the third which yields the final K^* distribution. We can see that this method works, because when we look at the π nucleon mass distributions there is some Δ present. But on subtraction the final mass distribution shows essentially no more Δ, as expected for an $I = 0$ state, which proves that we eliminated the $I = 1$ part. Figure 10 shows what the angular distributions do. Again we add the first two angular distributions and subtract the third one and that gives the resultant angular distribution for $I = 0$. Here θ is the K^* production angle in the overall center of mass, α is the decay angle of the K^* (Jackson angle), and φ is the Treiman-Yang angle of the K^* decay both in the $K\pi$ center of mass. In the communication by Hirata et al.[14] we show these distributions for three momenta 1200, 1360, and 1585 MeV/c. What is relevant to our discussion today is the question of the angular momentum state of K^* production. You see that the production angle is extremely sharp even at 1200 MeV/c. We must remember that 1200 MeV/c is just slightly above the K^* threshold (K^* threshold is at 1080 MeV/c or so). As soon as we can produce it, the K^* is strongly forward peaked which means that in terms of s channel K^*N production there are many angular momentum states involved. An analysis requires up to $\ell = 4$. I won't go into the details, only enough to say that it looks like the K^* is produced by pion exchange in this process even near threshold. Furthermore, there is a great similarity between the $I = 1$ K^* production cross section and the $I = 0$ K^* production cross section. We show that if you put a curve through the $I = 1$ K^* production cross section the same curve scaled up by a factor of 2.5 fits rather well to the $I = 0$ K^* production cross section. The physical significance of this is still under study.

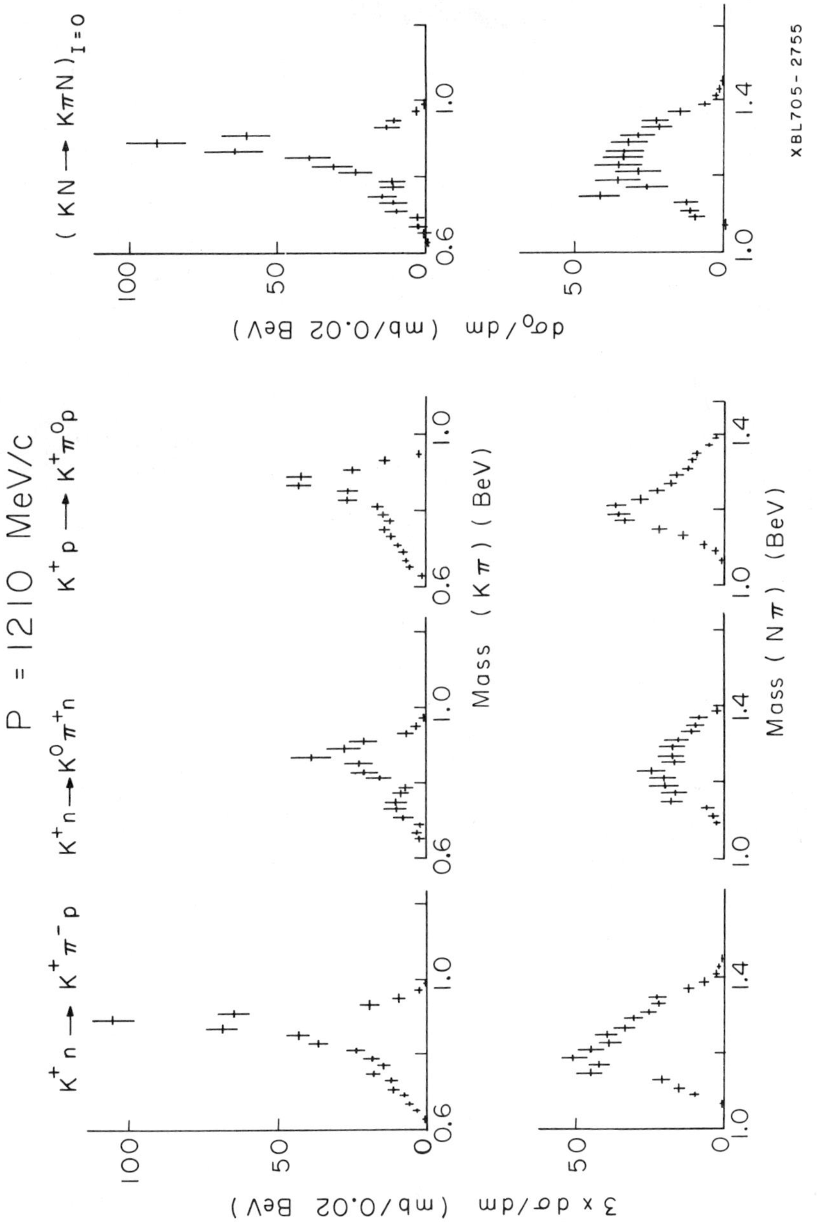

Fig. 9. Data of Hirata et al. The I=0 K^* production evaluated according to Eq. (1) from the algebraic sum of 3 cross sections.

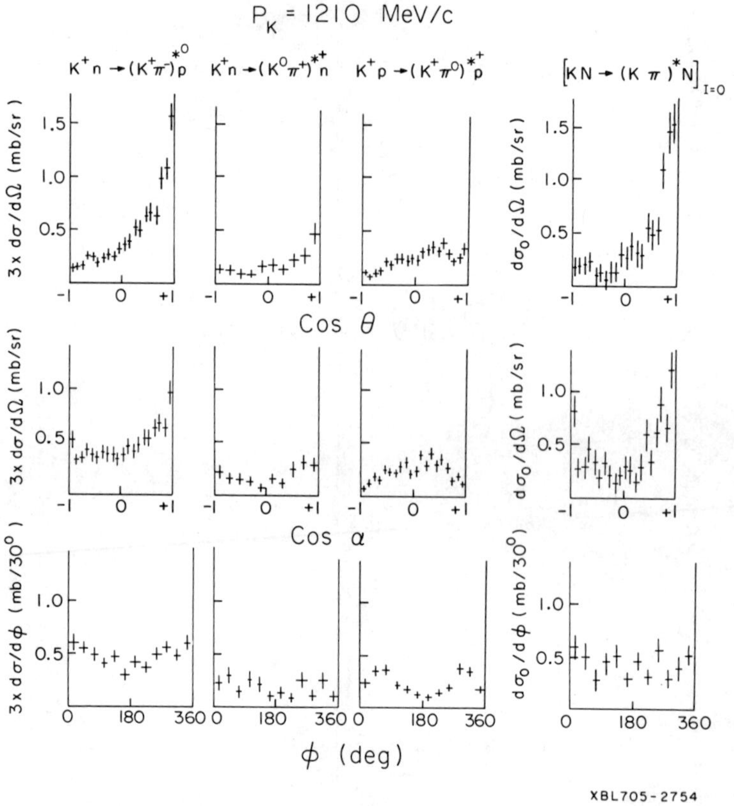

Fig. 10. Construction of the corresponding angular distributions.

IV. THE QUESTION ABOUT THE EXISTENCE OF A Z_0^*

In 1968 we showed our data for the $I = 0$ inelastic cross section (Hirata et al.[15]; see Fig. 11). If we compare this with the $I = 0$ total cross section deduced at that time by Carter from a linear combination of the results of Cool et al. and Bugg et al. which were available then (the Jenkins data was not yet available) then we found that subtracting the inelastic from the total you get a rather sharply falling elastic cross section in the $I = 0$ state. There is one other interesting point which was known from earlier work in collaboration with Stenger et al.[16] at UCLA and that is that the $I = 0$ scattering length at zero energy is very small; essentially it was deduced to be 0.04±0.04 Fermis. This means that the S wave cross section at zero momentum must be dropping down very rapidly.

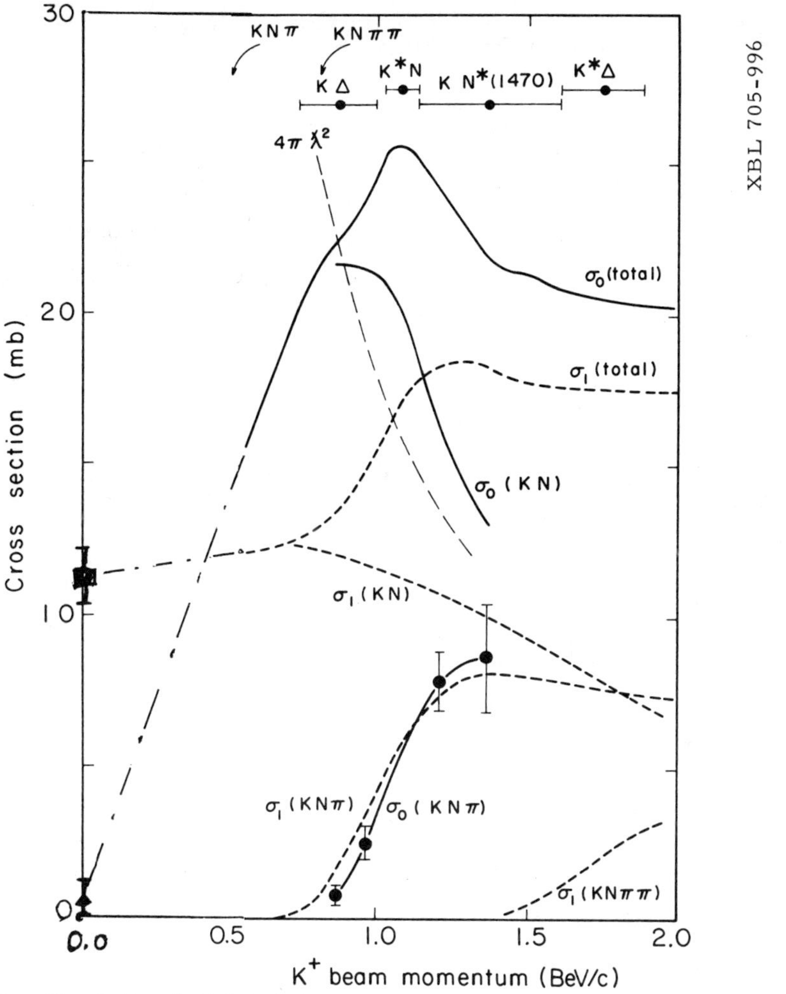

Fig. 11. The I=0 and I=1 total inelastic and elastic cross sections. The I=0 cross section was obtained by Carter as an interpolation between the data of Cool et al. and Bugg et al. This curve is shown extrapolated to zero energy to the scattering length value deduced by Stenger et al.▲ The I=1 cross section is similarly extrapolated to the scattering length value of S. Goldhaber et al.■

We pointed this feature out, and point out a distinction between the $I = 0$ and $I = 1$ cross sections: the $I = 0$ cross section drops off rather sharply, while the $I = 1$ cross section, which comes from the data I showed earlier remains more or less constant at threshold. The threshold limits for $I = 0$ and $I = 1$ are deduced as ≈ 0.2 mb and ≈ 11 mb respectively. This, of course, means that for the $I = 0$

state there occurs a rather rapid rise essentially from 0 up to some high value and then the elastic cross section drops off again. In fact with the numbers we had then the elastic cross section rises to $\approx 4\pi\lambda\!\!\!^{-2}$. The Brookhaven group (Abrams and co-workers[17]) analyzed this same data and made the point that this rapid rise together with the fall of the elastic cross section is a very striking feature and could be a Z_0^*. Figure 12 shows the most recent analysis by the Brookhaven group which I obtained from Dr. Cool (it includes the results of the University of Arizona experiment at the Bevatron by Jenkins et al.). This figure was already shown by Dr. Cool in his talk. There is at present some confusion as to exactly how the

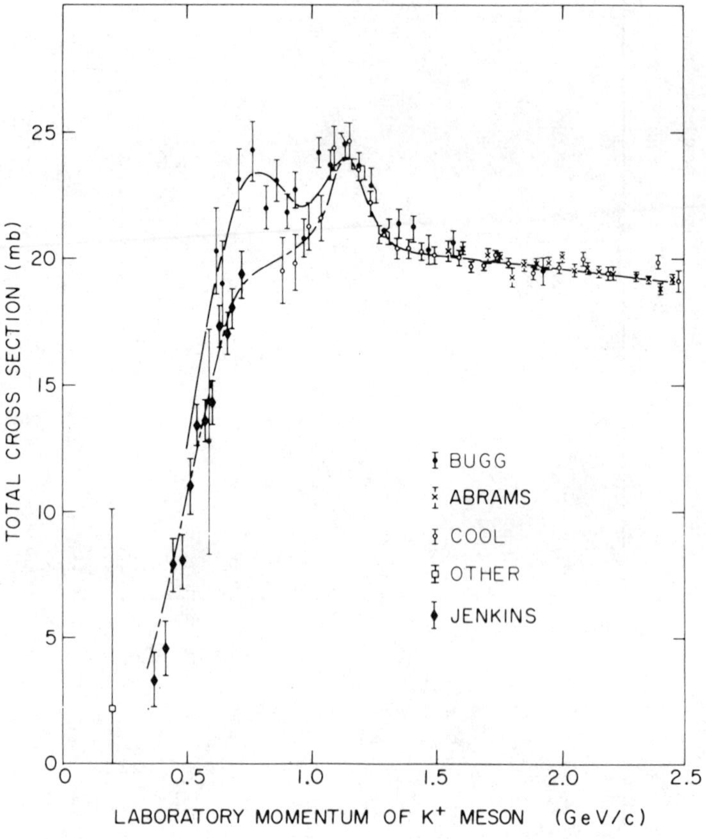

Fig. 12. The I=0 total cross section evaluated by Cool et al. including the new data by Jenkins et al.

curve goes through the points and whether we are dealing with one or two peaks, but there is no doubt that there is a rise from zero through a peak near $4\pi\lambda\!\!\!^{-2}$ followed by a drop off of the elastic cross section. Figure 13 gives two possible alternative interpretations of the I = 0 elastic cross section depending on just how you treat

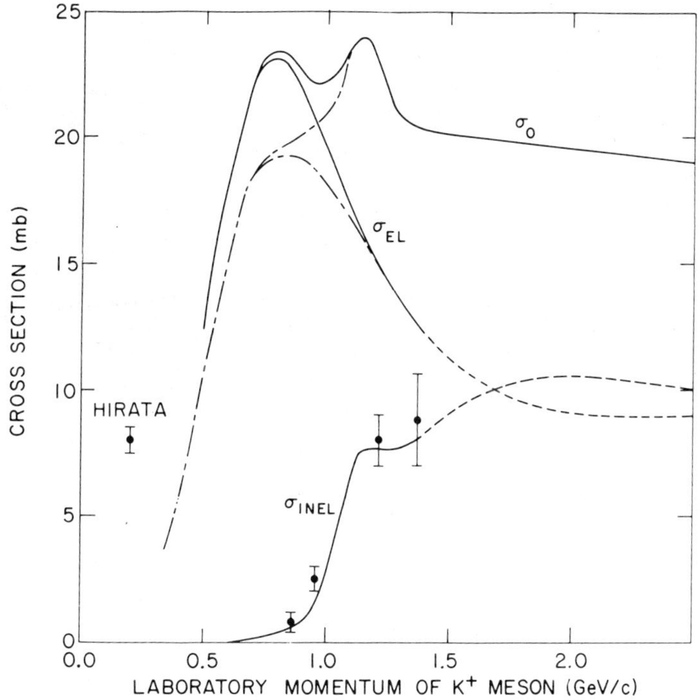

XBL 705-1000

Fig. 13. The resulting I=0 elastic cross section showing 2 possible interpretations of the data indicative of the current uncertainties in the 0.8-1.2 GeV/c region.

the data in the unfolding procedure and Dr. Lynch discussed this in his talk at this Conference. I won't go into this, but one of these curves should be right and for the present a decision between them is not that critical. Finally, Fig. 14 is a contribution by Dowell[18] to the Conference which again emphasizes the features of the Jenkins and Bugg data; he independently points out that this very large I = 0 peak occurs. I think the main difference in the analyses are the approximately 3 millibarns difference in how high the peak goes and the subsequent question as to whether this should be considered as two structures or a single structure.

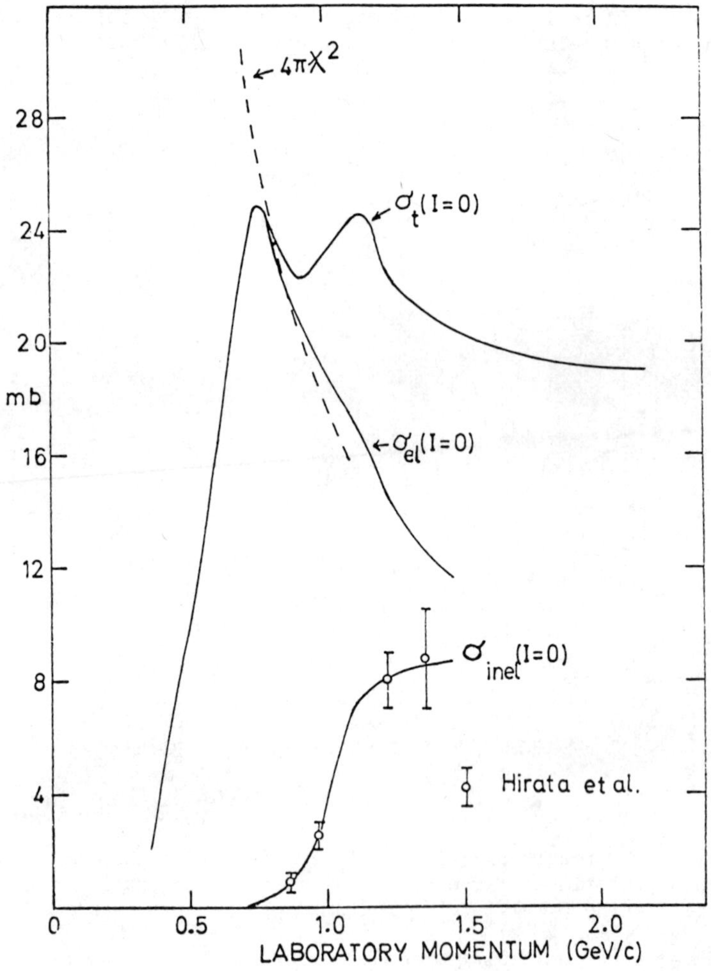

XBL 705-998

Fig. 14. Similar results deduced by J. Dowell.

V. SUMMARY ON EXISTENCE OF Z^*'S

This leads to the final point, what can we say as a summary to all the discussion we have heard?

(1) Is there a Z^* (I = 1) at the Cool bump? I think what we have learned is that _if_ the P_{13} curve is indeed a resonant circle on the Argand diagram then the answer is yes, there is a resonance there and its spin and parity would be $3/2^+$ with a mass somewhere in the region of 1900 MeV and width not yet determined. There is however the other possibility that there is a phase shift solution which is not resonant, namely that all solutions are heading for the center of the Argand diagram which would be the onset of diffraction processes. I feel that at the moment both possibilities still exist.

(2) The second question is does the Z_0^* exist and the answer is: what we know at the moment is that there is a large peak in the elastic cross section, $M(KN) \cong 1700$ MeV, $\Gamma \cong 500$ MeV; however, we need a phase shift analysis in order to be sure that this is a resonance. It has been suggested as a resonance by Abrams et al.[17] and also by John Dowell.[18] My feeling is that we will have to wait until we can do a phase shift analysis in the I = 0 system to be sure. We have done some phase shift analysis at lower momenta[16] and they show that the one wave which is large and positive is the P_1. This solution is favored by polarization measurements on the charge exchange reaction $K^+d \rightarrow K^0pp$.[19] If there is a resonance, it probably will be a $J = 1/2^+$ (a P_1 state).

Now why are we so cagey on this question since we have very readily accepted many, many resonances in the Y^*'s? The reason is really related to the point of view that the Z^* is a different animal from the Y^*; namely, in terms of the quark model in the case of the Z^* we need more than the usual qqq structure characteristic of a Y^*. Here we are dealing with an exotic structure such as $qqqq\bar{q}$. We could ask if three quarks bind why not five? But then we have to ask if five why not seven, nine, and finally why not 137? The main point to realize is that the Z^* _is_ a completely different structure. Although we have established the qqq structure, there is a need to establish this new one as well. That is why one has to be more rigorous in looking at all these questions than if we were just examining one additional Y^*. Now what I would like to pose as a question to the theorists here is the following: we do know that there is a positive P_{13} phase shift. Thus we are dealing with an attractive force. The question I have is, in what sense is the world different (a) if you have just a positive force which does not lead to resonance or (b) you have a positive force which is strong enough to lead to a resonance? In other words, what other features of our world would be different for these two cases. The positive force seems to be there in any case. The phenomenon has the qualitative features of an attractive force but what we are arguing about is the question of whether there is actually a resonance.

A second question is what should one do experimentally to really pin this down completely. We had a discussion between the members of the group giving the various talks in this session today. We thought that if very precise differential cross section and polarization measurements could be made at one intermediate momentum at least it would help in removing ambiguous solutions. Secondly, if this exotic resonance exists, and assuming the SU(3) picture applies here, this is only one out of 27; one should look for the other 26! Thirdly, there is the K^-p backwards scattering experiment which I believe is being done by the Stony Brook group at Brookhaven and also by various groups at CERN. If you have K^--proton backward scattering, the exchange particle would have to be a Z_1^*. As far as I know at present most differential cross section measurements either have only a limit in the backward direction for K^-p scattering or it is more or less flat. If one could measure $d\sigma/du$ as a function of u (the very backward scattering) and if one really observes the distinct characteristics of an exchange particle at sufficiently high energy so that one is out of the s channel resonance regions, I think that would be a very excellent demonstration of Z^* exchange. So far, to my knowledge, none of the exotic resonances has been observed in exchange processes at high energy.

REFERENCES

1. M. Gell-Mann and Y. Ne'eman, The Eightfold Way (W. A. Benjamin Inc., New York, 1964), p. 85-90.
2. R. L. Cool, G. Giacomelli, T. F. Kycia, B. A. Leontic, K. K. Li, A. Lundby, and J. Teiger, Phys. Rev. Letters 17, 102 (1966).
3. D. V. Bugg, R. S. Gilmore, K. M. Knight, D. C. Salter, G. H. Stafford, E. J. N. Wilson, J. D. Davies, J. D. Dowell, P. M. Hattersley, R. J. Homer, A. W. O'Dell, A. A. Carter, R. J. Tapper and K. F. Riley, Phys. Rev. 168, 1466 (1968).
4. R. W. Bland, M. G. Bowler, J. L. Brown, G. Goldhaber, S. Goldhaber J. A. Kadyk, V. H. Seeger, and G. H. Trilling, Nuclear Physics B13, 595 (1969) and B18, 537 (1970).
5. R. W. Bland, G. Goldhaber, and G. H. Trilling, Phys. Letters 29B, 618 (1969).
6. R. J. Abrams, R. L. Cool, G. Giacomelli, T. F. Kycia, B. A. Leonti K. K. Li, and D. N. Michael, Phys. Rev. Letters 19, 259 (1967).
7. J. Tyson, J. S. Greenberg, V. W. Hughes, D. C. Lu, R. C. Minehart, S. Mori, and J. E. Rothberg, Phys. Rev. Letters 19, 255 (1967).
8. S. Andersson, C. Daum, F. C. Erne, J. P. Lagnaux, J. C. Sens, F. Udo, and F. Wagner, Phys. Letters 30B, 56 (1969); J. G. Asbury, J. D. Dowell, S. Kato, D. Lundquist, T. B. Novey, A. Yokosawa, B. Barnett, P. F. M. Koehler, and P. Steinberg, Phys. Rev. Letters 23, 194 (1969); S. Kato, P. Koehler, T. Novey, A. Yokosawa,

and G. Burleson, ANL/HEP 6926, to be published; B. Barnett, J. Goldman, A. Laasanen, and P. H. Steinberg, Partial Wave Analysis of K^+p Elastic Scattering, Contribution to this Conference. See also the new results presented by V. Hughes at this Conference.

9. P. C. Barber, T. A. Broome, B. G. Duff, F. F. Heymann, D. C. Imrie, G. J. Lush, B. R. Martin, K. M. Potter, D. M. Ritson, L. A. Robbins, R. A. Rosner, S. J. Sharrock, A. D. Smith, R. C. Hanna, A. T. Lea and E. J. Sacharides, K^+p Elastic Scattering Between 1.4 and 2.3 GeV/c, Contribution to this Conference.
10. B. Carreras and A. Donnachie, DNPL/P 26, submitted to Nuclear Physics B.
11. A. T. Lea, B. R. Martin and G. C. Oades, Phys. Rev. 165, 1770 (1968).
12. B. Carreras, A. Donnachie, and R. G. Kirsopp, DNPL/P 31, Contribution to this Conference.
13. S. Goldhaber, W. Chinowsky, G. Goldhaber, W. Lee, T. O'Halloran, T. F. Stubbs, G. M. Pjerrou, D. H. Stork and H. K. Ticho, Phys. Rev. Letters 9, 135 (1962).
14. A. A. Hirata, G. Goldhaber, V. H. Seeger, G. H. Trilling, and C. G. Wohl, The Reaction $K^+N \rightarrow K^*N$ in the Isospin-0 Channel Near Threshold, UCRL-19774, Contribution to this Conference.
15. A. A. Hirata, C. G. Wohl, G. Goldhaber, and G. H. Trilling, Phys. Rev. Letters 21, 1485 (1968).
16. V. J. Stenger, W. E. Slater, D. H. Stork, H. K. Ticho, G. Goldhaber, and S. Goldhaber, Phys. Rev. 134, B1111 (1964).
17. R. J. Abrams, R. L. Cool, G. Giacomelli, T. F. Kycia, K. K. Li, and D. N. Michael, Phys. Letters 30B, 564 (1969).
18. J. D. Dowell, Contribution to this Conference.
19. A. A. Hirata, C. G. Wohl, R. W. Bland, G. Goldhaber, B. H. Hall, J. A. Kadyk, V. H. Seeger, and G. H. Trilling, K^+d Interactions Near 1 BeV/c, Contribution to XIVth International Conference on High-Energy Physics, Vienna, 1968, UCRL-18322; A. K. Ray, R. W. Burris, H. E. Fisk, R. W. Kraemer, D. G. Hill, and M. Sakitt, Phys. Rev. 183, 1183 (1969).

DISCUSSION

Kim: I would just like to make a comment. In the K^-p system, for instance, if we look at the Legendre coefficients for the elastic scattering, say between 600 and 800 MeV it's almost monotonically increasing like K^+p, yet there are so many inelastic resonances. So just looking at the elastic Legendre polynomial, I think that one doesn't gain very much if the elasticity is small. Now you have never mentioned there the measuring for instance of the inelastic cross sections accurately. You have the experience, so I would like to get your comment.

Goldhaber: We have done extensive measurements on the inelastic cross sections and the difficulty you run into is the following -- you have a Dalitz plot where you cross $K\pi$ mass squared against $p\pi$, for instance where you have $K^0\pi^+$ proton. Below the K^* threshold, you just have $p\pi^+$, a delta production and that can be analyzed, but then when you get above K^* threshold you have both delta and K^* production. And what's more they decide to interfere quite violently and we have not been able to disentangle this thing and get a unique phase shift analysis from the inelastic results.

Derrick: Did the Donnachie analysis that didn't show the $P_{3/2}$ going around deal with just the elastic scattering; did it include the polarization data?

Goldhaber: It included all polarization data even up to 6 GeV and particularly emphasizing high momentum data.

Derrick: What I was going to say was that perhaps one of the first things to do would be to have someone analyze all the available data that we have heard about.

Goldhaber: I believe that Wagner has analyzed all that was available to him.

Barnett: I have several comments with some pictures which I can't show but could possibly draw if I come up front.

First of all there was a question about the coefficient analysis at lower energies. We have done this analysis and I have figures of the results here with me if anyone is interested. In general, however, the A's of the differential cross section expansions do not show anything indicating a resonance. For instance, they should show a bump, or you might expect some sort of bump, if there were a resonance. None is observed although the error bars are large enough that it could be obscured if it were there. The B's, which are the coefficients of the expansion of the differential cross sections are polarizations, should show an inflection point at the resonance if the non-resonant background is slowly varying, which may not be true here because of the several thresholds. I didn't have any of the data from University College when these curves were drawn so we have good results only up to 1.45 GeV/c. Above that there was no momentum where there were both differential cross section and polarization measurements until you get up to about 1.9 GeV/c, where we combined our polarization data at 1.89 with differential cross section data at 1.95. This data point did not fall on a smooth continuation of the B_1 curve, which might indicate something, but this may change now that better data has come out. None of the other B coefficients showed anything striking.

Second, in our phase shift analysis at Maryland we have included all the published data which were available, whereas the other analyses have used only a partial list of data. When we combined our solutions from different momenta we used a least distance requirement on the square of the speed of the amplitudes because we noticed that the speeds of the partial wave amplitudes in the previous analyses at CERN and Argonne were very erratic. The speed of a resonant amplitude without a background should be proportional to the imaginary part of the amplitude. The actual relationship is $s = (2/\Gamma)\ \mathrm{Im}\ (f_{\ell,J})$. The P_3 amplitudes from our analyses have speeds which maximize near the peak in the Argand plot, and, in general, obey the above expression. Since the speeds of the P_3 amplitudes in the CERN and Argonne analyses around Christmas time did not peak near the resonant energy (they tended to be larger in the 1.45 to 1.9 GeV/c region than in the resonant region), I would like to ask if their P_3 speeds look more resonant-like and less erratic now?

My final comment is related to the suggestion made earlier by Dr. Goldhaber that K^-p backward scattering should be studied for resonance effects. Some work was done by Carroll et al. in which they found that $d\sigma/du$ $(u = 0)$ had approximately a s^{-10} dependence, which means that it is decreasing more rapidly than any other known backward cross section at comparable energies. They successfully fit their data with a direct channel resonance model and also with a Regge-pole (Z_1^*) exchange model. In the latter case they found $\alpha = -3.73 + 1.1u$, and a Chew-Frautschi plot of this trajectory shows that it passes very near to the proposed P_3 resonance at 1.9 GeV. That is all.

Tuan: What is the mass of your Z^*?

Goldhaber: Cool mentioned this yesterday. I think it may be 1760 and very wide, about 500 MeV.

Garfinkel: I am wondering if Donnachie couldn't tell whether the s-wave was attractive or repulsive. How can you tell in the p-waves without the Coulomb interference?

Goldhaber: I think all phase shift analysis which have been performed to my knowledge have assumed this negative s-wave and that sets the sign for the polarization of everything else of all the others. So none of the others is determined directly with Coulomb interference. Now whether or not one can do a Regge fit with a negative s-wave I don't know.

O'Halloran: When you showed your I=0 K^* nucleon cross section, the cosine α distribution was not symmetric for the K^*. Is there another effect coming in there?

Goldhaber: I don't know.

Kim: When you change the sign of the s-wave amplitude, the real part, don't you have to change the p-wave too?

Goldhaber: Yes.

THE REACTION $K^+N \rightarrow K^*N$ IN THE ISOSPIN-0 CHANNEL NEAR THRESHOLD‡

A. A. Hirata, G. Goldhaber, V. H. Seeger, G. H. Trilling, and C. G. Wohl*

Department of Physics and Lawrence Radiation Laboratory

University of California, Berkeley, California 94720

In an earlier Letter,[1] partial cross sections for K^+d and isospin-0 KN reactions at K^+ momenta of 865, 970, 1210 and 1365 MeV/c were presented. In this paper we present a discussion of the isospin-0 $K^+N \rightarrow K\pi N$ channel based on further analysis of K^+d data at 1210 and 1365 MeV/c plus recently analyzed data at 1585 MeV/c from the same exposure of the 25-inch bubble chamber at the Bevatron. Our results show that above the K^*N threshold at 1080 MeV/c most of the $I = 0$ $K\pi N$ production proceeds via the K^*N channel. The K^* production and decay angular distributions indicate the presence of several partial waves even near threshold and suggest that the main production mechanism is t-channel pion exchange. Thus there is no apparent support for an s-channel resonance interpretation of the peak in the $I = 0$ KN total cross section around 1200 MeV/c.[2]

(1) <u>Methods</u>.--(a) As shown earlier the isospin-0 $KN \rightarrow K\pi N$ cross section is given by

$$\sigma_0(KN \rightarrow K\pi N) = 3[\sigma(K^+n \rightarrow K^0\pi^+n) + \sigma(K^+n \rightarrow K^+\pi^-p) - \sigma(K^+p \rightarrow K^+\pi^0p)] \quad . \qquad (1)$$

Since the same relation applies to angular distributions and invariant-mass spectra, the K^*N cross section could be obtained by making the appropriate fit to the $K\pi N$ Dalitz plot obtained from Eq. (1).

(b) Alternatively, the K^*N cross section can also be determined from

$$\sigma_0(KN \rightarrow K^*N) = 3[\sigma(K^+n \rightarrow K^{*+}n \rightarrow K^0\pi^+n) + \sigma(K^+n \rightarrow K^{*0}p \rightarrow K^+\pi^-p) - \tfrac{1}{2}\,\sigma(K^+p \rightarrow K^{*+}p \rightarrow K^0\pi^+p)] \quad , \qquad (2)$$

in which each appropriate deuterium channel is individually separated into K^*N and the $K^0\pi^+p$ hydrogen channel rather than the $K^+\pi^0p$ state is used to subtract out the $I = 1$ $K^{*+}N$ part of the cross section.

(2) Results.--In Fig. 1 we give the energy dependence of the isospin-0 and isospin-1 $KN \rightarrow K^*N$ cross sections from threshold to 3.0 BeV/c. The isospin-1 cross sections at 1210, 1365 and 1585 MeV/c are taken from the work of Bland et al.[3] Higher momentum data are taken from the compilation of Price et al.[4] The isospin-0 cross sections have been calculated using Eq. (2). At 1210, 1365, and 1585 MeV/c we have used the $K^+p \rightarrow K^{*+}p \rightarrow K^0\pi^+p$ data of Bland et al.[3] and at 3.0 BeV/c we have used the published K^+d data of Bassompierre et al.[5] to calculate the isospin-0 cross section. A smooth curve has been drawn through the isospin-1 $KN \rightarrow K^*N$ cross sections and has been scaled upward by a factor 2.5 to give the smooth curve passing through the isospin-0 $KN \rightarrow K^*N$ cross sections. The data indicate that the energy dependency of the isospin-0 and isospin-1 $KN \rightarrow K^*N$ cross sections are quite similar in this energy region.

In Fig. 2 we give the production and decay angular distributions for the isospin-0 $KN \rightarrow K^*N$ reaction at 1210, 1365, and 1585 MeV/c. For the 1210 and 1365 MeV/c data procedure (a) was used, whereas at 1585 MeV/c, the availability of only the $K^0\pi^+p$ final state in hydrogen permitted only procedure (b). Cosine θ is the K^* production angle cosine $\hat{K}^+ \cdot \hat{K}^*$ in the K^*N center of mass, cos α is the polar-decay-angle cosine $\hat{K}^+ \cdot \hat{K}$ in the K^* c.m., and angle φ is the Treiman-Yang azimuthal decay angle.

The K^* angular distributions in Fig. 2 are strikingly different from those produced in the isospin-1 state.[6] In the isospin-0 channel, the production angular distributions peak sharply in the forward direction, the polar decay angular distributions vary mainly as $\cos^2\alpha$, and the azimuthal decay angular distributions are flat. In the isospin-1 channel (Ref. 6), the production distributions peak much less sharply, the polar decay distributions vary mainly as $\sin^2\alpha$, and the azimuthal decay distributions vary mainly as $\sin^2\varphi$. In the first case, the distributions are characteristic of exchange in the t-channel of a light pseudoscalar meson, whereas in the second case they are characteristic of exchange of a heavier vector meson. All the main features point to π exchange being the dominant K^* production mechanism in the isospin-0 channel, and ω, ρ exchange being dominant in the isospin-1 channel.

As is the case with $K\Delta$ and K^*N production in the $I = 1$ channel, the major inelastic process in the $I = 0$ channel in the region of the total cross-section bump, namely K^*N production, exhibits behavior which can be interpreted in terms of t-channel exchanges. Furthermore as is evident from Fig. 2 this process is highly peripheral even very close to the bump and therefore does not suggest dominance of one partial wave. There is therefore no evidence of resonant behavior near 1200 MeV/c at least insofar as the inelastic channels are concerned.

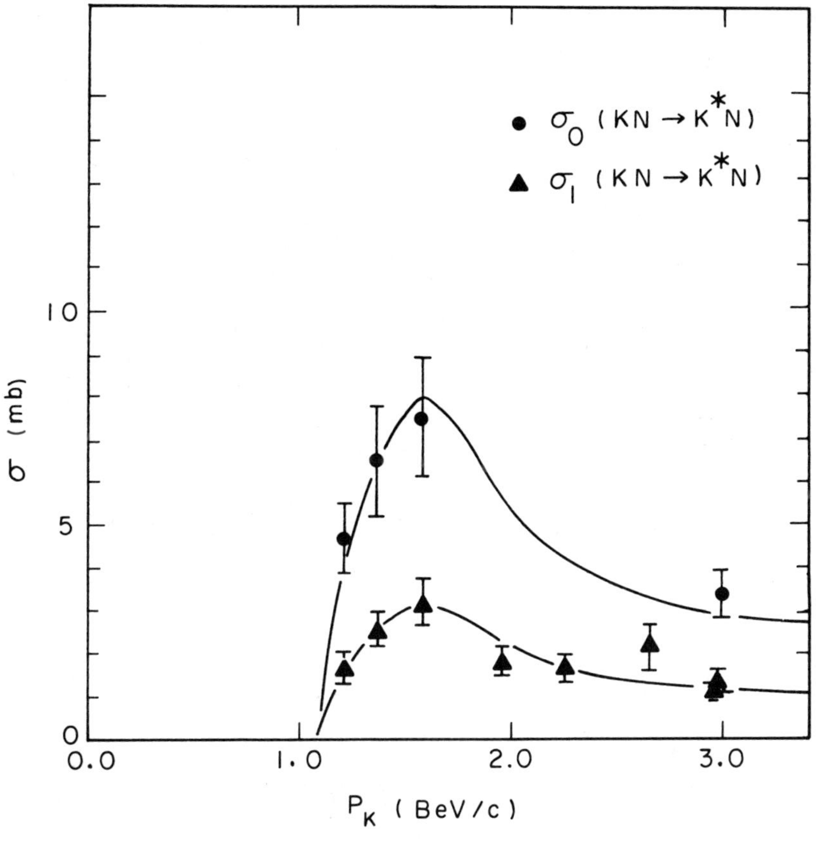

Fig. 1. Energy dependence of the isospin-0 and isospin-1 KN → K^*N cross sections.

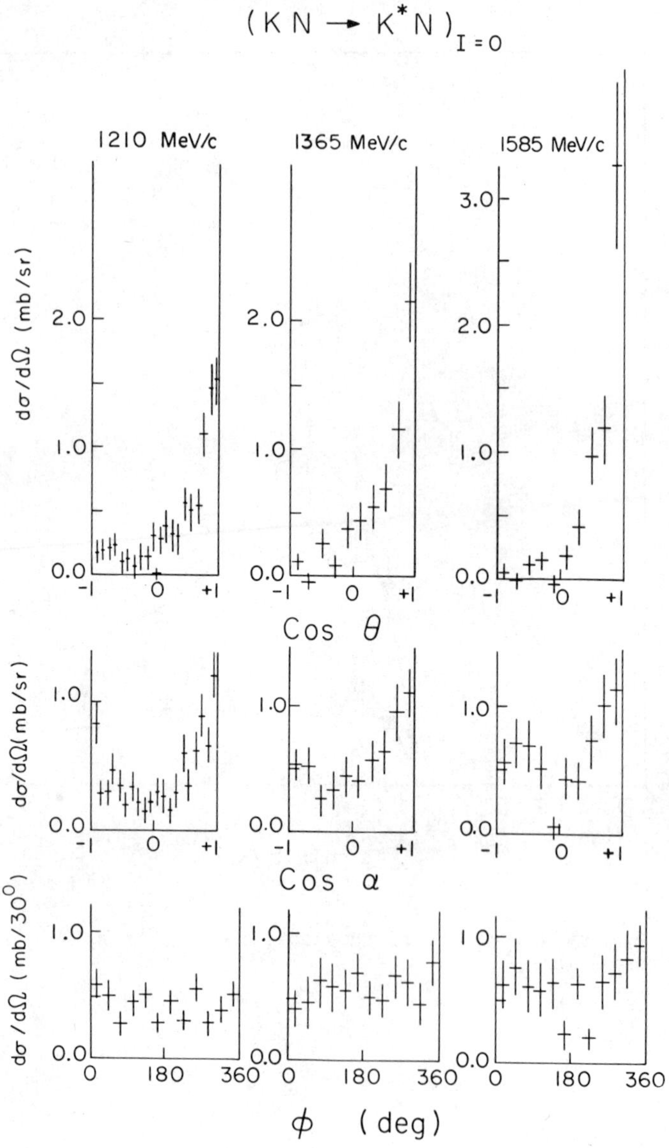

Fig. 2. Production and decay angular distributions of the isospin-0 $KN \rightarrow K^*N$ reaction at 1210, 1365, and 1585 MeV/c. Angles are defined in text.

We gratefully acknowledge the contributions made to this experiment by the Bevatron and bubble-chamber crews, by the F.S.D. staff under the direction of Howard White, and by our scanning and programming groups. We especially thank Dr. Roger W. Bland for data and numerous discussions.

REFERENCES

*Present address: Nuclear Physics Laboratory, Oxford, England.
‡Work supported by the U. S. Atomic Energy Commission.
1. A. A. Hirata, C. G. Wohl, G. Goldhaber, and G. H. Trilling, Phys. Rev. Letters 21, 1485 (1968), Errata, Phys. Rev. Letters 21, 1728 (1968).
2. R. L. Cool, G. Giacomelli, T. F. Kycia, B. A. Leontic, K. K. Li, A. Lundby, and J. Teiger, Phys. Rev. Letters 17, 102 (1966); D. V. Bugg et al., Phys. Rev. 168, 1466 (1968).
3. R. W. Bland, M. G. Bowler, J. L. Brown, J. A. Kadyk, G. Goldhaber, S. Goldhaber, V. H. Seeger, and G. H. Trilling, Nucl. Physics B13, 595 (1969).
4. L. R. Price, N. Barash-Schmidt, O. Benary, R. W. Bland, A. H. Rosenfeld, and C. G. Wohl, UCRL-20000 K^+N, September 1969.
5. G. Bassompierre et al., Nucl. Physics B16, 125 (1970).
6. R. W. Bland, M. G. Bowler, J. L. Brown, G. Goldhaber, S. Goldhaber, J. A. Kadyk, V. H. Seeger, and G. H. Trilling, Nucl. Physics B17 (in press).

K^+p PARTIAL WAVE ANALYSIS AT 860 TO 1210 MeV/c[‡]

B. H. Hall, R. W. Bland,[*] G. Goldhaber, and
G. H. Trilling

Department of Physics and Lawrence Radiation Laboratory

University of California, Berkeley, California 94720

In this note we report revised results of a partial wave analysis of K^+p elastic scattering at 860, 970, and 1210 MeV/c,[1] based on new polarization data[2] as well as the previously used differential cross section data.[3] We have also included the partial inelastic cross sections as determined by Bland et al. from an analysis of the $K\Delta$ final state at the two lower momenta.[4] The values of these cross sections for the K^+p initial state are shown in Table I. At 1210 MeV/c we were unable to use this analysis due to the additional presence of the K^*N final state and here we have used only the total inelastic cross section. The real parts of the forward scattering amplitude as calculated by Carter[5] were also used; this effectively fixes the s wave to be negative, as determined at lower momenta.[6]

The solutions we have obtained at 865 and 970 MeV/c are shown in Table I. Inclusion of the new polarization data eliminated our previous B^- type solution and slightly decreased the errors on the phase shifts of the remaining solution (A^- type). We feel that this solution is the only one consistent with both the elastic data and the partial wave makeup of the inelastic cross section as determined by Bland et al.

In Table II we show solutions obtained at 1207 MeV/c. The solutions fall into two classes, the fits equally good for each: the A^- type, continuous with the lower momenta, and the C^- type, distinguished by a negative $P_{3/2}$ wave.

Presently existing data, when fit with a Legendre series, do not seem to require the inclusion of f waves. However, SPDF solutions give partial wave inelasticities in better agreement with the results of the KΔ final state analysis of Bland et al., i.e., a fairly inelastic $P_{3/2}$ wave and an almost elastic $S_{1/2}$ wave. For this reason we show both our SPD and SPDF solutions. Figure 1 shows an Argand plot of the solutions at 865 and 970 MeV/c and of our preferred solution at 1210 MeV/c, the A^- SPDF solution.

REFERENCES

*Present address: Ecole Polytechnique, Paris, France.

‡Work supported by the U. S. Atomic Energy Commission.

1. B. H. Hall, R. W. Bland, G. Goldhaber, and G. H. Trilling, Lawrence Radiation Laboratory Report UCRL-19231 (1969).
2. S. Andersson, C. Daum, F. C. Erne, J. P. Lagnaux, J. C. Sens, F. Udo, and F. Wagner, Contributed paper to Lund International Conference on Elementary Particles, 1969.
3. R. W. Bland, G. Goldhaber, and G. H. Trilling, Physics Letters 29B, 618 (1969).
4. R. W. Bland, et al., Nucl. Physics B17 (in press). When fitting, errors in the partial cross sections were increased to 25% to account for possible uncertainties in the theory.
5. A. A. Carter, The Argand Diagrams of the KN and $\overline{K}N$ Forward-Scattering Amplitudes, Cavendish Laboratory Report HEP 68-10, March 1968.
6. S. Goldhaber, W. Chinowsky, G. Goldhaber, W. Lee, T. O'Halloran, T. F. Stubbs, G. M. Pjerrou, D. H. Stork, and H. K. Ticho, Phys. Rev. Letters 9, 135 (1962).

Table I. Phase shift solutions at 865 and 970 MeV/c.

		865 MeV/c	970 MeV/c	Partial inelastic cross sections, input values from Ref. 4 in millibarns 865 MeV/c	970 MeV/c
$S_{1/2}$	δ	-38.8±2.4	-40.5±2.6	--	--
	η	$0.987^{+0.013}_{-0.015}$	$0.996^{+0.004}_{-0.045}$		
$P_{1/2}$	δ	-13.5±2.7	-15.3±3.5	0.96±0.11	1.7±0.2
	η	0.915±0.014	0.833±0.026		
$P_{3/2}$	δ	11.9±1.4	15.3±1.4	0.57±0.07	1.5±0.15
	η	0.974±0.006	0.903±0.015		
$D_{3/2}$	δ	-3.7±0.5	-3.9±0.7	0.05±0.01	0.21±0.06
	η	0.998±0.001	0.988±0.005		
$D_{5/2}$	δ	-2.1±0.8	-1.2±1.1	0.05±0.01	0.23±0.07
	η	0.9985±0.0004	0.992±0.003		
Re(kf) in cm		-0.54	-0.44		
$\sigma_{inelastic}$ (mb)		1.72	3.65		
σ_{total} (mb)		13.64	15.36		
χ^2		38.5	38.7		
Degrees of freedom		38	36		
Probability		0.45	0.35		

Table II. Phase shift solutions at 1210 MeV/c.

Solution type		A^-	A^-	C^-	A^-	C^-
$S_{1/2}$	δ	-72.5±6.2	-59.9±7.9	-29.7±0.8	-54.0±16.2	-32.4±4.0
	η	0.68±0.07	0.67±0.22	0.90±0.04	$0.92^{+0.08}_{-0.54}$	0.84±0.15
$P_{1/2}$	δ	-13.1±2.7	-28.4±6.4	-2.0±0.9	-17.7±6.8	-9.1±3.3
	η	$1.0^{+0.0}_{-0.03}$	0.90±0.08	$1.0^{+0.0}_{-0.62}$	$0.95^{+0.05}_{-0.37}$	$0.98^{+0.02}_{-0.14}$
$P_{3/2}$	δ	2.5±2.7	2.8±3.8	-28.6±0.9	4.2±8.4	-28.1±8.8
	η	0.70±0.05	$0.94^{+0.06}_{-0.12}$	0.82±0.04	0.78±0.16	0.64±0.04
$D_{3/2}$	δ	-8.9±1.2	-7.0±2.6	-4.2±0.6	-10.0±2.7	-7.8±2.1
	η	$1.0^{+0.0}_{-0.03}$	$0.98^{+0.02}_{-0.06}$	$1.0^{+0.0}_{-0.06}$	$1.0^{+0.0}_{-0.10}$	0.92±0.02
$D_{5/2}$	δ	-1.8±1.1	1.6±2.0	13.7±0.5	0.6±3.1	12.7±2.7
	η	0.89±0.02	0.79±0.02	0.74±0.04	$0.94^{+0.06}_{-0.18}$	$1.0^{+0.0}_{-0.12}$
$F_{5/2}$	δ				0.7±2.9	-1.3±0.6
	η				$0.95^{+0.05}_{-0.19}$	0.94±0.03
$F_{7/2}$	δ				2.3±1.1	0.1±2.4
	η				0.93±0.04	$1.0^{+0.0}_{-0.51}$
Re(kf) in cm		-0.73	-0.75	-0.75	-0.73	-0.72
$\sigma_{inelastic}$ (mb)		7.45	7.42	7.42	7.42	7.41
σ_{total} (mb)		18.32	18.32	18.32	18.32	18.32
χ^2		33.7	29.5	29.4	24.4	25.3
Degrees of freedom		32	32	32	28	28
Probability		0.39	0.60	0.61	0.66	0.61

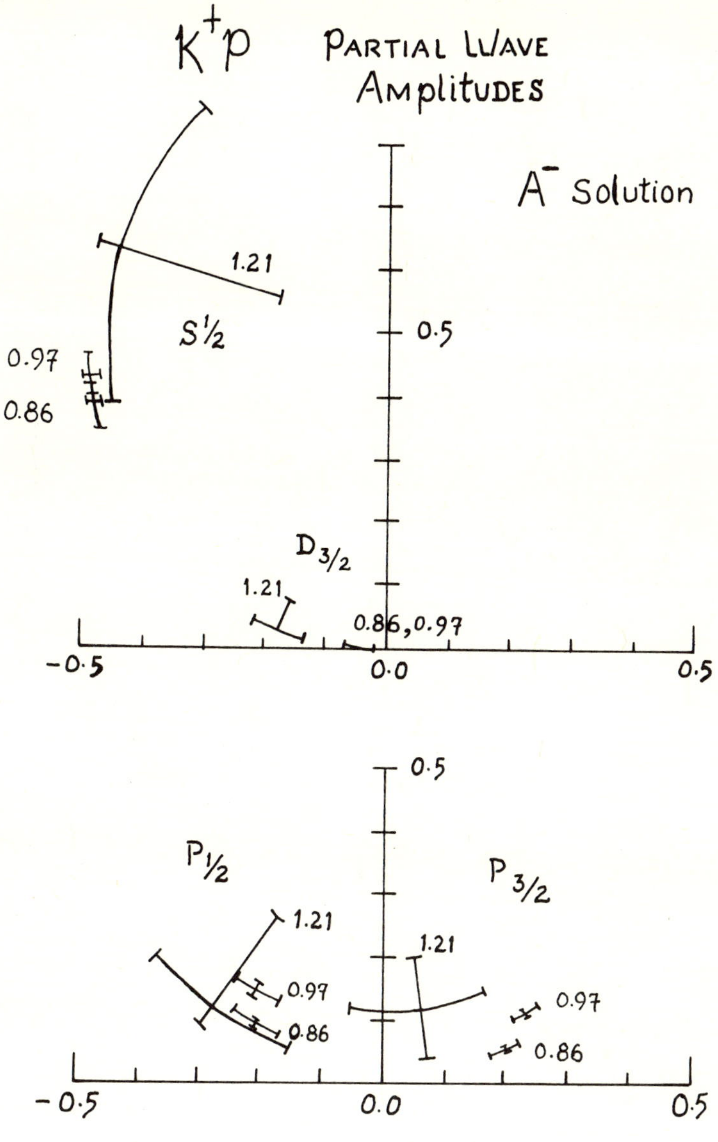

Fig. 1. Argand diagram for K^+p partial wave amplitudes at 865, 970 and 1210 MeV/c.

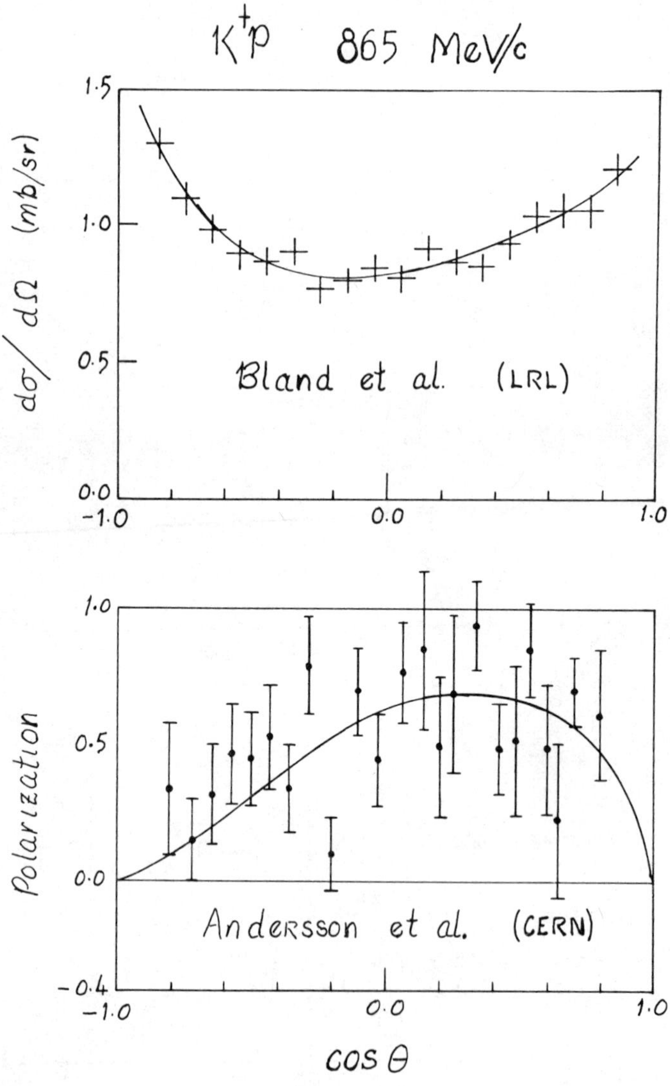

Fig. 2a. Differential cross section and polarization data. The curves correspond to the best fits.

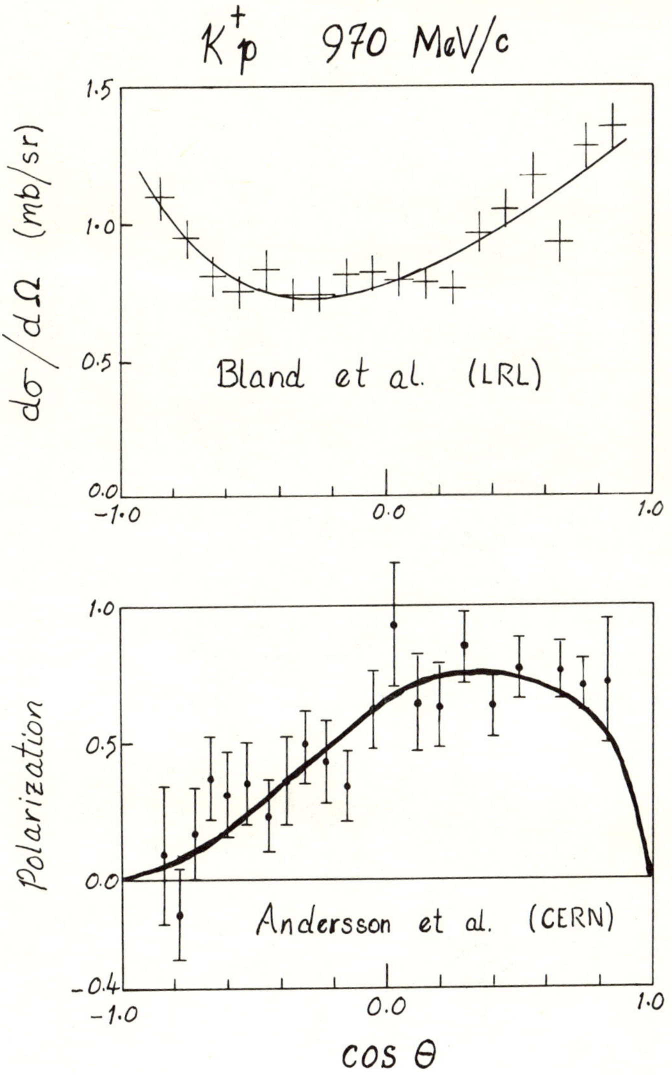

Fig. 2b. Differential cross section and polarization data. The curves correspond to the best fits.

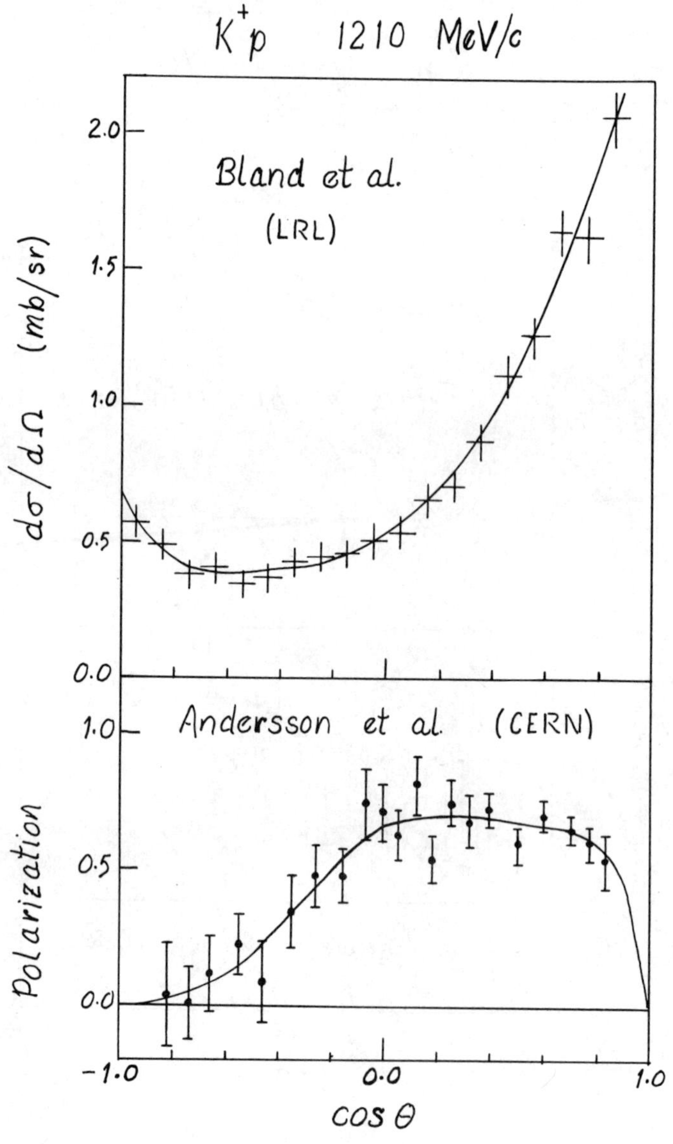

Fig. 2c. Differential cross section and polarization data. The curves correspond to the best fits.

PARTIAL WAVE ANALYSIS OF K^+p ELASTIC SCATTERING

B. Barnett, J. Goldman, A. Laasanen, and P.H. Steinberg

University of Maryland

Reported here are some of the results of a partial wave analysis of elastic K^+p scattering in the vicinity of the bump in the total cross section at 1.2 GeV/c. [1,2] The analysis is carried out at momenta; 0.78, 0.87, 0.97, 1.09, 1.17, 1.21, 1.37, and 1.45 GeV/c, and makes use of all cross section and polarization data available at these momenta. [3] A minimization program is used to find sets of partial waves (S_1, P_1, P_3, D_3, D_5) such that χ^2 is "flat" and the χ^2 probability is greater than 1/1000. [4] By "flat" we mean that the gradient of χ^2 in the parameter space of partial waves should be no larger than 10^{-4}. Starting points for the searches for solutions, (as defined above), were taken in the vicinities of solutions reported by Kato *et. al.*, [5] and Anderson *et. al.*, [6]. Typically over fifty solutions were found at each energy and these were grouped into distinct clusters of solutions wherever possible. The grouping into clusters is based upon the distribution of distances between all solution pairs where the distance between solution α and β at the same energy is defined as

$$D^{\alpha\beta} = \sum_{pw} \left| T_{pw}^{(\alpha)} - T_{pw}^{(\beta)} \right| \sqrt{j + 1/2} \quad .$$

The linking of solutions at various energies is carried out in two stages. In the first stage a central member of each cluster is assigned and these are linked subject to the requirement that the average speed between adjacent energies should *not* in any partial wave exceed that of pure Breit-Wigner resonance where $(j + 1/2)\ x = 0.3$ and $\Gamma = 100$ MeV.

It is also required that the linkages continue the low energy S-wave behavior of solution $\bar{A}_1$ of Bland *et. al.* [5] Based upon these criteria four distinct linkages were found.

In the second stage, a single member of each cluster in a linkage is selected such that the squared speeds summed over energy is minimum. Specifically, solution set $\alpha_1, \ldots \alpha_8$ is sought, which minimizes the quantity

$$L(\alpha_1, \ldots \alpha_8) \equiv \sum_{i=1}^{7} \left\{ \sum_{pw} (j + 1/2) \left| \frac{T_{pw}^{(\alpha_i)}(E_i) - T_{pw}^{(\alpha_{i+1})}(E_{i+1})}{E_i - E_{i+1}} \right|^2 \right\}$$

The resulting four solutions can be briefly described as follows. Solutions I and IV are respectively very similar to Solutions II and I of the CERN group and will not be discussed any further. Solution II bears a qualitative resemblance to Solution II of the ANL group although there are significant differences at some energies. Solution III appears to be different from the existing published results. It undergoes a counter clockwise motion in the Argand plot of the $P_{3/2}$ partial wave amplitude as can be seen in the figure. If this behavior is interpreted as a Z^* resonance the parameters of the corresponding Briet-Wigner amplitude can be estimated from the magnitudes of the partial cross sections and the speed, both of which are maximum at the resonance energy. One obtains $M \sim 1840$ MeV, $x = .09$ and $\Gamma = 80$ MeV.

One notes, however, that the entire effect is encompassed at several of the energies by the extension of the clusters from which the solutions are selected. If errors assigned to the partial wave amplitudes are derived from the sizes of the clusters, then the tight counter clockwise effect observed in Solution III is barely above the noise level.

We are indebted to Dr. A. Yokosawa of the Argonne National Laboratory for making available to us the phase shift and minimization program.

REFERENCES

[1]University of Maryland Technical Report No. 70-101 (1970). A complete listing of the results along with a more detailed description of the methods used in the analysis are contained in this report.

[2]R. L. Cool, <u>et</u>. <u>al</u>., Phys. Rev. Letters <u>17</u>, 102, (1966); D. V. Bugg, <u>et</u>. <u>al</u>., Phys. Rev. 168, 1466 (1968).

[3]A Compilation of K^+N Reactions, UCRL-20000 K^+N, Sept. 1969; B. Barnett, Ph.D. Thesis (University of Maryland Technical Report No. 70-071, unpublished). The cross section data of Carroll, <u>et</u>. <u>al</u>. have been renormalized wherever necessary to make them consistant with the bubble chamber data.

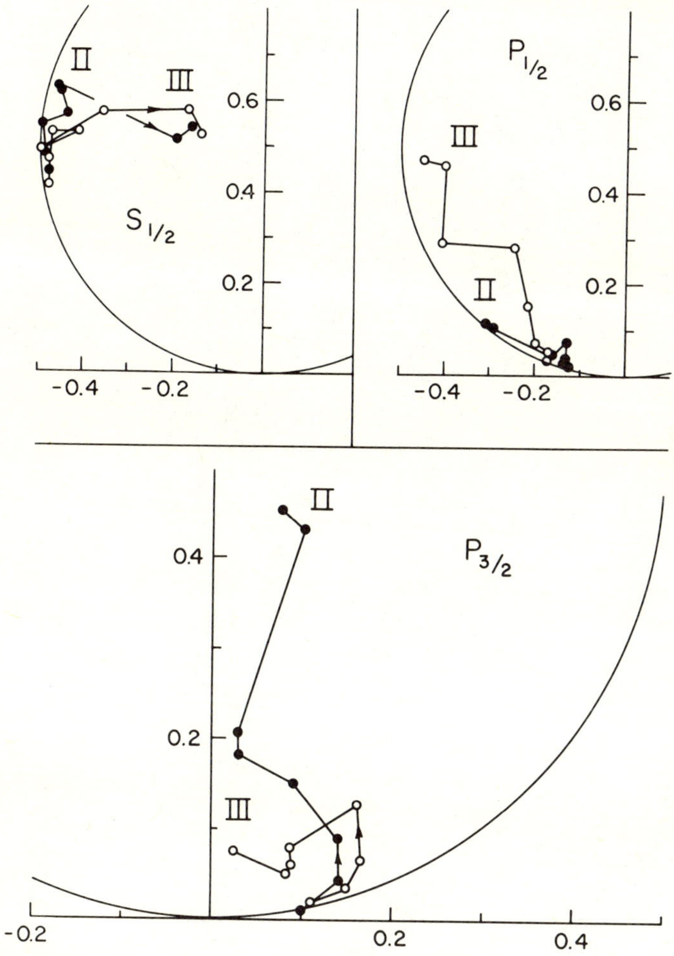

Figure. Argand plots.

[4] William C. Davidon - Argonne National Lab. Report-ANL-5990 (Rev.2), 1966.

[5] R. W. Bland, et. al. UCRL-18323.

K^+P PHASE SHIFT ANALYSIS: AN ALTERNATIVE SOLUTION

B. Carreras*

Daresbury Nuclear Physics Laboratory

A. Donnachie

Department of Theoretical Physics, University of Manchester, and Daresbury Nuclear Physics Laboratory

R. G. Kirsopp

Tait Institute of Mathematical Physics, University of Edinburgh

Recently an energy dependent study of K^+P elastic scattering was presented(1), based on t-channel and u-channel regge exchange. The results obtained differ in one significant respect from those of previous analyses(2,3) namely in the sign of the real part of the S-wave, which is positive. These results are now used as the input to an energy independent phase shift analysis of K^+p elastic scattering, taking them as starting values and constraining the phases and elasticities to them.

Of the two solutions obtained in ref.(1) we have made use of solution II, which was the more compatible with high energy KP and $\bar{K}P$ data. The analysis was performed in two steps. First of all in laboratory momentum range 0.75 to 2.0 GeV/c, making use of the theoretical solutions as indicated above, and secondly from 0.75 GeV/c down to threshold using as initial values for the parameters at each energy those values obtained in the final solution at the previous (higher) energy, and constraining the parameters to the initial values. This approach was adopted here because of the increasing unreliability of the regge pole extrapolations towards

---

*On leave from Junta de Energia Nuclear, Madrid

threshold. By construction, continuity of the solution was guaranteed and since the object of the analysis was to find a solution as close to the regge pole model as the data would allow, general random searches in parameter space were not performed.

From the energy-dependent study it was clear that the important partial-waves below 2.0 GeV/c are the S, P, D and F waves, G-waves being completely negligible below 1.5 GeV/c and contributing very little in the region up to 2.0 GeV/c. It was also possible to judge when to "switch-on" the higher partial waves, and in this analysis we have included S- and P-waves from threshold, D-waves from 0.3 GeV/c and F-waves above 0.7 GeV/c. At no point are the data sufficient to support G-waves as parameters, and they have not been included. This in turn implies that the precise values for the F-waves may be suspect at the higher energies (although not, of course, the general features) and consequently we have restricted ourselves to quoting results only for the S-, P- and D-waves.

Below 0.3 GeV/c laboratory momentum it was found possible to obtain a complete spectrum of solutions, with different combinations of S-waves and P-waves, the most general kind corresponding to an attractive S_{11}-wave and repulsive P_{11}-wave of approximately the same magnitude, with small contributions from P_{13}. This kind of solution is completely different from the repulsive S-wave dominant solution proposed by Goldhaber et al.[(2)], and which has been the basis of all previous phase shift analyses. We have studied this low energy region in detail and it is transparently clear that with the level of data presently available there, it is only by the assumption of either complete S-wave dominance or complete P-wave dominance (i.e. the two situations considered by Goldhaber et al.[(2)]) that a clear statement can be made. In our situation, with approximately equal contributions coming from both S- and P-waves almost any combination is possible, and much more detailed experimental information is required.

Since the fit to the experimental data was a constrained fit it is very difficult to make a reasonable statement about the real number of degrees of freedom we have. Certainly it will be smaller than the number of data points but larger than the number of data points less the number of parameters being varied. For this reason, in the table of results we simply give the total χ^2 at each laboratory momentum and the total number N of data points at that momentum.

Of the twenty one momenta given in the table, four stand out as having χ^2 appreciably greater than N, namely 0.90 GeV/c, 1.25 GeV/c, 1.32 GeV/c and 1.38 GeV/c. However in each of these cases the excess is coming from only one datum point which is contributing between 10 and 20 to χ^2 and which looks, particularly in two cases,

TABLE I

P_{LAB} GeV/c	S_{11} $\delta°$	S_{11} η	P_{11} δ	P_{11} η	P_{13} δ	P_{13} η	D_{13} δ	D_{13} η	D_{15} δ	D_{15} η	N	χ^2
.355	19.8±1.4	1.0	- 6.5±1.7	1.0	4.9±0.8	1.0	1.2±1.4	1.0	0.8±1.1	1.0	5	1.3
.520	30.0±2.4	1.0	-10.2±2.2	1.0	5.6±1.8	1.0	0.7±1.2	1.0	2.7±0.4	1.0	11	6.2
.642	37.2±3.7	1.0	- 5.7±9.9	1.0	2.8±2.1	1.0	-0.3±1.4	1.0	1.8±1.7	1.0	5	0.4
.735	40.8±1.9	1.0	- 8.5±4.7	1.0	3.9±1.1	1.0	-0.6±0.6	1.0	1.9±0.8	1.0	10	2.7
.778	42.6±1.7	0.96±0.04	-11.0±3.0	1.0	4.7±0.8	1.0	1.0±0.8	1.0	3.2±0.6	1.0	19	13.7
.810	43.3±2.0	0.99±0.01	-11.5±3.6	1.0	3.5±1.2	1.0	-0.9±1.2	1.0	3.3±0.5	1.0	8	0.6
.864	46.3±0.3	0.94±0.02	-12.3±0.6	0.99±0.02	3.9±0.2	1.0	0.3±0.1	1.0	4.9±0.2	1.0	80	62.4
.900	47.6±1.1	0.92±0.03	-13.8±0.7	0.97±0.03	4.4±0.5	1.0	0.1±0.1	0.98±0.01	4.6±0.3	1.0	30	38.7
.970	62.9±1.0	0.61±0.03	-14.1±0.7	0.86±0.02	1.7±0.4	1.0	-0.7±0.1	0.99±0.01	5.9±0.4	1.0	95	91.0
1.060	59.5±1.7	0.60±0.04	-22.5±1.6	0.89±0.04	3.3±0.6	0.95±0.02	-1.0±0.1	0.90±0.02	7.9±1.1	1.0	20	20.0
1.087	60.7±0.4	0.58±0.02	-17.9±0.8	0.70±0.02	2.8±0.3	0.96±0.02	-1.5±0.1	0.86±0.02	4.3±0.3	1.0	61	30.7
1.124	64.5±0.9	0.59±0.06	-23.3±2.4	0.67±0.03	0.9±0.9	0.91±1.02	-1.7±0.1	0.94±0.04	8.0±1.0	1.0	20	15.8
1.170	56.8±1.4	0.67±0.09	-18.6±3.3	0.68±0.05	8.7±1.2	0.87±0.02	-1.9±0.1	0.85±0.04	7.8±0.7	1.0	19	13.9
1.215	64.3±0.6	0.78±0.03	-14.7±1.3	0.59±0.03	5.5±0.6	0.85±0.02	-2.2±0.2	0.83±0.02	5.5±0.6	0.98±0.04	82	82.4
1.250	62.9±0.7	0.69±0.03	-14.9±1.7	0.63±0.04	12.0±0.7	0.86±0.01	-2.3±0.2	0.83±0.02	7.3±0.8	0.98±0.01	21	34.0
1.320	63.5±0.8	0.64±0.05	-24.5±2.8	0.52±0.04	12.3±0.9	0.84±0.01	-2.3±0.1	0.88±0.03	10.4±1.2	0.98±0.01	21	36.0
1.380	76.3±1.1	0.57±0.02	-18.6±0.9	0.78±0.02	9.2±0.8	0.57±0.01	0.8±0.1	0.84±0.02	6.6±0.4	0.99±0.01	87	102.1
1.450	90.6±0.9	0.47±0.02	-22.3±1.1	0.90±0.02	- 1.4±0.6	0.50±0.01	-2.3±0.2	0.89±0.03	3.3±0.5	0.93±0.02	110	99.1
1.480	92.1±2.0	0.48±0.02	-24.3±1.3	0.80±0.05	- 0.4±1.3	0.50±0.01	-2.3±0.1	0.89±0.03	6.7±0.8	0.99±0.02	21	15.9
1.710	118.0±5.2	0.27±0.03	-35.1±5.7	0.80±0.04	- 6.4±1.6	0.28±0.03	-0.4±3.3	0.99±0.02	6.5±2.1	0.99±0.02	25	5.3
1.970	110.9±3.0	0.39±0.02	-20.0±2.1	0.50±0.02	5.2±2.0	0.50±0.01	-5.6±0.1	0.78±0.02	8.2±0.5	0.71±0.03	29	27.0
											779	699.2

quite incompatible with neighboring data.

The final results, with errors, which we have obtained for the phase shifts δ and elasticities η for S_{11}, P_{11}, P_{13}, D_{13}, and D_{15} are given in the table. For the reasons discussed above we have not included any results below 0.30 GeV/c. We simply note that, for a range of possible solutions, χ^2 has a value of about 35 for the 30 data points in that region, with a disproportionate contribution to this coming from only two points. Apart from the S-wave, the results are qualitatively similar to those obtained in other analyses(3).

This analysis shows the compatibility of the present K^+p elastic scattering data with an attractive S-wave. An interesting feature of this S-wave is that it has a remarkable similarity to the S_{11} partial wave in the πN elastic scattering, and for those who are willing to consider exotic resonances here is another candidate (or candidates?). The P_{13} amplitude obtained in this solution is very similar to that found in other analyses(3) and it can also be interpreted as having resonance behaviour. However, in both cases the amplitudes obtained are very close to the regge-pole solution II of ref.(1), which specifically did not include any resonance behaviour. For this reason, and quite apart from any built-in prejudicies, we are unwilling to claim that either of these states should be associated with a resonance.

REFERENCES

(1) B. Carreras and A. Donnachie, DNPL/P26.

(2) S. Goldhaber, W. Chinowsky, G. Goldhaber, W. Lee, T. O'Halloran, T. F. Stubbs, G. M. Pjerrous, D. H. Stork, and H. K. Ticho, P.R.L. 9, 135 (1962).

(3) B. H. Hall, R. W. Bland, G. Goldhaber, and G. H. Trilling, UCRL-19231 (1969).

J. G. Asbury, J. D. Dowell, S. Kato, D. Lundquist, T.B. Novey, A. Yokosawa, B. Barnett, P. F. M. Koehler, and P. Steinberg, P.R.L. 23, 194 (1969).

S. Anderson, C. Daum, F. C. Erne, J. P. Lagnaux, J. C. Sens, F. Udo and F. Wagner. Contributed paper to Lund International Conference on Elementary Particles, 1969.

A. T. Lea, B. R. Martin, and G. C. Oades, Phys. Rev. 165, 1770 (1968).

G. Giacomelli, P. Lugaresi-Serra, G. Mandrioli, A. M. Rossi, F. Griffiths, I. S. Hughes, D. A. Jacobs, R. Jennings, B. C. Wilson, G. Ciapetti, V. Constantini, G. Martellotti, D. Zanello, E. Castelli, and M. Sessa. University of Glasgow preprint (1970).

K^+p ELASTIC SCATTERING BETWEEN 1.4 AND 2.3 GeV/c

P.C. Barber, T.A. Broome, B.G. Duff, F.F. Heymann,
D.C. Imrie, G.J. Lush, B.R. Martin, K.M. Potter,
D.M. Ritson*, L.A. Robbins, R.A. Rosner**, S.J. Sharrock,
and A.D. Smith.

University College London

and

R.C. Hanna, A.T. Lea and E.J. Sacharides

Rutherford High Energy Laboratory

* Permanent address: Stanford Linear Accelerator Center, Stanford, California 94305, USA

** Presently at: Rutherford High Energy Laboratory, Berks., England.

The differential cross section for K^+p elastic scattering has been measured at 26 momenta between 1.4 and 2.3 GeV/c, with approximately 4000 events in each angular distribution. The experiment was performed with the 7 GeV proton synchrotron NIMROD at the Rutherford High Energy Laboratory , Chilton, England, using an unseparated K^+ beam with an intensity of approximately 1000 kaons per burst. Incident positive kaons were identified with the aid of a DISC Cerenkov counter, a threshold gas Cerenkov counter and time-of-flight. Particles scattered from a liquid hydrogen target were detected by a core-readout wire spark chamber system connected on-line to a PDP-8 computer. Elastic events were selected on a basis of coplanarity and angular correlation.

The angular distributions have been normalised to a smooth interpolation of existing total elastic cross section measurements. The absolute normalisation of the data has not been finalised, but preliminary calculations at selected momenta indicate agreement to better than $\pm$ 12%.

The shape of the differential cross section varies smoothly throughout the momentum region, the process clearly becoming more diffractive as the momentum increases. The backward contribution remains significant over the whole momentum range, suggesting that baryon exchange should be included in any model of this process. The smooth variation of the cross section is confirmed by Legendre Polynomial fits. Individual Legendre coefficients as a function of momentum show no significant structure. The order of fit required varies from two at the lowest momentum to six at the highest.

The smooth behaviour of the differential cross section suggests that a simple Regge model involving both t- and u-channel exchange might produce an acceptable fit to the data. Such a model has recently been developed by Carreras, Donnachie and Kirsopp [1,2] using the five t-channel meson trajectories of Dass et al[3], together with two u-channel baryon trajectories Λ_α and Λ_γ. This model is in excellent agreement with the present measurements at 1.48 GeV/c and 1.97 GeV/c, where other angular distribution data exist, but predicts a significantly deeper minimum in the vicinity of 1.7 GeV/c than is observed in our data.

We have carried out a phase-shift analysis incorporating the new data with all other published data. Two continuous solutions have been obtained from 0 to 2.2 GeV/c. In the first of these the analysis used an energy-dependent solution from 0 to 0.8 GeV/c (similar to that of Lea, Martin and Oades [4]) and beyond 0.8 GeV/c, energy independent searches were carried out. No resonant structure is observed in any partial wave. All S and P wave amplitudes are moving slowly into the centre of the Argand diagram with increasing momentum.

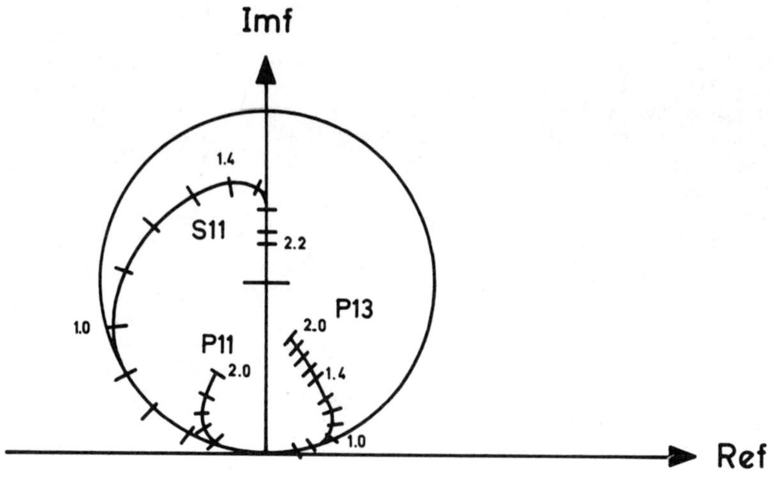

Scattering amplitudes of S and P waves for non-resonant solution.

The second solution was obtained using the Argonne phase shift solutions [5] as a starting point and a stable solution exists which coincides with the Argonne I solution from 1.4 to 1.8 GeV/c and is similar to the Argonne IV solution from 1.8 to 2.2 GeV/c. This solution joins smoothly on to the energy dependent analysis below 0.8 GeV/c and shows the same P_{13} "resonant" behaviour obtained by the Argonne group.

REFERENCES

1. B. Carreras, A. Donnachie, Daresbury Nucl.Phys. Lab.Preprint 26
2. B. Carreras, A. Donnachie, R.G. Kirsopp, Daresbury Nucl.Phys.Lab.Preprint 31
3. G.V. Dass, C. Michael, R.J.N. Phillips, Nucl.Phys. B9 (1969), 549
4. A.T. Lea, B.R. Martin, G.C. Oades, Phys.Rev. 165 (1968), 1770
5. S. Kato, P. Koehler, T. Novey, A. Yokosawa, G. Burleson, Phys.Rev.Letts. 24, 615 (1970)

K^+p INTERACTIONS IN THE RANGE 0.9-1.5 GEV/C AND ELASTIC SCATTERING PHASE SHIFT ANALYSIS

Bologna-Glasgow-Roma-Trieste Collaboration

A bubble chamber experiment is presented in which observations were made of the elastic and inelastic interactions of positive kaons with protons. Total, total elastic and inelastic cross sections, partial inelastic cross sections and elastic scattering angular distributions were measured at 9 kaon momenta in the range 0.9 to 1.5 Gev/c. Fig. 1 shows the measured cross-sections as function of the laboratory momentum. Fig. 2 shows the angular distributions. We have carried out a phase shift analysis using k^+ p total, total inelastic, elastic differential cross sections and recoil proton polarization in elastic scattering. We have used the method of Lea, Martin and Oades [1] in which the energy dependence of the phase shifts and inelasticities is parameterised in terms of 5 parameters. All the data available in the literature at the end of 1969 have been used in the analysis. We assumed S_{11} to be negative. The best four solutions found (see Fig. 3) point to the main inelasticity being in the P_{13} and D_{15} waves, with a possible need for smaller inelastic contributions from another wave. The introduction of F waves has negligible effects on the determination of the parameters of the other waves.

The four best solutions are quite similar. The solid lines in Figs. 1 and 2 represent the results of the phase shift analysis fit.

The preferred solutions show resonance-like behaviour of the P_{13} amplitude, as seen in Fig. 4. The peak in the total cross-section seen by Cool et al.[2] and Bugg et al.[3] can be accounted for by the behaviour of the P_{13} amplitude.

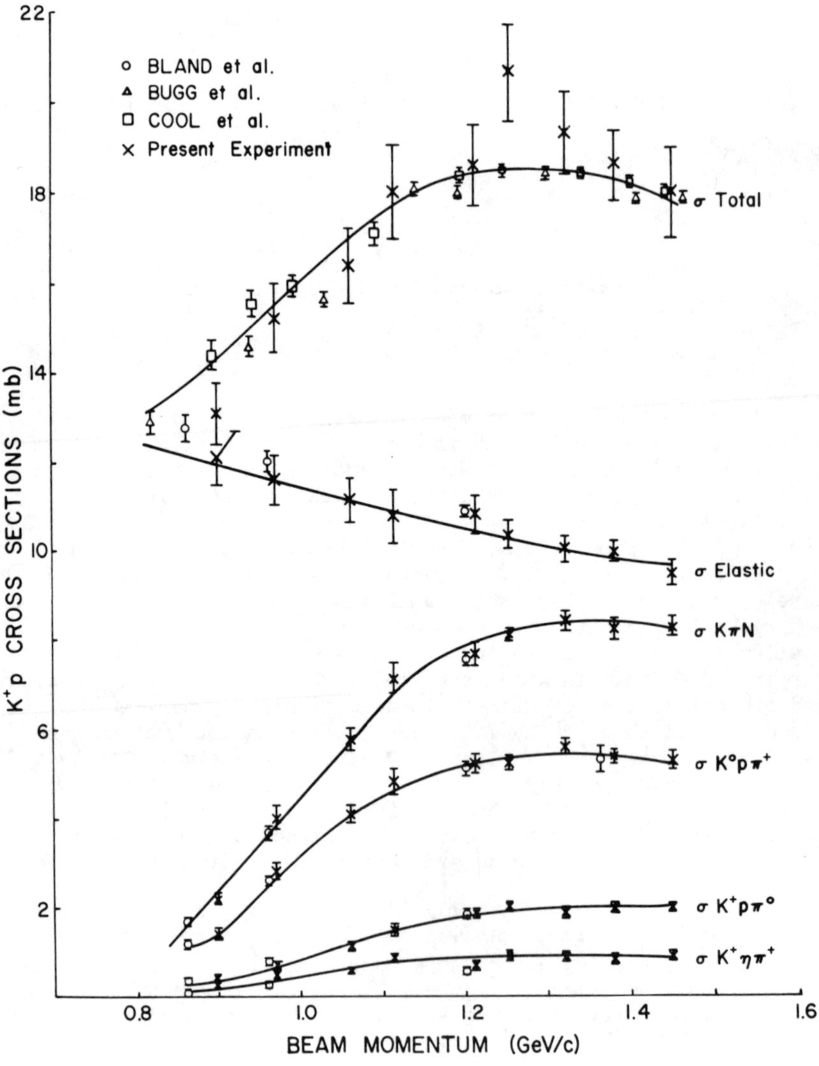
BLAND et al.
BUGG et al.
COOL et al.
Present Experiment
σ Total
σ Elastic
σ KπN
σ K°pπ⁺
σ K⁺pπ°
σ K⁺ηπ⁺
K⁺p CROSS SECTIONS (mb)
BEAM MOMENTUM (GeV/c)
22
18
14
10
6
2
0.8
1.0
1.2
1.4
1.6

Fig. 1

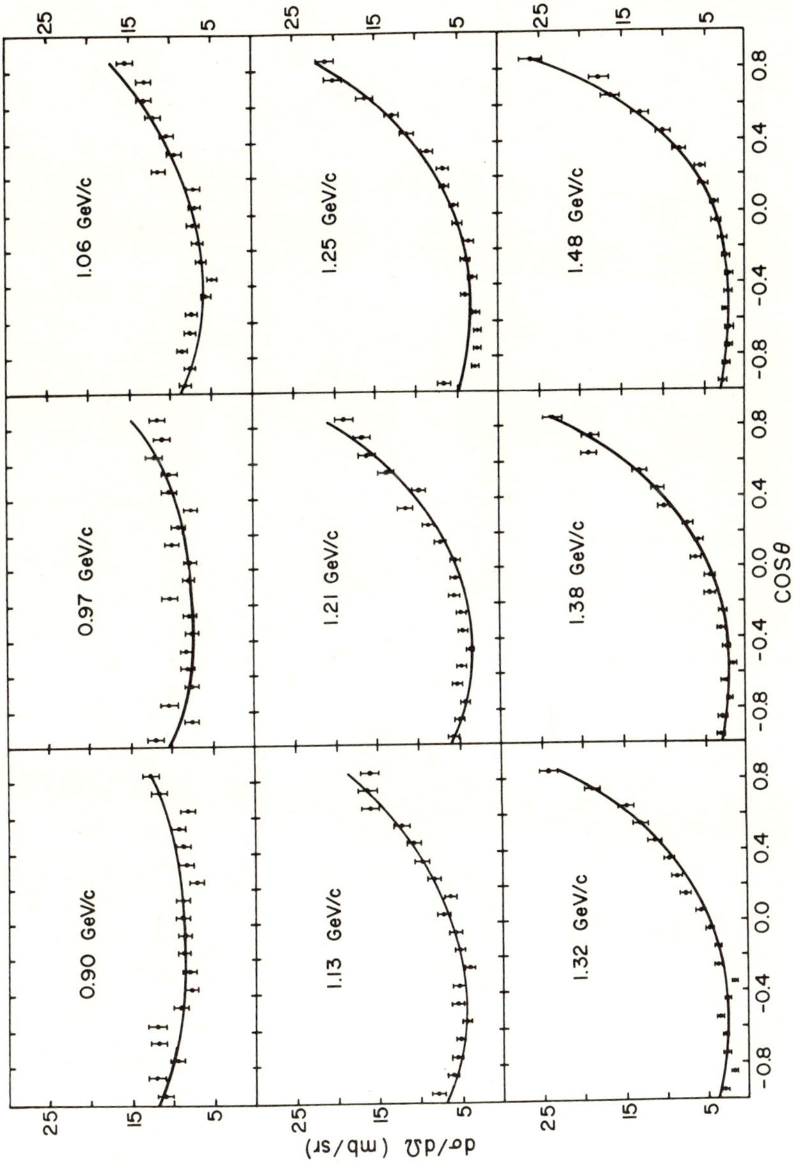
0.90 GeV/c
0.97 GeV/c
1.06 GeV/c
1.13 GeV/c
1.21 GeV/c
1.25 GeV/c
1.32 GeV/c
1.38 GeV/c
1.48 GeV/c
dσ/dΩ (mb/sr)
COSθ
25
15
5
-0.8
-0.4
0.0
0.4
0.8

Fig. 2

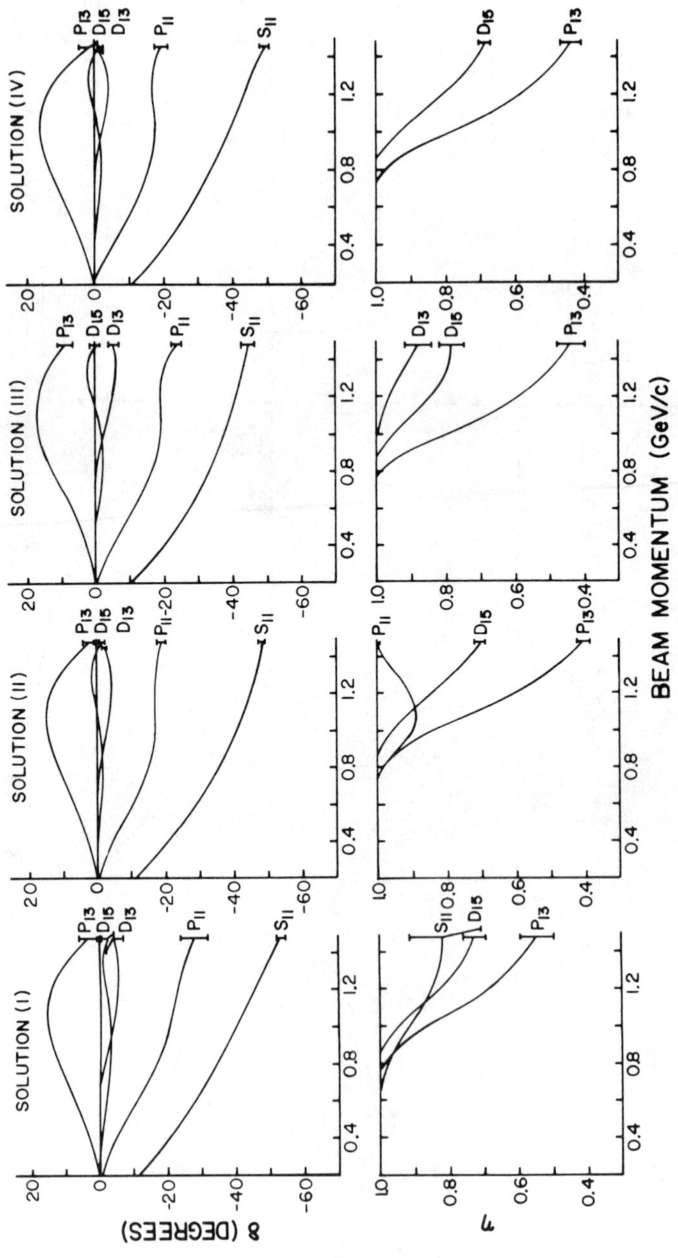
SOLUTION (I)
SOLUTION (II)
SOLUTION (III)
SOLUTION (IV)
δ (DEGREES)
η
P13
D15
D13
P11
S11
BEAM MOMENTUM (GeV/c)

Fig. 3

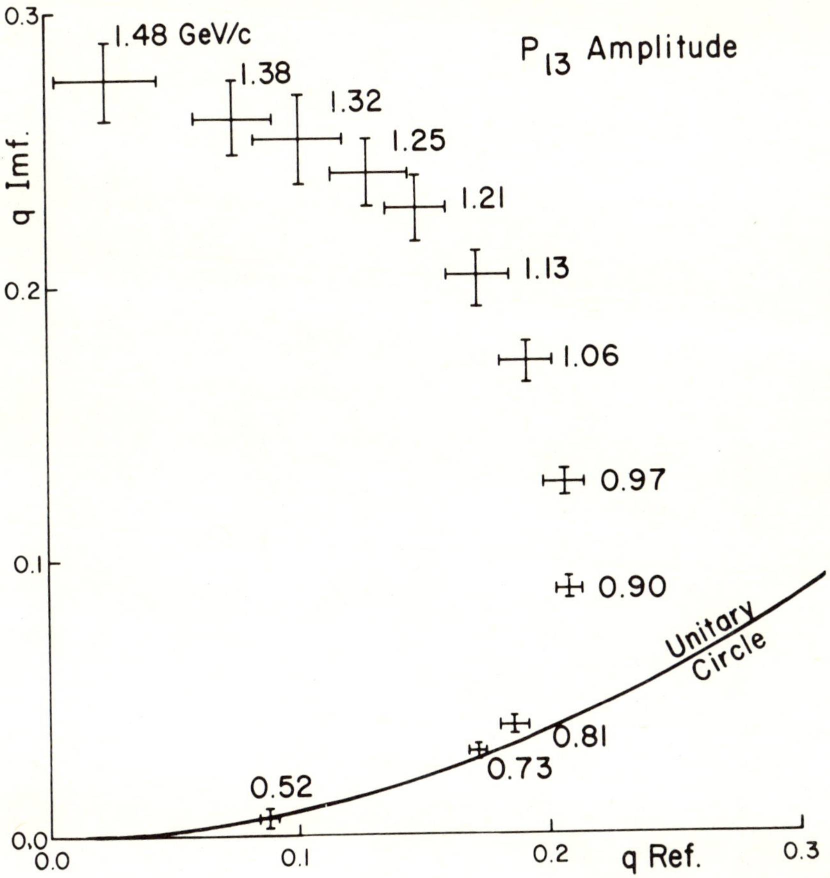

Fig. 4

REFERENCES

1. A.T.Lea, B.R.Martin and G.C.Oades, Phys. Rev. 165 (1968) 1770.

2. R.L.Cool, G.Giacomelli, T.F.Kycia, B.A.Leontic, K.K.Li, A.Lundby and J.Teiger, Phys. Rev. Lett. 17 (1966) 102.

3. D.V.Bugg, R.S.Gilmore, K.M.Knight, D.C.Salter, G.H.Stafford, E.J.N.Wilson, J.D.Davies, J.D.Dowell, P.M. Hattersley, R.J. Homer, A.W.O'Dell, A.A.Carter, R.J.Tapper and K.E.Riley, Phys. Rev. 168 (1968) 1466.

SESSION VIII

BARYON SPECTROSCOPY: THEORETICAL STANDPOINT

Discussion Leader: Bernard Feld

$$35 \times 35 = 405 + 189 + 35 + 35 + 1 + 280 + 280$$

DUALITY AND THE BARYON SPECTRUM

Richard H. Capps

Purdue University, Lafayette, Indiana

I want to discuss briefly an idea that is not new, but is quite relevant to the modern activity of hyperon hunting. It is the result of applying a simple duality principle to meson-baryon states that are exotic, (contain no resonances). A well-known implication of the principle is that the sum of the imaginary parts of the baryonic Regge-pole contributions to an exotic meson-baryon amplitude must vanish; the odd-signature terms much cancel the even-signature terms.

Let us combine this principle with the assumption that the residues are SU(3)-invariant. In nature we know of a baryon decuplet (of spin-parity $3/2^+$) that is nearly exchange-degenerate with an octet (of spin-parity $5/2^-$). The result of the duality principle for such a system is $g(\Lambda^* KN) = 0$, i.e., the F/D ratio of the coupling of the octet trajectory is such that the Λ^* is decoupled from the $\bar{K}N$ state. This result is easy to derive. Since the KN states of both isotopic spins are exotic, the residue cancellation must occur separately for Λ^* trajectories and Σ^* trajectories. Since there is no Λ^* in the decuplet, the octet Λ^* must be decoupled from $\bar{K}N$ states.

Another possible baryon system is a singlet of one parity, exchange-degenerate with an octet of the other. For such a system, the duality principle predicts that $g(\Sigma^* KN) = 0$ for the Σ^* of the octet. This follows because the singlet contains no Σ^*.

In nature, there is no definite evidence for such a singlet-octet pair without additional multiplets. On the other hand, there is very good evidence for a singlet plus an octet that is approximately exchange-degenerate with an octet. This is the nucleon octet and $3/2^-$ nonet. For such an $8 \oplus 1$ -- 8 combination, the conclusion $g(\Sigma^* \bar{K}N) = 0$ no longer follows. This conclusion remains, however, if one extends the duality principle by adding one of the following two assumptions: (a) The two Λ's of the nonet are sufficiently different in mass that the duality principle must be applied separately to the two trajectories through these two Λ's.[1] (b) The principle applies to states that are exotic in the t channel also.[2] With either of these assumptions, both the Σ^* particles of the two octets should be decoupled from $\bar{K}N$ states.

How well do these predictions work for the two most common cases? The $\Lambda^*(5/2^-, 1830)$, approximately exchange-degenerate with a decouplet, is not decoupled from $\bar{K}N$ states, but the $\bar{K}N$ branching ratio is only about 10%, or about 1/3 of the $\pi\Sigma$ branching ratio. In the case of the $8(1/2^+)$ -- $1(3/2^-) \oplus 8(3/2^-)$ combination, it is believed that the $\Sigma(1192)$ is weakly coupled to the $\bar{K}N$ state. There is uncertainty as to whether the $\Sigma^*(3/2^-, 1660)$ is one or two particles, but nobody disputes the fact that the $\bar{K}N$ width is small. Thus, the predictions are pretty good, but not perfect. It is as if nature wanted to break the predictions just enough to give the resonance-formation experimenters a chance to see these resonances.

Recently, I have been struck by the number of Y^*'s that have been observed that are weakly coupled to $\bar{K}N$ states. The above results can be helpful for sorting these out. In a case of a weakly coupled Λ^*, one should look for a decouplet of the other parity, approximately exchange-degenerate with the Λ^*. The other possibility, the Σ^* weakly coupled to $\bar{K}N$ states, appears more common experimentally. In such a case, one should look for an octet of the other parity, approximately exchange-degenerate with the observed Σ^*, one should see if the Σ^* of the new octet is also weakly coupled, and should see if one of the two octets is accompanied by a singlet. Such observations could help bring order to the hyperon spectrum.

REFERENCES

1. R. H. Capps, Phys. Rev. D1, 254 (1970).

2. V. Barger and C. Michael, Phys. Rev. 186, 1592 (1969).

BRIEF SUMMARY OF BARYON SPECTRUM AND ITS INTERPRETATION ACCORDING TO THE QUARK MODEL

B.T. Feld

Massachusetts Institute of Technology

It may be interesting to summarize the current status of the spectrum of baryonic states, taking into account the new data and identifications presented at this Conference. The following tables represent such a summary, in which an attempt is made to classify the levels according to the predictions of a quark model. It is assumed that the appropriate states are given by the SU(6) classification of 3-quark combinations, with either radial excitation or internal orbital angular momentum. It is also assumed that the order of levels is initially that which obtains for a system of three quarks in a harmonic oscillator potential. Even so, a number of choices are possible for the SU(6) multiplets to be associated with the observed levels. We have, on the basis of economy and simplicity, adopted the following scheme: For the positive parity states, we assume that only the 56-plet representation is required, either in the ground state, with L=0 (56-0), or as radially excited levels (56*-0 and 56**-0) or as Regge recurrences of the ground state (56-2 and 56-4). All the known level assignments fit into this scheme and, indeed, the $\Delta(2420)$ - $J = 11/2^+$ is strong evidence for this scheme. There is, however, one outstanding problem: this is the presence of two $7/2^+$ - Λ's, which cannot be accomodated by this, or indeed by any other simple 3-quark model assignment.

The negative parity levels are easily and most simply accomodated in the 70, L=1 supermultiplet and it's Regge recurrence, 70-3.

A summary is given, at the bottom of the second table, of the number of levels observed but unidentified as to spin-parity, which could account for the unfilled spaces in our tables. It is noteworthy that the only serious discrepancy is in the aforementioned extra Λ- state with $J^P = 7/2^+$.

Positive Parity Baryon Resonances [a]

SU6-L	SU3	J^P	N	Δ	Λ	Σ	Ξ	Ω
56-0	8_2	$1/2^+$	940		1115	1190	1320	
	10_4	$3/2^+$		1236		1385	1530	1672
56*-0	8_2	$1/2^+$	1470		(1680)	1620	—	
	10_4	$3/2^+$		(1690)		—	—	—
56**-0	8_2	$1/2^+$	1780		1800	1900	—	
	10_4	$3/2^+$		(2160)				
56-2	8_2	$3/2^+$	1861		1885	2070		
		$5/2^+$	1688		1815	1915	(1960)	
	10_4	$1/2^+$		1910		—	—	—
		$3/2^+$		—		—	—	—
		$5/2^+$		1890		—	—	—
		$7/2^+$		1950		2030	—	—
56-4	8_2	$7/2^+$	1990		{1870, 2015} [b]	2250	—	
		$9/2^+$	—		—	—	—	
	10_4	$5/2^+$		—		—	—	—
		$7/2^+$		—		—	—	—
		$9/2^+$		—		—	—	—
		$11/2^+$		2420		—	—	—

a) Underlined means new; in parenthesis means identification uncertain.

b) There is room for only one $7/2^+$ Λ in this scheme. Two is too many!

Negative Parity Baryon Resonances

SU6-L	SU3	J^P	N	Δ	Λ	Σ	Ξ	Ω
70-2	1_2	$1/2^-$			1405			
		$3/2^-$			1520			
	8_2	$1/2^-$	1535		1670	1750		
		$3/2^-$	1520		1690	1670	1820	
	10_2	$1/2^-$		1650			–	–
		$3/2^-$		1670		(1660)	–	–
	8_4	$1/2^-$	1700		1850		–	
		$3/2^-$	(1700)		2030	1940		
		$5/2^-$	1670		1830	1765	(1930)	
70-3	1_2	$5/2^-$						
		$7/2^-$			2100			
	8_2	$5/2^-$			–		–	
		$7/2^-$	2190		–	(2130)	–	
	10_2	$5/2^-$		(1960)		–	–	–
		$7/2^-$		–		–	–	–
	8_4	$3/2^-$	2040		–	–	–	
		$5/2^-$	–		–	–	–	
		$7/2^-$	–		–	–	–	
		$9/2^-$	–		–	2250	–	

	N	Δ	Λ	Σ	Ξ	Ω
Missing (numbers)						
$M \lesssim 2000$ MeV/c^2	0	1	0	7	14	8
$M \lesssim 2500$ MeV/c^2	6	5	8	23	30	20
possible candidates	0	0	-1	4	2	0

COMMENTS ON BARYON SPECTROSCOPY

Sydney Meshkov *

National Bureau of Standards

Washington, D. C. 20234

1. RESONANCES

Year after year people give summary talks at conferences at which they ritualistically devote themselves to the question of what is a resonance, and I've been asked to give my opinions on this subject, too. In a sense, we're like the patient who runs from doctor to doctor in order to find the diagnosis which pleases him. I can always tell you that a resonance is a pole on the second Riemann sheet of a scattering amplitude. That's very nice but probably entirely useless. In the long run, we seem to be led back to looking either for bumps in a cross section curve or for a loop to be made in an Argand plot when we graph partial wave amplitudes obtained from a phase shift analysis. The ramifications of these possibilities and others have been exhaustively explored by Haim Harari [1] at the Vienna Conference, and by Wally Greenberg [2] in the Lund Conference, and I doubt that I'll shed much more light, but let me voice a few comments.

Sometimes it seems that a resonance lies in the eyes of the beholder. Fortunately, the other viewers of the resonance would like to have some macroscopic manifestation of the divine revelation. At some point, we probably do want a bump in some cross section. It doesn't really have to be a bump in the particular cross section that we have started out with, but there has to be some place where we are going to see a bump and from which we are going to see particles coming off such that we can go back to the original bump and label its J^P, baryon number, isospin, Y, i.e., its quantum numbers. We also would like to have a mass, a total width and various partial widths. For example, perhaps we find, via a phase shift analysis that there exists a resonance. The

* Work supported in part by Office of Naval Research Contract No. NAonr-14-70.

total cross section may look rather smooth in that region. Nevertheless, most of us won't be convinced of its existence until we find a peak in the cross section for <u>some</u> inelastic process.

2. CLASSIFICATION CRITERIA

Why are we interested in the study of resonances in the first place? We're interested in them because they seem to have a key role in our attempt to unravel the mysteries of elementary particle interactions. We'd like to have some way of classifying them according to their spins and parities, and to see if we can say something about the strength of the various couplings. If we have a symmetry scheme like SU(3), we're able to relate the strengths of the various couplings for particular decay modes. A baryon resonance, R, may decay into any of the decay modes given below. B denotes members of the $J^P = 1/2^+$ octet, B^* denotes members of the $J^P = 3/2^+$ decuplet. P and V denote the pseudoscalar and vector mesons. SU(3) relates the couplings within one of these four modes; a higher symmetry like SU(6) relates the relative strengths of the four modes to each other.

$$
\begin{aligned}
R &\rightarrow B \times P \quad 8^2 \times (8^1 + 1^1) \\
&\rightarrow B \times V \quad 8^2 \times (8^3 + 1^3) \\
&\rightarrow B^* \times P \quad 10^4 \times (8^1 + 1^1) \\
&\rightarrow B^* \times V \quad 10^4 \times (8^3 + 1^3)
\end{aligned}
$$

In trying to classify, what criteria do we use to say that a multiplet of some higher symmetry, say, like isospin, SU(3), SU(6), or SU anything is filled? Classification is a very subjective business. We say things fit, when some group of us decides that we like the classification as it stands. There may always be some mavericks who decide that they don't agree with you but in some way they are outvoted (Democracy is a great thing). We shall assume that the first and foremost criterion is that the candidates for a multiplet obey a mass relation like the Gell-Mann Okubo formula for particles with a given spin and parity. Sometimes that works very nicely, as, for example, in the baryon octet, and in the $J^P = 3/2^+$ decuplet. When we get to some of the other multiplets, some deviations set in; and we often ascribe these differences to configuration mixing.

3. CONFIGURATION MIXING AND ITS EFFECT ON MASSES

If you write a mass matrix, then you find that it is very stable to even relatively large amounts of configuration mixing. In the past I have always used a rough rule of thumb that says that all configuration mixing is 10%. You may laugh, but it's actually not too bad. In any case, it's very hard to move around a mass matrix's eigenvalues by small amounts of configuration mixing. But it is very easy to change properties, like decay widths, using just a very small amount of mixing. This is particularly vexing in dealing with Ξ's. So an attempt to really pin down the coupling constants to very precise levels and to use them as a strong test for classification is probably not correct.

Let me give some examples of the role of configuration mixing. Consider the first 2 x 2 mass matrix shown in Table I, where the intensity of the admixed state, B, is 0.01; this gives an amplitude of 0.1. The original unperturbed state, A, will have an amplitude of the square root of 0.99, which is 0.995. If the coupling constant g^2, is proportional to $|\psi_1|^2$ then the cross term, which is the coefficient of 2AB is important. Its value for the first matrix is 0.2, even though the eigenvalue shifts by only 0.01. If the admixture has an intensity of 0.04 as in the second matrix the admixed amplitude is 0.2 and the cross term contribution is 0.39. If the intensity that is admixed is 0.09, as in the third matrix, the amplitude corresponding to that is 0.3 and the cross term which is mixed in is 0.57. In φ-ω mixing analogous to the fifth matrix the squares of the amplitudes are 2/3 and 1/3, respectively. The square root of 2/3 is 0.816, the square root of 1/3 is 0.577; that corresponds to a tremendous amount of configuration mixing, far more than, I think, we are likely to run across in most of the other situations where we can have it. Yet the effect on the mass eigenvalues is small. The moral of all of this is that widths and consequently the coupling constants extracted from them may be tremendously affected by configuration mixing and so a detailed comparison of widths can't be used as a clincher for assignments.

Before getting too deeply immersed in configuration mixing and its ramifications (as we'll do shortly), we'd best see what our theoretical framework will be and delineate what configurations will come in.

4. SU(3) AND SU(6) CLASSIFICATIONS

In the beginning there was SU(3). As you all know, this started with mesons in octets and baryons in octets.

A	B	E_1	$\psi_1 = \alpha A - \beta B$	β^2	$2\alpha\beta$
2	0.1	2.01	$\psi_1 = 0.99A - 0.1B$	0.01	0.20
0.1	1				
2	0.2	2.04	$\psi_1 = 0.98A - 0.2B$	0.04	0.39
0.2	1				
2	0.3	2.083	$\psi_1 = 0.95A - 0.3B$	0.09	0.57
0.3	1				
2	0.6	2.28	$\psi_1 = 0.91A - 0.42B$	0.18	0.77
0.6	1				
2	0.8	2.44	$\psi_1 = 0.806A - 0.59B$	0.35	0.95
0.8	1				

Table I. 2x2 Matrices. The largest eigenvalue E_1, for each matrix is listed. ψ_1 is the corresponding eigenfunction. It is given in terms of the unperturbed state, A, and the admixed state, B. Also listed are the intensities, β^2, of the state, B, and the cross term $2\alpha\beta$.

Meson baryon products (or resonances) could, in principle, be in

$$8 \times 8 = 27 + 10 + \overline{10} + 8_D + 8_F + 1 \quad , \tag{4.1}$$

and similarly for the mesons. By the middle sixties no baryon candidates for the 27, and $\overline{10}$ were known, nor were there any candidates for meson 27, 10, $\overline{10}$. This led Gell-Mann [3] and Zweig [4] to invent quarks, because simple qqq products only produced

$$3 \times 3 \times 3 = 10 + 8 + 8 + 1 \quad . \tag{4.2}$$

$q\bar{q}$ products gave

$$3 \times \bar{3} = 8 + 1 \quad . \tag{4.3}$$

To obtain baryons in 27's you need qqq $q\bar{q}$; to obtain mesons in 27's you need $qq\bar{q}\bar{q}$.

Once quarks were invented, attaching spin indices to them was immediate and SU(6) followed. The SU(6) analogs of (4.1), (4.2), (4.3) are easily written down. (See Figure I.) These expressions all involve quarks in s-waves. An obviously nice feature of SU(6) was that the baryon representation 56 contained $10^4 + 8^2$, and the meson 35 contained the $8^3 + 1^3 + 8^1$, which accommodated the observed vector and pseudoscalar mesons.

The SU(6) x O(3) scheme for classification was introduced by Mahanthappa and Sudarshan [5], by Borchi and Gatto [6], and by Gatto, Maiani and Preparata [7] and has been used in conjunction with the simple quark model by Dalitz [8] and with the shell model of Greenberg [9] to provide what is the classification scheme most in vogue today, for both mesons and baryons.

The SU(6) x O(3) approach has been extensively studied and excellent reviews by Dalitz [10], Harari [1], and Greenberg [2] with this as a basis have been given in past conferences. The positive parity baryons are placed in 56, [L = 0, 2, 4,. . .] and the negative parity baryons are in 70, [L = 1, 3, 5,. . .]. Nobody knows what has happened to the 70,20,[L = 0, 2,. . .] or to the 56,20, [L = 1, 3, 5,. . .].

SU(3)	SU(6)
$8 \times 8 = 27 + 10 + \overline{10} + 8_D + 8_F + 1$	$35 \times 56 = 1134 + 700 + 70 + 56$
	$35 \times 35 = 405 + 189 + 35_D + 35_F + 1 + 280 + \overline{280}$

QUARKS

$qqq \rightarrow 10 + 8 + 8 + 1$	$qqq \rightarrow 56 + 70 + 70 + 20$
$q\bar{q} \rightarrow 8 + 1$	$q\bar{q} \rightarrow 35 + 1$
$SU(3)^{SU(2)}$ breakdown of SU(6):	$35 = 8^3 + 1^3 + 8^1$
	$56 = 10^4 + 8^2$
	$70 = 10^2 + 8^2 + 1^2 + 8^4$
	$20 = 8^2 + 1^4$

Baryon	27	$qqqq\bar{q}$	700, 1134
Meson	27	$qq\overline{qq}$	405, 189

Figure I. SU(3), SU(6) and Quark Algebra

5. Z^*

A. Classification

Note that the presence of exotic resonances doesn't bother the SU(6) x O(3) approach. Recall that Z_1^*'s lie in SU(3) 27 representations, and that Z_0^*'s lie in SU(3) $\overline{10}$ representations. Low spin negative parity 27's can come from qqq $q\bar{q}$ 700^-, L = 0 or from 1134, L = 0. Positive parity 27's could come from 700, L = 1, 3, . . ., or from 1134, L = 1, 3, . . .;they could also arise from qqqqq $\overline{qq}$. In the preceding session you heard the various arguments for and against the existence of the Z^*'s. What emerges for the Z_1^*'s seems to be the result that if there is a Z_1^* it should be a positive parity P_{13} system [11]. The P_{13} solutions could come 1) from the 27's in the 700, L = 1 (the 700 contains a 27^4 and a 27^2); 2) from the (27^6, $27_A{}^4$, $27_B{}^4$, $27_A{}^2$, $27_B{}^2$) in the 1134, L = 1, or 3) from the $q^5\bar{q}^2$ representation (252, $\overline{21}$) → 4536 + 700 + 56. There is less evidence for a bump in the phase shift solution D_{35} for K^+P scattering. This could come from the 27^6 of the 1134, L = 0.

B. Branching Ratios

Tables II, III, and IV display the reduced matrix elements obtained by Joseph Coyne and me for the decay of Z_1^*'s and Z_0^*'s into various final states from various initial U(6) x U(6) and $SU(6)_S$ assignments. The calculation is carried out using $SU(6)_W$. (As mentioned before, a higher symmetry like SU(6) relates decays into different modes.)

Decay matrix elements squared from the (126,$\bar{6}$) 700^- and from the (210,$\bar{6}$) 1134^- are listed. The same results hold for positive parity 27's and $\overline{10}$'s obtained from 700, L = 1 and 1134, L = 1.

In the <u>700</u> $\quad 27^4 \rightarrow J = 5/2, 3/2, 1/2$

$\quad\quad 27^2 \rightarrow J = 3/2, 1/2$

Therefore in the 700, L = 1 there are two J = 3/2 27's. In principle, the physical Z^* states can be any linear combination of these two J = 3/2 27's. Since the KN coupling is weak experimentally, we shall assume that the "observed" Z^* P_{13} is primarily from 27^4, L = 1. From Table IV we note that the KN coupling is zero [12]. This gives $K\Delta/K^*N/K^*\Delta$ = 5/3/16. The $K^*\Delta$ channel isn't open below 2.120 GeV. A slight admixture of 27^2 L = 1 gives a small amount of KN and alters the K^*N somewhat. In this picture there should be another Z_1^* state with lots of KN, very little $K\Delta$, and some K^*N. A similar but more complicated state of affairs holds for decays from the 1134, L = 1 because there are more states. A

(126, $\overline{6}$) 700^-

hel 3/2 = hel 1/2

	$\underline{27^4}$	$\underline{27^2}$
K N	0	3/6 A^2
K Δ	5/24 A^2	0 A^2
K^*N	3/24 A^2	1/6 A^2
K^*Δ	16/24 A^2	2/6 A^2

Also true for $[700, L = 1]^+$

700, L = 1 $27^4 \rightarrow J = 5/2, 3/2, 1/2$
$27^2 \rightarrow J = 3/2, 1/2$

Table II. Z_1^* decay amplitudes from 700^-, L = 0 coming from the qqqq$\overline{q}$ representation (126,$\overline{6}$) and calculated using $SU(6)_W$. The entries shown are the squares of decay amplitudes obtained by averaging over initial states and summing over final states.

	$\underline{27^6}$	$\underline{27_A^{\ 4}}$	$\underline{27_B^{\ 4}}$	$\underline{27_A^{\ 2}}$	$\underline{27_B^{\ 2}}$
K N	0	0	0	1/4 B^2	3/12 B^2
K Δ	0	5/8 B^2	1/4 B^2	0	0
K^*N	0	0	1/3 B^2	3/4 B^2	1/12 B^2
K^*Δ	1 B^2	3/8 B^2	5/12 B^2	0	8/12 B^2

Also true for [1134, L = 1]

Table III. Z_1^* decay amplitudes from 1134^-, L = 0 coming from the qqqq$\overline{q}$ representation (210, $\overline{6}$) and calculated using $SU(6)_W$.

700			
	$\overline{10}^2$	K^*N	3/4 A^2
		K N	1/4 A^2
1134			
	$\overline{10}^4$	K^*N	1 B^2
	$\overline{10}^2$	K^*N	1/4 B^2
		K N	3/4 B^2
Also true for L = 1, + parity			

Table IV. Z_o^* decay amplitudes from the 700^-, L=0 and 1134^-, L=0 calculated using $SU(6)_W$.

comparison of the experimental cross sections in the region of the possible P_{13} and D_{35} resonances with the branching ratios given above should be carried out. A similar study can be undertaken for the I = 0 system. It should be pointed out that if this picture of large SU(6) multiplets is invoked then the full complement of particles in these representations should accompany the Z^*'s.

C. How to Look for Z^*'s

At this point let me say something about how to look for states that are rare, like Z^*'s. We know that they don't like to couple to KN very much. Therefore, I would claim that a good way to make it is spectroscopically, i.e., first, make a state that is relatively easy to make; then let it decay into a Z^* plus something; then let the Z^* decay into modes that it likes to decay into, like KΔ. In this way we don't fight the coupling strength but make use of phase space to give us a fair share of Z^*'s. An example, undoubtedly with many drawbacks, is suggested in Figure II.

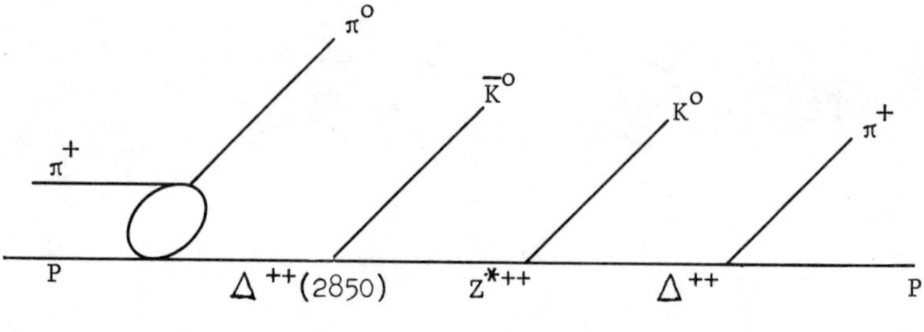

Figure II. Spectroscopic Production of Z^*

We are stuck in analyzing the Z^*, because we can't use SU(3) for decay width comparisons for 2 body decays, but must go to SU(6) for such. Similar things can be done for the Z_0^* which can only fit in $\overline{10}$'s.

D. Significance of Z^*'s

You may ask, so what if there are Z^*'s? What will change? For one thing, anything that depends on the absence of 27's to give a condition on amplitude relations will be affected:

a. Because this condition cannot be invoked, and

b. By the magnitude of the cross term between the 27 and the non-27 amplitudes in any cross section evaluation.

E. Tuan Analysis

Another approach to the Z^* problem has been made by Tuan [13], in a contribution to this conference. He assigns the Z^*'s to the (1/2,1/2) representation of a broken non-chiral SU(2)xSU(2) subgroup of global symmetry. This leads to doublets (Z_0^*,Z_1^*) which are nearly degenerate with a splitting which is typically of the order of the Σ-Λ mass splitting, and which have the <u>same spin</u> and <u>parity</u>. As Tuan points out, his predictions are distinct from those of strong coupling theory [14] which place the Z_1^* (1910) and Z^* (1865) in $P_{3/2}$ and $P_{1/2}$, respectively.

6. SU(6) x O(3) CLASSIFICATION

A. General Comments

The SU(6) x O(3) classification of even and odd parity baryons is not really very much different at present from that proposed by Dalitz [15] in the 1966 Berkeley Conference. Tables V and VI list the more common SU(3) singlet, octet, and decuplet assignments used in this classification as currently made by Tripp [16,17], Levi-Setti [18], and most recently by Flaminio, Metzger, Samios, and Goldberg [19]. Similar assignments by Tuan [20] and D. E. Plane, <u>et al</u> [21] have been discussed at this conference.

The need for the use of configuration mixing to untangle the structure and to extract coupling constants is evident. Let's delineate the kinds of configuration mixing which we might expect to encounter in the usual SU(6) x O(3) approach.

B. Even Parity

The various states that can be formed from 56, L = 0, 2, 4 ... are listed in Table VII. The contribution that each configuration

	$1/2^-$	$3/2^-$	$5/2^-$	$7/2^-_{FMSG}$	$1/2^+$	$5/2^+$
N	1525	1515	1675	2190	938	1688
Λ_8	1670	1690	1830	2350	1115	1815
Σ	1780	1660	1765	2280	1185	1940(1910)
Ξ	1825*	1820	1930	2430	1321	2030
Λ_1	1405	1520	(1600-1800?)**	2100***		

* Levi-Setti leaves out

** FMSG only

*** All

Table V. SU(3) Octet and Singlet Assignments.

	$3/2^+$	$7/2^+$
Δ	1236	1900
Σ	1385	2020
Ξ	1530	2140
Ω	1672	2260

Table VI. Decuplet Assignments

56	L=0	L=2	L=4
$8^{1/2}$	$8^{1/2}$	$8^{3/2,5/2}$	$8^{7/2,9/2}$
$10^{3/2}$	$10^{3/2}$	$10^{1/2,3/2,5/2,7/2}$	$10^{5/2,7/2,9/2,11/2}$

		[radial excitation]
$J = 1/2^+$	$8^{1/2}(L=0) + 10^{3/2}(L=2) +$	$n = 2\ [8^{1/2}(L=0)]$
$J = 3/2^+$	$10^{3/2}(L=0) + 8^{1/2}(L=2) + 10^{3/2}(L=2)$	$n = 2\ [10^{3/2}(L=0)]$
$J = 5/2^+$	$8^{1/2}(L=2) + 10^{3/2}(L=2) + 10^{3/2}(L=4)$	$n = 4\ [8^{1/2}(L=2)] + [10^{3/2}(L=2)]$
$J = 7/2^+$	$10^{3/2}(L=2) + 8^{1/2}(L=4) + 10^{3/2}(L=4)$	$n = 4\ [10^{3/2}(L=2)]$
$J = 9/2^+$	$8^{1/2}(L=4) + 10^{3/2}(L=4) + 10^{3/2}(L=6)$	$n = 6\ [8^{1/2}(L=4)] + [10^{3/2}(L=4)]$

Table VII. SU(6) x O(3) Even Parity Baryon Configuration Mixing.

makes to states of a particular J for J = 1/2, 3/2, 5/2, etc., are written in order. A pattern for the kind of mixing permissible in this scheme emerges. We see immediately that $J = 1/2^+, 5/2^+, 9/2^+ \ldots$ states form one group; $J = 3/2^+, 7/2^+ \ldots$ form another.

For the purpose of describing the mixing in a qualitative way, let's take the spacing between multiplets in units of δ defined by:

$$10^{3/2}(L = 0) - 8^{1/2}(L = 0) = \delta \text{ (pure)}$$

$$8^{1/2}(L = 2) - 8^{1/2}(L = 0) = 2\delta \text{ (very pure)}$$

or $$8^{1/2}(L = 0)\text{radial} - 8^{1/2}(L = 0) = 2\delta \text{ (very pure)}$$

$$8^{1/2}(L = 2) - 10^{3/2}(L = 0) = \delta \text{ (pure)}$$

Some observations are:

1. The $J = 1/2^+$ octet is anomalously pure and is low. (The nearest mixing partner is the first radial excitation.)

2. $1/2^+, 5/2^+, 9/2^+$: N, Λ are very pure because the nearest state that can mix is 2δ away.

3. $3/2^+, 7/2^+, 11/2^+$: Δ, Ω are very pure because the nearest state that can mix is 2δ away.

4. For all spins other than $1/2^+$, Σ, Ξ are pure, mixing with states one δ unit away.

$5/2^+, 9/2^+$ states allow mixing with states of the same L.

$3/2^+, 7/2^+$ states allow mixing between states of different L.

Because of this mixing, the Σ, Ξ should be slightly depressed relative to the other members of their multiplets which only have 2δ mixing.

5. We expect $3/2^+, 7/2^+, \ldots$ octets to be high, especially the N, Λ members, because there is nothing with which they can mix.

C. Negative Parity Resonances

The same mixing analysis can be done for the negative parity members of the 70^-(L = 1, 3, 5. . .). These states are listed in

Table VIII. Configuration mixing is much more important for the negative parity states than for the positive parity resonances. Looking at the pattern in Table VIII leads to the following observations:

1. The $J = 1/2^-$, $5/2^-$, $9/2^-$ states form one group, qualitatively different from the $J = 3/2^-$, $7/2^-$,. . . subset.

2. The $1/2^-$ states have an anomalous behavior. They properly belong to the sequence $5/2^-$, $9/2^-$,. . . but because there are no $L = -1$ states they seem to behave like $3/2^-$, $7/2^-$,. . . but with one less configuration.

3. The purest states are the $5/2^-$, $9/2^-$,. . . octets. We don't expect any low lying singlets to go with them.

4. The $3/2^-$ and $7/2^-$ multiplets should be strongly mixed, slightly more so than the $1/2^-$. The purest states of these should be Δ,Ω in $1/2^-$, $3/2^-$, $7/2^-$. Λ,Σ,Ξ should have three components, N only two, for $J = 1/2^-$.

Λ :	1	8	8	
Σ :		8	8	10
Ξ :		8	8	10
N :		8	8	

D. New Resonances and Results

1. Kim [22] has reported two new even parity resonances:

P_{03} 1.710 GeV

P_{11} 1.660 GeV

2. The CHS group [23] has one new odd parity resonance:

S_{01} 1.851 GeV

3. At higher energies Barbaro-Galtieri [24] and Butterworth [25] have presented phase shift analyses which led to the proposed resonances listed in Table IX. Also listed there are some remarks summarizing the aforementioned authors' likes and dislikes.

70	L=1	L=3	L=5
$1^{1/2}$	$1^{1/2,3/2}$	$1^{5/2,7/2}$	$1^{9/2,11/2}$
$8^{1/2}$	$8^{1/2,3/2}$	$8^{5/2,7/2}$	$8^{9/2,11/2}$
$10^{1/2}$	$10^{1/2,3/2}$	$10^{5/2,7/2}$	$10^{9/2,11/2}$
$8^{3/2}$	$8^{1/2,3/2,5/2}$	$8^{3/2,5/2,7/2,9/2}$	$8^{7/2,9/2,11/2,13/2}$

$$J = 1/2^- \quad \left\{1^{1/2} + 8^{1/2} + 10^{1/2} + 8^{3/2}\right\}(L=1)$$

$$J = 3/2^- \quad \left\{1^{1/2} + 8^{1/2} + 10^{1/2} + 8^{3/2}\right\}(L=1) + \left\{8^{3/2}\right\}(L=3)$$

$$J = 5/2^- \quad \left\{8^{3/2}\right\}(L=1) + \left\{1^{1/2} + 8^{1/2} + 10^{1/2} + 8^{3/2}\right\}(L=3)$$

$$J = 7/2^- \quad \left\{1^{1/2} + 8^{1/2} + 10^{1/2} + 8^{3/2}\right\}(L=3) + \left\{8^{3/2}\right\}(L=5)$$

$$J = 9/2^- \quad \left\{8^{3/2}\right\}(L=3) + \left\{1^{1/2} + 8^{1/2} + 10^{1/2} + 8^{3/2}\right\}(L=5)$$

Table VIII. SU(6) x O(3) Odd Parity Baryon Configuration Mixing.

			B - G		B
$1/2^+$	P_{11}	$\Sigma(1900)$	x	Litch-field	x
$3/2^+$	P_{13}	$\Sigma(2070)$	x		x
$3/2^-$	D_{13}	$\Sigma(1940)$	x $(\Sigma\pi)(\Lambda\pi)$		Likes $\Lambda\pi$ Not $\Sigma\pi$
$3/2^-$	D_{03}	$\Lambda(2040)$	x		Can't tell
$7/2^+$	F_{07}	$\Lambda(2015)$	x		Dislikes

Table IX. New High Mass Resonances Discussed by A. Barbaro-Galtieri and I. Butterworth.

	No.	SU(4) Rep.	Eiv. of SU(4) Casimir Oper.	Strange Spin
$\Delta^{1/2},N^{1/2},N^{3/2}$	20	20	39/4	0
$\Sigma^{1/2},\Lambda^{1/2},\Lambda^{3/2}$	12	6	5	1/2
$\Sigma^{1/2},\Lambda^{1/2},\Sigma^{3/2}$	20	10	9	1/2
$\Xi^{1/2},\Xi^{3/2}$	12	4	15/4	1
$\Xi^{1/2}$	4	4	15/4	0
$\Omega^{1/2}$	2	1	1	1/2

Table X. SU(4) x SU(2) Breakup of 70 Representation.

4. The CHS 1.851 GeV resonance gives us the three $1/2^-\Lambda$'s which we might expect from the configuration mixing analysis mentioned above:

$$1/2^-: \Lambda(1405),\ \Lambda(1670),\ \Lambda(1851).$$

Is the $\Lambda(1851)$ associated primarily with 8^4?

5. The D_{03} $\Lambda(2040)$ gives three $3/2^-\Lambda$'s.

$$3/2^-: \Lambda(1520),\ \Lambda(1690),\ \Lambda(2040).$$

6. It is possible that $3/2^-\Lambda(2040)$ and the $7/2^-\Lambda(2100)$ both come from $8^{3/2}$(L = 3).

7. An interesting system is the J = $3/2^-\Sigma(1660)$. As we have heard [26,27] there seem to be two different J = $3/2^-$ resonances, which like to decay into $\Sigma\pi$ amd $\Sigma\pi\pi$.

$$\Sigma(1660) \xrightarrow{56x35} \Sigma\pi$$

$$\Sigma(1660) \xrightarrow{70x35} \Lambda(1405)\ \pi \rightarrow \Sigma\pi\pi$$

The first decays into 56x35. The second is an example of a different SU(6) decay mode, namely 70x35. J.C. Carter and J.J. Coyne [28] have shown that the following representations lie in the product 70x35

$$70 \times 35 \rightarrow 20 + 56 + 70 + 70 + 540 + 560 + 1134$$

It is likely that the decaying object lies in a 70, and is mixed with that which decays into $\Sigma\pi\pi$, but it could even be an example of a <u>20</u>, something which doesn't arise in the product, 56 x 35, or of an <u>1134</u>.

8. It looks like the $\Sigma(1915)$ should be upgraded from a C to an A in Levi-Setti's notation.

9. Dalitz mentioned that states arising from 8^4 in 70,L = 1 have a narrow spread.

S_{11}	1.760	S_{01}	1.850	D_{13}	1.660
D_{15}	1.770	D_{05}	1.830	D_{11}	1.670

Nick Samios at the Texas Conference suggested the possibility of a degenerate decuplet.

$\Delta(1670)$, $\Sigma(1660)$, $\Xi(1640)$ and an Ω^- over the present one, all with $J^P = 3/2^-$.

10. We should have a large number of Ξ's in the mass region ~1.7 - 2.0 GeV. Every octet and decuplet should have one. As we have heard, the experimental situation is a mess and very few resonances stand out.

If we are to believe that there really exists a Ξ(1635) [29], its mass alone makes it interesting. Yodh and I have the following argument. Assuming only that Ξ (1635) is part of a multiplet, say an octet, which has normal mass ordering, with a spread of ~240 MeV, then there must exist an N, Σ and Λ below it.

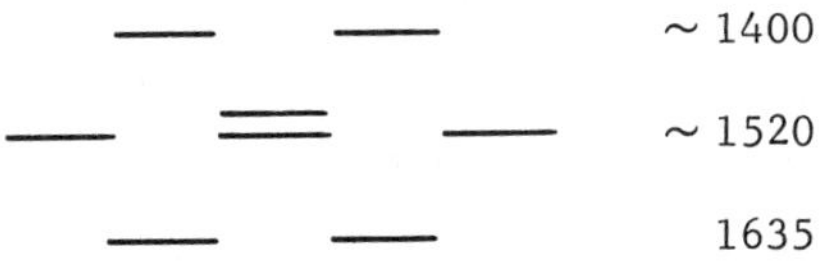

This would imply a nucleon resonance around 1400 MeV (possibly the Roper?) and a center of mass of 3Λ + Σ at about 1520 MeV. Tripp mentioned the possibility of a P_{11} hyperon at 1520-1540, but it didn't seem to be on firm ground.

E. Theoretical Guide to Mixing Analysis

In general, in the analysis of the data, particularly for negative parity baryons, most people have invoked a 1,8 mixing. Greenberg and Divgi [2,30] and Faiman and Hendry [31] have written down how to mix the different amplitudes in a more general way, but with some dynamical assumptions.

We would like to find a simple way to predict these mixings ahead of time, guided by some principles, whether they stem from quark models or from particular subgroups of higher symmetry groups. The simplest example we know of where a principle like this works is in the φ - ω mixing problem. We have found out through the years that the φ and the ω can be classified according to the strange spin. In quark model language the φ is a pure $\lambda\bar{\lambda}$ system (strange spin one) and the ω is a pure $p\bar{p}$ and $n\bar{n}$ system with strange spin 0. Attempts to use a similar classification scheme in explaining the f_o and the f^* have been done. When you go to the baryons, there is in principle, no reason why you shouldn't try such an approach.

In a very early paper in SU(6) days, Kuo and Yao [32] discussed mixing in an SU(6) 70. They decomposed the SU(6) 70 according to SU(4) x SU(2), where the SU(2) is that of strange spin. The breakup is shown in Table X. They found that the Λ members of $8^{1/2}$ and $1^{1/2}$ are mixed; that the Σ members of $8^{1/2}$

and $10^{1/2}$ are mixed, and that the Ξ members of $8^{1/2}$ and $10^{1/2}$ are mixed; and that in each case the mixing angle was 45°. They could carry out no complete analysis because 1) there was no data, 2) the idea of a 70, L = 1 hadn't been formulated.

In a modern context, for the Σ's and Λ's, the ordering might be in the order given by the eigenvalues of the SU(4) Casimir operator; i.e., the lowest (Casimir operator eigenvalue 9) $(\Sigma_8 + \Sigma_{10})^{1/2}$ combination would combine with $\Sigma^{3/2}$, all with L = 1. Similarly, the $\Lambda^{1/2}$ of the SU(4) 10 representation might be the low one. An immediate problem is that the experimental 1,8 mixing angle is measured as $\sim 20^o$ for $1/2^-$ and as $\sim 23^o$ for $3/2^-$, so clearly the upper state is somewhat admixed. For the Ξ system there may be a natural grouping according to strange spin. In any case, a complete analysis along this line may enlighten us.

7. HIGHER SPIN STATE CLASSIFICATION

A. Introduction

Now that we have analyzed various SU(3) multiplets up through spin $5/2^{+,-}$ and even $7/2^-$, it is time to ask a serious question. How do we describe the higher spin states like 7/2, 9/2, 11/2, 13/2, ..? Is SU(6) x O(3), the proper scheme? Are all higher spin even parity states to be looked at as 56, L = even states, and all odd parity states as 70, L = odd? Do we include 700, 1134 and still higher multiplicity representations?

One problem with this approach is that it's hard to predict accurately where higher spin states will fall. For this purpose, i.e., predicting where high spin states fall, we might use Regge trajectories. This approach has the difficulty that most trajectories contain at most 2 states at present. (Δ_α possibly has 3), and as we go higher in energy, configuration mixing may become more and more important. Furthermore, predicting spins differing by 2 means that each trajectory just won't have many states in regions of mass below 5 GeV. Are there any other approaches one can take to grouping particles and predicting their spins?

B. Rotational Model

1. General Motivation. I would like to suggest that perhaps there is. In particular, the patterns of the baryon spectra for individual states such as Λ, Σ, Ξ, N, Δ, as functions of spin and parity, look to me very much like those that correspond to rotational bands in nuclear physics. These rotational bands are based on the idea of the nucleus as being a deformed object which is made

to rotate. The fact that in baryon spectroscopy we start with or often have available spins of 1/2 leads us to the idea that most of our present particle states are members of what might be called K = 1/2 bands. K is defined as the projection of the total angular momentum along an axis of symmetry. In nuclear physics the assumption is generally that one is dealing with deformed axially symmetric nuclei, so the particular axis of symmetry might be the long axis of a football.

Another bit of motivation is that in nuclear physics, the rotational model has been a good representation when there is lots of configuration mixing, and we have seen that in particle physics we seem to be encountering considerable configuration mixing.

A whole formal apparatus has been built up in low energy nuclear physics to treat rotational systems [32-34]. In particular the one that is pertinent to us involves the description of odd A nuclei, namely, those with half integral spin. Since our time is limited, I won't go into great detail, but I'll present only the formalism that I need.

2. Mass Relations. The energy spectrum of a given rotational band, assuming that band mixing is neglected is

$$E_K(J) = A + \frac{\hbar^2}{2\mathcal{J}}\left[J(J+1) + a(-)^{J+1/2}(J+1/2)\delta(K,1/2)\right]$$

↑ characteristic rotator spectrum

$\mathcal{J}$ = moment of inertia

$\vec{J}$ = total spin angular momentum

$= \vec{j} + \vec{R}$

$\vec{j}$ = intrinsic angular momentum

$\vec{R}$ = rotational angular momentum

K = projection of the total angular momentum on the axis of symmetry (Z')

$\mathcal{M}$ = projection of the total angular momentum on the space fixed axis, (Z)

a = decoupling parameter (due to Coriolis force term$(\hbar^2/2\mathcal{J})[\vec{J}\cdot\vec{j}]$)

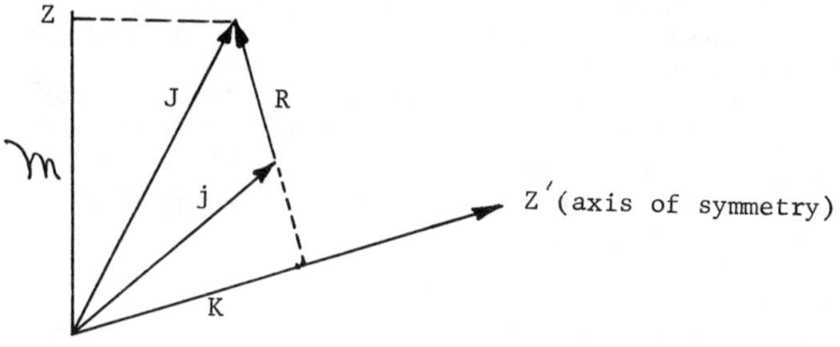

Figure III. Angular Momentum Diagram for a Deformed Rotator

If j lies along the Z′ axis then $J \cdot j = j^2$ and this term is independent of J. Therefore the term $J \cdot j$ is a measure of the decoupling of the intrinsic motion j from the symmetry axis Z′.

This term is called the Coriolis coupling or the rotational particle coupling. It corresponds to the coupling between the angular momentum of the particle, j, and the angular velocity of the rotational motion $= \frac{J}{\mathcal{J}}$. It is a term which is due to the rotation of the deformed potential and the amount of coupling varies with the J of each state in a band.

3. <u>Mass Patterns</u>. Enough of philosophy. Let's see whether this is of any use to us. Figure IV is a plot of energy differences, (or for our purposes) of mass^2 differences, plotted as a function of a. It is taken from notes of J. P. Elliott [36] of SU(3) fame. At a = 0 we get the pure rotational spectrum with spacings J(J+1). As a takes on different values the relative spacings of levels with different spin changes. By a = -1 the lowest state is no longer spin 1/2, but the spin 3/2 takes over. Also shown in Figure IV are the expressions for various energies and their differences. Each band involves three parameters A, B, a.

We treat Δ, Σ, Ξ, N, Λ separately, and shall form separate rotational bands for each parity. To evaluate all possible splittings within a band we must use 3 states as input. From then on we can predict the positions of all other states in a band. Effectively, we determine a, $B = \hbar^2/2\mathcal{J}$ and A.
Through trial and error, I have found that using M^2 seems to be best.

The rotator model bands differ from Regge trajectories in that 1) there is a state for <u>every</u> spin in a K = 1/2 band (There may be bands starting with spin 3/2, etc. which also go up by 1 unit of spin.); 2) it is not necessary that J = 1/2 be the lowest state

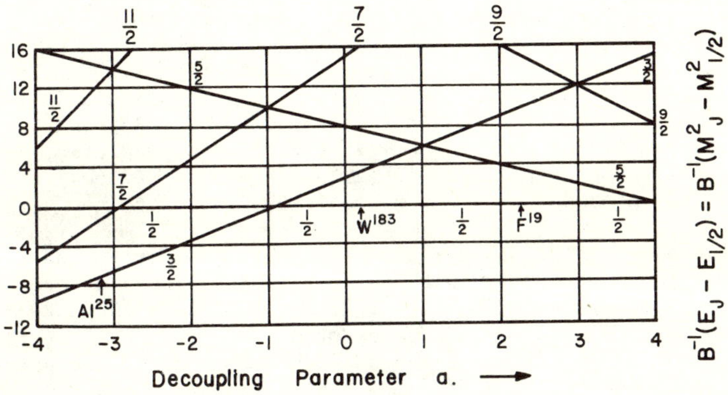

$E(1/2) = A + B\,[3/4 - a]$	
$E(3/2) = A + B\,[15/4 + 2a]$	$3/2 - 1/2 = B\,[3 + 3a]$
$E(5/2) = A + B\,[35/4 - 3a]$	$5/2 - 1/2 = B\,[8 - 2a]$
$E(7/2) = A + B\,[63/4 + 4a]$	$7/2 - 1/2 = B\,[15 + 5a]$
$E(9/2) = A + B\,[99/4 - 5a]$	$9/2 - 1/2 = B\,[24 - 4a]$
$E(11/2) = A + B\,[143/4 + 6a]$	$11/2 - 1/2 = B\,[35 + 7a]$
$E(13/2) = A + B\,[195/4 - 7a]$	$13/2 - 1/2 = B\,[48 - 6a]$

Figure IV. Energy Level (or Mass^2) Differences for an Axially Symmetric Rotator. M^2 differences are plotted for each spin as a function of a, the decoupling parameter.

of a band; 3) $M^2 \sim J(J+1)$ (for a = 0).

4. Band Structure for Each Particle. Proceeding in an admittedly ad hoc fashion, we try to accommodate the known resonances in rotational bands. There may be more than one K = 1/2 band, since we know that there are at least three $1/2^-$ Λ's and hence three $1/2^-$ SU(3) multiplets. Table XI displays a Σ^+ rotational band in which the $1/2^+$ Σ(1.189), $3/2^+$ Σ(1.385) and $5/2^+$ Σ(1.910) are used as input. A $7/2^+$ Σ(2.056) is predicted, to be compared with an observed $7/2^+$ Σ(2.030). A $9/2^+$ Σ(2.507) is predicted, to be compared with a possible $9/2^+$ Σ(2.455).

Table XI also lists the two bands which may be formed for Σ^-. The first one presents us with the first case of an inverted order. We take as input $3/2^-$ Σ(1.670), $1/2^-$ Σ(1.730) and $7/2^-$ Σ(2.280), yielding an a = -1.27. This value of a leads to a $5/2^-$ Σ(2.380) above the $7/2^-$ at 2.280. A second Σ^- band may be formed using the second $3/2^-$ Σ(1.670) and $5/2^-$ Σ(1.765) as input. If a = 0, then a $1/2^-$ state at 1.610 would be predicted; it could be identified with the Σ(1.616) state of Crennell, et al [37]. This band could be a K = 3/2 band, in which case there should be no $1/2^-$ state.

Table XII lists the rotational bands which may be constructed for Λ^-. Using $1/2^-$ Λ(1.405), $3/2^-$ Λ(1.518) and $5/2^-$ Λ(1.835) as input gives a decoupling parameter a which doesn't indicate any violent scrambling of the usual rotational ordering. A $7/2^-$ Λ(2.034) is predicted; experimentally there is a $7/2^-$ Λ(2.100). An $11/2^-$ Λ(2.696) is predicted, compared to a possible Λ(2.600). A second Λ^- band has three known states but nothing to compare with predictions. These states yield a = -.92; a glance at Figure 4 shows that such a decoupling parameter leads to pairs of states close together, i.e., ($1/2^-$, $3/2^-$), ($5/2^-$, $7/2^-$), etc.

No calculations were carried out for Λ^+, since there are not enough states to use as input. Only the $1/2^+$ Λ(1.115) and the $5/2^+$ Λ(1.815) are certain. [After the conference, I found that if I let a = 0, and predict a $3/2^+$ Λ, the predicted mass is at 1.736 to be compared with Kim's new state at 1.710 GeV. A $7/2^+$ Λ(2.259) is predicted.]

Table XIII lists three possible bands for the Ξ system. Really the only firm states are the $1/2^+$ Ξ(1.321) and $3/2^+$ Ξ(1.530). If we form a band with a = 0 using the latter two states as input, we predict a $5/2^+$ Ξ(1.823), a $7/2^+$ Ξ(2.170) and a $9/2^+$ Ξ(2.540), all of which states have been suggested in the past. We could include the $7/2^+$ Ξ(2.140) as input; this perturbs things only slightly and gives an a = 0.0652. We can try to form a Ξ^- band with a = 0, using as input a $3/2^-$ Ξ(1.820) and a $5/2^-$ Ξ(1.930), leading to a $7/2^-$ Ξ(2.072).

	Expt	Theory
$1/2^+$	1.189	<u>1.189</u>
$3/2^+$	1.385	<u>1.385</u>
$5/2^+$	1.910	<u>1.910</u>
$7/2^+$	2.030	2.056
$9/2^+$	2.455?	2.507
$11/2^+$	(2.600)?	2.849

a = -.156

B = .197

A = 1.24

	Expt	Theory
$3/2^-$	1.670	<u>1.670</u>
$1/2^-$	1.730	<u>1.730</u>
$7/2^-$	2.280	<u>2.280</u>
$5/2^-$	2.455?	2.380
$11/2^-$		
$9/2^-$		

a = -1.27

B = .255

A = 2.48

	Expt	Theory
$1/2^-$?	?1.616	1.610?
$3/2^-$	1.670	<u>1.670</u>
$5/2^-$	1.765	<u>1.765</u>
$7/2^-$		1.890

Pick a = 0

B = .0652

A = 2.54

Is this K = 3/2 band?

Table XI. Σ^+ and Σ^- Rotational Bands. Input states are underlined.

	Expt	Theory
$1/2^-$	1.405	<u>1.405</u>
$3/2^-$	1.518	<u>1.518</u>
$5/2^-$	1.835	<u>1.835</u>
$7/2^-$	2.100	2.034
$9/2^-$	?	2.462
$11/2^-$	2.600?	2.696
$13/2^-$	?	3.170

a = -.32
B = .162
A = 1.80

	Expt	Theory
$1/2^-$	1.670	<u>1.670</u>
$3/2^-$	1.690	<u>1.690</u>
$5/2^-$	?	2.316
$7/2^-$	2.350	<u>2.350</u>
$9/2^-$		3.170

a = -.92
B = .262
A = 2.35

Table XII. Λ^- Rotational Bands. Input states are underlined.

	Expt	Theory
$1/2^+$	1.321	1.321
$3/2^+$	1.530	1.530
$5/2^+$	1.820	1.823
$7/2^+$	2.140	2.170
$9/2^+$	2.500	2.540

Let a = 0

B = .197

	Expt	Theory
$1/2^+$	1.321	1.321
$3/2^+$	1.530	1.530
$5/2^+$	1.820	1.790
$7/2^+$	2.140	2.140
$9/2^+$	(2.500)?	2.476

a = .0652

B = .185

	Expt	Theory
$1/2^-$	?	1.750?
$3/2^-$	1.820	1.820
$5/2^-$	1.930	1.930
$7/2^-$	2.030	2.072

Let a = 0

B = .0825

Is this a K = 3/2 band?

Table XIII. Ξ^+ and Ξ^- Rotational Bands. Input states are underlined.

I have tried to incorporate the N's and Δ's into similar bands, but my results depend strongly on how much I work at the mass range limits given in the latest Particle Properties Compilation. Many of the N's and Δ's have ill-defined masses; a glance at Figure IV shows that near a = -1, the sensitivity to precise masses is large, and that is the region where shifts of 10 or 20 MeV render fits acceptable or impossible. For example, the D_{13} N(1.520) has a listed spread from 1.510 to 1.540 and the S_{11} N(1.535) has a spread from 1.500 to 1.600. By making an appropriate choice I either do or don't get a fit. Similar difficulties hold for the higher mass states and for the Δ's as well.

I am grateful to J. Joseph Coyne, Gaurang B. Yodh, Nicholas P. Samios, Angela Barbaro-Galtieri, and Robert D. Tripp for many useful suggestions and conversations.

REFERENCES

1. H. Harari, in Proceedings of the Fourteenth International Conference on High Energy Physics, Vienna, Austria, September 1968.

2. O. W. Greenberg, in Proceedings of the Lund International Conference on Elementary Particles, Lund, Sweden, 25 June - 1 July, 1969.

3. M. Gell-Mann, Phys. Letters 8, 214 (1964).

4. G. Zweig, CERN Report, 1964 (unpublished).

5. K. T. Mahanthappa and E. C. G. Sudarshan, Phys. Rev. Letters 14, 163 (1965).

6. E. Borchi and R. Gatto, Phys. Letters 14, 352 (1965).

7. R. Gatto, L. Maiani and G. Preparata, Phys. Rev. 140, B1579 (1965).

8. R. Dalitz, in Proceedings of Oxford International Conference on Elementary Particles (Rutherford High Energy Laboratory, Chilton, England) January 1966.

9. O. W. Greenberg, Phys. Rev. Letters 13, 598 (1964).

10. R. Dalitz, See Ref. 8, and also see R. Dalitz, in Proceedings of International Conference on Symmetries and Quark Models, Wayne State University, Detroit, Michigan, June 1969; R. Dalitz, Irvine Conference, December 1967.

11. S. Kato, P. Koehler, T. Novey, A. Yokosawa, and G. Burleson, Phys. Rev. Letters 24, 615 (1970).

12. This argument complements the discussion at the end of D. Horn, H. J. Lipkin and S. Meshkov, Phys. Rev. Letters 17, 1200 (1966).

13. S. F. Tuan, in "The Role of Z^*'s in a Theory of Clashing Internal Symmetries," (contributed to this conference).

14. C. J. Goebel, Phys. Rev. Letters 16, 1130 (1966).

15. R. Dalitz, in Proceedings of the Thirteenth International Conference on High Energy Physics, Berkeley, California, 1966.

16. R. Tripp, in Proceedings of Third Topical Conference on Particle Physics, Honolulu, Hawaii, August 1969; also UCRL-19361 (preprint).

17. R. Tripp, in Proceedings of the 14th International Conference on High Energy Physics, Vienna, 1968.

18. R. Levi-Setti, in Proceedings of the Lund International Conference on Elementary Particles, 1969.

19. E. Flaminio, W. Metzger, N. P. Samios and M. Goldberg, Rev. Mod. Phys. (to be published).

20. S. F. Tuan, contribution submitted to this conference.

21. D. E. Plane, et al., talk at this conference and CERN-D.Ph.II/Phys. 70-9; 31.3 1970.

22. J. J. Kim, talk at this conference.

23. D. E. Plane, et al., talk at this conference.

24. A. Barbaro-Galtieri, talk at this conference.

25. I. Butterworth, talk at this conference.

26. R. Ross, talk at this conference.

27. D. Bassano, talk at this conference.

28. J. C. Carter and J. J. Coyne, J. Math. Phys. 10, 1204 (1969).

29. S. Apsell, et al., Phys. Rev. Letters 23, 884 (1969).

30. O. W. Greenberg and D. R. Divgi, Phys. Rev. 175, 2024 (1968) and erratum, Phys. Rev. 178, 2487 (1969).

31. D. Faiman and A. W. Hendry, Phys. Rev. 173, 1720 (1968); D. Faiman and A. W. Hendry, Phys. Rev. 180, 1572 (1969).

32. T. K. Kuo and T. Yao, Phys. Rev. Letters 13, 415 (1964).

33. A. Bohr and B. R. Mottelson, Dan. Matt. Fys. Medd. 27, 16 (1953).

34. S. G. Nilsson, Dan. Matt. Fys. Medd. Dan. Vid. Selsk. 29, 16 (1955).

35. A. K. Kerman, Matt. Fys. Medd. Dan. Vid. Selsk. 30, 15 (1956).

36. J. P. Elliott, Collective Motion in Nuclei, University of Rochester preprint NYO-2271 (1958).

37. D. J. Crennell, et al., Phys. Rev. Letters, 26, 648 (1968).

PREDICTION OF SLOPE CHANGES IN KAON-NUCLEON SCATTERING

L. M. Libby

University of Colorado, Boulder, Colorado

The relativistic relation between cross section and matrix element for elastic scattering may be written,

$$|M|^2 = |\int H_\theta \, dt|^2 = 2 v^2 c^2 \hbar^2 (d\sigma/dt) . \tag{1}$$

If the effect of Lorentz contraction in a high energy scattering is taken into account by reducing the matrix element for the effect of Lorentz contraction, as follows,

$$|M| = |M_0| \gamma^{-1}; \text{ and } (M/v) = M_0 \, m/p , \tag{2}$$

then the relation between cross section and uncontracted matrix element becomes,

$$|M_0| = (2^{\frac{1}{2}} \hbar c/m) \, p \, (d\sigma/dt)^{\frac{1}{2}} \tag{3}$$

where m is the mass of a colliding particle.

Substituting momentum exchange t where $t = 2p^2(1-\cos\theta)$, this relation becomes,

$$| M_0 | = (\hbar c/m)\ (1 - \cos\theta)^{-\frac{1}{2}}\ t^{\frac{1}{2}}\ (d\sigma/dt)^{\frac{1}{2}}$$
$$| M_0 | \sim t^{\frac{1}{2}}\ (d\sigma/dt)^{\frac{1}{2}} \text{ at fixed } \theta . \tag{4}$$

The derivative with respect to t is,

$$(d/dt)\ | M_0 | \sim (t)^{-\frac{1}{2}}\ (1 + 2t\ (d/dt)\ \ln\ (d\sigma/dt)^{\frac{1}{2}})\ (d\sigma/dt)^{\frac{1}{2}} .$$

For all high energy cross sections (except annihilation) the logarithmic dependence on t has a negative slope, so that the parenthesis has a form $\sim$ (1-bt). This condition insures that the matrix element vs t has a minum, occurring at t $\sim$ $(b)^{-1}$. These considerations are discussed at some length in Ref. 1.

To study $| M_0 |$ analytically, we put for the cross section its parametrization from optical model,

$$d\sigma/dt = (kR_0)^2\ (J_1(x)/x)^2\ F(\theta) \tag{5}$$

which can be made to fit elastic scattering by adjusting $F(\theta)$.

The condition that a minimum exist in $| M_0 |$ now becomes,

$$J_0(x) = 0;\ \ x = 2kR_0 \sin\ (\theta/2)$$
$$t - 2\hbar^2 k^2 (1-\cos\theta) - x^2\ (c\hbar/R_0)^2 \tag{6}$$

This condition is satisfied for values of x, x^* = 2.49, 5.52, 8.64, 11.8, 14.9, 18.0, ... Assuming R_0 lies between 0.65 and 0.75 fermi, appropriate for K-nucleon scattering, the corresponding values of t^* are $\sim$ 0.43, 1.9, 4.4, 8.6, ... $(GeV/c)^2$.

It is very interesting that, using experimental K-nucleon scattering data [2], plots of $\ln(d\sigma/dt)$ versus t show a dependence consistent with changes in slope at the predicted values of t^*=0.4, 1.9, and 4.4, see Fig. 1 and 2.

Similar predictions [1] of observed slope changes in pp elastic scattering have been previously reported. The values of t^* predicted for the pp system are $t^* \cong 0.31, 1.54, 3.80, 7.0, \ldots$, corresponding to $R_0 \cong 0.83$ fermi, appropriate for the pp interaction. Slope changes have been observed for all these values of t^* in 90° center-of-mass scattering, that at ~ 3.8 having been observed [3] after the prediction was made.

Predicted slope changes [4] for pion-nucleon scattering are shown by the experimental data $t^* \cong 0.3, 1.5, 3.8, 7.0, \ldots$, corresponding to $R_0 \cong 0.8$ fermi, appropriate for the πp system.

REFERENCES

1. L. M. Libby, Physics Letters 29B, 345 (1969).

2. A Compilation of KN Reactions, by the Particle Data Group, Lawrence Radiation Laboratory, UCRL-20000 KN, September, 1969.

3. B. Brabson, R. Crittenden, R. Heinz, R. Kammerud, H. Neal, H. Paik, R. Sidwell, and K. Suen, Phys. Rev. Letters 23, 1306 (1969).

4. L. M. Libby, "Prediction of Slope Changes in Pion-Nucleon Scattering" in preparation for publication.

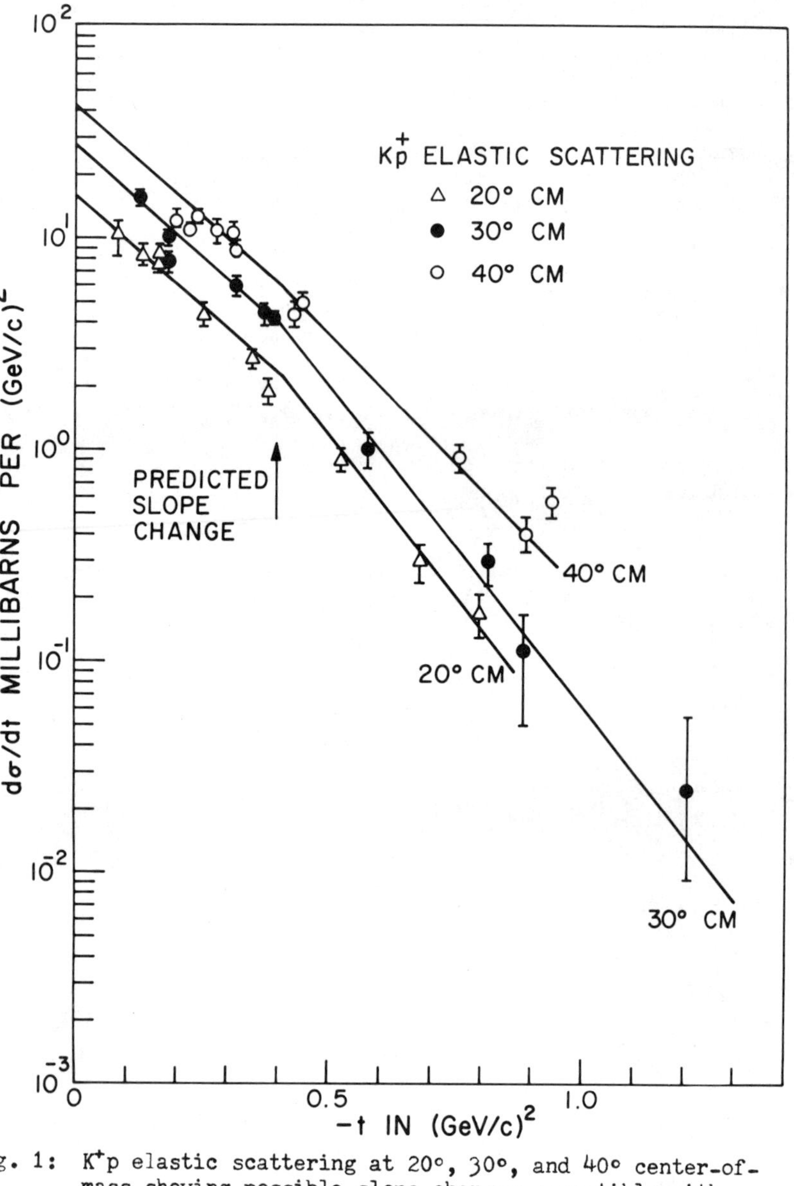

Fig. 1: K^+p elastic scattering at 20°, 30°, and 40° center-of-mass showing possible slope changes compatible with predictions.

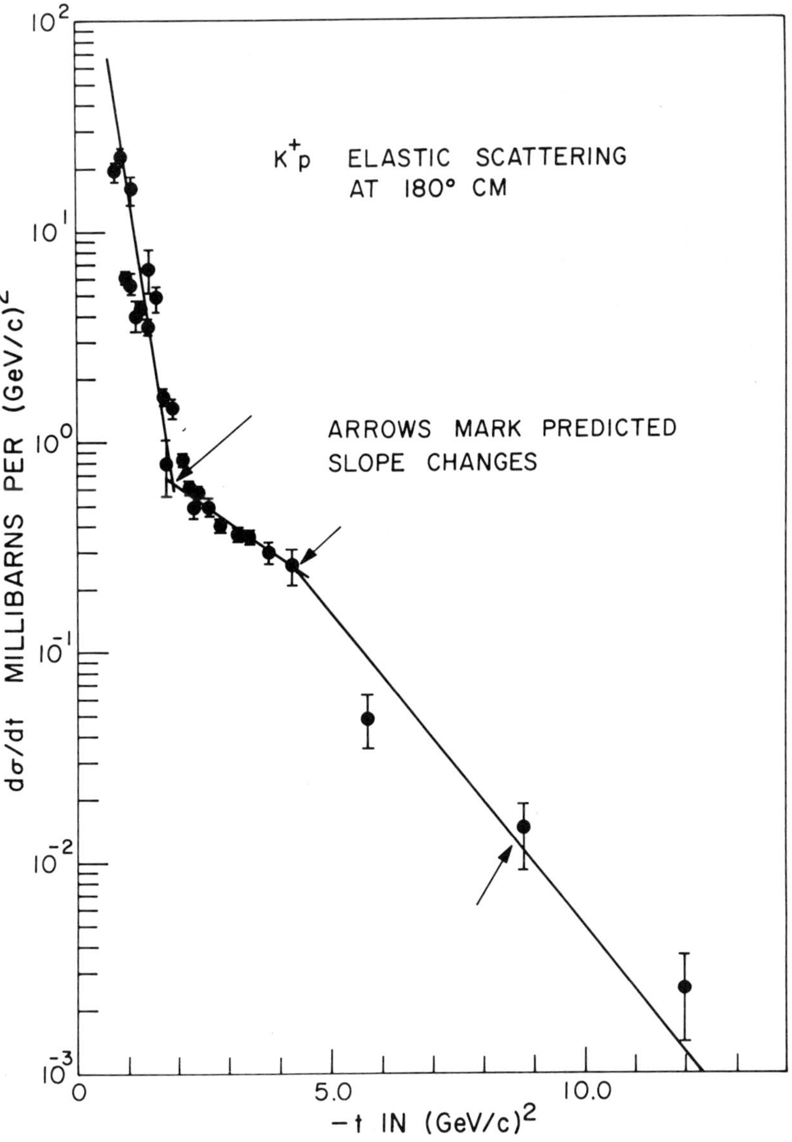

Fig. 2: K^+p elastic scattering at 180° center-of-mass showing slope changes compatible with predictions.

PARTICIPANTS

Bangerter, R.	University of California - Berkeley
Barbaro-Galtieri, L.	Lawrence Radiation Laboratory
Barloutaud, R.	C.E.N. Saclay
Barnes, V. E.	Purude University
Barnett, B.A.	University of Maryland
Bassano, D.	Brookhaven National Laboratory
Berley, D.	Brookhaven National Laboratory
Biedenharn, L.C.	Duke University
Binkley, M.E.	Duke University
Brucker, E.B.	Rutgers University
Bugg, W.	University of Tennessee
Burnstein, R. A.	Illinois Institute of Technology
Butterworth, I.	Rutherford Laboratory
Button-Shafer, J.	University of Massachusetts
Capps, R.	Purdue University
Carmony, D.D.	Purdue University
Carpenter, D.W.	Duke University
Chang, C.Y.	University of Maryland
Cline, D.B.	University of Wisconsin
Cool, R.L.	Brookhaven National Laboratory
Coyne, J.	National Bureau of Standards
Dalitz, R.	University of Oxford
Damerell, C.J.S.	Rutherford Laboratory
Davis, R.	Northwestern University
Derrick, M.	Argonne National Laboratory
Engler, A.	Carnegie-Mellon University
Feld, B.	Massachusetts Institute of Technology
Ferro-Luzzi, M.	CERN
Fickinger, W.	Case Western Reserve University
Fortney, L.R.	Duke University
Fowler, E.C.	Duke University
Friedman, J.	Lawrence Radiation Laboratory
Fung, S-Y.	University of California - Riverside
Garfinkel, A.	Purdue University
Gershwin, L.	Columbia University
Goldberg, M.	Syracuse University
Goldhaber, G.	University of California - Berkeley
Haberler, P.L.F.	Duke University
Han, M.-Y.	Duke University
Harrison, C.	Harvard University
Hart, E.L.	University of Tennessee
Hemingway, R.	University of Maryland
Hertzbach, S.	University of Massachusetts
Hildebrand, B.	U.S. Atomic Energy Commission
Hughes, V.W.	Yale University
Jenkins, E.W.	University of Arizona
Kalbfleisch, G.R.	Brookhaven National Laboratory
Kalmus, G.	Lawrence Radiation Laboratory

Kendall, B.N.	University of California-Santa Barbara
Kim, J.K.	Harvard University
Kirsch, L.	Brandeis University
Kraemer, R.W.	Carnegie-Mellon University
Lannutti, J.E.	Florida State University
Lasinski, T.	University of Chicago
Lassila, K.E.	Iowa State University
Levi-Setti, R.	University of Chicago
Lipshutz, N.R.	Duke University
London, G.	Brookhaven National Laboratory
Lynch, G.R.	Lawrence Radiation Laboratory
Ma, Z-M.	Michigan State University
Marshall- Libby, L.	University of Colorado
Martin, H.J.	University of Indiana
Mast, T.	Lawrence Radiation Laboratory
Meadows, B.	Syracuse University
Merrill, D.	University of Chicago
Meshkov, S.	National Bureau of Standards
Miller, D.H.	Purdue University
Moneti, G.	Syracuse University
Muir, R.	Univ. of North Carolina - Greensboro
Mulvey, J.H.	University of Oxford
Novey, T.	Argonne National Laboratory
Oberlack, H.	University of Heidelberg
O'Halloran, T.A.	University of Illinois
Plane, D.E.	CERN
Rahm, D.	Brookhaven National Laboratory
Rebka, G.	Yale University
Robbins, L.A.	University College - London
Robinson, D.K.	Case Western Reserve University
Roper, L.D.	Virginia Polytechnic Institute
Rose, C.M.	Duke University
Rosenfeld, A.	Lawrence Radiation Laboratory
Ross, R.	University of California - Berkeley
Schneps, J.	Tufts University
Schultz, P.F.	University of Illinois
Segar, A.	University of Oxford
Sharrock, S.J.	University College - London
Sims, W. H.	Brookhaven National Laboratory
Smart, W.	Stanford Linear Accelerator Center
Steinberg, P.H.	University of Maryland
Thews, R.	University of Rochester
Thompson, J.	Brookhaven National Laboratory
Tripp, R.D.	Lawrence Radiation Laboratory
Tuan, S.F.	University of Hawaii
Wagner, F.	Max Planck Institute for Physics
Wangler, T.P.	Argonne National Laboratory
Ward, C.	Argonne National Laboratory
Watkins, P.M.	University of Birmingham
Whitmore, J.	University of Illinois

Willis, W.J.	Yale University
Wolsky, G.	Tufts University
Yodh, G.	University of Maryland
Yokosawa, A.	Argonne National Laboratory
Zdanis, R.A.	Johns Hopkins University
Zeller, M.	Yale University

AUTHOR INDEX